Copper Interconnect Technology

Tapan Gupta

Copper Interconnect Technology

Springer

Tapan Gupta
Radiation Monitoring Devices, Inc.
Watertown, MA
USA
TGupta@rmdinc.com

ISBN 978-1-4419-0075-3 e-ISBN 978-1-4419-0076-0
DOI 10.1007/978-1-4419-0076-0
Springer Dordrecht Heidelberg London New York

Library of Congress Control Number: 2009928021

Cover illustration: Figures on front cover reproduced with permission from IBM

Printed on acid-free paper

Springer is part of Springer Science+Business Media (www.springer.com)

Dedicated to the memory
of my father, Dr. Gopeswar Gupta,
and mother, Dr. Kanak Lata Gupta
&
To The Godfather of Modern Electronics,
Dr. John F. Bardeen

Preface

The frontiers of knowledge are advancing at an ever-increasing rate. Today's discoveries will tomorrow be part of the toolbox of every researcher. By analogy, the process of advancing the line of settlements, and cultivating and civilizing new territory, takes place in stages, and the semiconductor industry, which has brought a revolution in modern civilization, is a vivid example. The intellectual foment that marked its beginnings has changed the way we view the physical world forever. Since the creation of the integrated circuit in 1960 there has been a continuous reduction in the size of circuits and thus the feature size of devices.

Following Moore's famous prediction, the number of devices manufactured on a single chip has exceeded the expectation of very large-scale integrated (VLSI) circuits. As a result, new materials are being introduced to meet the challenge of 21st century IC technology. The increasing device count accompanied by a shrinking minimum feature size, which was expected to be smaller than 1 μm before 1990, has reached 0.1 μm at the beginning of the 21st century. Progress in the development of new materials is marching in tandem and with much the same speed as circuit density (number of devices per unit square area).

The entire field of VLSI circuits depends upon circuit design, layout of electronic circuitry, process development, and synthesis of new materials. As a result, traditional materials (e.g. copper) are sometimes renewed in a fashion necessary to accommodate them to modern technology. Mankind has known about copper and its processing for different uses since the prehistoric age. Now, at the beginning of the 21st century, the use of copper has opened a new era in the IC manufacturing industry.

The 1990s were the decade in which copper as an interconnecting material came to the forefront and gained much attention from microelectronics engineers and scientists. The metallic conductivity and resistance to electromigration of bulk copper (Cu) are better than aluminum (Al). But as the feature size of the Cu-lines forming interconnects is scaled, the resistivity of the lines is seen to increase. At the same time, the electromigration and stress induced voids due to increased current density become reliability issues. Innovative ideas like the use of a cobalt-tungsten cap layer and alloying of copper have worked well, but both come with an increased *RC* product line.

The use of Cu-interconnect has introduced the additional burden of integration of the barrier layer in Cu-damascene architecture. The barrier layer affects the resistance of the Cu-lines too. The PVD-Cu/PVD-Ta interface is "magical" and the use of precursors in the ALD/CVD (Atomic Layer Deposition/Chemical Vapor Deposition) system has raised several questions, although the advantages of ALD in comparison to the PVD system in producing perfect step coverage with very thin layers is clear.

In the field of low-*K*, time-dependent dielectric breakdown (TDDB) and interlayer dielectrics (ILDs), reliability is becoming more important. As the dimensions shrink, the stress gradients increase because the same amount of stress is confined within a reduced geometry. Moreover, these low-*K* materials are softer than SiO_2 and susceptible to failure.

There are many challenges and difficulties with copper interconnect but not without any gain. Most of the challenges are overcome partially by using advanced instrumentation and photolithography materials. Photolithography has come a long way from 365 nm exposure technology to 157 nm exposure technology. Use of advanced laser systems with lower radiation wavelengths, additional hardware, and enhancement techniques, have all contributed to these improvements.

In 1997 IBM and Motorola introduced new approaches to Cu-interconnecting technology and in September 1997 IBM demonstrated their complementary metal oxide semiconductor (CMOS) device with six layers of copper lines. At the end of the decade, copper damascene processes were introduced and the century-old chemical mechanical polishing (CMP) procedure was renewed to integrate copper in deep sub-micron level circuitry.

This book was developed from a series of lecture notes prepared for graduate students of different universities. The notes are based on the research and publications of countless scientists and engineers engaged in this field. I express my sincerest thanks and gratitude for their indirect help.

Copper Interconnect Technology is the first book on the subject to treat materials, technology, and applications comprehensively, and is a product of my 25 years of research and teaching experience in different universities and research organizations. It is written for professionals as well as graduate students, and belongs on the bookshelf of workers in several microelectronics disciplines.

The chapters of the book are arranged sequentially following the sequence of the damascene process. In Chapter 1 basic properties of the materials used in copper interconnect are presented. Chapter 2 deals with the low- and high-*K* dielectric materials (dealing with the physical, chemical, and structural properties) that are of potential interest for scaled-down, high-speed devices. The diffusion of Cu in silicon is well known, so Chapter 3 is devoted to the search for new barrier materials and metal complexes to minimize diffusion of Cu from Cu interconnects. Some of the promising barrier materials, their physical and chemical properties, and the interpretation of binary and ternary phase diagrams, are also discussed.

Chapter 4 addresses different resist materials (DUV and EUV) and lithography techniques that are being used or are in the development stage. Pattern generation technologies together with different etching systems applied in the modern IC industry are also considered.

As the state of the art of modern integrated circuit technology has changed from the subtractive aluminum metallization process to via and trench filling additive Cu, the deposition technology has also evolved. Chapter 5 covers different deposition technologies that are frequently used in the modern Cu-damascene process. Chapter 6 deals with the damascene procedure and the chemical mechanical planarization (CMP) process.

Cu-interconnect replaced Al-interconnect because of its higher bulk-conductivity and electromigration resistance. But as the feature size of the conducting lines is shrinking, the thin metal lines can no longer retain the bulk properties of the metal. In Chapter 7 the conductivity and electromigration properties of Cu-interconnecting lines are discussed. Chapter 8 deals with the routing design of the Cu-interconnects together with the reliability issues of the scaled Cu-lines forming interconnects.

Watertown, Massachusetts Tapan K. Gupta

Acknowledgments

A book of such diversity would not have been possible without the indirect help of many research workers, engineers, and teachers. Virtually all of the information presented here is based on my lectures to graduate students of different universities. The notes were prepared from the research work of countless scientists and engineers engaged in this field. Their contributions are recognized to a small degree by citing some of their work in references at the end of the chapters. I also wish to acknowledge the work of the people who are not cited directly but who have contributed indirectly to the development of the book. My sincerest thanks go to them.

The entire manuscript has been read by Dr. Rafael Reif, Provost, and former Professor and Head, Electrical Engineering Department, MIT, Cambridge, Massachusetts and Dr. K.N. Tu, Professor of The Henry Samueli School of Engineering and Applied Science, and former Head of the Department of Materials Science and Engineering, University of California, Los Angeles, California (UCLA). The author is grateful to them for their help.

The author also expresses his gratitude for the encouragement received from Professor Carl Thompson, Rickey/Nelson Professor, MIT, Cambridge, Massachusetts, Professor Krishna Saraswat, ECE Department and Professor Reinhold Dauskardt, Materials Science, Stanford University, CA, and Professor P.S. Ho, Professor, University of Texas, Austin, TX. I am grateful to the anonymous reviewers for their comments and criticism that have helped in shaping the book.

I have taken many illustrations and materials from different journals and magazines. I want to express my sincerest thanks to all the authors and the staff members for their permission to reproduce these in my book.

Most of all I want to thank my wife Arundhati, my daughter Atreyee, and my son-in-law Jesse, for the love, understanding, patience, and impatience that made the preparation of this book possible.

Tapan K. Gupta

Contents

Author Biography

Tapan K. Gupta received his Master of Science degree in Physics from the Indian Institute Technology, a PhD degree in Physics from Boston College, and was a Post-Doctoral Fellow in the Electrical Engineering and Communication Department at Lehigh University, Pennsylvania. A former Analog Devices Career Development Professor in the Electrical Engineering and Computer Science Department at Tufts University, an International Rotary Foundation Scholar, and recipient of a Teaching Excellency Award, Dr. Gupta is currently a Chief Materials Scientist performing research in the field of Nuclear Medicine in collaboration with different universities and industries at Radiation Monitoring Devices, Massachusetts. Dr. Gupta has 25 years of teaching and research experience at different universities and industries in the USA and 75 peer reviewed articles in the fields of physics, materials science, semiconductor physics, and nuclear medicine. Dr. Gupta is the author of the book *Handbook of Thick and Thin Film Hybrid Microelectronics* published by John Wiley and has authored a book chapter on solar cells and materials for Allied Publication, New Delhi.

Chapter 1
Introduction

The word *interconnect* (interconnection) in very large-scale integrated circuits (VLSIs) means a *metal line* of *low resistivity* (*high conductivity*) which connects the various electronic devices to carry current or to transport charge [1]. The *interconnecting metal lines* are separated from the substrate by *insulating layers* (*dielectric material*), except on the contact area. Since the creation of the integrated circuit (IC) in 1960, *aluminum* (Al) or its *alloy* (Al+Si+Cu) has become the primary material for interconnecting lines, and *silicon dioxide* (SiO_2) has become the *insulating layer* (dielectric material) to separate the *interconnects*. Besides being an insulating material for interconnecting lines, SiO_2 has been used also as a gate material in metal oxide semiconductor (MOS) devices. As a matter of fact, Al coupled with SiO_2 has become the workhorse of IC technology.

Figure 1.1 shows a scanning electron microphotograph (SEM) of an integrated circuit (IC) with local (which is used within a cell or a block using the finest pitch

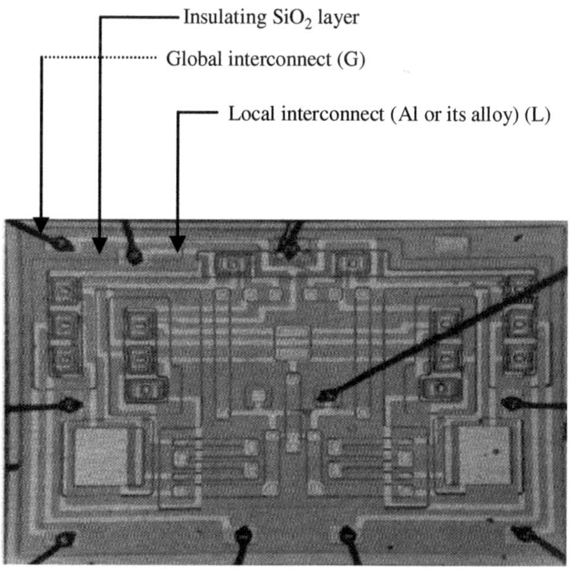

Fig. 1.1 SEM of the interconnecting lines (G & L) and the insulating SiO_2 layer inside an IC chip (Courtesy, ICE Corp., AZ)

Tapan K. Gupta, *Copper Interconnect Technology*,
DOI 10.1007/978-1-4419-0076-0_1, © Springer Science+Business Media, LLC 2009

metal allowed by the design rules) and global (which is used for inter-block communication using the upper metal layers) interconnecting lines. The fine interconnecting lines are fabricated on the deposited metal film by *photolithography* and selective etching.

IC technology remains the dominant force in semiconductor device fabrication, and is likely to retain this position for the foreseeable future. As a result, the number of devices manufactured on a chip exceeded the expectations of VLSI, and by the mid-1970s it exceeded 100,000 devices per chip. The increasing device count accompanied by the shrinking feature size resulted in the historical prediction of Moore that the number of transistors per integrated circuit would double every two years [2]. Figure 1.2 shows the advancement of integrated circuit technology since 1960. From 1960 to 1970, complexity was the number of components as initially described by Moore. After 1970, it was often cited as the number of bits in a DRAM (dynamic random access memory) or the number of transistors in an MPU (microprocessor unit).

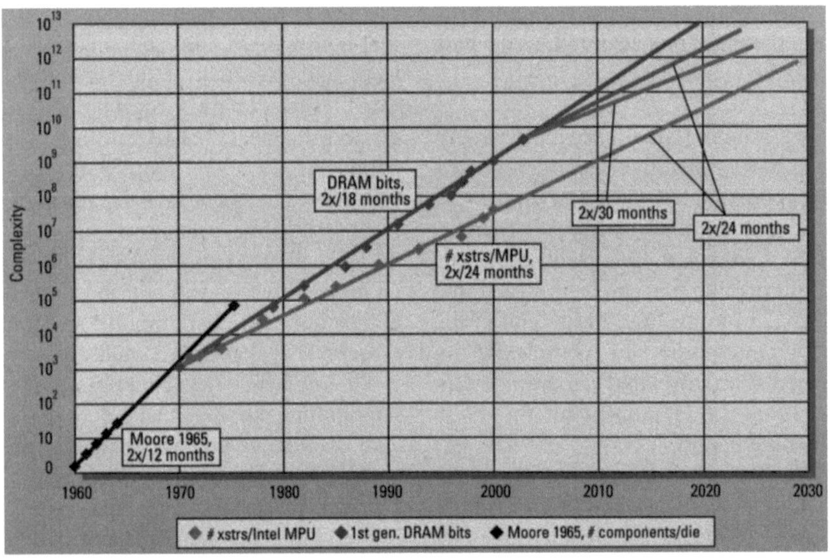

Fig. 1.2 The lines drawn after the prediction of Moore about the first 70 years of IC development (Courtesy, M. L. Hammond, TMX International, Cupertino, CA)

1.1 Trends and Challenges

After 1990, progress in VLSI manufacturing technology continued to proceed by further reductions in unit cost per unit function, and in power-delay products of VLSI devices. As a result, at the end of the 20th century, a production level MOS device with a 0.35 μm gate appeared on the market (Fig. 1.3).

Fig. 1.3 A scanning electron microphotograph of a MOS device (Reprinted from P.A. Packan's article in MRS Bull. June 2000, p. 18, with permission from MRS, Pittsburgh, PA)

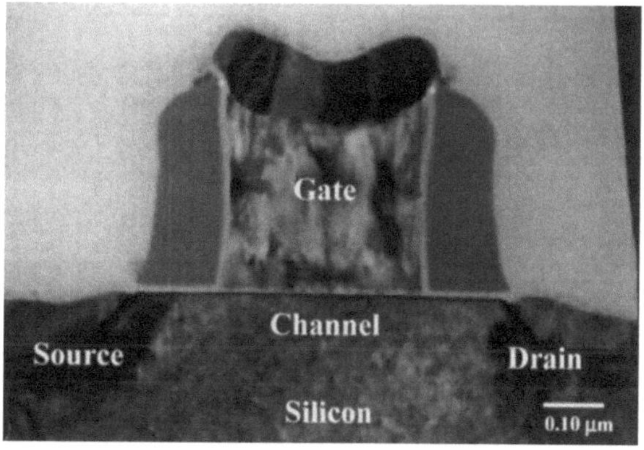

Now the gate oxide thickness (T_{ox}) of CMOS transistors has been scaled down constantly with the reduction in gate length. According to the 2006 ITRS projection, the gate length for a 65 nm device should be ~25 nm which will require 1.5 nm gate oxide thickness. In this thin gate oxide regime, the direct tunneling current will increase exponentially with decreasing tunneling oxide thickness [7].

Figure 1.4 presents the projections versus simulations of direct tunneling gate leakage current (J_g) density for low standby power (LSP) logic [3]. From the graph we can see that every 0.2 nm reduction of the gate oxide (T_{ox}) causes a 10× increase in J_g. As a rule of thumb, if we calculate the oxide thickness for 1.2 nm the respective gate length is 20 nm; then for a 350 nm gate length we can assume the gate oxide to be ~21 nm. Thus for a 65 nm node, the ITRS (2006) expectation for the gate oxide is 1.5 nm. Thus when the oxide thickness changes from 21 to 1.5 nm, the increase in J_g will be enormous. Thus it is expected that when the feature size lies in the submicron region, Al-SiO_2 technology will not be able to provide speed and will not be able to prevent leakage current and tunneling in MOS and other devices [4–8]. Because SiO_2 enables defect charge densities on the order of 10^{10}/cm^2, mid-band gap interface state densities of ~10^{10}/cm^2-eV, and hard breakdown fields in excess of 10 mV/cm, these are not adequate for sub-100 nm\ devices.

Again according to 2006 ITRS requirements CV/I delay have been projected to 0.65 ps. From Fig. 1.5a we can see that for that small CV/I value the gate length should be 50 nm and the corresponding $E \times D$ product will be 0.6×10^{-27} J-sec (Fig. 1.5b).

Figure 1.5a shows CV/I gate delay versus transistor gate length. From the figure we can see that a transistor with 60 nm gate-length will approach a (CV/I) value close to 1 ps. Figure 1.5b shows the estimated energy-delay ($E \times D$) product of an NMOS and the data are collected from a large amount of published research [10]. The $E \times D$ product is seen to vary almost linearly with the gate length.

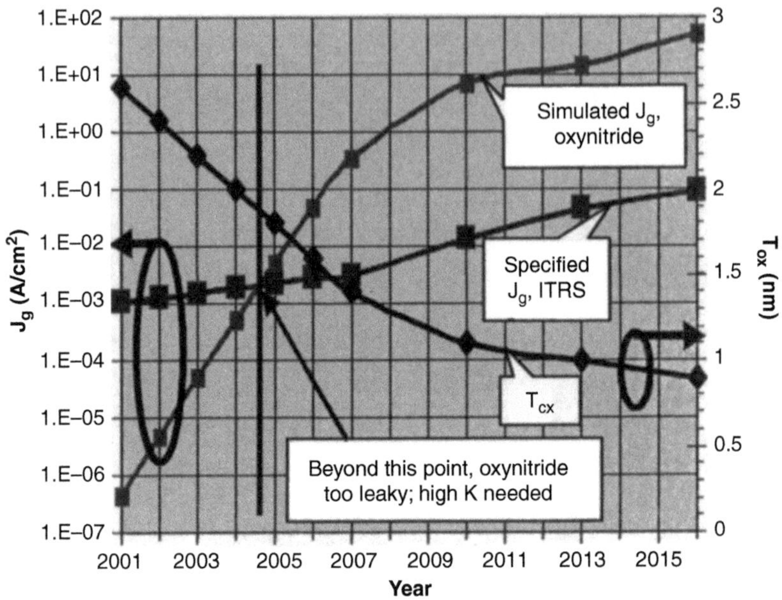

Fig. 1.4 2001 ITRS projections versus simulations of direct tunneling gate leakage current density for low standby power logic.J_g is the direct tunneling gate leakage current and T_{ox} is the thickness of the oxide (Courtesy, ITRS, 2001, Austin TX)

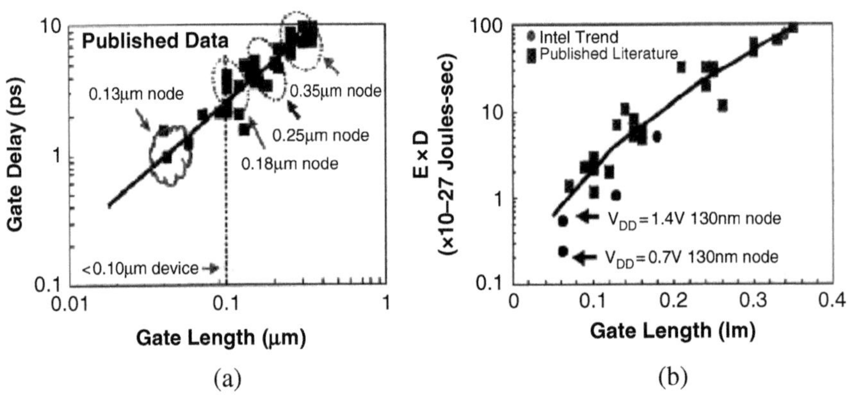

Fig. 1.5 (**a**) CV/I gate delay versus gate length and (**b**) energy-delay product versus gate length (Reprinted with permission, Intel Inc.)

Besides the *RC* effect, scaled technology has another disadvantage, which is known as *circuit-noise*, and it arises from the total wire capacitance. The *noise* can be classified as *cross-talk noise* and noise due to *dynamic delay*. The first type arises when one or more of the neighboring lines act upon a line and yields an unwanted voltage spike. The second one refers to a situation where the victim line is in a switching rather than static mode [11].

The capacitive effect, which affects the *RC* (delay) time constant and the ultimate speed of a device mostly comes from inter-layer capacitance. As a result, silicon dioxide (SiO_2), with dielectric constant (*K*) ~3.9, is difficult to accommodate in sub-100 nm devices. So the sub-100 nm device needs a low-*K* dielectric material to improve the speed of the device.

It has been estimated that vertical scaling requirements of the gate stacks (Fig. 1.6) and shallow junctions of MOS devices will require 1–2 nm equivalent oxide thickness (EOT), which is not possible to fulfill with SiO_2 [12] (Table 1.1).

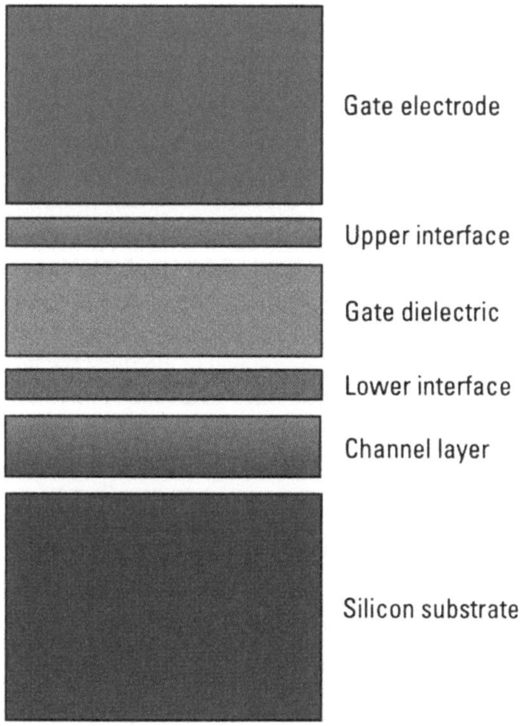

Fig. 1.6 The gate stack and the interfacial layers between gate electrode and substrate together with the scaled *interlayer dielectric* (ILD)

1.2 Physical Limits and Search for New Materials

The international Technology Road Map for semiconductors (ITRS) projected that by 2007 one billion transistors will be integrated into a monolithic die [9]. The wiring system of the billion-transistor die will deliver power to each transistor, provide a low-skew synchronizing clock to latches and dynamic circuits, and distribute data and control signals throughout the chip. As a result, 21st century

Table 1.1 Microprocessor chips at different technology nodes

Year	2006	2007	2008	2009
Dram 1/2 pitch Contacted	70 nm	65 nm	57 nm	50 nm
MPU/ASIC metal 1½ pitch	78 nm	68 nm	59 nm	52 nm
MPU physical gate length	28 nm	25 nm	22 nm	20 nm
EOT: extended planar bulk (EPB)	11 Å	11 Å	9 Å	7.5 Å
Gate poly depletion and inversion layer (EPB)	7.4 Å	7.4 Å	2.9 Å	2.8 Å
J_g limit (EPB)	5.3E+02	8.0E+02	9.09E+02	1.10E+02
V_{dd} power supply voltage	1.1 V	1.1 V	1.0 V	1.0 V
V_t sat (EPB)	168 mV	165 mV	160 mV	159 mV

Source: Adopted from International Technology Road Map for semiconductors ITRS, 2005, with permission.

opportunities for giga-scale integration (GSI) will be governed by a hierarchy of *physical limits* on interconnects whose levels are codified as: *material*, *device*, *circuit*, and *system* [10]. It is true that the physical limits of the interconnects have reached a point where Al-SiO_2 will not be able to provide the requirements of giga-scale integration, and therefore, ongoing research is focused on *materials*, *process technologies*, *circuit dimensions* and *the design of high performance architecture*.

To adequately describe the needs of interconnects, near term and long term technology requirements and potential solutions are addressed for two specific classes of products: high performance microprocessors (MPU) and dynamic memory (DRAM) circuit. For MPUs, local, intermediate and global wiring pitches/aspect ratios (ARs) are differentiated to highlight the hierarchical scaling methodology [3]. Figure 1.7 shows the picture of a 486DX4 100 MHz microprocessor with 1.6 million transistors.

The ideal scaling rules for interconnects basically serve to provide sufficient packing density. *Gates* are being rapidly *scaled down* in size and more and more wires are needed for communication. Needless to say, for better communication and speed the *RC* (*delay*) time constant of the local interconnects is reduced. In order to achieve that, one has to decrease both the line width and spacing of the local interconnects. In particular, the physics of *resistivity* behavior in *extremely fine conductor lines* represents a daunting and potentially insurmountable challenge that needs to be understood and resolved. On the other hand, both the size of a functional unit and the size of the entire chip, which may employ a constant dimension type of scaling, set the global wire lengths [10–11].

1.3 Challenges

The shrinking circuit dimensions have been the primary drivers behind the development of integrated circuits (ICs). Figure 1.8 shows the shrinkage in dimensions when 180 nm technology is changed to 130 nm technology. It is expected that by the year 2007, for 65 nm node technology [9] we will need to fabricate over one billion transistors on a single chip with a minimum interconnect thickness of ~50 nm [10]

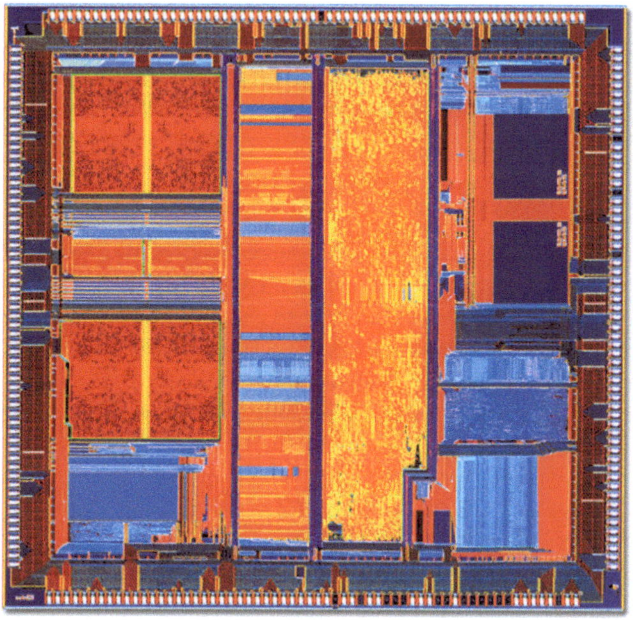

Fig. 1.7 486DX4 microprocessor (Photo courtesy AMD integrated circuits)

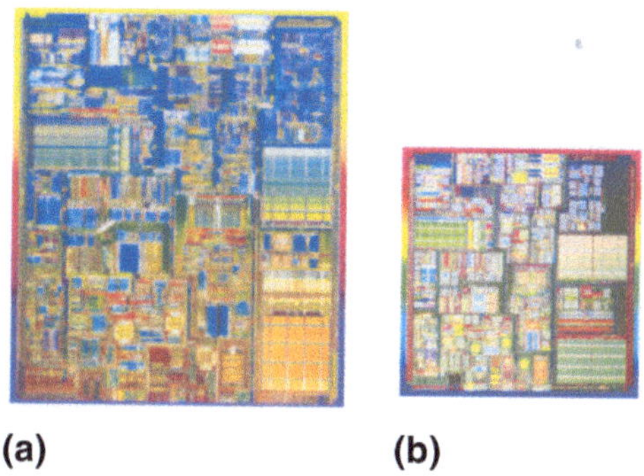

Fig. 1.8 The dimensions of a circuit when it is fabricated by (**a**) 180 nm technology and (**b**) 130 nm technology (Photo courtesy, Intel Corporation, Intel Tech. J. 6 (2), May 16, 2002)

and equivalent oxide thickness (EOT) of ~1.5 nm [12]. This means that the scaled Al-SiO_2 system will not be able to fulfill the requirements of giga-scale integrated circuits (ICs).

1.4 Choice of Materials

1.4.1 Why Copper (Cu) Interconnects?

One of the major reasons for the interconnect to limit the performance of a circuit is the resistance-capacitance (*RC*) effect [13]. As the chip area increases, parasitic interconnect capacitance dominates the gate capacitance and the speed improvement expected from simple scaling does not apply to circuits that drive global communication lines. In addition to large capacitance loads resulting from long interconnections, the resistance of the lines also becomes a major concern [14–15].

1.4.1.1 Line Resistivity

The advantage of using copper (Cu) as an interconnect in VLSI is mainly due to its lower bulk resistivity compared to aluminum (Al). Figure 1.9 shows the comparative analyses of the resistivities of copper and aluminum interconnects with respect to line width and thickness of the films. Experimental studies reveal that at room temperature, copper interconnect shows almost 35% lower resistivity than Al film having the same length, thickness, and width (when the film thickness is above 100 nm) [16–19].

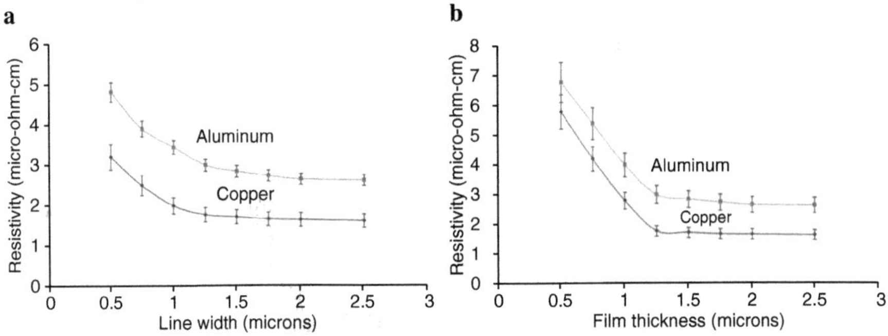

Fig. 1.9 Variation of line resistivities of aluminum and copper (**a**) with line width and (**b**) with line thickness (Data are collected from different sources including unpublished data of the author)

Figure 1.10 shows a schematic of a Cu-damascene architecture where a Cu-interconnect is embedded within the dielectric layer. In Fig. 1.10, we have assumed that the dielectric layer thickness (H_{d}) is equal to the interconnect height (H_{int}), and P is the pitch of the line, equal to 2 W.

Figure 1.11 shows the schematic of a multilayer stack of Cu-interconnects over a barrier layer embedded within the dielectric layer. The RC time constant of the generic multilayer stack can be calculated as [20–21]

$$\begin{aligned} RC &= 2R\,(C_{\mathrm{V}} + C_{\mathrm{L}}) \\ &= \left[(2\rho L)/(WT)\right]\left[\left\{(\varepsilon_0\,\varepsilon_r\,WL)/(H)\right\} + \left\{(\varepsilon_0\,\varepsilon_r\,TL)/(X)\right\}\right] \\ &= \left[\left\{(2\rho\varepsilon_0\varepsilon_r L)/(WT)\right\}\left\{(WL)/(H) + (TL)/(X)\right\}\right] \\ &= 2\rho\varepsilon_0\varepsilon_r\,(L^2/TH + L^2/WX) \end{aligned} \tag{1.1}$$

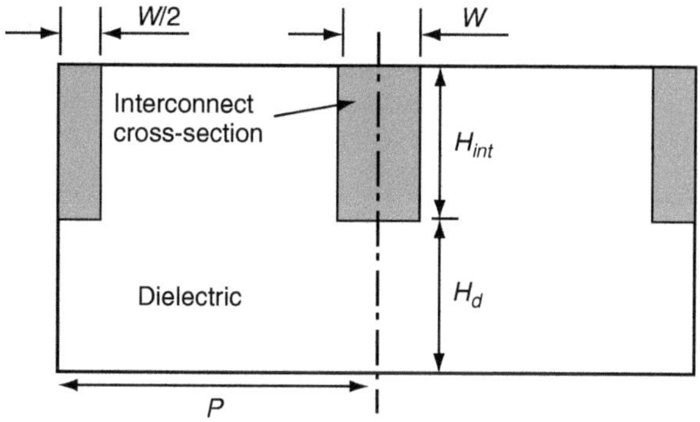

Fig. 1.10 Cu-damascene architecture with Cu-interconnects embedded within the dielectric layer

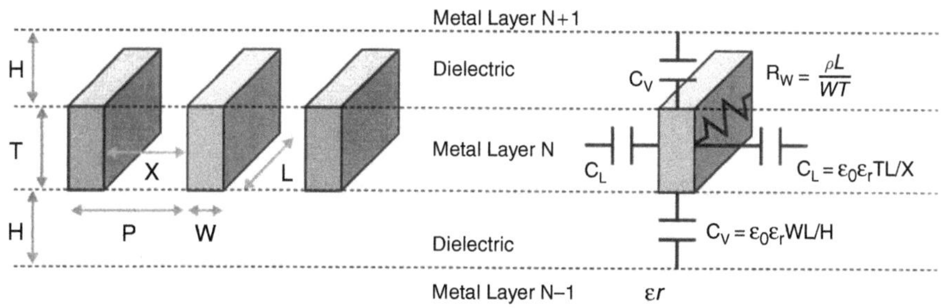

Fig. 1.11 Generic multilayer stack of Cu-interconnect and dielectric layers (Courtesy, A. Pratt, Advanced Energy Indus. Inc.)

where,

R = Resistance of the interconnect
C_V= Interlayer capacitance
C_L = Intra-layer capacitance
ρ = Resistivity of the interconnect metal
L= Length of the interconnect
W = Width of the interconnect
T = Thickness of the metal layer = H_{int}
ε_0 = The permittivity of space
ε_r = The relative permittivity of the dielectric
H = Height of the dielectric layer = H_d
X = Spacing between adjacent interconnect
P = Pitch of the line

If we assume that $(H_d) = H_{int} = H = T$, and $P = 2\ W$, (i.e. $X = W$), Equation (1.1) transform into:

$$RC = 2\rho\varepsilon_0\varepsilon_r L^2 (1/T^2 + 1/P^2) \quad (1.2)$$

Equation (1.2) clearly shows that *RC* – the *time* constant or the *delay* time (also related to the speed of a device) – depends upon the resistivity (ρ) of the Cu-interconnecting line, and the dielectric property of the material (ε_r).

1.4.1.2 Efficiency of Cu-Lines versus Node Technology

Since Cu film has almost 35% lower bulk resistivity than Al, it is expected that the number of interconnecting lines with Cu will be less than the number required with Al. It has been estimated that a reduction of almost 40% in the *RC* time constant can be achieved by using Cu rather than Al; or in other words, we can say that the *speed* of the circuit will be faster with Cu-interconnects compared to Al-interconnects.

It is also true that the interconnect *RC delay* can be reduced by increasing the thickness ofthe metal line (*T*) i.e., increasing the aspect ratio (AR). The ITRS (International Technology Road map for Semiconductors) predicted that for 65 nm node technology, the aspect ratio (AR) for local and global wiring should be 1.7 and 2.2, respectively [21–22]. At the same time when the capacitive reactance of the circuit is lowered by introducing low-*K* inter-layer dielectric (ILD), the number of interconnecting lines will be even smaller (Fig. 1.12). As a matter of fact, to reduce the capacitive effect alternate dielectric material other than the conventional silicon dioxide (SiO_2) is already at the manufacturing level. In addition, the Cu-CMP (chemical mechanical planarization) process has been adopted in Cu-damascene technology. The process will not only avoid dry and wet chemical etchings (which are not environmentally friendly) but it will be cost-effective too.

1.4.1.3 Heat Dissipation

Heat transfer issues significantly impact chip performance and reliability of deep submicron interconnects. Elevated temperatures degrade device performance due to

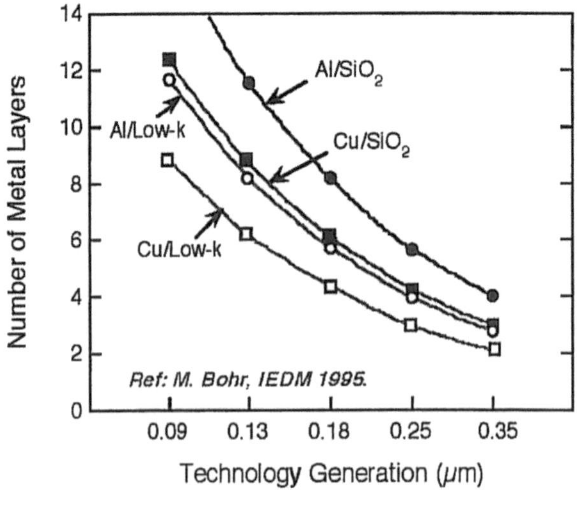

Fig. 1.12 The effect of the combination of metal and interlayer-dielectric layers on the number of required interconnect levels (Reprinted with permission from IEEE, IEDM Proc. 1995)

increased junction leakage and reduce interconnect reliability by accelerating *electromigartion* [23]. Experimental observation shows that a typical emitter coupled logic (ECL) gate dissipates less than 10 mW, but when these numbers become 10, 000 or more heat dissipation can bring the total power consumption to more than 20 W. As a result, heat generated by the IC components must be dissipated efficiently because virtually all failure mechanisms are enhanced by increased temperature [24]. In IC packages, thermal conduction is the only mechanism of practical significance, and one can expect that heat dissipation due to the current in resistive elements will be better with a material having high thermal conductivity. Figure 1.13 shows that Cu has higher thermal conductivity than Al, and it is expected that Cu-interconnecting lines will dissipate more heat than Al-interconnecting lines under the same conditions. As a matter of fact, it has been observed that high conductive (thermal) material dissipates more heat when it is used as the base material [25]. This feature contributes a factor to the development of high power, high performance, and fine pitch devices using thinner diameter Cu-wire without compromising performance and reliability.

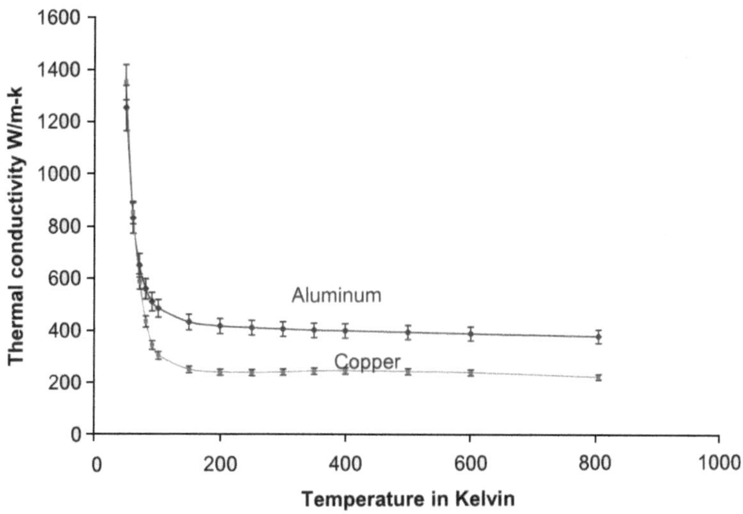

Fig. 1.13 Comparative analysis between the thermal conductivity of aluminum and copper as a function of temperature (Data are collected from different sources)

1.4.1.4 Electromigration (EM) Failure

Experimental evidence shows that grain boundaries/triple points account for electromigration (EM) in Al-interconnects. So, in a conventional integrated circuit (IC) Al-Cu alloy film is used to create Al-interconnecting lines, because Cu itself has a reduced susceptibility combined with reduced EM. Experimental data show that in median time to failure (t_{50}) the performance for copper damascene line is higher than Al-Cu without plug [26–29]. Usually, the t_{50} value of a metal interconnect

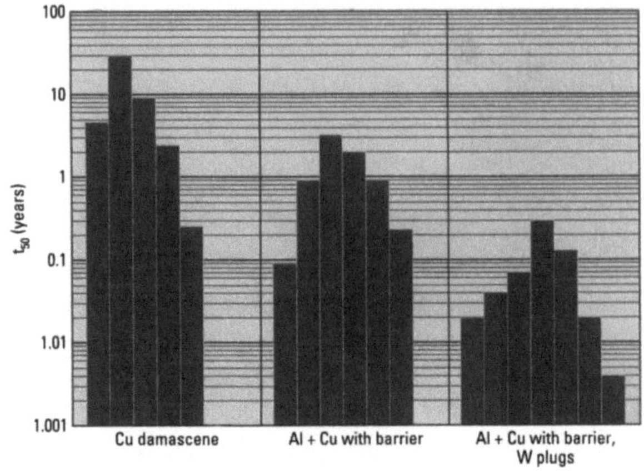

Fig. 1.14 The failure rates of copper and aluminum (Reprinted with permission, Semiconductor International, April 2000, p. 98)

depends on the temperature and current density [30]. Figure 1.14 is a plot of the failure rates of Cu and Al after Black's equation, when the current density and the temperature during the experiment are kept at ~5 × 10^6 A/cm^2 and 125 °C respectively [31].

1.4.1.5 Thermo-Mechanical Properties

The study of thermo-mechanical properties of a thin metal film can predict the amount of *stress* a metal line will experience during thermal cycling. The *stress* induced by thermal cycling can be *tensile* or *compressive* or the stress may arise due to differences in the coefficient of thermal expansion (CTE) between the substrate and the film [32]. Stress induced due to thermo-mechanical properties of a metal line is important especially when the circuit has to be built with multi-layer metal lines [33].

Some of the thermo-mechanical properties of Al and Cu are listed in Table 1.2 to testify to the technological superiority of copper (Cu) compared to aluminum (Al) as interconnects in microelectronics devices [34–37].

From Table 1.2 we can see that the thermo-mechanical properties of Cu are superior to Al. On the basis of stress calculations it has been found that the shear stresses

Table 1.2 Thermo-mechanical properties of Al and Cu interconnects

Thermo-mechanical properties	Copper (Cu)	Aluminum (Al)
Young's modulus (GPa)	128	71.1
Yield strength (Mpa)	69	10.30
Tensile strength (Mpa)	220	44.90
Coefficient of thermal expansion (CTE) ppm/C	17	24

and the propensity of shear induced interface delamination are actually lower in Cu-interconnect compared to Al-interconnect. Another superior metallurgical property of Cu over Al is its ultimate tensile strength, which determines the critical stress and failure site of a metal wire. It is particularly important in the case of ball bonding. The higher the tensile strength, the lower will be the deformations in wire diameter and failure site. Figure 1.15b shows that copper (Cu) wire has the highest breaking load compared to gold (Au) and aluminum (Al) wires of the same diameter.

It is beyond doubt that ultra-fine pitch (<25 μm) ball-bonding will be needed on Cu-pads deposited on the chip for global interconnection that requires smaller-diameter wire with sufficient stiffness to ease handling and to avoid molding related defects due to wire sweep [38–39].

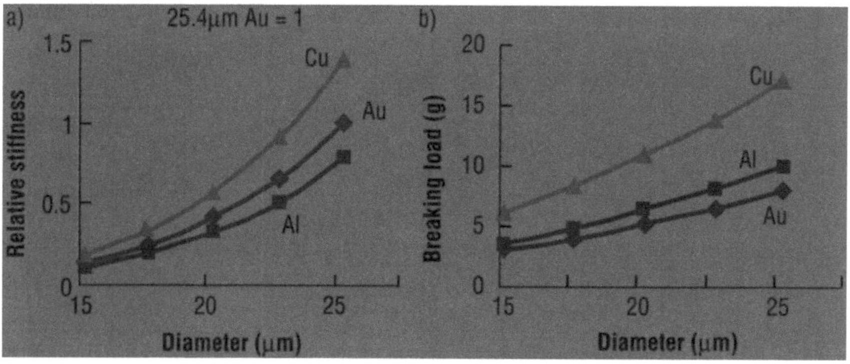

Fig. 1.15 (**a**) Relative diameter of a metal wire versus stiffness of the wire and (**b**) relative diameter of a connecting wire versus breaking load (Courtesy Kullick and Soffa, PA)

Figure 1.15a shows that Cu wire has twice the strength and 40% higher stiffness than gold. This means that Cu will be a better bonding metal than gold and will eliminate the formation of intermetallic compounds. Moreover, the process technology will be cost-effective (Fig. 1.16) [40].

Cu-Cu bonding has almost no intermetallic growth and it results in lower contact resistance and less generated heat. Figure 1.17 shows a scanning electron micrograph (SEM) and the intermetallic penetration in a Cu wire bond versus a gold wire bond. From the figure it is clear that the intermetallic penetration of the Cu ball bond is significantly less than that of the gold bond.

Needless to say, a multi-layer stack with layers of different materials should have matched coefficient of thermal expansion (CTE), otherwise the system will experience a stress. It has been found that the CTE of Cu is close to the CTE values of some of the low-*K* dielectric materials (14–17 ppm/°C) [41–42].As a result, the metal lines in Cu-low-*K* dielectric systems will experience smaller tensile hydrostatic stresses than traditional Al-SiO_2 systems during thermal cycling.

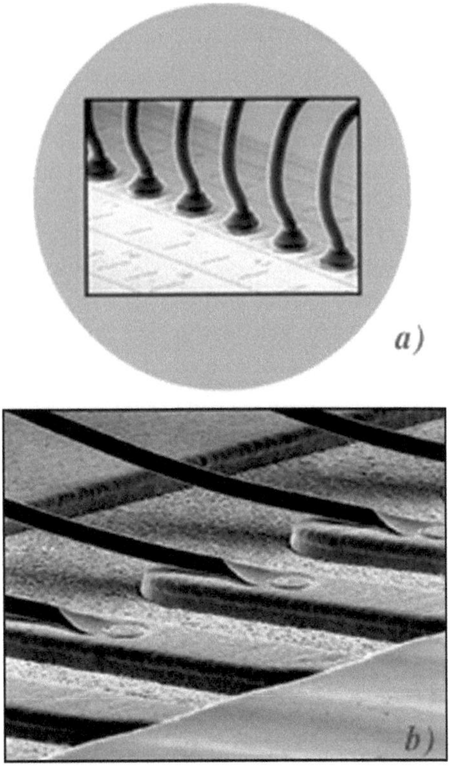

Fig. 1.16 (**a**) Cu-ball bonding on Cu-pads and (**b**) second bond on the package (Courtesy, Kullick and Soffa industries, PA)

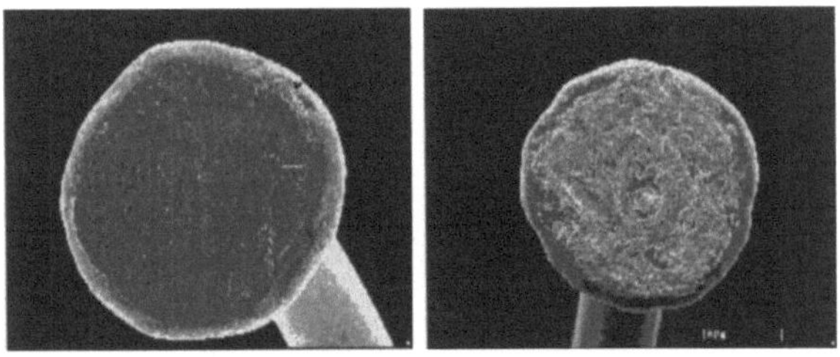

(a) (b)

Fig. 1.17 Intermetallic penetration: (**a**) Cu-wire ball and (**b**) Au-wire ball (courtesy Kullick and Soffa)

1.4.1.6 Stress Migration Effect

Deposited thin films always have different grain sizes and grain boundaries which are the paths of impurity diffusion during device processing. As a result, diffusional mass transport along grain boundaries in thin films leads to the formation of a nobledefect identified as a *diffusion wedge* [43–45]. It has been observed that the grain boundary diffusivity at a given temperature in Cu is much lower than that in Al [46–47]. As a result, Cu shows a better *stress migration resistance* than Al, which is related to the generation of a discontinuity in a metal line due to atomic migration induced by stress [48–49]. However, for a particular metal line the atomic migration

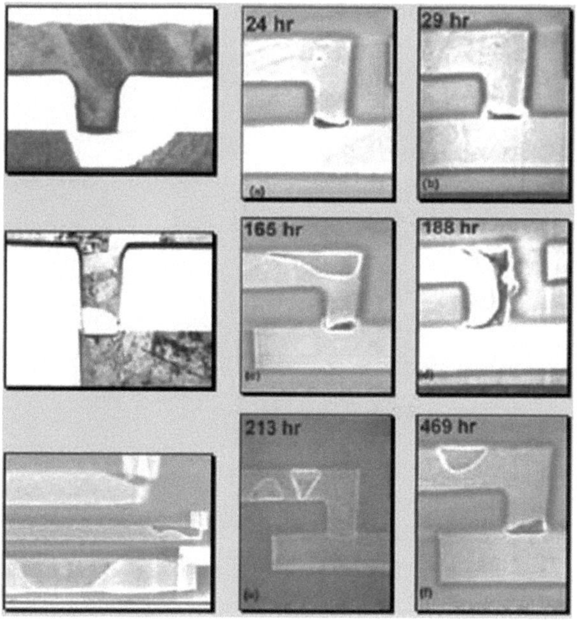

Fig. 1.18 Stress migration as the geometry of the Cu-interconnect shrinks (Reprinted with permission, Semiconductor International. July 2004 p. 95)

induced stress is related to the dimension of the line. Figure 1.18 shows the stress and electromigration in Cu-interconnect as the geometry of the interconnect shrinks.

1.5 New Technologies

1.5.1 Multilayer Metal Architecture

Interconnect delays are increasingly dominating IC performance in spite of copper-low-*K* dielectric systems, and it is expected that the interconnect delay will be substantial below 130 nm node technology [10,50]. Therefore the need exists for an alternative technology to overcome the problem. One such promising technique is 3-D ICs with multiple active silicon layers. The multilayer metal architecture will reduce the chip footprint leading to a significant reduction in *RC* delay and will greatly enhance the performance of logic circuits [51]. It is expected that to keep the *RC* delay at a respectable level, nine levels of metal will be needed for 130 nm node devices. Therefore we do need an alternative multilevel interconnection architecture in the dual Cu-damascene process [49, 52–53]. Figure 1.19 shows *multilevel interconnect processing* performed on a 300 mm silicon (Si) wafer with six levels

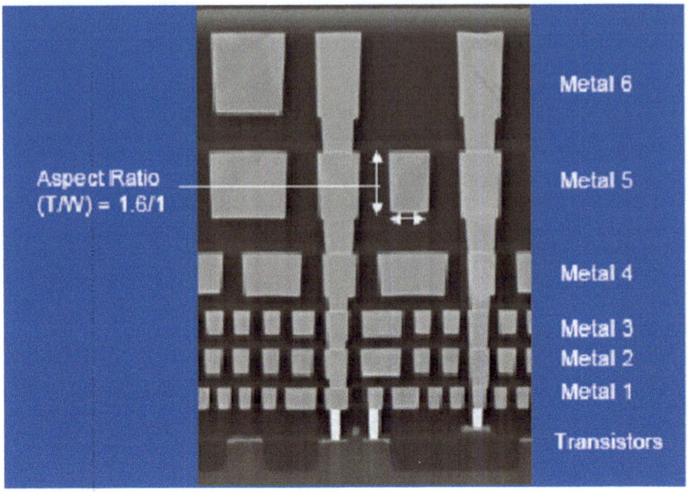

Fig. 1.19 90 nm interconnect technology on 300 nm wafer with six levels of metal (Photo courtesy, Intel, CA)

of metal following 90 nm node technology. The metal lines are connected via holes to minimize wire lengths and interelectrode resistances and capacitances.

In the multilevel metallization process (Fig. 1.19) interconnecting lines are routed through small holes, and they are so small and thin that conventional aluminum based interconnecting lines are inadequate for better and faster chip functions. Silver (Ag) has a highest *metallic conductivity* (~0.629 micro-mho/cm) [54–55], but suffers from *electromigration* [56–57]. On the other hand, copper (Cu) has higher conductivity (0.595 micro-mho/cm) than aluminum (Al) (0.377 micro-mho/cm) and it offers higher resistance to electromigration compared to both Ag and Al. Thus it is expected that Cu-interconnects will minimize *RC* delay, cross-talk and power dissipation in a microcircuit [58–61].

1.5.2 Substrate Engineering

The continued miniaturization of devices has reached a point where Cu-low-*K* coupling alone will not be able to support the manufacturing needs of sub-100 nm node technology. Thus, some industrial houses have already introduced a change in the substrate material (such as: silicon-on-insulator (SOI), silicon-germanium (Si-Ge), etc.) to meet the challenges of device scaling [62–65].

Figure 1.20 shows the cross-section of a CMOS device fabricated on a partially depleted SOI substrate. The advantage of this type of structure is that the device operation and design are very close to the operation and design of a bulk CMOS device. Compared with traditional bipolar approaches, SOI considerably reduces cross-talk between RF analog and digital logic elements, and enable easy integration with passive elements. Recently, some industrial houses are considering switching over to *ultrathin fully depleted* (FD) SOI to improve electrostatic device

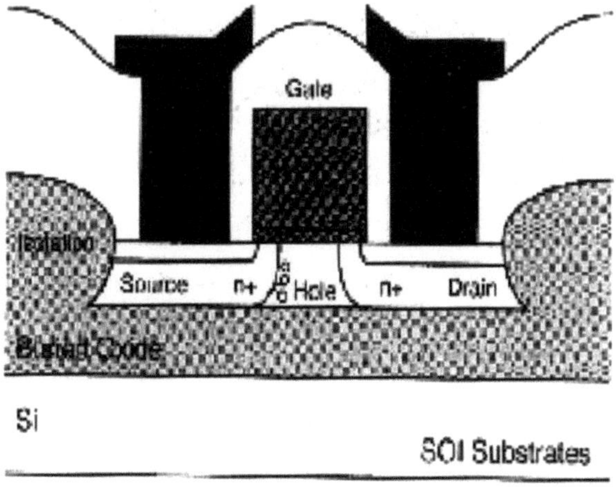

Fig. 1.20 SOI cross-section showing the device structure and the holes in the channel region (Courtesy, Wiley Interscience, New York, Modern Semiconductor Device Physics, S.M. Sze, p. 161)

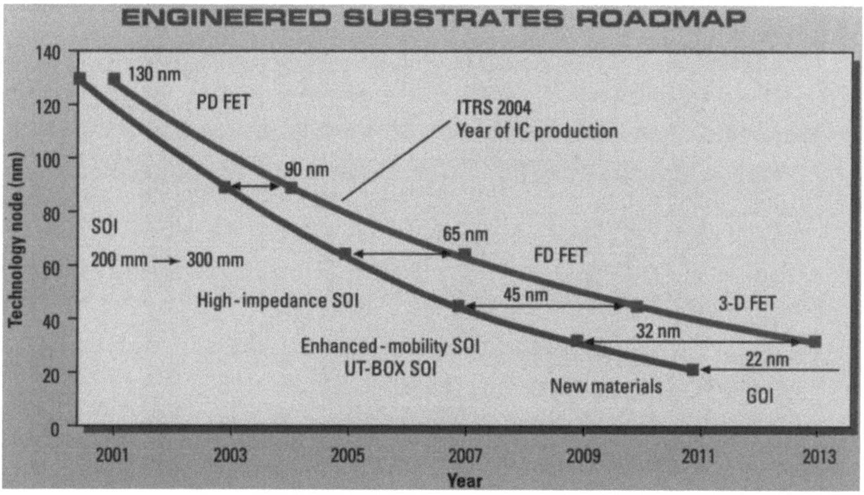

Fig. 1.21 Substrate readiness roadmap compared to the 2004 ITRS (Reprinted with permission, Semiconductor International. March 2006)

characteristics. However, the substrate industry must anticipate its final choices, and offer an increasing number of options well in advance of production implementation (Fig. 1.21).

Figure 1.22a shows a schematic of a nine level Cu-BEOL 130 nm technology on a SOI substrate and (b) shows a schematic for an AMD Opteron processor at 130 nm on SOI. Silicon carbide (SiC) and SiCOH have been used as low-*K* dielectric and etch stop layer respectively. A tantalum (Ta) barrier has been used in the 130 nm process, which has shown good reliability, low via resistance and excellent in-line defectivity.

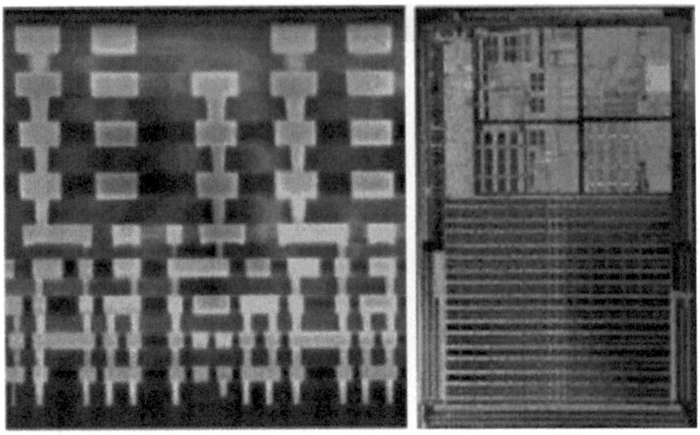

(a) (b)

Fig. 1.22 (**a**) Schematic of a nine level Cu-BEOL 130 nm technology on a SOI substrate and (**b**) schematic for AMD Opteron processor at 130 nm on SOI (Reprinted with permission, SST, January, 2004)

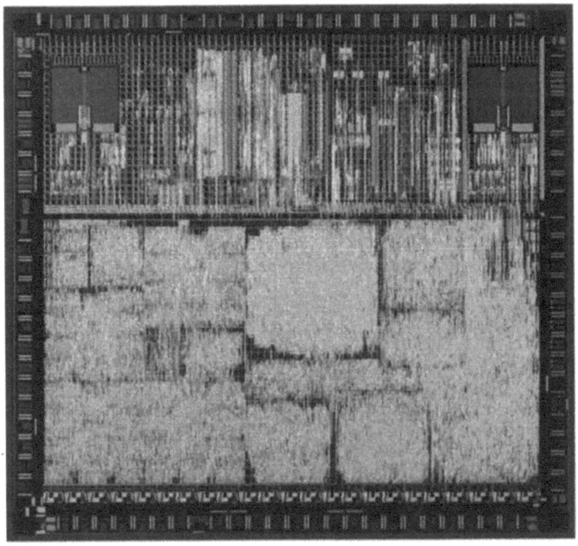

Fig. 1.23 A high speed BiCMOS device designed in IBM's 0.25 μm technology (Photo courtesy, IBM Research)

Another recent development in *substrate engineering* to meet the challenges of the sub-100 nm device structure is band-gap engineering in silicon to increase the hole flow to the source. This is accomplished by implanting germanium (Ge) into the silicon (Si) layer, forming a Si_{1-x} Ge_x region [66–69]. Figure 1.23 shows a 120 GHz BiCMOS device, which is based on Si/Ge technology to address low power 10 Gb/s applications[67–69].

Classical scaling is *running up* as the device dimensions have come down to 45 nm. With increasing circuit density and mobility, enhancements become key to maintaining device performance as supply voltages are scaled to manage active power dissipation at the chip level. As a result, effective scaling has taken over, and with it the technology of *strain engineering* is being widely used to speed carrier mobility (holes in pFETs and electrons in nFETs).

Strain engineering is a new kind of engineering which involves straining of silicon crystal to increase the mobility of charge carriers in the channel (electrons in MOSFETs and holes in PMOSFETs). An added benefit is that it reduces the source drain (S/D) series resistance. Compressive strain is induced in PMOS transistors, typically using epitaxially grown SiGe S/D and/or a compressively strained nitride layer over the gate. On the other hand, a tensile strained nitride layer is used in NMOS transistors. The process has been proven cost effective at the 90 nm node. Another method that has been used to enter manufacturing of the32 nm node is called a biaxial or global strain where hybrid orientation schemes (transferring a <110> silicon layer onto a standard <100> substrate) have been devised to take full advantage of geometrical preferences.

Figure 1.24 shows four stress techniques – dual stress liners, stress memorization (SM) and an embedded SiGe S/D – that were fully integrated on a partially depleted SOI structure.

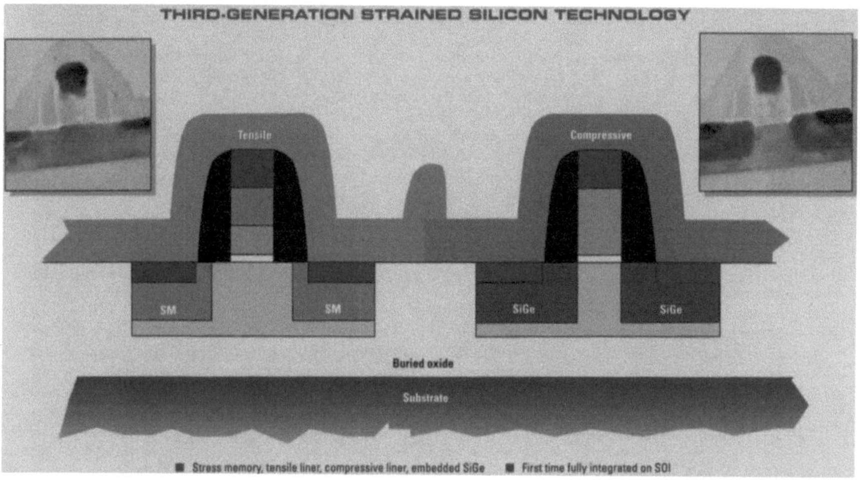

Fig. 1.24 Latest 45 nm device structure with strained engineering on SOI substrate (Reprinted with permission, Semiconductor International Jan. 2006, p. 37)

1.6 An Alternate Technology for Interconnects

Optical interconnects are likely to play an important role in the near future when all the other avenues with new exotic materials and architecture will be exhausted. It is expected that optical interconnects will reduce delay, increase bandwidth in signal wires and will reduce jitter, skew, and power in clock distribution. Figure 1.25 shows the expected signal delay when electrical interconnect is replaced by optical interconnect.

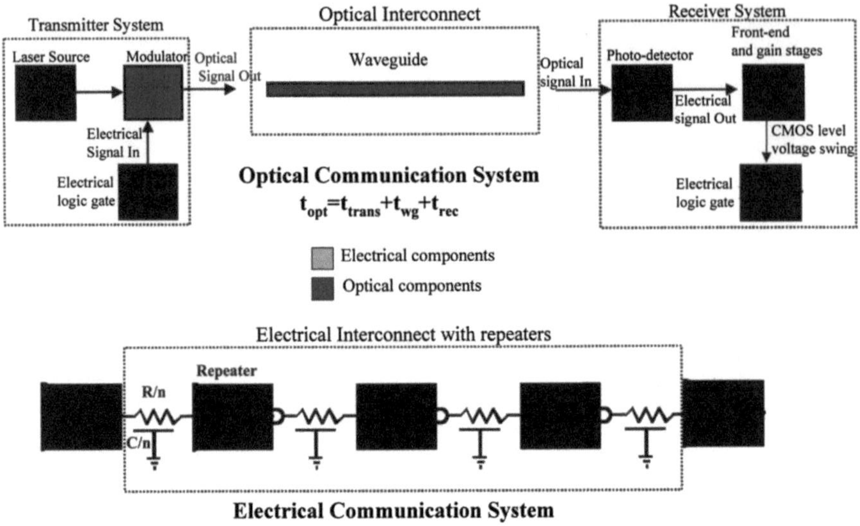

Fig. 1.25 Signal delays in optical versus electrical interconnects (Courtesy, Dr. K. Saraswat, Stanford University)

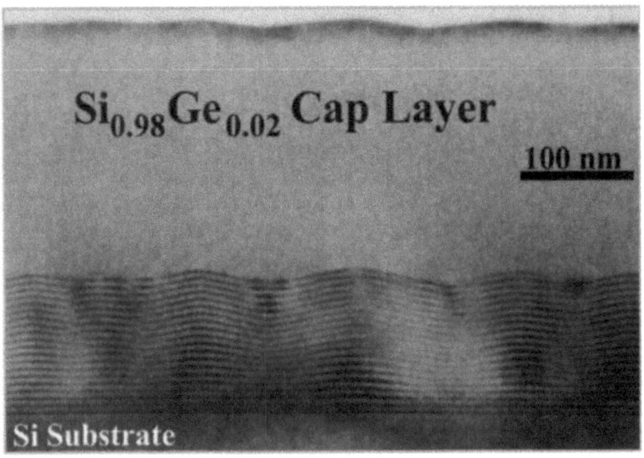

Fig. 1.26 Transmission electron micrograph of the Si/Ge unprocessed diode structure (Reprinted with permission Elsevier, UK, Material's Today, Dec. 2003, p. 9)

The real problem with optical interconnect is the lack of a proper light source. Two approaches are being taken to develop on-chip light sources: (a) use of erbium (Er)-doped silicon (Si) nanocrystals; and (b) fabrication of silicon (Si)/germanium (Ge) superlattices (Fig. 1.26) [70–71]. The second approach needs molecular beam epitaxial (MBE) technology, which is sophisticated and expensive. However, according to some experts, on-die optical distribution will not bring any significant advantages over scaled and non-scaled Cu-interconnects. Moreover, the realization of on-chip optical interconnects needs the development of high-speed low-capacitance CMOS compatible modulators and detectors [72].

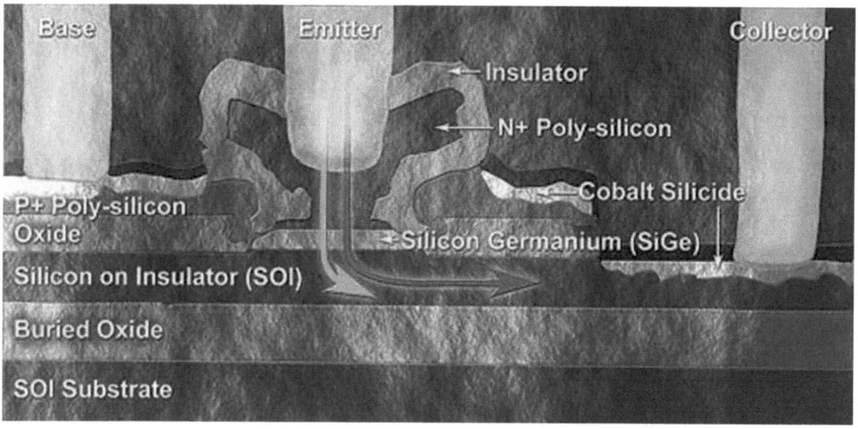

Fig. 1.27 A thin SiGe bipolar transistor on a SOI substrate (Reprinted with permission Semiconductor International, Nov. 2003, p. 30)

The excitement surrounding new materials (e.g. SOI, GeOI, Si/Ge) has never been greater [73]. High hopes are being placed on rapid technological developments of the *new materials* through advancements in the field of *integrated circuits* (ICs) [74–76]. Figure 1.27 shows a thin SiGe bipolar transistor on a SOI substrate.

We are going to discuss some of these materials, the processes to integrate them in ICs, new developments in the field of VLSI circuits, and some of the engineering maneuvers to handle copper interconnects.

1.7 Materials Used in Modern Integrated Circuits

Sub-micrometer devices with higher levels of speed and performance require proper materials and process technologies to integrate them into the integrated circuit (IC) [77–78]. The materials that are being used very often in integrated circuits (ICs) are (a) metals [79–81],(b) insulators/dielectric materials [82–83], (c) polymers and their derivatives [84–86], and (d) semiconductors [87–91].

(a) Metals: The most important property of the metal used as an interconnect in ICs is that it should offer minimum resistance to current flow, or in other words, the resistivity of the bulk metal should be low.

From Table 1.3, we can see that silver (Ag), copper (Cu) and gold (Au) have resistivities lower than aluminum (Al). Au has a resistance to electromigration (EM) but its resistivity is higher compared to Cu and Ag. Moreover, diffusion of Au in silicon (Si) will degrade the device properties [92]. Ag has the lowest resistivity, but shows EM and diffuses in Si and silicon dioxide (SiO_2) [93]. Moreover, the diffusion barrier of Ag is difficult to make [94–95]. Therefore, copper with resistivity 1.68 Ω-cm compared to 3.0 Ω-cm for Al-0.5% Cu alloy, and electromigration on the order of

Table 1.3 Materials of interest for microelectronics circuits*

Metal	At. number	At. weight	Structure	Lattice constant, nm	Atomic radius, nm	Density g/mL	Specific resistance μΩ-cm	Melting point (°C)
Aluminum (Al)	13	26.98	Fcc	0.2862	0.1431	2.70	2.65	660.32
Silver (Ag)	47	107.88	Fcc	0.2888	0.1444	10.5	1.59	961.78
Gold (Au)	79	197.0	fcc	0.2882	0.1441	19.3	2.04	1064.18
Copper (Cu)	29	63.54	fcc	0.2556	0.1278	8.96	1.68	1084.62
Lead (Pb)	82	207.21	fcc	0.3499	0.1750	11.3	19.0	327.3
Platinum (Pt)	78	195.09	fcc	0.2775	0.1386	21.45	9.81	1768.4
Tungsten (W)	74	183.86	bcc	0.2734	0.1367	19.3	4.9	3422
Chromium (Cr)	24	52.01	bcc	0.2498	0.1249	7.1	12.7	1907
Tantalum (Ta)	73	180.95	fcc/ bcc	0.33026	0.143	16.6	12.6	3017
Titanium (Ti)	22	47.90	hcp (α), bcc (β)	hcp: $a = 0.2953$ $c = 0.4729$ bcc: 0.333	hcp: 0.1448 bcc: 0.1432	4.5	5.0	1668
Silicon (Si)	14	28.09	diamond	0.54282	0.118	2.4	$\sim 10^{14}$	1410
Molybdenum (Mo)	42	95.95	bcc	0.314737	0.1362	10.2	5.2	2623

* Data are collected from different sources. The author acknowledges the help of the individuals.

ten times better than Al-alloy, will be the most appropriate material for interconnections [96]. Although Cu forms a deep level impurity in Si, several materials have been identified as its potential diffusion barrier [97].

1.7.1 Properties of Copper

Copper is a metallic element with atomic number 29, and atomic weight 63.54. The density of Cu is 8.96 g/cm^3. The atomic radius of copper is 1.4 Å with a packing factor (PF) of 0.74 cm^3/cm^3. Copper has a single s electron outside the filled 3d shell. The d shell is filled. Since the electrons of the d shell are involved in metallic bonding, the heat of sublimation and the melting point of copper are much higher than those of the alkali metals. These factors indicate the noble character of copper [98–99].

Copper can exist in two forms: cuprous, (valency 1), and cupric, (valency 2). However, the cuprous state of the metal complex is readily oxidized to the cupric state. So the cupric state is more stable. Copper melts at ~1084 °C (1983 °F). Most cuprous (Cu^{I}) compounds are oxidized to the cupric (Cu^{II}) state and can be presented in terms of the electrical potential [100–101]:

Copper metallized wafers can be ball-bonded very easily with ultra-fine pitch (pitch below 60 μm) copper wire of diameter 25 μm or less which is difficult with <25 μm-diameter gold wire [102–103]. The high-speed monometallic system (Cu-Cu bonding) shows longer term stability than the intermetallic system (Cu or Al-gold) and provides low manufacturing cost [104].

Starting from hull protection for sea-going vessels, to modern energy efficient motors and transformers, to high T_c superconducting materials, copper has been used either as a mainstream or as a supporting metal. The metal is malleable, durable, non-magnetic, recyclable, formable and easily machineable.

The introduction of copper as a metal interconnect in an integrated circuit has changed the fabrication procedure from subtractive metal etching to additive filling up of trenches or via holes with Cu. Before Cu-filling, the trenches or via holes are covered with a thin lining of a barrier layer to prevent out-diffusion of Cu to silicon. The technology is known as the *damascene process.* Figure 1.28 shows a schematic of a Cu-damascene architecture.

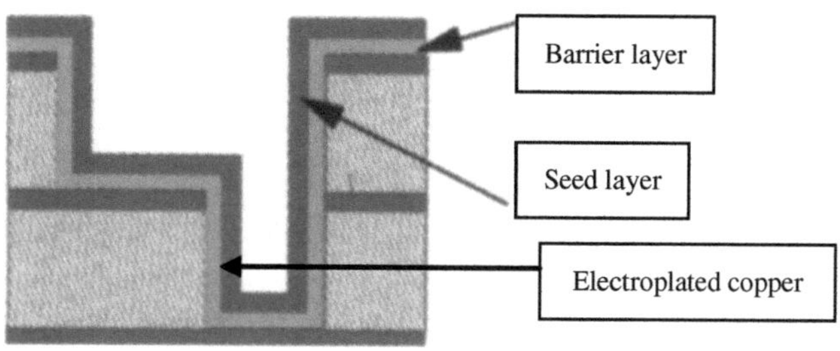

Fig. 1.28 The conventional Cu-damascene architecture

Copper is deposited by an electrochemical method over a Cu-seed layer. The Cu-film grows along the <111> direction with smaller grains but with time the grains grow larger even at room temperature. Studies have shown that electromigration (EM) performance, measured by the median time to failure (MTTF) can be improved by controlling the *texture* of the Cu-film. The strong <111> texture minimizes the presence of high angle grain boundaries along the interconnect line, thus minimizing the fast diffusion path for mass transport due to EM [105–106]. Thus a quantitative understanding of the nucleation and morphology that develops during the initial deposition time of the film is important for both fundamental and practical reasons [107–109]. A high volume production system used in commercial houses for copper electrofill is shown in Fig. 1.29.

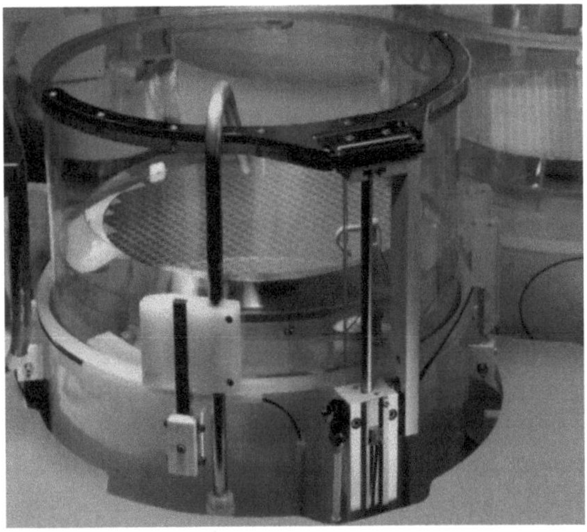

Fig. 1.29 A high volume production apparatus for copper electrofill from Novellus Corporation (Reprinted with permission, Semiconductor International, May 2000, p. 5)

1.7.2 Grain Size

Besides crystallographic orientation (texture), the performance and reliability of metal interconnects are dependent on the grain size as the line widths of the interconnecting lines approach $\leq$100 nm. As a matter of fact, during post depositions annealing electrical properties of thin films (including skin effects) are influenced considerably by the change in grain size and grain boundary characteristics (Fig. 1.30). On the other hand, the mean free path and the Fermi velocity of the charge carriers in traveling distances between two scattering events on grain boundaries seem to influence the resistance of the line when it is less than 100 nm. It is also true that EM in Cu-interconnect depends on the distribution of grain structures [110–111].

In Fig. 1.30 there are four groups and each group has three series, 1, 2, and 3. The series represents grain size, line width, and resistivity of the Cu-line in sequential

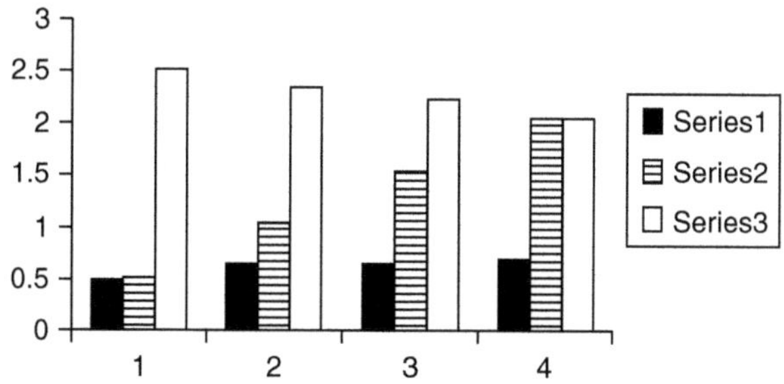

Fig. 1.30 The relation between grain size (in microns, series 1), line width (in microns, series 2) and resistivity (in ohm-cm, series 3) of Cu-interconnect

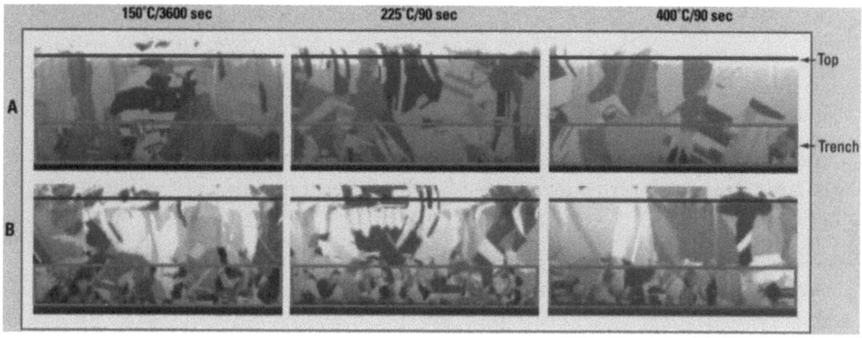

Fig. 1.31 Post annealing grain growth in electroplated copper inside a 5:1 aspect ratio trench (**A** and **B** are two films with different plating chemistries) (Reprinted with permission Semiconductor International, May 2006, p. 39, Courtesy: Novellus Systems)

order. The *Y*-axis represents numbers. The bar graph indicates how the grain size, line width and resitivity of a Cu-line are related to each other.

Figure 1.31 shows a comparison of two different plating chemistries on grain structure inside a 5:1 aspect ratio trench. For both chemistries, the grain size of Cu in the field is many times larger than the mean free path of the electrons, consistent with no grain boundary scattering in the copper after a post plating anneal.

1.7.3 Melting Temperature

The melting temperature (melting point, mp) of a metal plays an important role in determining the grain size of a metal during deposition. Aluminum (Al) has a low melting temperature compared to copper (Cu), which will show higher atomic mobility and larger grain size of an Al-atom compared to Cu-atom during deposition. As a result, thermal cycling can bring more grain growth in Cu-film, than in Al-film. As the grain size increases, the total grain boundary area and the resistance of the film are seen to decrease monotonically.

The melting temperature of a metal also influences the amount of stress that can develop within a film during temperature cycling. It has been observed that once the

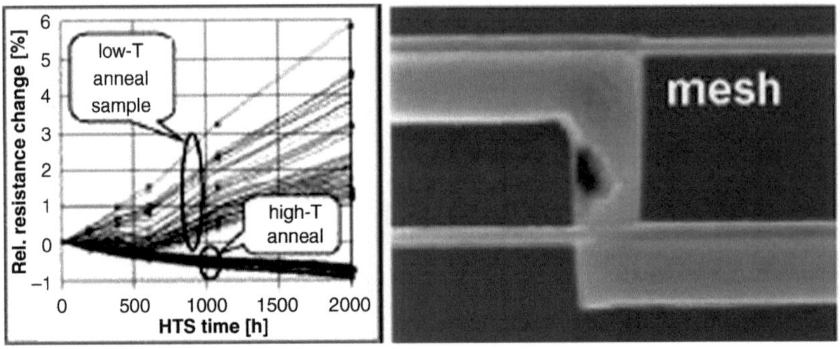

Fig. 1.32 *Left*: Resistance drifts during heating. *Right*: Thermal stress-induced voids (Courtesy, Infineon Technologies)

melting temperature of a metal has been exceeded, the maximum stress becomes very high near the top surface of the film. As a result, thermal stress-induced voids are formed that affect the resistivity of the film (Fig. 1.32) and current-induced Joule heating [112].

Table 1.4 shows the list of metals that are being used in *copper damascene technology* for interconnects and barrier layers (Fig. 1.28). Most of the metals used in the Cu-damascene process are *transition elements or their compounds*. The *transition metals* are highly conductive because of their incomplete sets of d electrons. Besides being highly conductive, the metals show high *cohesive energy*, which is the binding energy among metals. In general, they are malleable, ductile, and good conductors of heat and electricity. They form a large number of complex compounds and the rates at which these metals are attacked by oxidizing agents are small [113–114].

Table 1.4 Structure of the metals that are potential candidates for high density microelectronics circuits

Metals	Structure	Melting point (°C)
Titanium (Ti)	hcp, bcc	1668
Tantalum (Ta)	bcc	3017
Chromium (Cr)	bcc	1907
Molybdenum (Mo)	bcc	2623
Tungsten (W)	bcc	3422
Nickel (Ni)	ccp (fcc)	1455
Palladium (Pd)	ccp (fcc)	1554.9
Platinum (Pt)	ccp (fcc)	1768.4
Copper (Cu)	ccp (fcc)	1084.62
Silver (Ag)	ccp (fcc)	961.78
Gold (Au)	ccp (fcc)	1064.18
Aluminum (Al)	ccp (fcc)	660.32

Source: Printed with permission, CRC Handbook of Chemistry and Physics, 76th ed. 1995–1996, p. 4.

The transition metals are either face-centered cubic (fcc) or body centered cubic (bcc). During deposition as thin films, some of the transition metals change from fcc to bcc structure or vice versa. The structural change in the metals is always associated with a change in volume and strain in the film. The strain affects device performance especially when the device size is very small (on the order of a few nanometers) [115–118]. Thus during selection of a metal for interconnect or barrier layer one should consider the *physical, chemical, mechanical and electrical properties* of the bulk metal. We will discuss briefly some of the *properties of the transition metals* that are used as *barrier layer* in the copper damascene process.

1.8 Barrier Layer

Studies show that transition metals like tantalum (Ta), titanium (Ti), molybdenum (Mo), and tungsten (W) and their compounds are potential candidates as barrier layers for Cu-interconnects. However, an ideal barrier layer should effectively minimize copper diffusion, but should not introduce any additional resistance and capacitive reactance to the copper interconnects that will ultimately affect the speed of a device. Figure 1.33 shows two different processes to deposit barrier layers and their

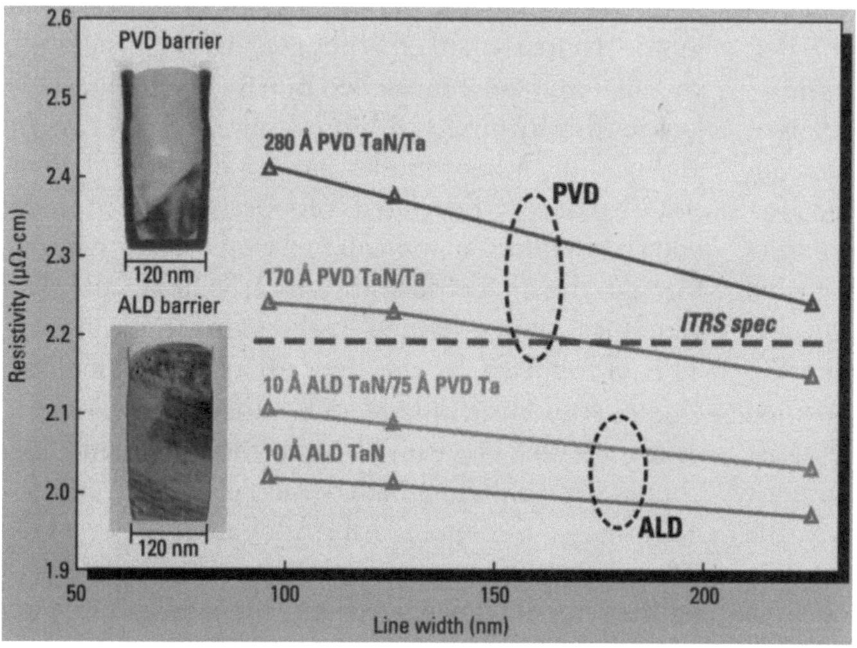

Fig. 1.33 The comparative analysis of the Cu-line resistances when the barrier layer is deposited either by PVD or ALD. The ITRS recommendation for line resistance is ~2.2 μm-cm (Reprinted with permission, Semiconductor International, May 2004, p. 42, Courtesy: Applied Materials)

consequences on the resistivity of copper lines. From Fig. 1.33, it is clear that compared to physical vapor deposited (PVD) barriers, atomic layer deposited (ALD) barriers will offer consistently lower line resistance.

Another important characteristic of a barrier layer in contact with an another metal or metal compound is its *electronegativity* – a measure of the degree of interaction between two intimate metals (or metal compounds) which can be a powerful concept in the selection of a barrier layer. In general one can expect that if two metals are in contact, the less electronegative metal will travel towards the more electronegative metal as part of the interaction between the two. However, in the presence of oxygen the rate of interaction (attraction of electrons) will be accelerated for the less electronegative element sitting on the top [119–121]. Electronegativity values of some metals of interest are given in Table 1.5:

Table 1.5 Metals of interest, with their electronegativity values*

Metals	Electronegative values
Copper (Cu)	1.75 (1.90) (1.9 Pauling)
Silver (Ag)	1.42 (1.93) (1.9 Pauling)
Aluminum (Al)	1.47 (1.61) (1.5 Pauling)
Gold (Au)	1.42 (2.54) (2.4 Pauling)
Platinum (Pt)	1.44 (2.28) (2.2 Pauling)
Nickel (Ni)	1.75 (1.91) (1.2 Pauling)
Tantalum (Ta)	1.33 (1.66) (1.0 Pauling)
Tungsten (W)	1.40 (2.36) (1.7 Pauling)
Titanium (Ti)	1.32 (1.54) (1.5 Pauling)
Molybdenum (Mo)	1.30 (2.16) (1.9 Pauling)
Chromium (Cr)	1.56 (1.66) (1.6 Pauling)
Zirconium (Zr)	1.22 (1.33) (1.4 Pauling)
Cobalt (Co)	(1.8 Pauling)

*Reprinted with permission from Cotton and Wilkinson [114].

1.9 Low-K Dielectric Materials

The resistance effect becomes serious when the interconnecting lines are small and thin. Since the birth of the integrated circuit (IC), the semiconductor industry has moved towards ultra-large-scale integration (ULSI). As a result, the device geometry has shrunk considerably and the circuit density (number of devices per unit area) has increased enormously. So to connect the maximum number of devices with short interconnecting lines (to reduce resistance and capacitive reactance, i.e. the *RC effect*) *multilevel metallization* (MLM) has been introduced [83–91, 122–123].

Figure 1.34 shows a scanning electron microphotograph (SEM) of a six level metal interconnect with a low-*K* material (SilK, $K{\sim}3$) as *inter-layer dielectric* (*ILD*).

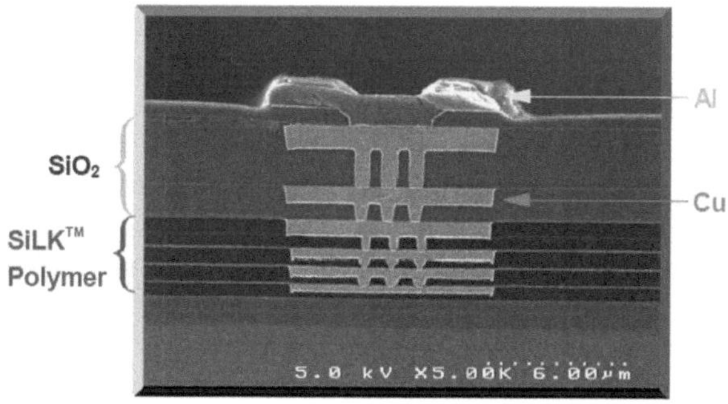

Fig. 1.34 Six level multilevel metal interconnects through via holes (Reprinted with permission, IBM Research)

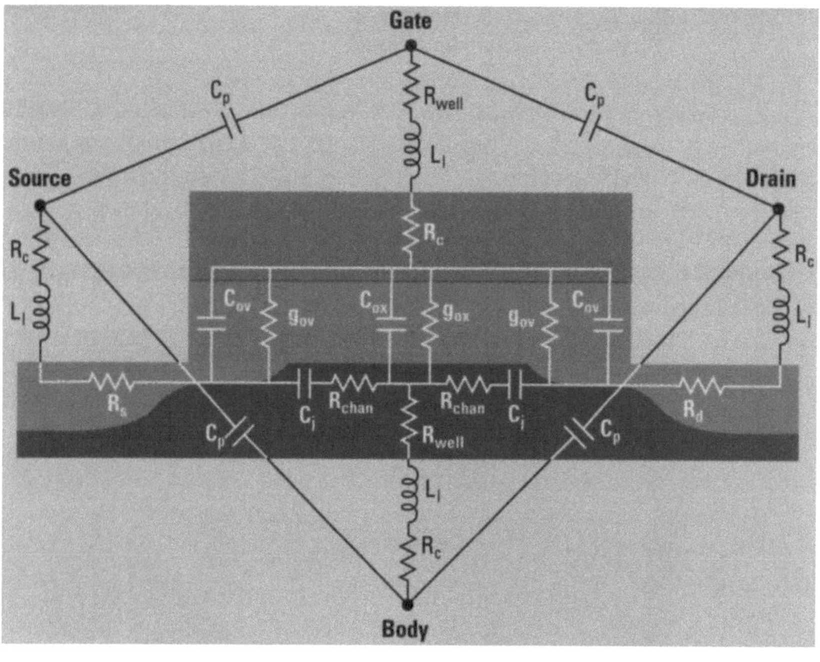

Fig. 1.35 A simplified circuit model of a MOSFET device. R, C, and L denotesthe resistance, capacitance and inductance, and ox, p, c, ov, chan, s, d, and l, represent oxide, pad, overlap, channel, source, drain, and lead, respectively (Courtesy, Semicond. Int.)

RC delay, which is linked to the capacitive reactance of the circuit (Fig. 1.35), can be minimized by the inclusion of a *low-K dielectric* material (*insulator*). Figure 1.35 shows a simplified model of a MOSFET device under test showing the equivalent circuit and the presence of parasitic *resistance and capacitance*. The model has been used to extract data to characterize equivalent oxide thickness (EOT) on thin SiO_2 and high-*K* dielectric in high performance logic devices. As a matter of fact, radio frequency (RF) measurement play an important role in accurate modeling of the dielectric by eliminating parasitic components that would prevent accurate representation of the C-V data by a traditional two-element model. It is expected that low-*K* interlayer dielectric (ILD) together with high-*K* gate material will minimize the parasitic capacitance effect.

Moreover, as the size of the gate length has scaled down to 10 nm with 1.5 nm oxide thickness, junction leakage, gate leakage, and off-state leakage became very much more prominent. Therefore, there is no doubt that the conventional dielectric SiO_2 will not be able to provide the required equivalent oxide thickness (EOT) for low power devices [124–125]. This has led to the search for a low-*K* dielectric material, which will be compatible with Cu-interconnects and will be a potential candidate for:

(a) inter-level dielectric materials (ILDs), and
(b) inter-metal dielectric materials (IMDs).

We will discuss the dielectric materials in a separate chapter.

There is an another class of *dielectric material* (oxide of metal), which is called a *ceramic* (e.g. alumina, Al_2O_3, beryllia, BeO, etc.). It is used mostly as substrate material in thick and thin film hybrid circuits [126]. The ceramic substrates are exclusively used in multilevel circuits because of the ease of creating holes through them. Figure 1.36 shows a low temperature co-fired ceramic (LTCC) tape with via holes drilled by laser [127–128]. The picture on the left-hand side shows a typical finished hybrid circuit.

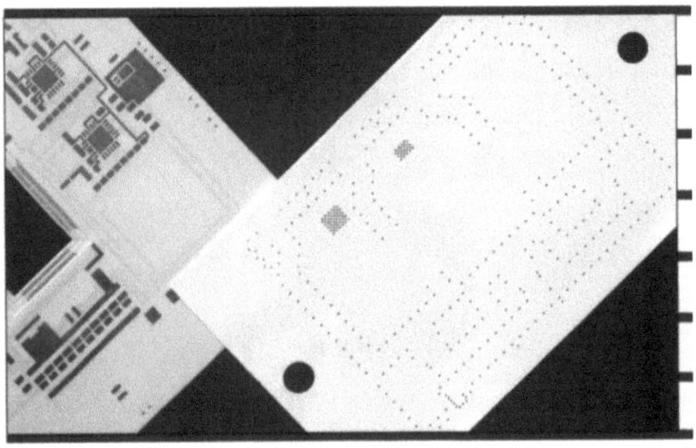

Fig. 1.36 A laser drilled ceramic substrate, and a built-in circuit on the substrate (Courtesy of Accu-Tech, San Marcos, CA)

1.10 Polymers

Polymers are plastic, non-metallic materials and are used as: substrates (in printed circuit boards, PCBs), [129], low-*K* dielectric materials (in *copper damascene processes*) [130], insulating materials (for forming crossover and capacitors in

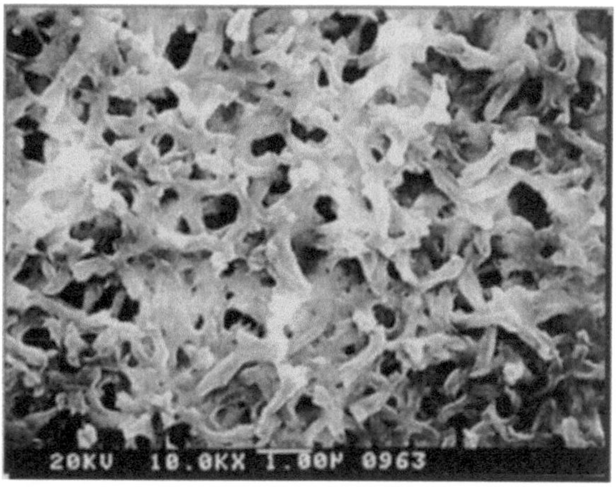

Fig. 1.37 Nylon 6,6 filtration membrane (Reprinted with permission SST, July 2000)

ICs) [133–135] and for creating high-resolution structures (*photolithography*) [131–132].

Development in the field of *polymer science* is so enormous that it is impossible for us to get through a day without using dozen or more synthetic organic polymers. Once a renowned polymer scientist remarked, *if you took away the man made polymer materials, you'd be utterly amazed how little you'd have left* [136]. By using supramolecular chemistry, scientists have been able to achieve control at the single molecule level, in the design of electro-active organic materials [137].

At the end of the twentieth century, it was difficult for Frazier to conceive that within 15 years 90 nm technology node will be possible in spite of fundamental physical limitations [138]. But the dream is going to come true because of the development of high molecular weight polymers which are known as *photoresist* [139]. However, resist slimming and profile control of the side wall have made critical dimension (CD) control more challenging. Although the acceptable 3σ scattering of the gate length is shared by lithography and etching at an optimum ratio, the tolerances in both technologies are approaching their limits. The isotropic filtration membrane supported by cage core and end caps provides an efficient filtering system for 193 nm lithography, which has enabled the production of advanced memory and logic devices with feature sizes of 100 nm and below. Figure 1.37 illustrates a typical nylon 6,6 membrane. Figure 1.38 represents the scanning electron micrograph (SEM) of a 180 nm feature created with DUV photoresist dispensed through polytetrafluoroethylene (PTFE). We will discuss photoresist materials in a separate chapter.

Recently, polymers are being used to fabricate light emitting devices (LEDs) and transistors [140–141]. Figure 1.39, shows the structure of an organic light emitting diode (OLED) where a thin layer of organic material (polymer) is sandwiched between anode and cathode layers.

Figure 1.40 shows a thin film transistor (TFT) fabricated on a flexible polymeric substrate, which is reported to achieve charge carrier mobilities in excess

Fig. 1.38 Top down scanning electron microscopic picture of a 40 nm feature created with DUV photoresist over a silicon wafer (Reprinted with permission, Solid State Technology (SST), July 200, p. 204)

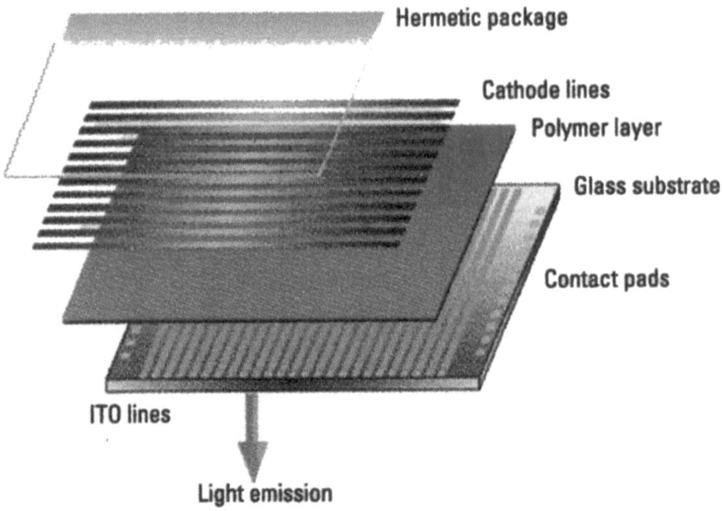

Fig. 1.39 The organic light emitting diode (OLED) display (Reprinted with permission from Semiconductor International. June 2003, p. 30, Courtesy, DuPont)

of 1 cm^2/V-s. It is expected that these transistors will be used in different commercial units, e.g. flat panel displays, light sensor arrays, and biochemical sensors [142–143]. Very recently use of synthesized organic conducting and superconducting polymers has been reported [144–145].

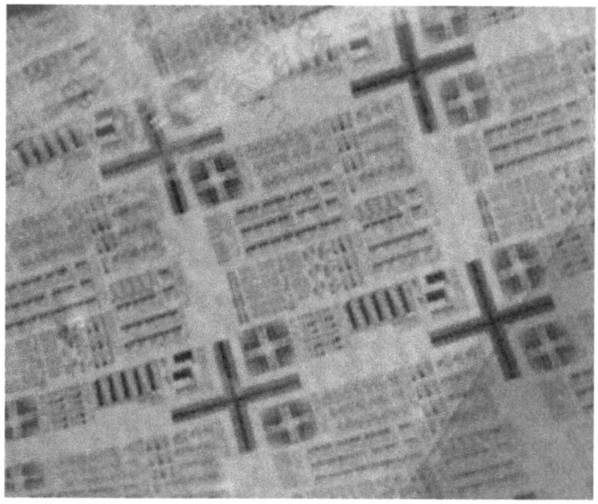

Fig. 1.40 Plastic thin film transistor used in an electronic circuit (courtesy, Infineon technologies, Munich, Germany)

1.11 Semiconductors

A semiconductor material does not have metallic cohesion, and therefore it has electrical conductivity smaller than metals (but higher than insulators). The conductivity of a semiconductor is a function of the energy gap. Dielectric materials and pure semiconductors are insulators at sufficiently low temperatures; however, dielectric materials retain their high electrical resistivities at high temperatures where the crystalline or semicrystalline phase ceases to exist, whereas semiconductors are reasonably good conductors at high temperatures [146–147]. A very distinctive property of semiconductors is that they possess negative temperature coefficient of resistance (NTCR) in some range of temperature [148].

1.11.1 Silicon (Si)

Silicon is the most extensively studied semiconductor, and modern VLSI technology is based almost entirely on silicon. In addition, integrated planar technology requires the capability of forming an insulating layer (SiO_2) on the semiconductor surface, which is accomplished very easily with Si. Germanium (Ge) and other semiconductors have difficulties in forming the oxides and /or the oxides formed are not reliable at different environmental conditions. These limitations make other semiconductors inferior materials for the fabrication of ICs, compared to *silicon* (Si).

The silicon industry has moved from a 100 mm-wafer fabrication process to a 150 mm-wafer fabricationprocess in six years, and to move from a 150-wafer fabricationprocess to a 200 mm wafer fabrication process it took another five years. Now the industry has gained enough experience of moving to larger wafer size (300 mm). Some industries are fully capable of handling 300 mm wafer at the production level (Fig. 1.41).

Silicon is an elemental semiconductor. It has a diamond crystal structure, which is characterized by extremely high stability. A single crystal of Si is grown mostly by

Fig. 1.41 The realization of a 300 mm silicon waferfabrication process from a 200 mm wafer fabrication process (Reprinted with permission Semiconductor International, June 2003, p. 47, Courtesy, Lam Research)

the Czocharlski method at high temperatures [149]. It is cut, polished, and cleaned. Generally two types of crystals are grown, <111> and <100>. The <111> direction grown single crystal is used for bipolar device fabrication andthe <100> direction grown is used for the fabrication of low-noise devices [150–151]. Figure 1.42 shows the range of electrical resistivities of pure and donor-doped silicon single crystals in comparison with metals and insulators [152–154].

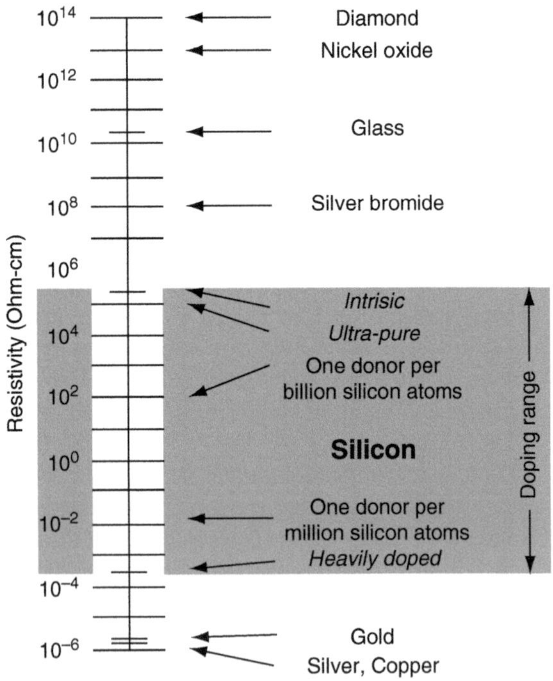

Fig. 1.42 Range of electrical resistivities of pure and doped silicon single crystals in comparison with metals and insulators (Reprinted with permission of AAAS, ref. [107])

1.12 Challenges and Accomplishments

1.12.1 Challenges

There are several challenges for the sub-100 nm level copper-damascene process. Some of them are identified as: introduction of low-*K* and high-*K* dielectric materials, deposition of very thin barrier and Cu-seed layers, and void free, continuous, and conformal deposition of Cu-interconnect inside high aspect ratio trenches and via holes. In addition to these, interfaces, contamination, adhesion, mechanical stability, electrical parametrics, thermal budget confounded by the number of wiring levels for interconnect, ground planes, and passive elements, have created *additional challenges* [155–157].

Although Cu-low-*K* is adequate for near term local interconnect in 65 nm node technology, global interconnect (whose length increases with scaling) performance at 65 nm node has posed several problems due to the resistance of the long wires. As a result, scaling has deteriorated all types of signal wire delays with respect to gate delay even with Cu-low-*K* (Fig. 1.43).

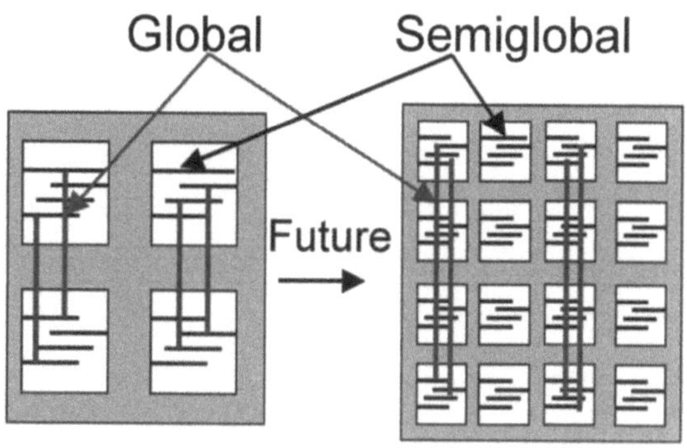

Fig. 1.43 The delay of signal wires due to the scaling of global wire length (Reprinted with permission, Dr. K. Saraswat, Stanford University, CA)

Increase in wire resistance due to scaling has caused more voltage drop and power consumption at constant voltage through equipotential wires to all V_{dd} bias points. Though ball-grid array packaging technology partially eliminates this problem, interconnect nets for distributing clock signals and power can lose 40–50% of the total power supplied to the chip. However, it is expected that the use of *superconducting* wires for global interconnects might be a possible avenue for reliable performance of global interconnects in the near future.

1.12.2 Accomplishments

The industrial research consortium in 1988 changed the technology node from 0.25 to 0.18 μm. This drove the industry to fabricate at production level the complementary metal oxide semiconductor(CMOS) device with six-level Cu-interconnects in 1997, which boasted several technical accomplishments of its own. In the same year the copper-based damascene process was adopted to fabricate a 0.20 μm CMOS logic device, which was operated at 1.8 V, and opened an avenue for commercialization of low power operated devices. Figure 1.44 shows a scanning electron micrograph (SEM) of a copper based CMOS circuit developed by using the *copper damascene process*.

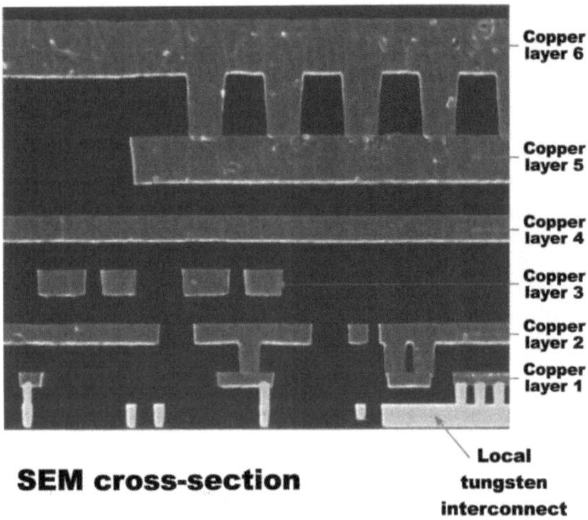

Fig. 1.44 SEM of copper-based CMOS device (7 s technology) developed at IBM (Courtesy of IBM research)

Soon the industry moved from 0.18 to 0.13 μm node technology and a CMOS device was fabricated with 18 Å gate oxide, adopting 248 nm lithography. Figure 1.45 shows a new 130 nm CMOS architecture designed with 18 Å gate-oxide [158–159].

The International Technology Road Map for Semiconductors (ITRS) 2001 evaluated Ti/TiN as a feature contact metal and barrier material for a high aspect ratio trench/via (e.g. DRAM stacked capacitor) filled with a tungsten (W) conductor. It seems that the ITRS evaluation might have stemmed from the success of fabricating

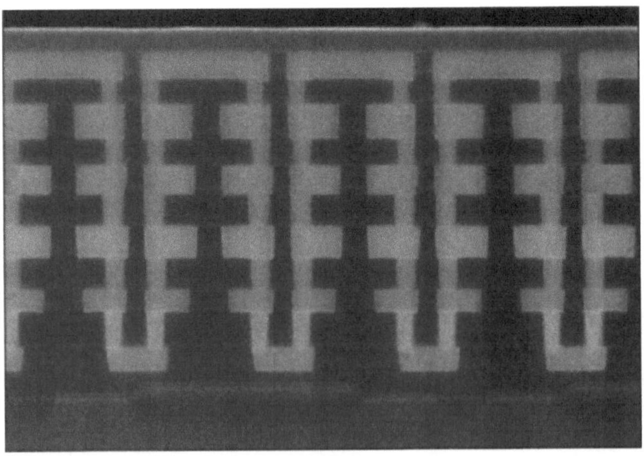

Fig. 1.45 130 nm CMOS architecture with nine levels of dual and single inlaid copper metal designed by Motorola (courtesy Free Scale Semiconductor, Inc.)

a copper based CMOS device with W/TiN contact plugs and collimated sputtered barrier layer in 1995 (Fig. 1.46) [160].

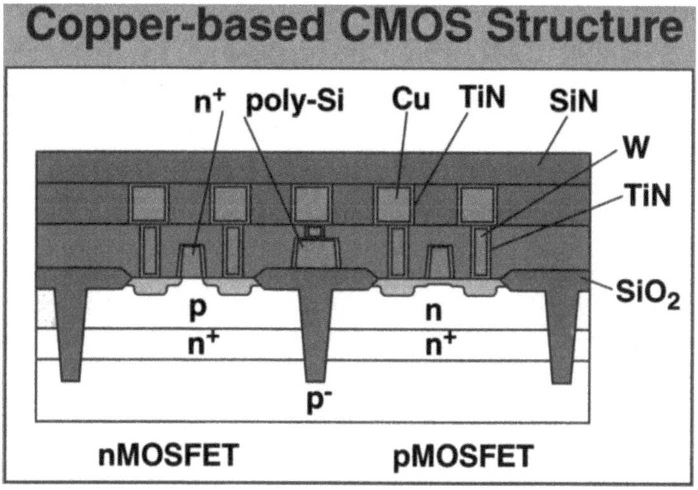

Fig. 1.46 A copper-based CMOS device using CVD W/TiN contact plugs and CVDCu/sputtered Cu/TiN 1st level metal (Reprinted with permission, Semiconductor International, July 1995, p. 68)

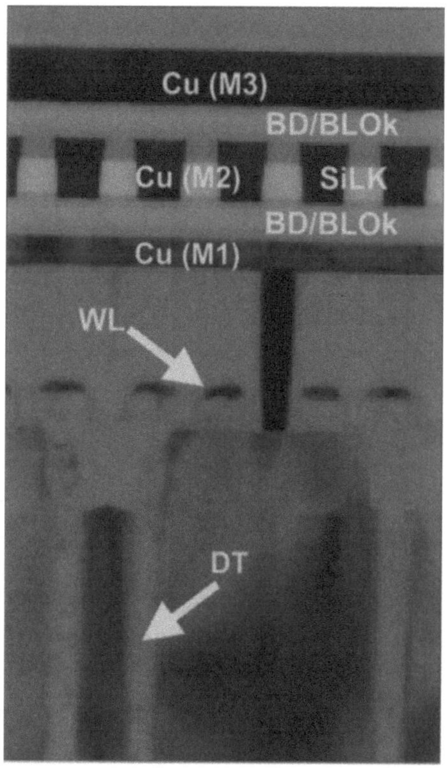

Fig. 1.47 An 8 Mb DRAM device with a three level Cu-interconnects. The cell size is 0.11 μm^2 (Reprinted with permission, Semiconductor International, Oct. 2003, p. 28)

The miniaturization of devices for high-speed circuits went on unabated, and by the end of 2003, a functional static random access memory (SRAM) chip was fabricated following 90 nm node technology. Subsequently, an 8-Mb dynamic read access memory (DRAM) device with 35 nm gate length and 0.57 um^2 cell size was fabricated by adopting the 65 nm process (Fig. 1.47).

1.13 Technologies of the 21st Century, and the Plan to Meet the Challenges

The last two centuries have seen an increasing flood of inventions and discoveries. These inventions and discoveries have reinforced one another and made possible new products, and new production methods. In the beginning of the 21st century we can see that many semiconductor industries have sprung up because of the discoveries of the fundamental research laboratories. According to the National Academy of Sciences in Washington, *the modern era of electronics has ushered in a second industrial revolution ... its impact on society could be ever greater than that of the original industrial revolution* [161].

Before the invention of the transistor, electronic devices including early computers depended on the use of vacuum tubes that were bulky and relatively power hungry. After 1957, the microelectronics industry witnessed miniaturization of the devices. Aggressive scaling of the feature size, like gate length in high performance device applications, makes it difficult to optimize device parameters. For example, the effect on *scaling* has pushed the gate length of a MOSFET transistor to 6 nm with less than 22 Å-gate oxide (Fig. 1.48).

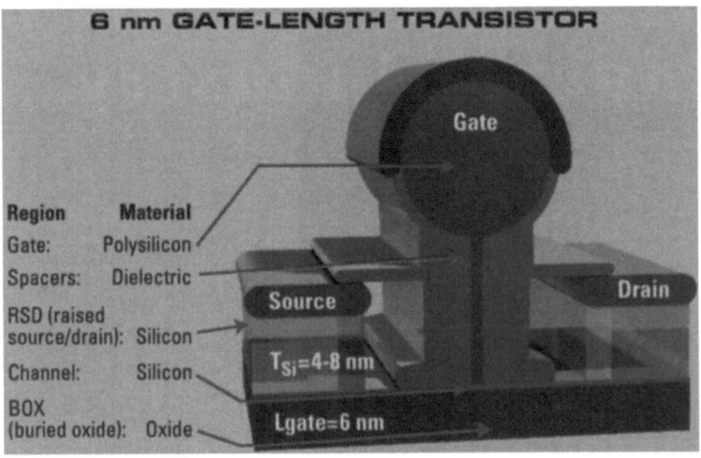

Fig. 1.48 A 6-nm gate length planar single gate MOSFET (Courtesy, IBM Research)

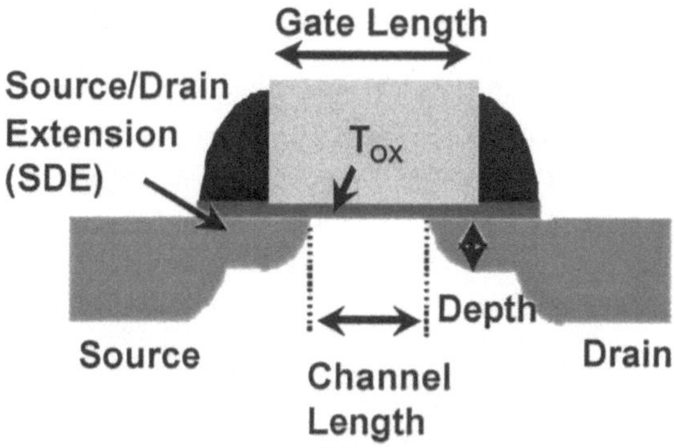

(a)

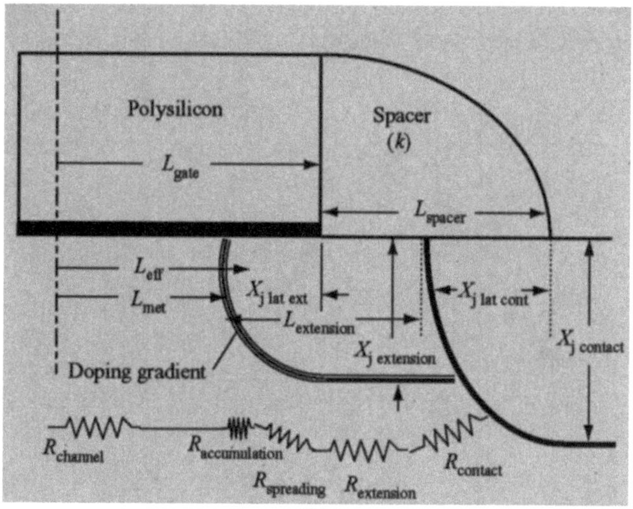

(b)

Fig. 1.49 (**a**) Cross-section of a CMOS transistor with dielectric spacing (Courtesy, Intel Tech.). (**b**) Drain extension structure for shallow junctions with the components of series resistance. The contacting junction separated by spacer is also shown in the picture (courtesy IBM Research)

In order to avoid the consequences of gate length scaling, a methodology for patterning ploysilicon at these extreme dimensions, called a *spacer gate*, has been used (Fig. 1.49) [162–163].

In a recent development the gate oxide has been scaled to 2 nm to achieve high drive currents and controllable short channel effects. As a result, for such a small gate length, one has to control the short channel effects and achieve sufficiently low external resistance and overlap capacitance by creating an *ultra-shallow junction (USJ)*.

1.14 Ultra-Shallow Junction (USJ)

Replacements of thin silicon oxide and depletion layers in doped polysilicon by high-K dielectric and metal gate electrodes to limit high leakage currents demand drastic junction depth (x_j) scaling of source/drain extension (SDE), as SDE is directly proportional to the gate length L_g ($x_j \propto L_g$). As a result, between 2001 and 2005, the shrinkage in junction depth became more than 40%. This aggressive scaling of the SDE demanded new implantation procedures of dopants including post annealing. Though these procedures have controlled the activities of the increased dopant levels, yet continuation of gate-length scaling, while controlling *short channel effects* in CMOS, was found to be difficult to achieve without incurring significant increase of *parasitic resistance*. Thus USJs (x_j <30 nm) required precise control of process, profile design, and careful monitoring of the electrical parameters, like sheet resistance, leakage current, and capacitance inside the USJ. Figure 1.50 shows the change in the doping profile and the sheet resistance inside a USJ.

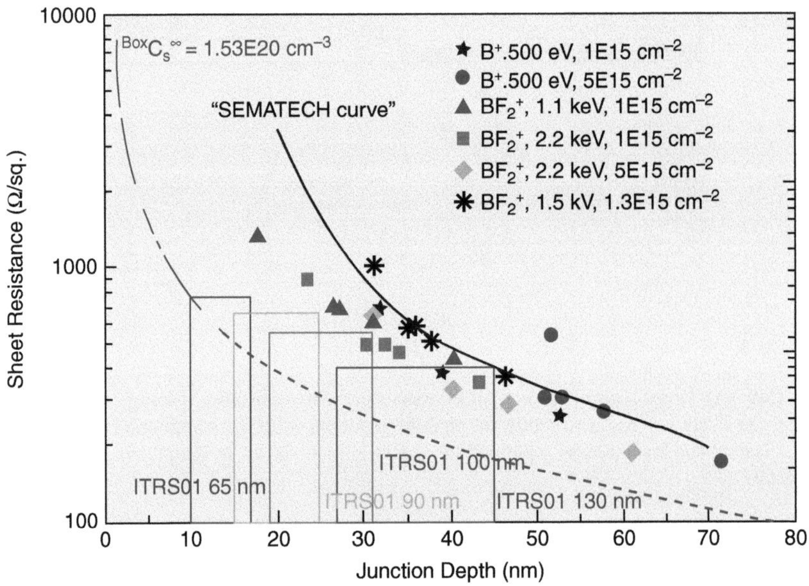

Fig. 1.50 Doping profile inside a USJ (Courtesy ITRS, SEMATECH, Austin TX)

Direct probe contact (four-point probe) to monitor the electrical parameters is difficult inside a USJ, and recently a non-contact optical probe method with modulated visible light has proven to be very successful (Fig. 1.51) [164]. As a rule of thumb, in a scaled USJ device the drain current is lowered to control the short channel effect. However, the solutions for these issues should be pursued concurrently with

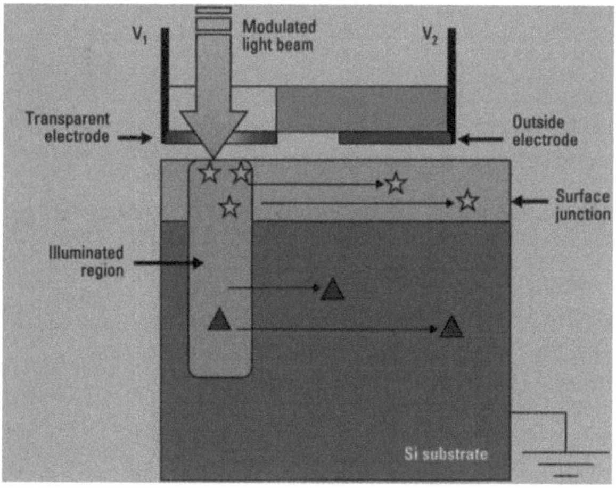

Fig. 1.51 A non-contact probe of p-n junction sheet resistance and leakage current (Reprinted with permission Semiconductor International, Nov. 2005, p. 38)

(a) circuit design and circuit-architecture improvements, (b) lowering power dissipation, and (c) integration of new materials and processes.

1.15 Circuit Design and Architecture Improvements

In order to reduce the *RC* delay especially in a high-speed device a typical multilevel metallization process might be a solution. The multilevel metallization has reached almost six to eight levels of metal with the lowest layer commonly acting as ground plane and the upper levels provide transmission lines, and routing traces. As a result, accurate and efficient modeling of interconnect *parasitics and delays* especially at high frequencies cannot be ignored [165,166]. At the same time, the 2-D and 3-D effects on interconnects must be considered with their statistical variations.

In the case of nanometer devices, front-end process modeling might be a solution. But the process seems to be challenging for the future IC industry. At the same time continued reduction of gate length demands a new high-*K* dielectric material which will be able to suppress the *tunneling current* while maintaining the drain current necessary for *low standby power* (LSTP). Thus for gate electrode, we need a new material and deposition process as oxynitride films have reached onan unacceptable level. On the other hand, in 65 nm node technology the basic transistor architecture will mainly be associated with processing and integration of the gate stack with a highly doped *ultra-shallow junction* (*USJ*) contacting the device.

Figure 1.52 shows a CMOS device with eight levels of Cu-interconnect. Low-*K* (SiLK) has been used as the inter-level dielectric (ILD). Tungsten (W) as the local interconnect level has increased wiring density with wire bond pitch (minimum C4 pitch on 8) of 60 μm in line.

Fig. 1.52 A CMOS with eight levels of copper interconnects with low-*K* (SiLK) ILD (Courtesy IBM Research)

1.16 Performance and Leakage in Low Standby Power (LSTP) Systems

In portable products where the power supply is low, suppression of the leakage current is achieved by a low standby power (LSTP) system. Unfortunately, it is difficult to maximize device performance under low operating power (LOP). Low power will challenge isolating phase locked loops (PLLs) due to the scaling of the supply voltage V_{DD}, enabling a wide range of input sources and output frequencies. It will also affect custom circuits, because analog devices and I/O drivers do not voltage-scale easily. Moreover, with low V_{DD}, design of traditional A/D converters will be challenging. On the other hand, a clocking subsystem should have features like low voltage, and low power PLL to generate a high-frequency clock (Fig. 1.53). It is hoped that improvements in frequency and circuitry will reduce the number of gates and average gate delay in a clock by 30% [167–168].

In CMOS devices, reduced operating voltage greatly reduces active power. As a result, the maintenance of acceptable off-state leakage with continually decreasing channel lengths will require increased channel doping levels. This will degrade short channel effects of extremely small (nano-size) devices. As a matter of fact, when the device size is less than 90 nm, time-dependent dielectric breakdown (TDDB) and electromigration (EM) phenomena will be the biggest challenges. Thus to maintain low gate leakage current and high performance of the device, we neednew materials and advanced processes to integrate these materials.

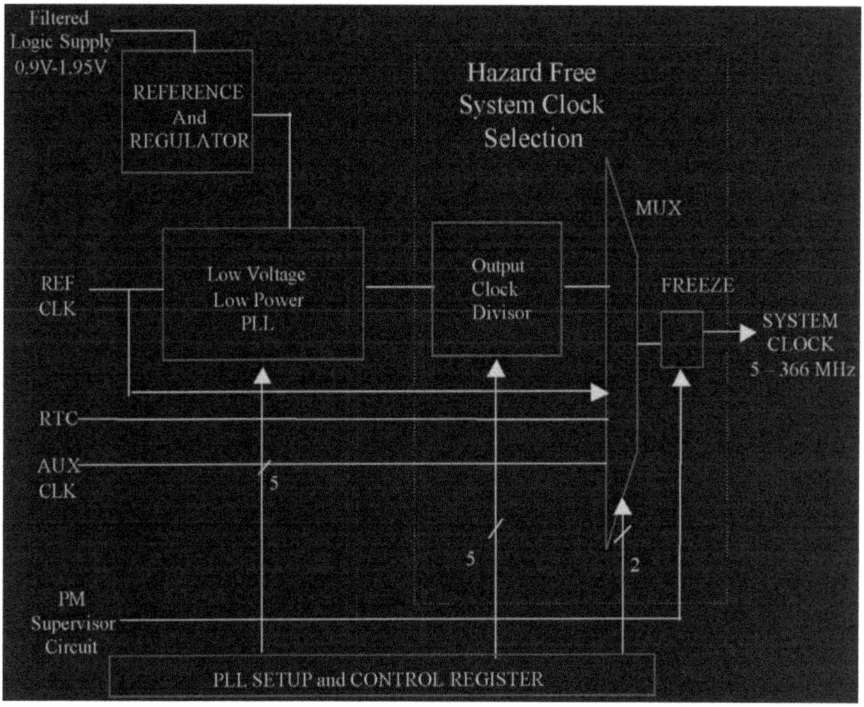

Fig. 1.53 Scalable clock generation subsystem (Reprinted with permission, IBM Research)

1.17 Introduction of New Materials and Integration Processes

From our discussions and presentations so far about so many research oriented results, it is clear that to minimize interconnect delay to increase device speed, successful integration of low-resistivity metal and low-dielectric material is essential. As a mater of fact, a Cu-low-*K* system is already at the production level. However, the deposition processes for these materials are still under investigation. For example, high temperature (>400°C) chemical vapor deposition (CVD) processes confront issues with wafer bowing [169–170]. But the challenges for physical vapor deposition (PVD) are somewhat different. It goes without saying that optimization of the PVD [171–172] source to obtain equivalent or better within-wafer uniformity is critical. It is true that compared to CVD or PVD, the low-temperature atomic layer deposition (ALD) process is better for very thin films (of the order of an ångstrom unit) but for thick films it is not a very cost-effective procedure [173–174]. On the other hand, the electroplating process (ECP) of copper, where deposition is performed at room temperature, has come a long way [175–178]. Recent developments of self-adjusting and self-monitoring electrolyte systems allow precise control of the plating chemistry. But still this process suffers from *terminal effects (higher rate of deposition of Cu on the edge).*

Integrated circuits (ICs) are typically classified in terms of the minimum feature size of the devices. The minimum feature size has come to 100 nm at the beginning of the 21st century with an expectation of a 65 nm production feature by the year 2009. But fact of the matter is that the impact of shrinking the feature size on silicon costs yield and reliability. For example, the resistivity of Cu interconnect below 100 nm size is expected to increase because of electron scattering at the barrier–metal interface. Thus it is likely that in the near future for nano-scaled (below 100 nm) devices we need to replace the copper interconnect. As a matter of fact, researchers are exploring nano-scale interconnect architectures, in organic and inorganic materials [179–180], and the use of *superconducting nano-wires* might be a future possibility.

1.17.1 Nano-Materials

Nano-technology [181] is one of the fascinating branches of science based mainly on the study of single-walled carbon nanotubes (CNTs) [182]. First identified by Iijim [183], carbon nanotubes are fabricated by several methods, e.g. arc discharge [184], laser ablation [185], and by chemical vapor deposition (CVD) [185]. A single walled carbon nanotube can have a diameter as small as 1.5 nm with length exceeding tens of microns [186]. By tailoring the electronic functionality, an atomic scale transistor has been fabricated from a 1.6 nm wide CNT [187]. Another example of a nano-scale organic device is a molecular diode, which is being fabricated from aromatic hydrocarbons and conductive polymers [188]. CNTs are promising candidates for single electron transistors [189–191], and field effect transistors (FETs) [192]. Thus materialization of 50 nm device node interconnect is a fair possibility in the very near future [193–195]. As a matter of fact, who could guess that the baby transistor, a few inches long in 1947, would be reduced to 100 nm at the beginning of the 21st century? Figure 1.54 shows a picture of a single nanotube field-effect transistor (FET).

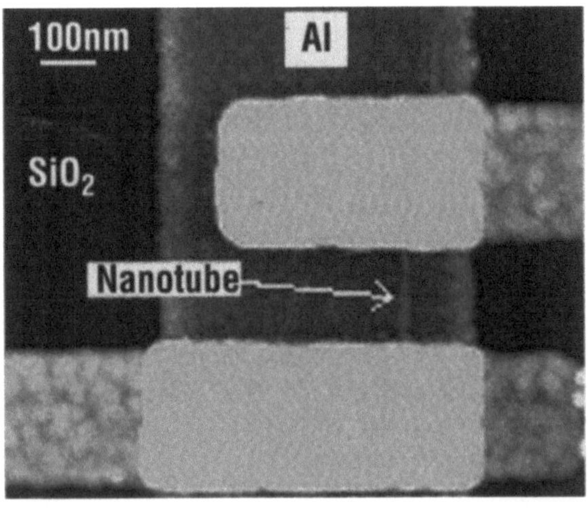

Fig. 1.54 A single nanotube transistor with an aluminum (Al) gate (Reprinted with permission, Solid State Technology, Feb. 2002, p. 24)

In the year 2004, researchers from International SEMATECH and the University of Texas at Austin, fabricated a transistor on an ultra-thin (<10 nm) silicon-on-insulator (SOI) substrate with 60 Å gate length. The features are at the nano-wire level and it will require reliable instruments to test the different parameters of the device.

Figure 1.55 shows a future transistor with gate length on the order of 60 Å. The features of the device are at the nano-level. The device has been fabricated on a thin SOI substrate. It is hoped that these nano-size devices will enable us to gather fundamental knowledge about device physics.

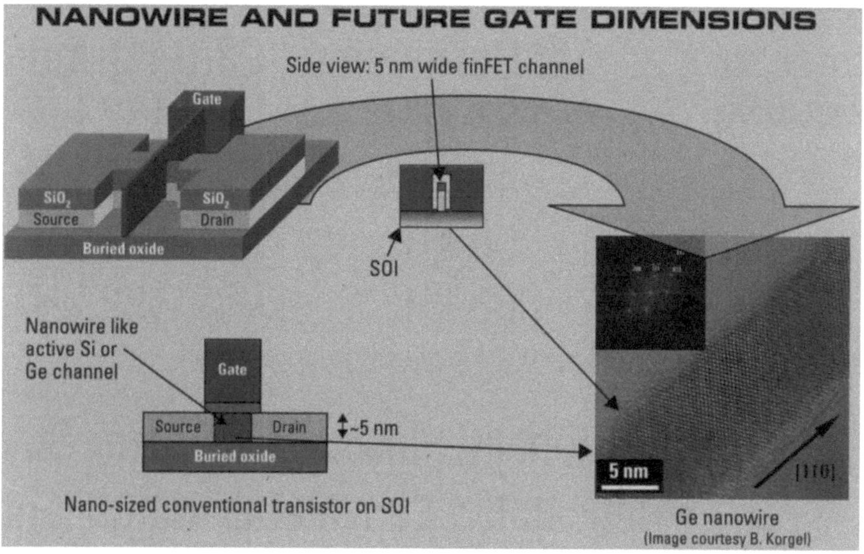

Fig. 1.55 The future transistor with 60 Å gate on an SOI substrate (Reprinted with permission, Semiconductor International, May 2004, p. 36)

1.17.2 Superconductors

In a nano-scale device, conduction through small interconnects creates several problems [196–197]. Potential solutions must clearly involve noble conducting material/materials, which will be able to eliminate the fundamental quantum mechanics problems. The idea of superconductivity and the invention of high-temperature superconductors (high T_c) in 1987 brought new hopes in the field of interconnect technology [198–202]. Bulk superconductivity based on a copper–oxygen array is an important structural feature (Fig. 1.56). But neither the metallic bond, nor copper–oxygen bonding guarantees superconductivity [203]. Figure 1.57 shows a picture of a single crystal of $Y_1Ba_2Cu_3O_{1-x}$ superconductor.

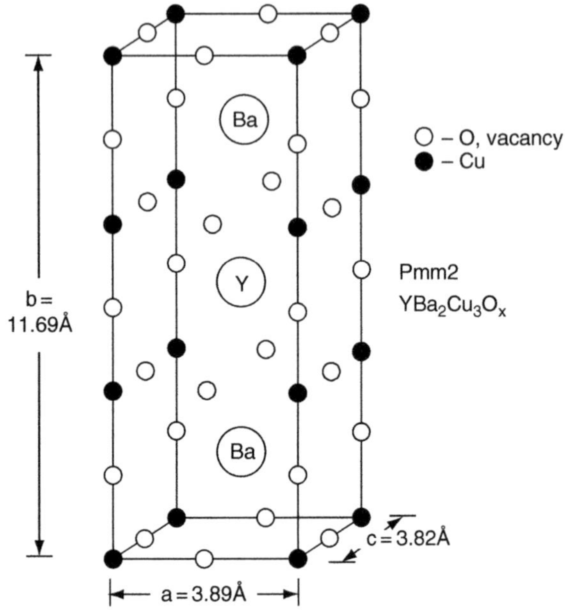

Fig. 1.56 The unit cell of the oxygen-defect perovskite $YBa_2Cu_3O_x$ (Photo courtesy, IBM Research)

Fig. 1.57 A single crystal of $Y_1Ba_2Cu_3O_{1-x}$ (Photo courtesy, IBM Research)

There is an another class of superconductors which is known as a *layered superconductor* [204] where we have two interpenetrating perpendicular vortex lattices consisting of Josephson vortices (JVs) and pancake vortices (PVs). It has been demonstrated by Savele'ev and Nori [205] that owing to the mutual JV-PV interaction and asymmetric driving, the ac motion of JVs and/or PVs can produce a net dc vortex current. The controllable vortex motion can be used for making pumps, diodes, and lenses of quantized magnetic flux [206–209].

From all the evidences there is no doubt that divalent copper assumes a distorted *Jahn–Teller* coordination where oxygen forms the sensitive Cu–O sublattice structure and the total oxygen stoichiometry is responsible for high T_c superconductors. Thus it will not be unfair to say that copper has proven its superiority over other metals as a conductor, and as a key material in superconductors. Benefiting from all its incredible ability to extend technologies, it goes without saying that this reddish-brown soft metal has established its superiority in every sphere of life.

1.17.3 Integration Processes

1.17.3.1 3-D Technology

Among all the integration processes described so far, the three-dimensional (3-D) integration process is thought to be one of the emerging technologies in Cu-interconnecting systems. According to some experts [210] the 3-D technology will lead the copper interconnecting paradigm shift due to its tremendous *benefits*.

The potential *benefits* of 3-D integration will vary depending on the approach. Some of the benefits are: (a) increased performance of the circuit, (b) reduced power, (c) multifuctionality, (d) small form factor, (e) reduced packaging, (f) flexible heterogeneous integration, (g) increased yield and reliability, and (h) cost [211].

Among various 3-D technologies [212–213] that are presently been implemented, two of them are worth mentioning: (i) fabrication of *nano-integrated circuit chips* (ICs), and (ii) packaging.

The *nano-integrated circuit chips* that are designed on 3-D technology can achieve the highest density of silicon transistors (because of *thinning* and *stacking*) with wafer level functioning using advanced photolithography. As a matter of fact, the electrical performance of CMOS devices is fundamentally determined by physical structure, e.g. gate length and 3-D activated dopant profiles within the devices [214]. It is expected that 3-D design will reduce the fabrication costs for high volume production of these nano-integrated circuit chips. However, processing constraints (3-D trench capacitors for SiP and atomic layer deposition (ALD) of multiple MIM layer stacks of high *K* dielectrics and conductive layers) especially the thermal budget, affect the properties of the transistors and limit the material choices mostly to silicon (Si) and tungsten (W) [215–217].

Through-silicon via (TSV) is the latest progression of technologies for making stacking silicon devices in 3-D. Different techniques may be used for TSV drilling and filling to allow for 3-D die and wafer stacking with a small form factor [218]. Stacking two wafers together, at the same time integrating them with vertical vias adopting 3-D technology, is a challenge because of the difference in design rules. As a result, the whole process becomes expensive [211]. However, there is no question that sooner or later TSV technology will be adopted, but the timing for mass production depends on how the TSV technology compares in terms of cost with the existing technologies.

As a matter of fact, 3-D integration of the wiring system and placing them properly within the circuit promises higher clock rates, lower power dissipation, and higher integration density. TSVs are already being put into production for CMOS image sensors, where the active silicon area is bonded into glass and contacted from the back side. For some applications, a high bandwidth memory interface to the logic has been the main driver for the development of TSV technologies. However, the current design tools used for 2-D ICs cannot be easily extended to 3-D ICs [219].

3-D integration and high-density (HD) packaging, such as system-in-package (SIP), package-on-package (PoP), die-to-die integration, die-to-wafer integration, wafer-level back end of line (BEOL)-compatible 3-D hyper-integration (hyper-integrate terahertz technology, mixed signal processing, wireless and optical systems) [220], have become important parts in the development of nano-technology and future telecommunication systems.

According to some pundits, 3-D back end of line (BEOL)-compatible 3-D hyper-integration is perhaps the most attractive 3-D technology because of its flexibility for heterogeneous integration of different materials, processing technologies and functional components with additional benefits in cost and performance.

Recent study reveals that 3-D integration for system-in-package modules requires a high resistivity silicon (HRSi) substrate. Functions such as broadband couplers, filters, matching networks, etc. are easily integrated on an interposer high resistive silicon substrate. On the other hand, AF-45 glass, which was previously used for MCM–D, has been transferred to an HRSi substrate (ρ>4 kΩ) allowing the integration of high quality passive components and circuits (e.g. performance of a 7 GHz MCM-D power splitter and a 50 GHz distributed bandpass filter). This study is a significant step forward in the development of 3-D thin film RF modules [221–223] and future generations of cellular RF transceivers that require a higher degree of integration using 3-D generation.

Taking into account all the viewpoints of the experts in this field, one can say that once the 3-D integration technology has matured and the manufacturing infrastructure (such as ECAD tools, fabrication equipment and standards) is in place, it is very likely that more ICs will be designed following 3-D integration because production of 3-D ICs would lower the production cost [220].

1.17.3.2 Air Gap Technology

Due to constant shrinkage of the critical dimensions in ultra-large-scale integration (ULSI) the performance of ULSI chips is increasingly limited by the capacitance of the SiO_2 dielectric. The capacitance of a circuit in Cu-interconnect ismostly related to the resistance and the capacitance of the circuit which influences the speed (*RC* delay), AC power (CV^2f) and cross-talk. Though current low-*K* materials like fluorinated SiO_2, amorphous carbon fluoride, and polymer carbon reduce the *K* value of the circuit, they require additional processing and pose many challenges of reliability, manufacturability, and integration. Additionally, as integrated circuits (ICs) continue to scale, the aspect ratio of metal lines increases so that the intralevel dielectric (IaLD) capacitance increasingly dominates the interlevel dielectric (IeLD) capacitance in determining the total interconnect performance [224–226]. To reduce the interconnect capacitance, the introduction of low-*K* with microscopic pores have been investigated aggressively [227].

Air gaps formed between metal lines during SiO_2 deposition address many of the concerns associated with low-*K* materials while offering comparable if not better capacitance reduction [228–229]. Air gaps reduce the dominating IaLD capacitance, while leaving the IeLD SiO_2 intact which provide the structural integrity of the interconnect stack. These air gaps can be produced by fine tuning the CVD oxide process.

Different air gap integration approaches have been applied to fabricate multilevel interconnects. The approaches can be classified mainly in two categories: (a) partial or complete material removal in between metal lines followed by non-conformal CVD deposition, because conformal CVD requires an additional dedicated air-gap lithography for each metal level;however, to overcome the additional processing and cost blanket dielectric etch processing has been suggested [230]; and(b) damascene integration of metal lines in a sacrificial material, which can be selectively removed through a dielectric cap [231–232]. However, the damascene process requires an additional lithography/etching step to pattern trenches [233]. Moreover, the aspect ratio of the trenches (closure point height, width and volume of the cavity) must be controlled.

The most important drawback of air gap engineering is the deterioration of the mechanical strength of the stack due to the presence of the air gap though the thermo-elastic stress around the air gap is not very high [234]. At the same time exact placement of air gaps in specific locations is difficult. As a result, mechanical integrity is not achieved fully. However, mechanical integrity is less problematic for non-conformal CVD approaches compared to the sacrificial material approaches, since the CVD dielectric contributes to the mechanical strength of the stack in wide spaces. In addition to the mechanical stability, the reliability of air gap engineering in nano-scale devices raises several questions regarding, via reliability, electromigration failures [235], degraded barriers, interline leakage and air gap breakdown [236–237].

Recently, IBM has introduced this air gap technology to computer chip manufacturing. The method deployed by IBM causes a vacuum to form in between the copper wires on a computer chip, allowing electrical signals to flow faster (35%), while consuming less power (15%). A vacuum is believed to be the ultimate insulator for what is known as wiring capacitance, which occurs when two conductors (in this case adjacent wires on a chip) sap or siphon electrical energy from one another, generating undesirable heat and slowing the speed at which data can move through a chip. The new technique adopted by IBM to make air gaps by self-assembly skips the photolithography and associated processes. Once the holes are formed, carbon silicate glass is removed, creating a vacuum between the wires. According to Dr. Douglas Paul, senior research associate at Cambridge University, *35% speed increase is quite a significant jump in performance just by changing one process in the fabrication of the chip.* The self-assembly process has already been integrated with IBM's state of the art manufacturing line in East Fishkinn, NY, and is expected to be fully incorporated in IBM's manufacturing line by 2009.

1.17.3.3 Fabrication Status of Sub-Micron (65 nm and Below) Devices and Reliability Issues

65 nm Node Technology

In January 2006, about two years after Intel shipped the first 90 nm Prescot, it began shipping 65 nm versions of its Pentium D Presler processor. According to the spokesperson of Intel, the gate length of the 65 nm device has been reduced to 40 nm from 65 nm which was used for their 90 nm Prescot product, without any alteration of the gate oxide thickness. It is worth mentioning that in 2006 Intel shipped 70 million 65 nm microprocessors [238]. According to Mark Bohr, Senior Intel Fellow, 65 nm transistors increase the drive current by 10–50% with enhanced strain, an improvement that can either provide a leakage reduction of $4\times$ or improve the CPU performance [239].

Texas Instruments (TI), on the other hand, has been quietly producing 65 nm products for third parties such as Nokia since March 2006 using ASIC design flow. The Nokia baseband processor has a small die size of 13.2 mm^2. According to TI, its 65 nm process boosts transistors' performance by 40% and reduces power consumption from idle transistors by a factor 1000 [240].

In December 2006, AMD entered the 65 nm node technology with an open market availability of its 110 mm^2 Athlon64 X2 dual core desktop processor. The process innovation includes a silicon-on-insulator (SOI) substrate, a unique mobility enhancement scheme and nine levels of Cu-metallization [228].

Xilinx Inc. San Jose, CA has delivered 12 devices across three of the four platforms (LX, LXT, and SXT) from the second half of 2006 (May 15, 2006) through the first half of 2007. According to the spokesperson of Xilinx, their Virtex-5 LX FPGAs can deliver unprecedented performance and density gains, with 30 percent faster speed and 65 percent increased capacity over previous generation 90 nm FPGAs [241].

By the end of 2006, TSMC, Taiwan, has manufactured the foundry's first functional 65 nm embedded DRAM. The 65 nm embedded DRAM process (80 GC) provides high bandwidth, low power consumption, and small cell and macro size, making it ideal for game consoles, high end networking, and processors for consumer and multi-media applications [242].

Semiconductor International Company, (SMIC) China, has been in the pre-production phase since June 2007, and is expected to deliver production from December 2007 of their 2 Gb Nand flash devices.

45 nm Technology

In the beginning of 2008, Texas instruments (TI) Inc. announced that their 45 nm transistor technology has now widening its way toward mass production in the second half of that year [243]. The biggest challenge TI's scientists and engineers are facing is the difficulty in controlling the leakage while scaling the silicon oxynitride (SiON) gate dielectric by several ångstroms and shortening the gate length. However, TI's engineers have been successful in creating abrupt junctions to control the leakage. There are continual improvements to the implant tools supported by millisecond annealing techniques. In addition, they have adopted immersion lithography for selected critical layers avoiding embedded silicon-germanium (Si-Ge) stressors or high-*K* dielectrics. As a matter of fact, hyper-NA immersion lithography has brought 45 nm and below imaging within reach [244]. TI reported their 45 nm device fabrication progress at the Solid State Circuits Conference (ISSCC) in San Francisco, in February 2008.

Intel technology, on the other hand, has been successful in implementing 45 nm node technology fortheir high-*K* gate transistors. The introduction of a high-*K* gate dielectric has enabled a 0.7× reduction in T_{ox}while reducing gate leakage 100× for PMOS and 25× for NMOS transistors. In addition to the high-*K* dielectric, Intel has introduced dual-band edge work-function metal gates eliminating polysilicon (poly-Si) gate depletion, which has greatly improved the compatibility with high-*K* dielectrics. In addition to the high-*K* gate, 35 nm gate length CMOS transistors have been integrated with a third generation of strained silicon that have demonstrated the highest drive currents to date for both NMOS and PMOS. They further reported that Intel has been successful in achieving an SRAM cell size of 0.346 μm^2 by using 193 dry lithography [245]. On November 12, 2007, Intel shipped 45 nm nanometer microprocessors using high-*K* metal gate technology [246].

Advanced Micro-Device's (AMD) researchers have developed a low aspect ratio FinFET-like transistor at the 45-nm node in the year 2007. The transistor's switching speed – expressed in CV/I, a measure of capacitance, voltage, and current – was 0.26 ps for NMOS and 0.45 ps for PMOS transistors. According to AMD those are the fastest transistors with 20 nm gate length so far reported by any manufacturers.

The gate surrounds a vertical channel, rather than the planar structure, which stacks the channel, gate oxide, and electrodes between the source and drain. AMD has used fully silicided (FUSI) metal gates, instead of electrodes made of polysilicon (poly-Si). Rather than depositing the nickel material, AMD has used the silicidation process to gradually replace polysilicon with nickel silicide to form the metal gate electrodes. AMD has also employed fully depleted SOI combined with a metal gate that creates a strain on the silicon in the channel delivering high mobility electrons and holes. According to the spokesperson of AMD, their multigate device demonstrates 50 percent better performance than other multigate devices available on the market. The structure has a lower aspect ratio than the conventional FinFET, which eases the burden on the lithographic tools and its depth of field. The multigate device was reported by AMD at the International Conference of Solid State Devices and Materials (SSDM) in Tokyo on Thursday, September 18, 2007.

32 nm Technology

There are no unsurmountable obstacles foreseen for scaling 45 nm devices to 32 nm node in 2009. However the transition poses several challenges that have remained unsolved. The fourth generation of uniaxial stress including films on planar transistors will debut with the 32 nm-logic technology node. At the same time, nitride stressors for both NMOS and PMOS will reach >3 Gpa compressive and over >2 Gpa tensile stress for blanket film depositions [247]. Since electrons and holes respond differently to the strain, 32 nm node device fabrication should need different process parameters to enhance electron and hole mobility. However, by careful integration these stresses can be transferred to the channel of the device [248–249].

At 32 nm node, metal 1 and metal 2 pitches will be nearly 100 nm with aspect ratios 1.7 and 2.3, respectively. As a result, an effective bulk dielectric constant of the low-*K* material will be ~2.2. High-*K* metal gate implementation will require new materials for 32 nm node with atomically smooth surfaces as most of the hafnium based oxides degrade the channel mobility [250]. The dielectric barriers should have reduced *K* value (low carbon) and film thickness for 32 nm node compared to 45 nm node.

The integration of the gate stack for the 32 nm node may require either of two unique methods, e.g. two different dielectrics and one electrode or two different metal work functions and one dielectric [251].

Double exposure with dry 193 nm ArF and single exposure water based 193 nm ArF immersion lithography will not meet the printing requirements. However, it is most likely that immersion lithography combined with either dual exposure or double patterning might meet the printing requirement.

As a matter of fact, no destructive technologies such as new substrates, biaxial stress, or double gates, will be needed; instead 32 nm will continue to use silicon uniaxial stress, and planar transistors. Tables 1.6 and 1.7 illustrate process enhancements that will enable semiconductor manufacturers to deliver significantly greater functionality in less area.

Table 1.6 SRAM cell sizes

Manufacturer/alliance	65 nm SRAM cell size μm^2	45 nm SRAM cell size μm^2	32 nm SRAM cell size μm^2
TSMC	Not reported	0.242	0.15
ST Micro, Free scale, NXP	Not reported	0.25	Not reported
Fujitsu	Not reported	0.255	Not reported
Intel	0.57	0.346	Not reported
IBM	0.54	0.37	Not reported
Texas Instruments	0.49	Not reported	Not reported
IBM, Chartered, Infineon, and Samsung	0.54	Not reported	Not reported
IBM, Toshiba, Sony, and AMD	0.65	Not reported	Not reported

Source: Real world technologies, Process technology advancements at IEDM 2007.

Difficult challenges:

Table 1.7 Interconnect difficult challenges for $\geq$ 32 nm node technology*

Difficult challenges $\geq$32 nm	Summary of issues
Introduction of new materials to meet conductivity requirements and reduce dielectric permittivity	Introduce integration and characterization challenges
Engineering manufacturable interconnect structures compatible with new materials and processes	Integration complexity, CMP damage, resist poisoning, degradation in *K* value. Lack of interconnect/packaging architecture design optimization tool
Achieving necessary reliability	New materials, structures, and processes create new chip-reliability (electrical, mechanical, and thermal) exposure. Detecting, testing, modeling and control of failure mechanism will be the key
Three-dimensional control of interconnect features (with its associated metrology) is required to achieve necessary circuit performance and reliability	Line edge roughness, trench depth and profile, via shape, etch bias, thinning due to cleaning, and CMP effects. The multiplicity of levels combined with new materials, reduced feature size, and end pattern dependent processes create this challenge
Mnufacturabilty and defect management that meet overall cost/performance requirements	As feature sizes shrink, interconnect process must be compatible with device roadmaps and meet manufacturing targets at the specified wafer size. Plasma damage, contamination, thermal budgets, cleaning high A/R features, defect tolerance processes, elimination/reduction of control wafers are the key concerns. Where appropriate, global wiring and packaging concerns will be added in an integrated fashion

*Adopted from The International Technology RoadMap for Semiconductors, 2006.

1.18 Summary

Integration of one billion transistors into a single chip will become a reality before 2010, although some industries are very much optimistic and hope to achieve this before 2009. The chip will be built on 45 nm CMOS technology with transistors having a gate length of around 25 nm. It seems that CMOS will remain the main driving horse as industry continues toward the scaling limit.

In the meantime, *130 nm* technology has become a volume production in some leading IC industries for high-performance and high-density products. However, *180 nm* technology is still used in many industries for production of many logic circuits.

Development of *65 nm* technology is in full swing in the industry and there are some reports in the market that some leading IC industries have already in production 65 nm logic circuits following 65 nm node technology. As a matter of fact, there is no sign of any slowing down in scaling the logic technologies to the 65 nm generation.

The trends of high chip integration densities, enhanced integrated circuit (IC) performance without propagation delays, cross-talk noise, and reduced cost per function have been the driving force in device scaling, with the consequence of increased interconnect complexity. Integration of copper (Cu) interconnect with low-*K* dielectric material in place of aluminum (Al) interconnect and silicon dioxide (SiO_2) dielectric material has resulted in the reduction in the number of interconnect layers and manufacturing costs.

The introduction of new materials and process technologies for Cu-interconnects has raised several questions due to additional work related costs. These include conformal Cu-interconnects with conformal barrier layer, chemical mechanical polishing (CMP) to form inlaid copper lines/plugs, adhesion at the interface, contamination, and reliability of the low-*K* materials.

The inlaid dual damascene Cu-interconnect structure can be fabricated by sequential depositions of PVD (physical vapor deposited) barrier and Cu-seed, and ECD (electrochemically deposited) copper. The processes are adequate for 0.18 μm node technology, but at 0.1 μm and beyond physical vapor deposited (PVD) barrier and seed layers will not be able to provide void free filling. However, it has been argued that low Cu-resistivity, and good barrier integrity can be achieved by metal organic chemical vapor deposition (MOCVD).

Lithography consumes a significant portion of the manufacturing cost especially when the radiation wavelengths are shortened. Today, almost everyone agrees that we have hopeful potential solutions for extending general CMOS scaling to at least the *hp 32 nm node*. At this node, minimum transistor gate lengths are projected to be within the 13–15 nm range, which is still within theoretical CMOS device limits. What makes people more nervous in this regime are things like affordable lithography.

In the field of ion implantation and annealing, monitoring of the precision steps to reduce device variability has become routine work as the number of implant steps is increasing with each device generation. For example, with system-on-a-chip (SOC) devices, the implantation steps can average ~38 for low and high voltage transistors. Implant micro- and macro-dopant precision and variation directly affect device threshold voltage variability for high energy and high current implantation. Therefore, implantation damage, amorphization, recrystallization, and silicidation must be accurately simulated.

The stress–strain behavior of silicon-on-insulator (SOI) structures in CMOS processing has also to be addressed. As a matter of fact, *strain engineering*, which has opted for 45 nm device structure, can be described as *gain with strain and some pain*. It is also true that control of high-aspect-ratio technologies such as damascene challenges all metrology methods. On the other hand, void detection in the copper seed layer, sidewalls of trenches and via holes, pore size distribution in patterned low-*K* materials, and proper CD measurements, are the areas of much concern. Most of all, the cost corresponding to customers' rapidly changing complex business requirements should not be too high.

References

1. T.K. Gupta, Aluminum alloy as an interconnecting material in the fabrication of integrated circuits, *Micron. Reliab.*, **19**, 337 (1979); T.K. Gupta, Integrated Circuit, Chapter-1, p. 3, in Hand book of Thick and Thin film Film Hybrid Microelectronics, T.K. Gupta, Wiley, NJ (2003); K.N. Tu, Recent advances in electromigration in very large scale integration of interconnects, *J. Appl. Phys.*, **84** (9), 5451 (Nov. 2003); J. Cong, L. He, C.K. Koh and P.H. Madden, Performance optimization of VLSI interconnects lay out integration, *VLSI J.*, **21**, 1 (1996); G. Ta, C. Ouvrad, H. Chauveau, and S. Nath, Experimental study of carrier transport in multi-layer structures, *Microelectron. Relib.*, **47** (5), 610 (2007)
2. G.E. Moore, Lithography and future Moore's law, Proc. VIIIth. Optical /Microlithography Conf. SPIE, Vol. 2440, pp. 2–17 (Feb. 1999); G.E. Moore, Cramming more components onto integrated circuits, *Electronics*, **38** (8), April 19 (1965); T. Kudo and N. Kimizuka, US Patent, 6853037 (2005); X. Ma, *Nano Technol.*, **19**, 275706 (May 2008)
3. *2001 International Technology Road Map for Semiconductors* (ITRS), H. Kamura et al., *J. Solid State Circ.*, **39** (6), 919 (2006); C. Tsui, R. Yi-c Au, and R.Y.K Choi, Integration, *VLSI J.*, **19**, 275706 (May 2008)
4. G.C.F. Yeap, Leakage current in low standby power and high performance devices: Trends and challenges, ISPD 02, April 7–10, 2002, San Diego, CA; J. Baliga, Power devices, in S.M. Sze (ed.), *Modern Semiconductor Devices*, Chapter 4, John Wiley, New York (1998); T.M. Pan, C.L. Chen, W.-W. Yeh and W.J. Lai, *Electrochem. Solid State Lett.*, **10** (3), H101 (2007)
5. N. Yang, W.K. Hensen, and J.J. Wortman, A comparative study of gate tunneling and drain leakage currents in N-MOSFETs with sub-2 nm gate oxide, *IEEE Trans. Electron. Dev.*, **47** (8), 1636–1644 (2000); R.S. Muller and T.I. Kamins, *Device Electronics for Integrated Circuits*, John Wiley, New York, Chapter 9 (1986); B. Mheen, Y.-J. Song, J.-Y. Kang, K. Hshim, and S. Hong, *Mater. Sci. Semicond. Proc.*, **7** (4–6), 374 (2004)
6. T. Ghani et al., Scaling challenges and device design requirements for high performance sub 50 nm gate Planar C-MOS transistors, Tech. Digest of 2000 Symp. VLSI Tech. pp. 174–175 (2000); S.M. Sze (ed.), *Modern Semiconductor Devices*, Chapters 3 and 5, John Wiley, New York (1998); N. Oda et al., *Jpn. J. Appl. Phys.*, **46**, 954 (2007)
7. C.-H. Choi et al., Impact of gate direct tunneling current on circuit performance: A simulation study, *IEEE Trans. Electron. Dev.*, **48** (12), 2823–2829 (Dec. 2001); J. Sunea, I. Placencia, E. Farreas, N. Barniol, and X. Aymerich, *Phys. Status Solidi (a)*, **109** (2), 479 (2006)
8. P.A. Pecan, Pushing the limits, *Science*, **285** (5436), 2079–2081 (1999); Z. Lu et al., *IEEE Trans. VLSI*, **15** (20), 159 (2007)
9. Table I, Semiconductor Industry Association, International Technology Road Map, 1999, Austin, TX: Int. SEMATECH (1999); K. Mistry et al., *IEEE IEDM Dig.*, p. 247 (2007)
10. J.A. Devis et al., Interconnect limits on gigascale integration in the 21st century, *Proc. IEEE*, **80** (3) (2001); S. Thompson et al., *Int. Tech. J.*, **6** (2), (May 16, 2002); G. Decher and J.B. Schlenoff (eds.), *Multilayer Thin Films*, Wiley-VCH, Weinheim, Germany (2002)
11. D. Silvester and C. Hu, Analytical modeling and characterization of deep-sub-micrometer interconnect, *Proc. IEEE*, **89** (5) (2001)
12. C.M. Osburn et al., Vertically scaled MOSFET gate stacks and junctions: How far are we likely to go? *IBM J. Res. Dev.*, **46** (2/3), 299 (March/May, 2002); F. Sacconi et al., *IEEE Trans. Electron. Dev.*, **51** (5), 741 (2004)
13. T. Edwards and M. Steer, *Foundations of Interconnect and Microstrip Design*, John Wiley, New York (2000); J. Meindl, Theoretical, practical andlogical limits in ULSI, IEEE Int. Dev. Meeting, (IEDM '83) pp. 8–13 (Dec. 1983); G. Moore, VLSI: Some fundamental challenges, *IEEE Spect.*, **16**, 30 (1979); R. Aghavani et al., *Semiconductor Fabrication Techniques*, 35th edition ICG Pub., UK, (2007)

14. A.K. Sinha, J.A. Cooper, and H.J. Levinstein, Speed limitations due to interconnect time constants in VLSI integrated circuits, *Electron Dev. Lett.*, **3**, 90 (1982); H.B. Bakoglu, *Circuits, Interconnections, and Packaging for VLSI*, Addison Wesley, Reading, MA, Chapters 4 and 5 (1993); J.W. Seight et al., Challenges and opportunities for high performance 32 nm NMOS Technology, IEEE IEDM Dig., p. 697 (Dec. 2006)
15. W. Zang et al., Surface and grain boundary scattering studied in beveled polycrystalline thin copper-film, *J. Vac. Sci. Technol. B*, **22**(4), 1830 (2004)
16. J. Huo et al., Characteristics of copper films produced by atomic layer deposition, *J. Mat. Res.*, **17** (9), 2397 (2002); J. van Olmen et al., IEEE IITC Tech. Dig., p. 49 (2007)
17. G.L. Gan, C.V. Thompson, K.L. Pey, and W.K. Choi, Experimental characterization and modeling of reliability of three terminal dual damascene Cu-interconnect trees, *J. Appl. Phys.*, **94** (2), 1222 (2003)
18. S.P. Hau-Reige and C.V. Thompson, *J. Appl. Phys.*, **88**, 2382 (2000)
19. R.H. Havemann and J.A. Hutchby, High performance interconnects: An integration overview, *Proc. IEEE*, **89**, 586 (2001); X. W. Lin and D. Pramanik, Future interconnect technologies and copper metallization, *Solid State Technol.*, **41** (10), 63 (1998)
20. R.A. Powel, A.S. Hurrus, and R. Hill, *Raising the RC Speed Limit by the Use of Copper Interconnects*, Novellus Systems Inc., San. Jose, CA (2000); T.N. Theis, The future of interconnect technology, *IBM J. Res.*, **44** (3), 379 (2000)
21. *International Technology Road Map for Semiconductors* (ITRS), SEMATECH, Austin, TX (2002); A. Pratt, *Overview of the Use of Cu-Interconnects in Semiconductor Industry*, Advanced Energy Industries Inc., Fort Collins, CO (2004)
22. R. Liu, C. Pai, H. Cong, W. Lai, and E. Martinz, Impact of interconnect architecture on chip size and die yield, IEEE Int. Intercomm. Tech. Conf., p. 21 (1999)
23. W.R. Hunter, *IEEE Trans. Electron. Dev.*, **44** (2), 304, (1997)
24. T.K. Gupta, *Handbook of Thick and Thin Film Hybrid Microelectronics*, John Wiley, Hoboken, NJ (2003)
25. J. Cizek et al., Thermal stability of ultra-fine grained copper, *Phys. Rev. B*, **66** (9), 195331 (2002); A.A. Volincky and W.W. Gerberich, Nonindentation techniques for assessing mechanical reliability at nano-scale, *Micron. Eng.*, **69**, 519 (2003)
26. C.K. Hu, L. Gignac, S.G. Malhotra, R. Rosenberg, and S. Boettcher, Mechanism for very long electromigration lifetime in dual damascene Cu interconnects, *Appl. Phys. Lett.*, **78** (7), 904 (2001); I.A. Blech and C. Herring, *Appl. Phys. Lett.*, **29**, 131 (1976)
27. S.P. Hau-Riege and C.V. Thompson, *J. Appl. Phys.*, **89**, 601 (2001); J.R. Lloyd, *J. Appl. Phys.*, **69**, 7601 (1991)
28. E.T. Ogawa, K.D. Lee, V.A. Blaschke, and P.S. Ho, Electromigration reliability issues in dual damascene Cu-interconnects, *IEEE Trans. Reliab.* **51** (4), 403 (Dec. 2002); L.J. Sham, *Phys. Rev. B*, **12**, 3142 (1975)
29. R. Rosenberg, D. Edelstein, C.K. Hu, and K.P. Rodbell, *Annu. Rev. Matter. Sci.*, **30**, 229 (2000); W.L. Schaich, *Phys. Rev. B*, **13**, 3350 (1976)
30. J.R. Black, Electromigration – a brief survey and some recent results, *IEEE Trans. Electron. Dev.*, **ED-16**, 338 (1969)
31. R. Streiter, H. Wolf, Z. Zhu, X. Xiao, and T. Gessener, Thermal and electrical simulation on deep submicron interconnecting system, *Microelectron. Eng.*, **60**, 39 (2002); J.A. Cunningham, improving Cu-interconnects, *Semicond. Int.* April 1 (2000)
32. S.P. Murarka and M.C. Peckerar, *Electronic Materials, Science and Technology*, Academic Press, San Diego, CA, p. 314 (1989); N. Ranganathan et al., *J. Micromech. Microeng.*, **18**, 075018 (2008)
33. M.J. Kobrinsky et al., On chip optical interconnect, *Intel Technol. J.*, **8**(2) 136, May (2004); Z. Lu et al., *IEEE Trans. VLSI Syst.*, **15**(2), 159 (2007)
34. E.S. Ege and Y.L. Shen, Thermomechanical response and stress analysis of copper interconnects, *J. Electron. Mater.*, **32**(6), 1000 (2001)
35. V. Weihnact and W. Bruckner, *Acta Mater.*, **49**, 2365 (2001)

36. Y.L. Shen and U. Ramamurty, *J. Appl. Phys.*, **93**, 1806 (2003)
37. P.M. Igic and P.A. Mawby, *Micro. Electron. Relib.*, **40**, 443 (2000)
38. S. Ouimet, and M. Paquet, Overmold technology applied to cavity down ultra fine pitch PBGA packages, 48th Proc. ECTC, p. 458 (1998); T.K. Gupta, *Handbook of Thick and Thin Film Hybrid Microelectronics*, John Wiley, Hoboken, NJ, Chapter 10 (2003); A.G.K. Visanath et al., *IEEE Trans. Adv. Packaging*, **30**(3), 448 (2007)
39. T. W. Ellis, L. Levine, and R. Wicen, Copper: Emerging material for wire assembly, *Solid State Technol.*, **43** (4), 71 (April 2000); T.K. Gupta, *Handbook of Thick and Thin Film Hybrid Microelectronics*, John Wiley, Hoboken, NJ, Chapter 8 (2003)
40. A. Bischoff and F. Aldinger, Reliability criteria of new low cost materials for bonding wires and substrates, 34th Proc. ECTC, p. 441 (1984);US Patent, 7328830, (Dec. 2008)
41. E.S. Ege and Y.L. Shen, Thermomechanical response and stress analysis on Cu-interconnects, *J. Electron. Mater.*, **32** (10), 1000 (2003)
42. G.M. Phar, D.S. Harding, and W.C. Oliver, in M. Nastasi et al., (eds.) *Mechanical Properties and Deformation Behavior of Materials Having Ultra-Thin Micro-Structures*, Kluwer, Dordrecht, p. 449 (1943); P.S. Tsao et al., IEEE Proc. 54th Electron Comp. and Technol., **1** (1–4), 767 (2003)
43. H. Gao, L. Zang, W.D. Nix, C.V. Thompson, and E. Artz, Crack like grain boundary diffusion wedges in metal films, *Acta Mater.*, **47**, 2865 (1999); M.J. Buehler, A. Hartmaier, and G. Huajian, *J. Mech. Phys. Solids*, **51** (11012), 420 (2003)
44. M.J. Buehler, A. Hartmaier, and H. Gao, Atomic and continuum studies of crack like diffusion wedges and dislocations in submicron thin films, *J. Mech. Phys. Solids*, **51**, 2105 (2003); Y. A. antipov and H. Gao, *Proc. R. Soc. Lond. A*, **458**, 1673 (2002)
45. M.J. Buehler, A. Hartmaier, and H. Gao, Hierachial multiscale modeling of plasticity of submicron thin metal films, *Model. Simul. Mater. Sci. Eng.*, **12**, S391 (2004)
46. M.R. Sorensen, Y. Mishin, and A.F. Voter, *Phys. Rev. B.*, **62**, 3658 (2000); J.C. Fisher, *J. Appl. Phys.*, **22**, 634 (1951); D. Zhang, J. Reng, and T. Liu, *Mater. Sci. Eng.*, **425** (1–2), 78 (2006)
47. Y. Mishin, 50 years of grain boundary diffusion, *Phil. Mag. A*, **72**, 1589 (1995); Z. Stemberg and M. Stupnisek, *Eur. Phys. Lett.* **71**, 757 (2005)
48. Y.J. Park and Y.C. Joo, Electromigration induced stress interaction between vias and polygranular clusters, *Scripta Matter.* **44**, 2497 (2001); P.S. Ho and J.K. Howard, Grain boundary solute electromigration in ploycrystalline films, *J. Appl. Phys.*, **45** (8), 3249 (1974)
49. Y.C. Joo, C.V. Thompson, S.P. Baker, and E. Artz, *J. Appl. Phys.*, **85**(4) 2108 (1999); S.H. Rhee, Y. Du, and P.S. Ho, Thermal stress characteristics of Cu/oxide and Cu/Low-*K* submicron structures, *J. Appl. Phys.*, **93** (7), 3926 (2003)
50. K.C. Saraswat and F. Mohammadi, Effect of interconnection scaling on time delay of VLSI circuits, *IEEE Trans. Electron. Dev.*, **ED-29**, 645 (1982); K.P. Cheung, *Microelectron. Relib.* **41**, 193 (2001)
51. S.J. Souri and K.C. Saraswat, Interconnect performance modeling for 3-D integrated circuits with multiple Si-layers, Int. Interconnect Tech. Conf. Proc., 24, (1999); J. Cong, L. He, C.K. Koh, and P.H. Madden, Performance optimization of VLSI interconnect layout, *Integration, VLSI J.*, **21** (1&2), 1 (1996) D.-H. Kim et al., *Science*, **320** (5875), 507 (April 2008)
52. S.P. Murarka and R.J. Guttman, Advanced multilayer metallization schemes withCu as interconnection metal, *Thin Solid Films*, **236**, 257 (1993); W.S. Young and S.H. Knickerbocker, Multilayer ceramics, in *Interconnects for Electronics*, R.C. Buchanan (ed.). Marcel Dekker, New York, pp. 489–526 (1986)
53. H.B. Bakoglu and J.D. Meindl, Optimal interconnection circuits for VLSI, *IEEE Trans. Electron. Dev.*, **ED-32**, 903 (May 1985)
54. G.W.A Dummer, *Origin of Metallic Conductivity*, Chapter 1, p. 3, Hayden Book Co., New York (1970); X. Oudet, Metallic conductivity at low temperature, *Annales de la fondation Louis de Broglie*, **25** (3), 275 (2000)

55. W. Jones and N.H. March, *Theoretical Solid State Physics*, Dover, New York, Vol. 2, p. 691 (1973)
56. A. Christou, *Electromigration and Electronic Device Degradation*, John John Wiley, New York (1993)
57. F.M. D'Heurle, Electromigration and failure in electronics: An introduction, *Proc. IEEE*, **59** (10), 1409 (Oct. 1971); E.T. Ogawa, K.D. Lee, V.A. Blaschke, and P.S. Ho, Electromigration reliability issues in dual damascene Cu interconnects, *IEEE Trans. Relib.*, **51** (4), 403 (2002)
58. D.K. Ferry, Interconnection lengths and VLSI, *IEEE Circ. Dev.*, **1**, 39–42, (July 1985); E.T. Ogawa et al., Direct observation of a critical length effect in dual damascene Cu/oxide interconnect, *Appl. Phys. Lett.*, **78**, 2652 (2001)
59. J.A. David and J.D. Meindl, Length scaling and material dependence of cross talk between distributed RC interconnects, Proc. Int. Interconnect Tech. Conf., pp. 227–229 (May 1999); K.C. Saraswat and F. Mohammadi, Effect of interconnection scaling on time delay of VLSI circuits, *IEEE Trans. Electron. Dev.*, **ED-29**, 645 (1982)
60. W.S. Wong, and L.A. Glasser, Power distribution techniques for VLSI circuits, *IEEE J. Solid State Circ.*, **SC-21** (1), 150 (Feb. 1986)
61. H.B. Bakoglu, *Circuits Interconnections and Packaging*, Addison-Wesley, Reading, MA, p. 447 (1990)
62. S.J. Hillenius, MOSFETs and related devices, in *Modern Semiconductor Device Physics*, S.M. Sze (ed.). John Wiley, New York (1998); A. Sadaka et al., Fabrication and operation of sub-50 nm strained – Si directly on insulator,(SGOI) CMOSFETs, IEEE Int. SOI Conf. p. 209 (2004)
63. R.H. Yan, A. Ourmazd, and K.F. Lee, Scaling of MOSFET: From bulk to SOI to bulk, *IEEE Trans. Electron. Dev.***ED-39** (7), 965 (1992); C.M. Osburn et al., Vertically scaled MOSFET gate stacks and junctions, *IBM J. Res. Dev.* **46** (2/3), 299 (2002); K. Saraswat et al., IEEE IEDM Dig., p. 659 (2006)
64. R. People, Physics and applications of GeSi/Si heterostructures, *IEEE J. Quantum Electron.*, **22**, 1696 (1986); J. Jung, M.L. Lee, S. Yu, E.A. Fitzgerald, and D. Antoniadis, Implementation of both high hole and electron mobility in strained Si/strained Si, Ge on relaxed Si_{1-x} Ge_x, *IEEE Electron. Dev. Lett.* **24**, 460 (2003)
65. R.G. Grimmeiss and J. Olajos, Physics and applications of GeSi/Si hetrostructures, *Phys. Scripta*, **T69**, 52 (1997); M. Myronov et al., *Appl. Phys. Lett.*, **91**, 082108 (2007)
66. H. Temkin, J.C. Bean, A. Antreasyan, and R. Leibenguth, Ge_xSi_{1-x} strained layer hetro structure bipolar transistors, *Appl. Phys. Lett.* **52**, 1089 (1988); S.J. Hillenius, MOSFETS and related devices,in S.M. Sze (ed.), *Modern Semiconductor Device Physics*, John Wiley, New York, p. 137 (1998); F.P. Gusev, V. Narayana, and M.M. Frank, *IBM. J. Res. Dev.*, **90** (4/5), 387 (2006)
67. E. Levy, 15 years of advanced materials: Past and future, *Adv. Mater.* **15**, 13 (2003); T. Ashley et al., *Electron. Lett.* **43** (14) (July 2007)
68. W.C.B. Peatman, H. Park, B. Gelmont, M.S. Shur, P. Maki, E.R. Brown, and M.J. Rooks, Novel metal/2-DEG junction transistors, in Proc. 1993 IEEE/ Cornell Conf. Cornell Univ. Press, Ithaca, New York, p. 314 (1993)
69. M. Mastrapasqua, C. King, P. Smith, and M. Pinto, Functional devices based on real space transfer inSi/SiGe structure, *IEEE Trans. Electron. Dev.* **ED-43** (10), 1671 (1996); J. Wang et al., Symp. VLSI Tech. Dig., p. 46 (2007)
70. N.D. Zakarov, V.G. Talalaev, P. Werner, A.A. Torkikh, and G.E. Cirlin, Room temperature light emission from a highly strained Si/Ge superlattice, *Appl. Phys. Lett.* **83** (15), 3084 (2003)
71. V.G. Talalaev, G.E. Cirlin, A.A. Torkikh, N.D. Zakharov, and P. Warner, Room temperature electroluminescence from Ge/Si quantum dots,*Phys. Stat. Solidi*, **198** (1) R4 (2003)
72. M.J. Kobrinski et al., On chip optical interconnects, *Intel Tech. J.* **8** (2), 129 (May 2004)
73. J.F. Shackelford, *Introduction to Materials Science for Engineers*, Macmillan, New York (1988); J. Wang et al., Symp. VLSI Tech. Dig., p. 40 (2007)

74. J.W. Mayer and S.S. Lau, *Electronic Materials Science*, Macmillan, New York (1990); A. Wei et al., Symp. VLSI Tech. Dig., p. 216 (2007)
75. A.B. Glaser and G.E. Subak-Sharpe, *Integrated Circuit Engineering, Design, Fabrication and Applications*,Addison Wesley, Reading, MA (1977); H. Kim et al., *Electrochem. Solid State Lett.*, **11**(5) 127 (2008)
76. P.R. Gray and R.G. Meyer, *Analysis and Design of Analog Integrated Circuits*, John Wiley, New York (1993); C.E. Royer et al., *Solid State Electron.*, **52**(9), 1285 (2008)
77. M.P. Anderson and S. Ling, Computer simulation of transport in thin films, *J. Electronic Mater.* **19** (11) p. 1161 (1990); T, Turner, Cu-line width resistivity measurements, *Solid State Technol.*, **43**, 89 (April 2000); D. Ingerly et al., IEEE IITC, June-4, San Francisco, CA (2008)
78. N.L. Michael, C. Kim, P. Gillepie, and R. Augur, Electromigration failure in ultra-fine Cu-interconnects, *J. Electron. Matter.* **32** (10) pp. 988–993 (2003); A. Zehe, A selection rule of solutes for void resistance crystalline metallic alloys exposed to electromigration, *Cryst. Res. Technol.*, **37** (8), 817 (2002); and K.N. Tu, Recent advances on electromigration in very large scale integration interconnects, *J. Appl. Phys.*, **94** (9), 5452 (2003); N. Oda, *Jap. J. Appl. Phys.*, **46**, 954 (2007)
79. G.W.A. Dummer, *Materials for Conductive and Resistive Functions*, Hayden Book Co., New York (1970); D. Raabe, *Computational Materials Science*, Wiley-VCH, Weinheim, Germany (1998); A. Sakata et al., IEEE IITC Proc. (2006)
80. N.B. Hanny and U. Colombo (eds), *Electronic Materials*, Plenum Press, New York (1972); D. Wood, *Material Development and Processing*, John Wiley, Hoboken, NJ (2000)
81. C.A. Harper (ed.), *Handbook of Materials and Processes for Electronics*, McGraw Hill, New York (1970); P.C. Dunn, *Gateway into Electronics*, John Wiley,Hoboken, NJ (2000)
82. B. Tareev, *Physics of Dielectric Materials*, MIR Publishers, Moscow (1973); G.D. Wilk, R.M. Wallace, and J.M. Anthony, *J. Appl. Phys.*, **89**, 5243 (2001); and K. Mistry et al., IEEE IEDM Tech. Dig., p. 247 (2007)
83. J.B. Birks, *Modern Dielectric Materials*, Heywood, London (1960); R.L. Opila and D.W. Hess, A century of dielectric science and materials, *J. Electrochem. Soc.* **150** (1), S1 (2003); M. Chudzik et al., VLSI Tech. Dig., p. 194 (2007)
84. A. Charlesby, *Atomic Radiation and Polymers*, Pergamon Press, Oxford (1960); A.D. Schluter, *Synthesis of Polymers*, Wiley-VCH, Weinheim, Germany, (1998); S. Narashimha et al., IEEE IEDM Tech. Dig., pp. 1–4 (2006)
85. J.K. Stille, *Introduction to Polymer Chemistry*, John Wiley, New York (1962); H.G. Elias, *Introduction to Polymer Science*, Wiley-VCH, Weinheim, Germany (1997); G. Gambino et al., IEEE IEDM Tech. Dig., p. 131 (2007)
86. F.W. Billmeyer, *Textbook of Polymer Chemistry*, 2nd edn. Interscience. New York (1950); L.H. Sperling, *Polymeric Multcomponent Materials*, Wiley-VCH, Weinheim, Germany (1997)
87. H.F. Wolf, *Semiconductors*, John Wiley, New York (1971); D.A. Miller, IEEE IITC June 2008, San Francisco, CA
88. D.K. Ferry, *Semiconductors*, Macmillan, New York (1991); P. Zimmerman et al., IEEE IEDM Tech Dig., p. 437, (2006)
89. W. Crawford, *An Introduction to Semiconductors*, John Wiley, New York (1957); D. Sing et al., *Mater. Today*, **9** (6), 26 (2006)
90. P.S. Kireev, *Semiconductor Physics*, Mir Publishers, Moscow (1975)
91. W.C. Dunlap, Jr., *An Introduction to Semiconductors*, John Wiley, New York (1957); A. Khakifirooz and D.A. Antoniadis, IEEE IEDM Tech Dig. 84 (9–10), p. 2047 (Sept 2008)
92. H.F. Wolf, *Semiconductors*, John Wiley, New York, p. 201 (1977)
93. J.D. McBrayer, R.W. Swanson, and T.W. Sigmon, *J. Electrochem. Soc.* **133**, 1242 (1986)
94. S.P. Murarka, *Tungsten and Other Advanced Metals for VLSI Applications*, MRS Publishers, Pittsburgh (1991)
95. J. Li, T.E. Sedal, and J.W. Mayer, *MRS Bull.*, **XIX** (8), 15 (1994)

96. M. Bohr, *IEEE Int. Electron. Dev. Meet.*, **95**, 241 (1995)
97. D. Edelstein et al., *IEEE Int. Electron Dev. Meet.*, p. 97 (1997) and T. Murata et al., *Jpn. J. Appl. Phys.*, **47**, 2488 (2008)
98. L.L. Bran, R.S. Phillips, and N.H. Dickey, *Funk and Wagnall's New Encyclopedia*, Vol. 7, p. 200, MCMLXXI, (1989)
99. I. Bremmev, Copper the smart choice, The International Copper Association, New York (2001); T.K. Gupta, *Handbook of Thick and Thin Film Hybrid Microelectronics*, John Wiley, Hoboken, NJ, p. 143 (1983)
100. F.A. Cotton and G. Willkinson, *Basic Inorganic Chemistry, Junior and Advanced*, John Wiley (1976); U. Schubert and N. Husing, *Synthesis of Inorganic Materials*, Wiley-VCH, Weinheim, Germany (2000)
101. W.E. Hatfield and R. Whyman, Coordination chemistry of copper, *Transit. Metal Chem.*, **5**, 47 (1969); G. Meyer, D. Naumann, and L. Wesemann, *Inorganic Chemistry Highlights*, Wiley-VCH, Weinheim, Germany (2002)
102. I. Hadar, Experimental investigation of wire diameter effect on the fine pitch ball bonding, Proc. SEMICON Test, Assembly and Packaging (April, 1996); T.K. Gupta, *Handbook of Thick and Thin Film Hybrid Microelectronics*, John Wiley, Hoboken, NJ, p. 258 (1983)
103. V. Shanon and D. Smith, Copper interconnects for high volume manufacturing, *Semicond. Intl.*, p. 93 (May 2001)
104. D. Simonaitis, IC Failure rate estimates from field data, *MIL-HDBK-217D*, **15**, 5 (1982); A.A. Shirzadi and E.R. Wallach, *Sci. Technol. Welding Joining*, **9** (1), 37 (2004)
105. L. Vanasupa, Y.C. Joo, and P.R. Besser, *J. Appl. Phys.*, **85** (5) (1999); A. Sekiguchi, J. Koike, and K. Maruyama, *Appl. Phys. Lett.*, **83** (10), 1962 (2003)
106. A.N. Campbell, R.E. Mikawa, and D.B. Knorr, *J. Electron. Mater.* **22**, 589 (1993)
107. K.A. Fichthorn, and M. Scheffler, *Phys. Rev. Lett.*, **84**, 5371 (2000)
108. P. Hyldgaard and M. Bersson, *J. Appl. Phys. Cond. Mater.* **12**, L-13 (2000)
109. J. Repp, M. Moresco, C. Meyer, and K.H. Rieder, *Phys. Rev. Lett.*, **85**, 2981 (2000)
110. C.S. Hau-Reige and C.V. Thompson, *Appl. Phys. Lett.*, **78** (22), 3451 (2001)
111. B.D. Knowlton, J.J. Clement, and C.V. Thompson, *J. Appl. Phys.* **81** (9), 6073 (1997)
112. D. Gan, P.S. Ho, R. Huang, J. Leu, J. Maity, and T. Scherban, *J. Appl. Phys.*, **97**, 103531 (2005)
113. B.H. Mahan, *University Chemistry*, Addison Wesley, Reading, MA (1975)
114. F.A. Cotton and G. Wilkinson, *Advanced Inorganic Chemistry*, John Wiley, New York, Chapter 20, p. 555 (1972)
115. L. Maissel, Thin film resistors, in L. Maissel and R. Glang (eds.), *Handbook of Thin Film Technology*, McGraw Hill, New York, Chapter 18, p-18-3 (1983)
116. M.P. Lepselter, *Bell Syst. J.* **45**, 233 (1966)
117. I. Blech, H. Sello, and L.V. Gregor, Thin films in integrated circuits, in L. Maissel and R. Glang (eds.), *Handbook of Thin Film Technology*, McGraw Hill, New York, p. 23–27 (1983)
118. G. Timp, *Nanotechnology*, Springer-Verlag, Berlin (1999)
119. F.A. Cotton and G. Wilkinson, *Basic Inorganic Chemistry*, John Wiley, New York, p. 50 (1976)
120. H.O. Pritchard and H.A. Skinner, Electronegativity scales, *Chem. Rev.* **55**, 745 (1955)
121. J.M. Mayer, and S.S. Lau, *Electronic Materials Science for Integrated Circuits in Si and GaAs*, Macmillan, New York, p. 411 (1990)
122. H.B. Bakoglu (author), Multilayer interconnections, in *Circuits Interconnections and Packaging for VLSI*, Addison Wesley, Reading, MA p. 211 (1990); and C.C. Chu et al., *Micrelectron Relib.*, **47** (9), 1560 (2007)
123. K. Otsuka (ed.), *Multilayer Ceramic Substrate Technology for VLSI Package/ MCM*, Elsevier, New York (1993); and N. Oda et al. *Jpn. J. Appl. Phys.*, **46**, 954 (2007)
124. N. Nang, W.K. Henson, J.R. Hauser, and J.J. Wortman, Modeling study of ultrathin gate oxides using direct tunneling current and capacitance, *IEEE Trans. Electron. Dev.*, **46**, 1464 (1999)

125. H. Yu, Y.T. Hou, M.F. Li, and D.L. Kong, Investigation of hole tunneling current through ultrathin oxynitride/oxide stack dielectrics in PMOSFETS, *Trans. Electron. Devices*, **49** (7), 1158 (2002)
126. T.K. Gupta, *Handbook of Thick and Thin Film Hybrid Microelectronics*, John Wiley, Hoboken, NJ (2003)
127. J.E. Sergent, The hybrid microelectronics and MCM technology, in C.A. Harper (ed.), *Electronics Packaging and Interconnections Handbook*, 3rd edn., McGraw Hill, New York (2000)
128. Y. Shimada, K. Utsumi, M. Suzuki, H. Takamizawa, N. Mitsura, and Y. Scihen, Low firing temperature multilayer glass ceramic substrate, *IEEE Trans. On Comp. Hybd. And Manuf.*, **6** (4), 382 (1983).
129. H. Comb, Jr. (ed.), *Printed Circuit Handbook*, 2nd edn., McGraw Hill, New York (1988)
130. S.P. Murarka, I.V. Verner, and R.J. Guttman, *Copper: Fundamental Mechanisms for Microelectronics Applications*, John Wiley, New York (2000)
131. R. Glang and L.V. Gregor, Generation of patterns in thin films, in L.I. Maissel and R. Glang (eds.), *Handbook of Thin Film Technology*, McGraw Hill, New York (1983); S.V. Sreenivasan et al., 23rd Int. VLSI/ULSI multilevel Interconn. Conf. Sept 26 (2006)
132. L.F. Thompson, C.G. Wilson, and J.M. J. Frecht, *Materials for Microlithography*, American Chemical Society Publications, Washington, DC (1984); and T. Schram et al., VLSI Tech Symp. (2008)
133. C.C. Ku et al., *Electrical Properties of Polymers*, Hanser, New York (1987)
134. T. Swager, Light from insulated organic wires, *Nature Mater.*, **1** (3), p. 151 (Nov. 2002)
135. F. Cacialli et al., Cyclodextrin-threaded conjugated polyrotaxes as insulated molecular wires with reduced interstrand interactions, *Nature Mater.*, **1** (3), p. 161 (Nov. 2002)
136. M. Good, J.C. Kotz, J.L. Wood, M.D. Joesten, and J.W. Moore, *The Technical World*, Saunders College Publications, New York, p. 579 (1994); H.L. Tsai, J.L. Schindler, C. R. Kannewurf, and M.G. Kanatzidis, Plastic superconducting polymer, *Chem. Mater.* **9**, 875 (1997)
137. H.A. Klok and S. Lecomanndoux, Supramolecular materials via block copolymer assembly, *Adv. Mater.* **13** (6), 1217 (2001); F. Cacialli et al., Cyclodextrin – threaded conjugated polyrotaxes an insulator molecular wires with reduced interstrand interactions, *Nature Mater.*, **1** (3), 161 (Nov. 2002); and H.J. Schneider and A. Yatsimirski, *Principles and Methods in Supramolecular Chemistry*, John Wiley, Chichester (2000)
138. J.G. Frazier, *An Ideology for Nanoelectronics in Computation*, Plenum Press, New York (1988)
139. T.K. Gupta, *Handbook of Thick and Thin Film Hybrid Microelectronics*, John Wiley, Hoboken, NJ, p. 116 (2003)
140. D.R.T. Zahn, T. Kampen, and R. Scholz, *Organic Molecular Semiconductors*, John Wiley, Chichester, (2004)
141. S. Zing and J. Shi, Novel polymers for light emitting diodes, in H.S. Nalwa and L.S. Rohwer (eds), *Handbook of Luminescence, Display Materials and Devices*, American Scientific Publishers, CA (2003); Y.G. Seol et al., *Thin Solid Films*, **515** (12), 5065 (2007)
142. C.D. Dimitrakopoulos and P.R.L. Malenfant, Organic thin film transistors for large area electronics, *Adv. Mater.*, **14**, p. 401 (2002); S.J. Hillenius, MOSFETs and related devices, in S.M. Sze (ed.), *Modern Semiconductor Device Physics*, John Wiley, New York,Chapter 3, pp. 165–168 (1998)
143. T. Mcnelly et al., High performance 0.25 μm SRAM technology with tungsten interpoly plug, in IEDM Tech. Dig., 927 (1995); and Y.G. Seol, J.G. Lee, and N-E Lee, *Organic electron.*, **8** (5), 513 (Oct 2007)
144. J.E. Katon (ed.), *Organic Semiconducting Polymers*, Marcel Dekker, New York (1968)
145. D. Jerome and K. Bechgaard, *Nature*, **410**, March 8, p. 162–163, (2001); P. Singer, Emerging technologies, Technical News, *Semicond. Int.* **14** (4), April, p. 36, (2001)

146. H.F. Wolf, *Semiconductors*, John Wiley New York, p. 12 (1971)
147. H. Herbs, *Semiconducting Materials*, H.K. Henisch (ed.), Butterworths Scientific, London, Chapter 22 (1952)
148. W. Crawford Dunlap, Jr., *An Introduction to Semiconductors*, John Wiley, New York, Chapter 11, p. 224 (1957)
149. W. Zuhlehner and D. Huber, *Czochralski Growth Silicon*, Springer-Verlag, Berlin, Germany (1982)
150. R.C. Jaeger, *Introduction to Microelectronics Fabrication*, Addison Wesley, Reading, MA (1988)
151. A.J. Khambata, *Introduction to Large Scale Integration*, John Wiley, New York (1969)
152. H.J. Queisser and E. E. Haller, Defects in semiconductors: Some fatal, some vital, *Science*, **281**, 945 (Aug, 1998)
153. R.B. Fair, Concentration of diffused dopants in silicon, in F.F.Y. Wang (ed.), *Impurity Doping Processes in Silicon*, North Holland, New York (1981)
154. E.E. Haller and F.S. Goulding, in C. Hilsum (ed.), *Handbook on Semiconductors*, 2nd edn, Vol. 4, pp. 937–963, Elsevier, New York, Chapter 11 (1993)
155. P.C. Andricacos, C. Uzoh, J.O. Dukovic, J. Horkans, and H. Deligianni, *IBM, J. Res. Dev.*, **42** (5), p. 567 (Sept. 1998); Z. Lu et al., *IEEE Trans. VLSI Symp.* **15** (2) 159 (2007)
156. P. Moon et al., Process road map and challenges for metal barriers, Proc. IEDM (2003); Nd X. Zhang, P.S. Ho, and T. Nakamura, IEEE IITC Conf. June 2008, San Francisco, CA
157. International Device Meeting (IEDM), Dec. 11–14, San Francisco Hilton Tower (2000)
158. VLSI Multilevel Interconnecting Conf. (VMIC), Santa Clara, June 26–30 (1995)
159. S. Thompson et al., 130 nm logic technology featuring 60 nm transistors, low-K dielectrics, and Cu-interconnects, *Intel Tech. J.* **6** (2), p. 5 (May16, 2002)
160. K.L. Lai, Ionized hollow cathode magnetron sputtering, in J.P. Hopwood (ed.), *Ionized Physical Vapor Deposition*, Academic Press, San Diego, CA, p. 115 (2000)
161. Microstructure Science; Engineering and Technology, National Academy of Science, Washington DC (1979)
162. Y.K. Cho et al., A spacer patterning technology for nano-scale CMOS, *IEEE Trans. Electron. Dev. Lett.*, **49**, 436 (2002)
163. B. Doyle et al. Transistor elements for 30 nm physical gate lengths and beyond, *Intel Tech. J.* **6** (2), 43 (May 16, 2002)
164. T. Clarysse, D. Vanhaeren, and W. Vandervorst, Impact probe penetration on electrical characterization of sub-50 nm profiles, *J. Vac. Sci. Technol.*, **B20** (1), 459 (2002); V.N. Faifer, M.I. Current, T.M. H. Wong, and V.V. Souchkov, Non contact sheet resistance and leakage current mapping for ultra shallow junctions, Proc, 8th Intl., Workshop on Fabrication, Characterization, and modeling of ultra shallow junctions in semiconductors, June 5–8 (2005)
165. T.K. Gupta, *Handbook of Thick and Thin Film Hybrid Microelectronics*, Chapter 8, John Wiley, Hoboken, NJ (2003)
166. H.B. Bakoglu, *Circuits, Interconnections, and Packaging for VLSI*, Addison Wesley, Reading, MA, Chapter. 3 (1990)
167. M. Mahalingam, J. Andrews, and J. Drye, Thermal pin grid array packages for high density LSI and VLSI logic circuits, *IEEE Trans. CHMT*, **6** (4), pp. 246 (Sept. 1983)
168. R.F. Pease and O.K. Kwon, Physical limits to the useful packaging density of electronic systems, *IBM J. Res. Dev.*, **32** (5), 636 (Sept. 1985)
169. R.C. Mehrotra, R. Bhora, and D.P. Gaur, *Metal Diketones and Allied Derivatives*, Academic Press, London (1978); E. Levy and C. Lopez, *J. Adv. Mater.* **15** (20), p. 1665 (Oct. 16, 2003)
170. A.E. Kaloyeros, A. Feng, J. Garhart, K.C. Brooks, S.K. Ghosh, A.N. Saxena, and F. Luehrs, Low temperature metal-organic chemical vapor deposition (LTMOCVD) of device quality copper film for microelectronics applications, *J. Electron. Mater.*, **19**, 271 (1990)
171. B. Chapman, *Glow Discharge Processes*, John Wiley, New York (1980)
172. S. Wolf and R.N. Tauber, *Silicon Processing*, Vol. I, Lattice Press, Sunset Beach, CA, Chapter 10 (1986)

173. J. Sundqvist, H. Hogberg, and A. Harsta, Atomic layer deposition of Ta_2O_5 using the TaI_5 and O_2precursor combination, J. *Adv. Mater.*, **15** (20), 245 (Oct. 16, 2003)
174. T. Suntola, and J. Anston, US patent, 4058430 (1977)
175. A.J. Bard and L.R. Faulkner, *Electrochemical Methods: Fundamentals and Applications*, 2nd edn. John Wiley, New York (2001)
176. J. Goodisman, *Electrochemistry: Theoretical Foundations, Quantum and Statistical Mechanics, Thermodynamics, The Solid State*, John Wiley, New York (1987)
177. D.A. MacInnes, *The Principles of Electrochemistry*, Dover, New York (1947, revised 1961)
178. P.C. Andricacos and L.T. Romankiw, Magnetically soft materials: Their properties and electrochemistry, in H. Gerischer and C.W Tobias (eds) *Advances in Electrochemical Science and Engineering*, Vol. 3, VCH, New York, pp. 227–321 (1994); and P.C. Andricacos, C. Uzoh, J.C. Dukovic, J. Horkans, and H. Deligiani, Damascene copper electroplating for chip interconnections, *IBM. J. Res. Dev.,* **42** (5), 567 (Sept. 1998)
179. H.S. Nalwa, *Handbook of Organic-Inorganic Hybrid Materials and Nanocomposites*, Vol. I, American Scientific Publishers, Steven Ranch, CA (2003)
180. A. Tiwari, A. Chugh, C. Chin, and J. Narayan, Role of self-assembled gold nanodots in improving the electrical and optical characteristics of zinc oxide films, *J. Nanosci. Nanotech.*, **3** (5), 368 (2003)
181. R. Saito, G. Dresselhaus, and M.S. Dresselhaus, *Physical Properties of Carbon Nanotubes*, Imperial College, London (1999); R. Chau et al., *Nano Mater.* **6**, 810 (Nov. 2007)
182. W.A. De Heer, Structure and transport in nanotubes, *Nature Mater.*, **1** (3), 153 (Nov. 2002)
183. S. Iijim, Helical nanotube of graphite carbon, *Nature*, **354**, 56–58 (1991)
184. T.W. Ebbessen and P.M. Ajayan, Large scale synthesis of carbon nanotubes, *Nature*, **358**, 220 (1992)
185. A. Thesis et al., Crystalline ropes of metallic carbon nanotubes (PCNTs), *J. Phys. Chem. Solids*, **54**, 1841 (1993)
186. S. Amelinckx et al., A formation mechanism for catalytically grown helix shaped graphite nanotubes, *Science*, **265**, 635 (1994)
187. M. Endo, Grow carbon fibers in vapor phase, *Chem. Tech.* **18**, 568 (1988)
188. R. Martel et al., *Appl. Phys. Lett.* **73** (17), 2447 (1998)
189. A. Aviram and M. Ratner, *Chem. Phys. Lett.*, **29**, 288 (1974)
190. S.J. Trans et al., Individual single wall carbon nano tubes as quantum wires, *Nature*, **386**, 474 (1977)
191. M. Bockrath et al., Single electron transport in rope of carbon nanotubes, *Science*, **275**, 1922 (1977)
192. S.J. Trans. A.R.M. Verschueren, and C. Dekker, Room temperature transistor based on a single carbon nanotube, *Nature*, **393**, 49 (1998)
193. Z. Yao, H. Postma, L. Balents, and C. Dekker Carbon nanotube intermolecular junctions, *Lett. Nat.*, **40** (2), 273 (1999); Y. Kamata, *Mater. Today*, **11** (1–2), 30 (2008)
194. A.E. Kaloyeros, E.T. Eisenbraun, J. Welch, and R.E. Geer, Exploiting nanotechnology for terahertz interconnect, *Semicond. Int.* **26** (1), 56 (Jan. 2003)
195. K. Takayanagi, Y. Kondo, and H. Ohnishi, Suspended gold nano wires: Ballistic transport of electrons, *Int. J. ISAP*, **3** (3) (Jan. 2001)
196. R.A. Webbs, S. Washborn, C.P. Umbach, and R.B. Laibowitz, *Phys. Rev. Lett.*, **54**, 2696 (1985)
197. S. Lijima, *Nature*, **354**, 56 (1991)
198. M. Bockrath et al., *Science*, **275**, 1922 (1997)
199. J. Bardeen, L.N. Cooper, and J.R. Schrieffer, *Phys. Rev.* **108**, 1175 (1957)
200. R.E. Joynson, Superconductive thin films and devices, in L.I. Maissel and R. Glang (eds), *Handbook of Thin Film Technology*, McGraw Hill, New York, Chapter 22 (1983)
201. M. Tinkham, *Documents on Modern Physics*, Gorden & Breach, New York (1990)

202. Superconductivity, APS Special Issue, American Physical Society Publishers, New York (1987); also D.K. Brock, E.K. Track, and J.M. Rowel, Superconductor ICs: The 1000-GHz second generation, *IEEE Spectrum*, **37** (12) (Dec. 2000); and K. Likharev, Superconductors: Computation, *Physics World*, **10**, pp. 39–43 (May 1997)
203. C.W. Chu, P.H. Hor, R.L. Meng, L. Gao, Z.J. Huang, and Y.Q. Wang, Evidence of superconductivity above 40 K in the La-Ba-Cu-O compound system, high temperature superconductivity, Special Issue, *Phys. Rev. Lett.* **58**, 405 (1987)
204. T.K. Gupta, Preparation and characterization of layered superconductors, *Phys. Rev. B*, **43** (7) (1991)
205. S. Savel'ev and F. Nori, Experimentally anisotropic superconductors, *Nat. Mater.*, **1** (3), 179 (Nov. 2002)
206. G.D. Anna, Controlling the motion of quanta, *Nat. Mater.* **1** (3), 143 (Nov. 2002)
207. G.D. Anna, Controlling the motion of quanta, *Nat. Mater.* **1** (3), 143 (2003)
208. R.D. Astumian, *Science*, **276**, 917 (1997)
209. J. Prost et al., *Phys. Rev. Lett.*, **72**, 2652 (1994)
210. R.J. Gutmann et al., A wafer level 3-D IC technology platform, Adv. Metal. Conf. October 2003, p. 19
211. P. Garrou, C.B. Power, and P. Ramm (eds.), *Handbook of 3-D Integration: Technology and Application of 3-D Integrated Circuits*, John Wiley, New York (2008)
212. A Klumpp et al., Integration technologies for 3-D systems, Int. Workshop of 3-D System Integration (Dec. 2003)
213. S. Pozder et al., Back end compatibility of bonding and thinning process for wafer level 3-D interconnect technology platform, IEEE IITC, June 2004, p. 102
214. K. Ishimaru, 45/32 nm CMOS Challenge and perspective, **52**(9), 1266 (2008)
215. F. Roozeboom et al., *Electrochem. Soc. Trans.*, **3** (15) 173 (2007)
216. J.H. Klootwjik, A. Kemmeren, R. Wolters, F. Roozeboom, J. Verhoeven, and E. van den Heuvel, in, E. Gusev (ed.), *Defects in High-K Gate Electric Stacks*, Springer, Dordrecht, pp. 17–28 (2006)
217. J. Vardaman, What's delaying the adoption of 3-D TSV, *Semicond. Int.* March 2008, p. 80
218. M.A. Blauw, P.J.W. van Lankvelt, F. Roozeboom, M.C.M. van de Sanden, and W.M.M. Kessels, *Electrochem. Solid State Lett.*, **10**, H-309 (2007)
219. Y. Xie, G. Loh, G. Black, and K. Bernstein, Design space exploration for 3-D architecture, *ACM J. Emerg. Technol. Comput. Syst.*, **2** (2), 65 (2006)
220. J.Q. Lu et al., Wafer level 3-D hyper integration, 20th Int. VLSI Multilevel Interconnect. Conf. 2003, p. 227
221. G. Posada et al., Low pass coupled line filters with transmission zeros in multilayer thin film MCM-D technology, IEEE MTTS-Digest 2004, p. 1471
222. G. Posada et al., Microstrip thin-film MCM-D technology on high resistivity silicon with integrated through substrate vias, *Eur. Microw. Week*, **37**, pp. 8–12 (October 2007)
223. G. Carchon et al., Multilayer thin film MCM-D for the integration of high performance RF and microwave circuits, *IEEE Trans. CPT.*, **24** (3), 29 (2001)
224. A. Stich, Z. Gabric, and W. Pamler, Potential of air gap technology by selective ozones/TEOS deposition: Effects of air gap geometry on dielectric constant, *Microelectron. Eng.*, **82** (3–4), 362 (2005)
225. B. Shieh, K. Saraswat, M. Deal, and J. Mcvittie, Air gaps lower K in interconnect dielectrics, *Solid state technology*, **42** (2), 57 (1999)
226. H.J. Lee et al., Structural characteristics of porous low-*K* thin films prepared by different techniques using x-ray prosimetry, *J. Appl. Phys.*, **95** (5), 2355 (2004)
227. J. Noguchi et al., Process and reliability of air gap Cu interconnect using 90 nm node technology, *IEEE Trans. Electron. Dev.*, **52** (3), 352 (March 2005)
228. B. Sheih et al., Air gap formation during IMD deposition to lower interconnect capacitance, *IEEE Electron. Dev. Lett.* **19** (1) 16 (1998)

229. M. Lin, C.Y. Chang, T.Y. Huang, and M.L. Lin, A multilevel interconnect technology with internal air gap for high performance 0.25 μm and beyond device manufacturing, *JJAP*, **38**, 6240 (1999)
230. J.P. Gueneau de Mussy, C. Bruynsereade, Z. Tokei, G.P. Beyer, and K. Maex, Proc. IEEE-IITC, 2005, p. 150
231. L. Gosset et al., *Microelectron. Eng.*, **82**, 240 (2005)
232. R.J.O.M. Hoofman et al., Self aligned multilevel air gap integration, *Microelectron. Eng.*, **83** (11–12), 2150 (2006)
233. V. Arnal et al., IEE IITC 2001, p. 143
234. H. Park et al. Simulation of electrical and mechanical properties of air bridge Cu-interconnects, 44th Annual Int. Reliab. Phys. Symp. San Jose (2006)
235. C.S. Hau-Riege, S.P. Hau-Riege, and A.P. Marathe, *J. Appl. Phys.* **96**, 5792 (2004)
236. J. Noguchi, K. Sato, N. Konishi, S. Uno, T. Oshima, and K. Ishikawa, *IEEE Trans. Electron. Dev.*, **52**, 352 (2005)
237. R. Daamen, J. Michelon, and V. Nguyenhoang, Alternative to low-*K* nanoporous materials: Dielectric air gap integration, *SST* (Aug. 2006)
238. E. Keyes, Intel's 45 nm device, *EE Times* pp. 2–5 (Dec. 2, 2007)
239. IEEE, IEDM, Dec. 13–15, San Francisco, CA.,2004
240. P. Singer, 65 nm technology details emerge, *Solid State Technology*, **27** (10), 30 (2004)
241. P. Homberg, *Xilinx*, Open Systems Publishing, 2008 Editorial Calendar
242. *TSMC eNewsletter*, Feb. 16, 2007
243. Industry news, *Semicond. Int.* March 2008, p. 17
244. E. van Setten et al., The flash memory battle: How low can we go? *Opt. Microlitho.* **XXI**. 6924, 692441, *Proc. SPIE* (2008)
245. C. Auth et al., Intel's 45 nm CMOS Technology, *Intel Tech J.* **12** (2) 1, June (2008)
246. D. Scansen, *Under hood: EE Times*, (News-magazine) CMP media LLC. Pub., NY, Nov. 14 (2007)
247. R. Argavani and H. M'Saad, Despite engineering and cost challenges, 32 nm node IC manufacturing within reach, *Solid State Technology*, **51** (5), 25 (2008)
248. H.S. Yang et al., Dual stress liner for high performance sub-45 nm gate length SOI CMOS manufacturing, IEDM Tech Dig. Dec., p. 1075 (2004)
249. Agravani et al., Strain engineering push to the 32 nm logic technology node, Semiconductor Fabtech, ICG Pub., UK, 32 edition, p. 15 (2007); L. Washington et al., PMOS FET with 200% mobility enhancement induced by multiple stressors, *IEEE Trans. Electron. Dev.* **27** (6), 511–513 (2006)
250. E.P. Gusev, V. Narayana, and M. Frank, Advanced high-*K* dielectric stacks with poly-Si and metal gates, *IBM J. Res. Dev.*, **90** (4/5), 387 (2006)
251. M. Chudzik et al., High performance high-*K* metal gates for 45 nm CMOS and beyond with gate first processing, in VLSI Tech. Dig., p. 194 (2007)

Chapter 2
Dielectric Materials

2.1 Introduction

Copper (Cu) has higher *conductivity* and resistance to *electromigration* (EM) than aluminum (Al) and has been the choice of the semiconductor industry for interconnecting metal in sub-100 nm devices. With rapidly *decreasing feature sizes* and more demand for circuit *speed*, low-*K* and passivation materials have been inserted with Cu-interconnects to address the additional *RC delay* reduction [1–2]. Unfortunately, as the thickness of the gate oxide becomes very thin because of the scaling down of channel length, quantum mechanical *tunneling* occurs for voltages below the Si/SiO_2 barrier height which is approximately 3.1 eV [3–4] (Fig. 2.1).

Figure 2.2 shows the simulated results of gate oxide scaling and the current density when the applied gate voltage is 1 volt. As the thickness of silicon dioxide (SiO_2) as the gate material becomes very thin (less than 1.5 nm), the tunnel current increases significantly.

Therefore, the need exists for a dielectric material with high *K*(*K* is the dielectric constant of the material), which will provide the required electrical properties. Calculations based on experimental data show that the threshold voltage and transconductance fluctuations become major issues when the gate oxide thickness is scaled down to less than 1 nm. Figure 2.3 shows that the thinner the gate oxide, the greater the shift in threshold voltage. As a matter of fact, instability is more severe for static negative bias temperature (SNBT) than dynamic negative bias temperature (DNBT) [5].

According to some experts, the NBTI is thought to be an electrochemical activity that involves the electric field, holes, Si–H bonds, and temperature. Interface traps are formed when hydrogen (H_2) is released from the Si–H bond. Figure 2.4 shows some of the possible chemical activities and probable reasons for instability due to negative bias temperature (NBT). However, it is expected that using a recessed SiGe source/drain (S/D) with an elevated S/D structure will minimize the effect [6].

This chapter will be devoted to dielectric materials (low-*K* and high-*K*) – the second most important material in the copper damascene process. Figure 2.5 shows the Cu-damascene architecture with low-*K* materials into the trenches and via holes.

Tapan K. Gupta, *Copper Interconnect Technology*,
DOI 10.1007/978-1-4419-0076-0_2, © Springer Science+Business Media, LLC 2009

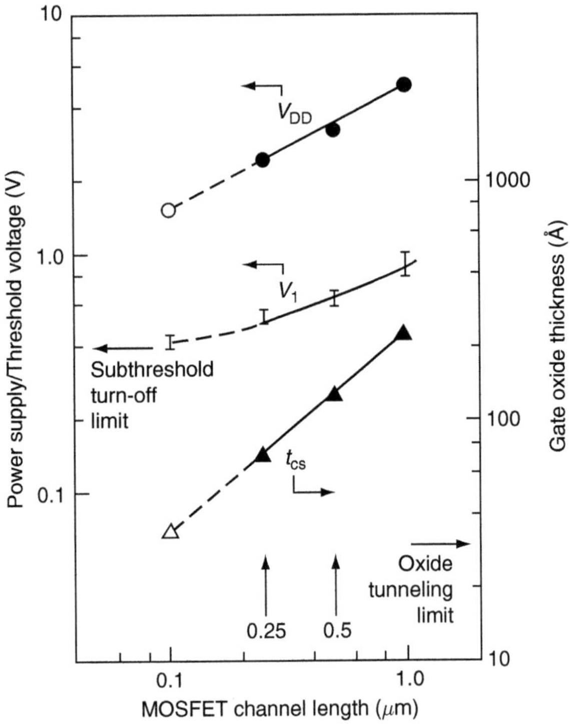

Fig. 2.1 The effect of gate oxide on channel length and its consequences on the oxide-tunneling limit (Reprinted with permission, IBM Research [3])

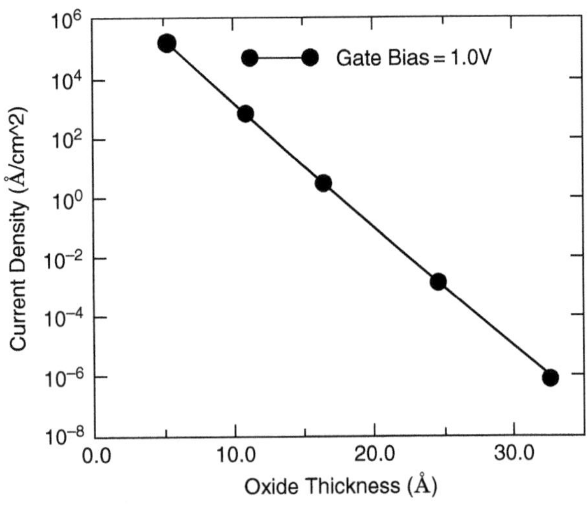

Fig. 2.2 Simulated direct tunneling current density versus gate oxide thickness for $V_{GB} = 1.0$ V

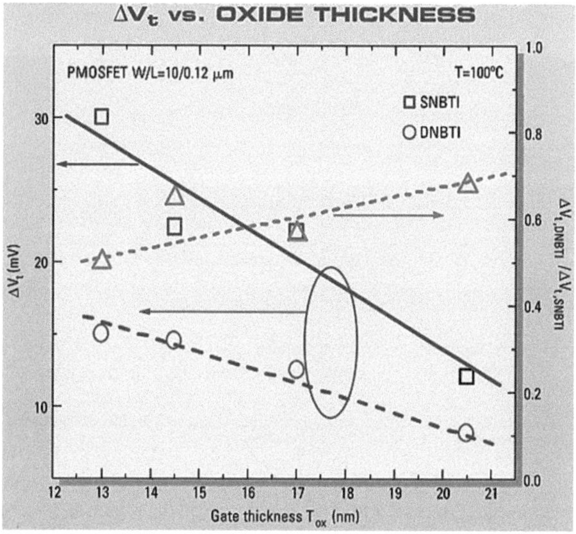

Fig. 2.3 The dependency of the threshold voltage (V_t) on the thickness of the gate (S = source, D = Drain, and NBTI = Negative bias temperature instability) (Reprinted with permission, Semiconductor International, March 2004, p. 48)

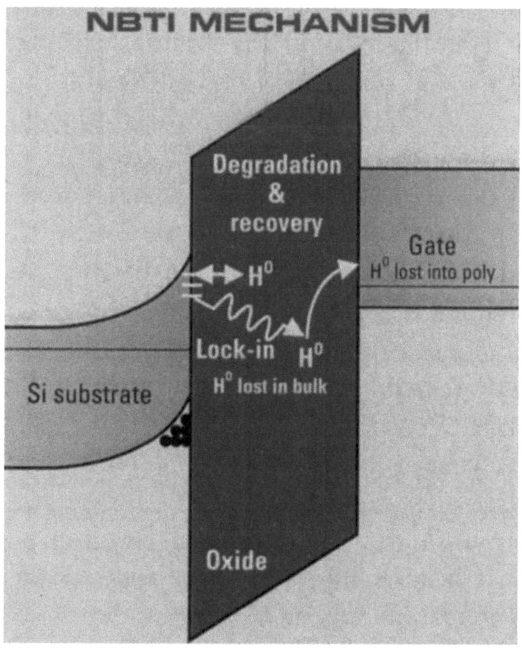

Fig. 2.4 Breaking up of the Si–H bond, release of hydrogen (H_2), re-passivation of the bonds, and lock-in when H_2 is no longer available (Reprinted with permission, Semiconductor International, March 2004, p. 48, Courtesy, Intel Corp.)

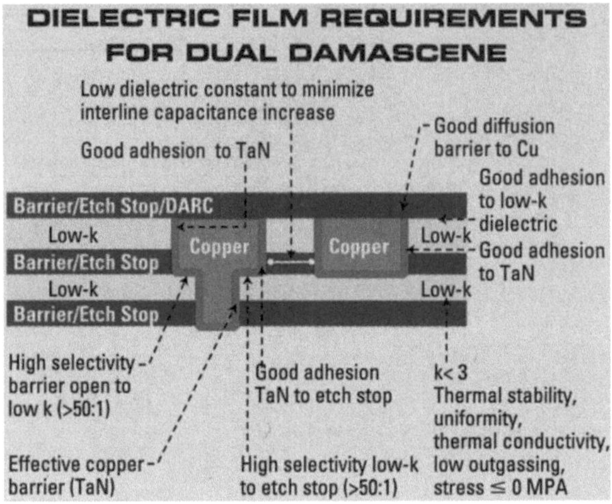

Fig. 2.5 The damascene architecture with low-*K* materials in the trench and via holes (Reproduced with permission, Applied Materials)

Dielectrics or electrically *insulated* materials are understood as materials in which electrostatic fields could persist for a long time. The materials offer a very high resistance to the passage of electric current, and have been used as thin films in electronic circuits. In conventional silicon planar devices, deposition of resistors, capacitors, and their connections are made on stable insulating substrates like silicon dioxide (SiO_2). Thin insulating dielectric films are used in a wide variety of components. The films are usually amorphous or near amorphous in nature and they show higher electrical resistance compared to polycrystalline or crystalline thin films. In integrated circuit (IC) dielectric materials are used as *insulating layers between conducting layers, diffusion and ion implantation masks, capping material for doped films to prevent loss of dopants, passivation layers* to protect devices from impurities, moisture and scratches, sandwich material between two electrodes to form *a capacitor*, and *gate oxide* [1–2].

The thin film dielectric materials that are most commonly used in exploratory *RC* (*R*, is resistance, and *C* is capacitance of a circuit) circuit applications are inorganic substances, mainly the oxides, and halides of metals and semiconductors. However, scaling down of the devices and the complexity of the integrated circuits (ICs) have stimulated the development of inorganic, organic and hybrid dielectric materials to deal with *RC* problems [7]. In comparing the properties of different dielectric materials it is convenient to use the capacitance density, which is defined as the capacitance per unit area, and is related to the *dielectric constant* (*K*) and the dielectric thickness of the material. The other important electrical properties of a dielectric material are its breakdown voltage, and dielectric strength. The breakdown voltage of a dielectric is dependent on the thickness and dielectric strength of the dielectric material. The dielectric strength for most of the dielectric film materials that have been used in conventional ICs is between 10^6 and $10^7 Vcm^{-1}$.

The inorganic dielectric materials that are now being used or are under investigation for future high-*K* dielectric materials are silicon monoxide (SiO, *K*~5.0), silicon dioxide (SiO_2, *K*~3.9), silicon nitride (SiO_3N_4, *K*~6), alkali halides (rubidium bromide, RbBr, *K*~4.7, lithium fluoride, LiF, *K*~9.2), barium titanate ($BaTiO_3$, *K* varies between 130 and 1000), lead titanate ($PbTiO_3$, *K* is between 200 and 400), and metal oxides (hafnium oxide HfO_2, *K*~40, tantalum oxide, TaO_5, *K*~ 27, tungsten oxide, WO_3, *K*~42, and zirconium oxide, ZrO_2, *K*~24.7).

Linear dielectric materials show a direct proportionality between the electric moment *p* (induced) acquired by the particle during the process of *polarization* and the intensity *E* of the electric field acting on the particle in question. The *polarizability* (χ_e) of a dielectric is defined as the electric dipole moment *p* per unit volume divided by the electric field (*E*) and the permittivity of free space ε_0. The dielectric materials having low polarizabilities are better in designing low-*K* materials. It has been observed that materials having single C–C and C–F bonds have the lowest electronic *polarizability* (≈0.531 and ≈0.555 $Å^3$ respectively), making fluorinated and non-fluorinated aliphatic macromolecules potential candidates for low-*K* applications [8–9]. Conversely, materials having double and triple bonds show higher electronic polarizability because of increased mobility of π-electrons (C=C ≈1.643, and C≡C≈2.036 $Å^3$). In general, materials with low polarizability having low dielectric constants show poor adhesion property. In contrast, aromatic π bonding configurations have higher bonding strength. Experimental observations show that increasing the bond length, bonding orientation, as well as discontinuing the chain by inserting single bond atoms or groups of atoms into the main structure, can also lower the *K* value of a material.

Advances in silicon ultra-large-scale integrated (ULSI) technologies have historically been made by scaling of the device dimensions. As a result, to minimize signal delay a reduced resistance, *R*, of the metal wiring and a reduced capacitance, *C*, of the encapsulating interlayer dielectric (ILD) line are required [1]. Thus the effort in the development of organic low-*K* dielectric materials has centered around inorganic dielectric materials. These organic dielectric materials offer highly desirable electrical and mechanical properties. At the same time for future sub-100 nm devices the thickness of the gate oxide will reach a value less than 1.5 nm, which will require high-*K* materials to lower the leakage current and tunneling. This chapter will be devoted to low-*K* and high-*K* dielectric materials, their characteristics, and deposition methods.

2.2 Interlayer Dielectric (ILD)

2.2.1 Introduction

Aluminum (Al) metal lines and silicon dioxide (SiO_2) as dielectric material have been used so far in conventional integrated circuits (ICs). But as the circuit lines changed to 0.18 μm in width from 0.25 μm, the limiting factor in computer processor speed shifts from the transistors′ gate delay to the interconnect delay caused by the Al-interconnects and SiO_2 (dielectric material). As a result, the interconnect *RC* delay (Fig. 2.6) becomes the major factor, which limits device performance [7].

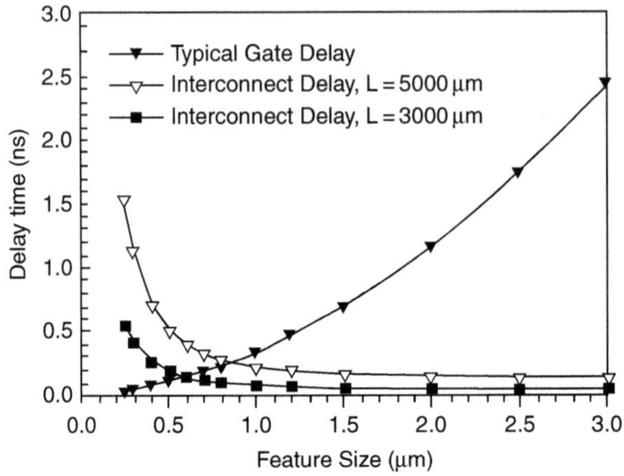

Fig. 2.6 The effect of feature size on the gate and interconnect delay (Reprinted with permission IEEE [7])

Thus the need to introduce advanced interconnect and dielectric materials becomes essential to limit the number of metal levels, die size and to reduce *RC* delay in *ultra-large-scale integrated (ULSI) circuits*. Figure 2.7 shows a multilayer metallization scheme with interlayer dielectric (ILD).

Technological innovation leads to the development of Cu-interconnect technology to replace Al- interconnecting lines since it will lower the line resistance (*R*) and improve electromigration reliability. Further requirements of the reduction of *RC* delay depend on the value of the stack and inter-electrode capacitance (Fig. 2.8).

The gate oxide thickness has reached a level where SiO_2 will not be able to provide adequate reliability (leakage current and tunneling). Therefore, there has been a continuous search for dielectric materials that will offer proper *K* values (both high-*K* and low-*K*), electrical and mechanical properties better than SiO_2, and better deposition processes to integrate the dielectric materials in sub-100 nm devices without much difficulty [1,7, 10–11].

The *dielectric constant*, *K*, of a specific material has three main contributions, namely *electronic*, *ionic (distortion)*, and *orientation*. The electronic contribution comes from the contribution of the electrons under an applied field and is related to the number of bonds per unit volume. Therefore for a particular class of dielectric material the electronic contribution is directly proportional to the density of the material. The ionic contribution represents the response of the atoms to an electric field and is dependent on the types of atoms (Si, C, H, N, F) present in the material. The last one, where the contribution is due to the orientation of the molecules under an applied field, is related to the structure of the material [12–13].

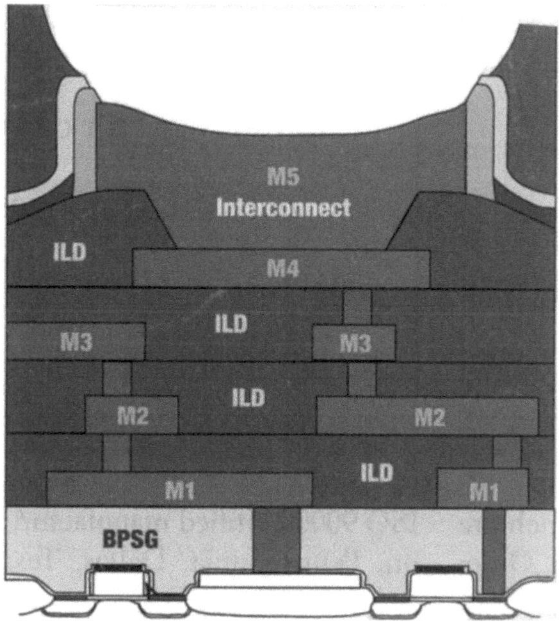

Fig. 2.7 Multilayer metallization with interlayer dielectric (ILD) (Reprinted with permission, Air Products and Chemicals, Inc., 2002)

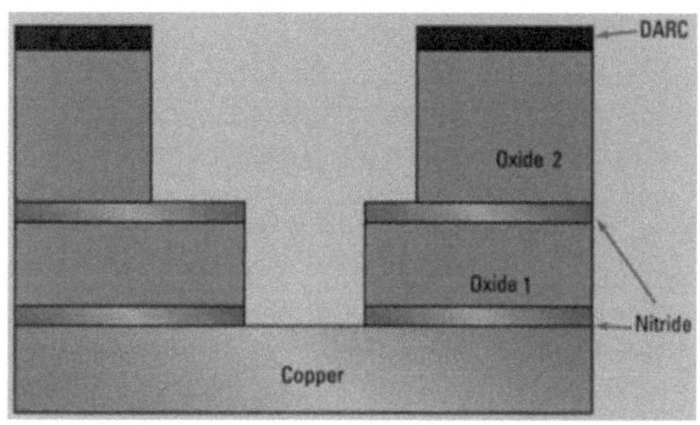

Fig. 2.8 A typical dual-damascene film stack of different dielectric materials including dielectric antireflecting coating (DARC)

2.2.1.1 Low-K

There are two primary approaches to achieve low-*K* dielectric materials. The first one is to lower the electronic contribution by the addition of fluorine (F) [14] and/or carbon (C) [15], which will provide the material with an inherently lower electronic polarizability. The second one is to lower the contribution due to the orientation and or the ionic contribution. This can be done by the introduction of a free volume in a material, which will decrease the number of polarizable groups per unit volume and will lower the atomic or dipolar contributions. Generally, the low-*K* materials fall under three categories, namely inorganic, organic, and hybrid (organo-silicates).

Due to their hydrophobic nature and low polarizability, organic dielectric materials show lower K values than inorganic materials. However, inorganic materials retain a SiO_2- like matrix, which help them to integrate easily into the existing SiO_2-like processes. Hybrid materials, on the other hand, are typically doped with carbon (C) to take advantages of both organic and inorganic regimes. Once the base material is selected (organic/inorganic/hybrid), the next step is the integration of the material in the sub-100 nm device.

2.2.2 Mathematical Model

The demand for higher speed and overall performance of the ULSI circuits has prompted extensive development of *low-K materials* for interlayer stacks (ILD/IMD). As the device down-sizes, the parasitic resistance and capacitance, i.e. *the RC effects*, become dominating factors for circuit performance. We can express the RC as: $RC = 2\rho K \varepsilon_0 \{(4L^2/P^2)+L^2/T^2)\}$, where ρ is the interconnect resistivity in Ω-cm, K the dielectric constant of interlayer dielectric (ILD), ε_0 the permittivity of free space, L the interconnect length in cm, P the interconnect pitch in cm, and T the interconnect thickness in cm (Fig. 2.9b). From the equation, we can see that by changing ρ and K, we can effectively minimize the value of RC [12].

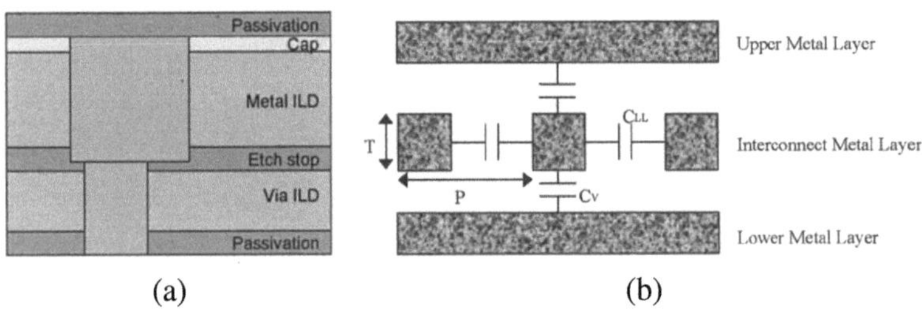

Fig. 2.9 (**a**) Schematic of ILD stack with many discrete layers (Reprinted with permission, AMD), and (**b**) equivalent circuit showing the effective capacitance within ILD layers (Reproduced with permission, Intel Corp.)

Figure 2.9a shows a schematic of an ILD stack with many discrete layers, and Fig. 2.9b represents the equivalent circuit showing the effective capacitance within ILD layers. The integrated K value of C_{LL} together with the K values of the other stacks will usually refer to the effective K value (K_{eff}) of the full stack assuming that the films are homogeneous.

The K_{eff} value is a strong function of the thickness of the individual stack, and the materials used in the cap and passivation layers. Considering the effective resistance ($R_{eff} = \{(2\rho L)/PT)\}$) and the total capacitance $C = (C_{LL} + C_V)$ we can write the total RC delay as:

$$RC = 2\rho K_{eff}\varepsilon_0 L^2\{(4/P^2 + (1/T^2)\} \tag{2.1}$$

From the above equation it is clear that as the chip geometry drivers (*P* and *T*) are decreased, and the interconnect length (as the device size shrinks) is increased, they will force the *RC* delay to increase. As a result the speed of the circuit will be affected severely. The dielectric constant (*K*) of the ILD layer comes indirectly in the factor *C* of the *RC* delay and thus the selection of an ideal dielectric material with very low-*K* will be a key to minimizing the *RC* effect. Figure 2.10 shows the effective capacitance effect of low-*K* on Cu-interconnects.

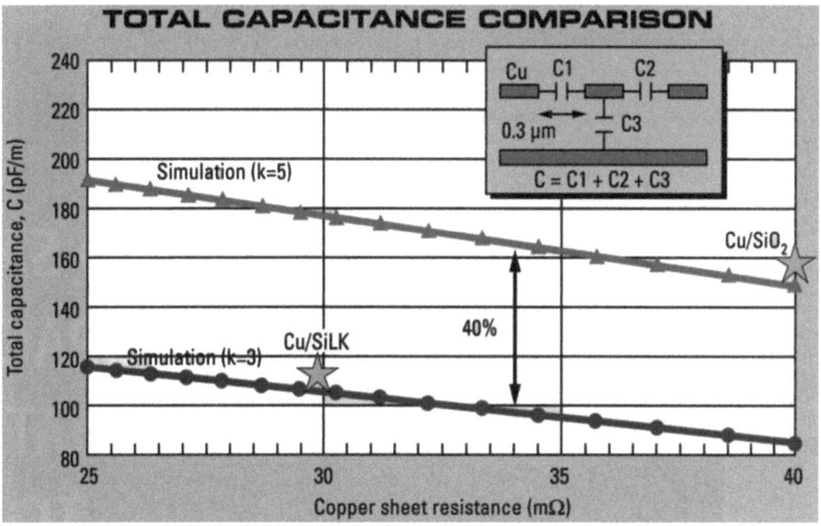

Fig. 2.10 The effective capacitance effect of low-K on Cu-interconnects (Reprinted with permission, Semiconductor International, June 2000, p. 118)

To extract K_{eff}, the capacitance between two inter-digitated combs is measured, and the dimensions of the combs (dielectric layer thickness and the metal line shape) obtained from cross-sections are entered into the model. The output of the model is matched to the measurement by varying the *K* value of the effective dielectric [12].

Figure 2.11 shows the results of capacitance simulations which have been used to extract the effective relative permittivities of interconnect structures by taking account of the bulk low-*K*, the dielectric thin films below and above the metal line (such as the etch stop, hard mask, and cap).

Capacitance simulations (Fig. 2.12) take care of the thickness and the *K* values of the damaged side walls (from electron energy loss spectroscopy measurements), especially when the thickness of the dielectric layer reaches $\leq$65 nm. The predictive K_{eff} simulations show that side-wall damage can no longer be neglected in the development work when the design node is less than 65 nm.

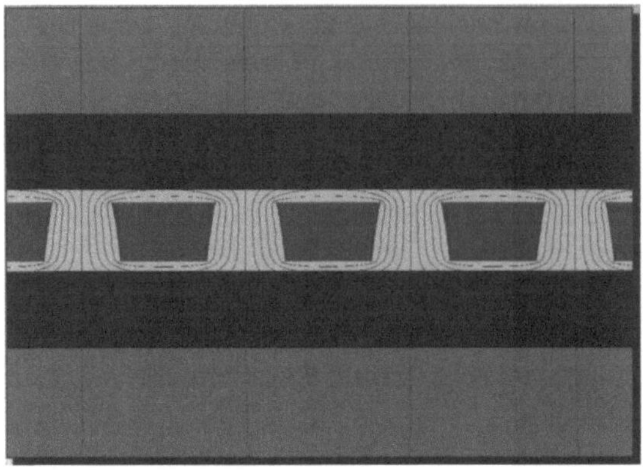

Fig. 2.11 The embedded metal inside the etch stop, bulk dielectric, hard mask, and cap. The combined structure gives rise to an effective dielectric (*broken lines* are equipotential lines) (Reprinted with permission Semiconductor International, July 2004, p. 87, Courtesy, SEMATECH, Austin, TX)

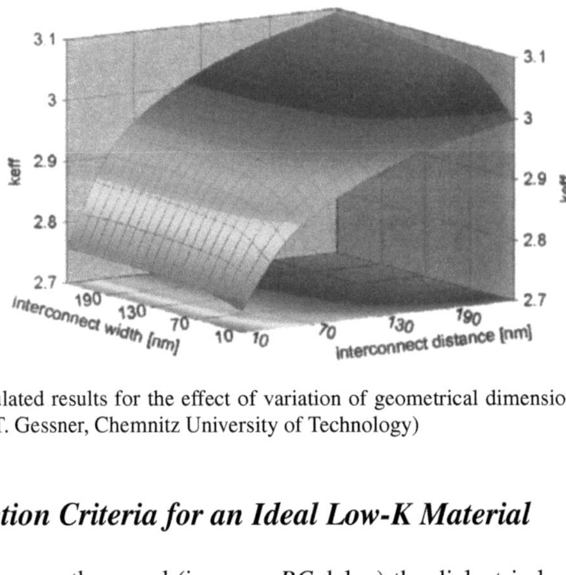

Fig. 2.12 Simulated results for the effect of variation of geometrical dimensions on K_{eff} (Photo courtesy, Prof. T. Gessner, Chemnitz University of Technology)

2.2.3 Selection Criteria for an Ideal Low-K Material

In order to increase the speed (improve *RC* delay) the dielectric layer should have low *K*, as well as reduce AC power dissipation (CV^2f, where *C* is the capacitance, *V* is the applied voltage, and *f* is the frequency of the applied voltage). The magnitude of the value of *K* is dependent upon the ability of the material to orient to the oscillations (frequencies) of an alternating electric field. As a matter of fact, the power consumed by the chip decreases as the interconnect capacitance is reduced.

The dielectric constant (*K*) is a physical measure of the electric polarizability of a material and can be expressed as $K = 1 + \chi_e, where\, \chi_e = M/E\varepsilon_0$, ε_0 being the permittivity of free space, *M* the induced dipole moment per unit volume of the dielectric, and E the applied field. Thus, the most obvious approach to synthesize a low-K material is to choose a material with few polar chemical groups with symmetry to cancel the dipoles of chemical bonds between dissimilar atoms.

Low-K dielectric materials have been targeted for better RC value to increase speed and to reduce interconnect cross-talk and bit-line capacitance for the memory chips [8]. It has been found that the devices are more susceptible to cross-talk at lower voltages and inductive coupling becomes more of an issue at higher operating frequencies.

The most important advantage of using the amorphous form of glassy silicon dioxide (SiO_2), in conventional silicon devices, is ease of forming silicon dioxide (SiO_2) by simple thermal oxidation [16]. But the disadvantages of SiO_2 in the copper damascene process are: (a) oxygen ion vacancies from positively charged defects that might oxidize the cap layer (barrier) or copper to form ionized $Cu^{+}/^{++}$ions; (b) non-uniformity and voids in loose disordered networks, which can form diffusion paths for the intermediate layers; and (c) most important of all is its high dielectric constant value (K of SiO_2 can vary between 3.9 and 4.9) [17].

Thus the need to introduce a low-K material compatible with the copper damascene process becomes essential to reducing RC delay in ULSI circuits. The liner material (ILD) should satisfy a number of diverse specifications, including good conformality and continuity, good adhesion, and good electrical and mechanical properties. But any process change in the semiconductor industry is difficult, yet, choosing a new dielectric material has been (and continues to be) an exercise in trying to find a low-K material with electrical, mechanical and thermal properties comparable to or better than SiO_2. Desired properties for these low-K dielectric materials will depend on the applications and the chip architecture. Some of the fundamental requirements that a processed dielectric material should show in the sub-100 nm level devices are listed bellow:

(i) dielectric constant should be below 3.0;
(ii) good thermal stability at least at the processing temperatures of the device, and low coefficient of thermal expansion;
(iii) good adhesion on the substrate and should be conformal after deposition;
(iv) should have low thermal shrinkage, ability to resist cracking and compatible with chemical mechanical polishing (CMP);
(v) should have an isotropic dielectric constant (K);
(vi) high dielectric breakdown, low charge trapping and leakage current;
(vii) low solubility in H_2O and low moisture absorption from the ambient;
(viii) materials should be friendly to the environment and safe during handling;
(ix) processing method should be simple and cost-effective;
(x) should be chemically inert;
(xi) should have high etch selectivity;
(xii) good mechanical properties, i.e. the film should have enough mechanical strength to prevent cohesive failure, and interfaces *delamination* (Fig. 2.13).

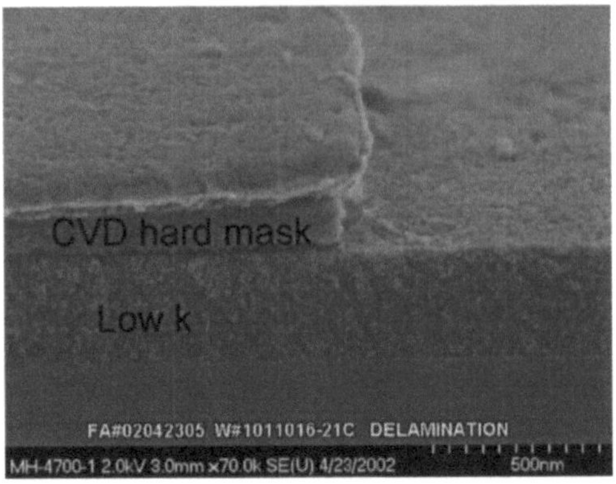

Fig. 2.13 Typical cap delamination failure (Reproduced with permission, AMD)

2.2.4 Search for an Ideal Low-K Material

The variation in the dielectric constant is attributed to the frequency dependence of the polarization mechanisms that contribute to the dielectric constant (*K*). The polarizability and *K* value of a dielectric material are generally results of the addition of three components (i.e. electronic + atomic + dipolar).

One approach to reduce the *K* value of a dielectric is to introduce carbon (C) or fluorine (F) atoms to increase the free volume of the matrix which will decrease the number of polarizable groups per unit volume. For example, in $SiO_2(K = 3.9)$, the introduction of C atoms to form SiCOH (*K* is between 2.7 and 3.3), and F atoms to form fluoro-silicate glass (*K* is between 3.2 and 4.0), and fluorinated polymides (*K* is in between 2.5 and 2.9), reduces its *K* value. On the other hand, hydroxyl and carbonyl groups are polar functional groups which can attract water via hydrogen bonding and thus drastically increase the dielectric constant (*K* of water $\sim$78.5). Thus to formulate a low-*K* material polar functionality containing elements like oxygen or nitrogen should be avoided.

The other approach that has been successfully implemented to reduce the *K* value is by introducing an air gap (*K* of air is 1) or pores. The addition of pores in a dielectric material is particularly challenging because the percentage of pores needed for low-*K* dielectric materials is not an absolute number that can be applied across the film. Thus the overall dielectric constant of a material can be varied from that of a dense material down to the value of air ($K = 1$) [18]. However, the porosity of a foam (static mixture) depends upon many factors, e.g. pore diameter, distribution of microstructure, and thickness of the pores (Fig. 2.16).

Based on this rationale, porous silica, referred to as silica xerogel, is synthesized by hydrolysis and condensation of tetra-ethylorthosilicate (TEOS). A great variety of formulas have been suggested for calculation of dielectric constant (*K*) of these static mixtures (doped and porous dielectric materials) [19]. It is easy to calculate the effective permittivity ε^* of a static mixture as:

$$\overset{*}{\varepsilon} = \Sigma_{i=1}^{i=m} y_1\, \varepsilon_i \text{ (homogeneous connected in series)}$$

and

$$\overset{*}{\varepsilon} = \Sigma_{i=1}^{i=m} (y_i/\varepsilon_i) \text{ (homogeneous connected in parallel).}$$

where y_i is the volume content of the components, ε_i is the permittivity of the components, and m is the number of the components. Figure 2.14 shows the effective permittivity of two components A and B versus their volume content in the mixture. ε_A and ε_B are the permittivities of material A and B, respectively. 1, 2, and 3 refer to parallel connection, series connection, and statistic distribution of the mixture, respectively. Our low-*K* dielectric materials (either doped or with pores) can be visualized as the static mixture of two materials.

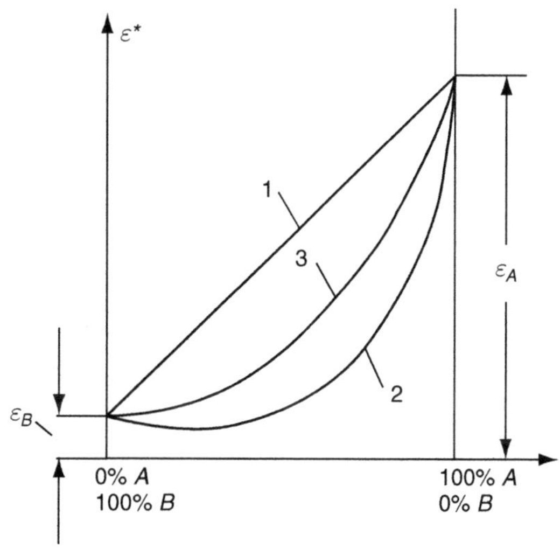

Fig. 2.14 The relative permittivity ε^* of a mixture of two components A and B mixed together vs. their volume content in the mixture (Reprinted with permission, MIR Publishers Moscow)

There are several formulas that have been developed for the calculation of permittivity of statistic mixtures. Among these, wide recognition has been given to the formula developed by Lichtenecker and Rother, which is known as the *logarithmic law of mixing*. The law can be mathematically expressed as:

$$\log \varepsilon^* = \Sigma_{i=1}^{i=m} y_i \log \varepsilon_i \tag{2.2}$$

For porous polymers with a vast number of small pores (porous low-*K*) the modified Lichtencker–Rother equation fits very well and can be represented mathematically as:

$$\log \varepsilon^* = (D^*/D) \log \varepsilon \tag{2.3}$$

where

$\varepsilon* =$ Permittivity of the foamed material = (mixture/matrix)
$\varepsilon =$ Permittivity of the solid material
$D* =$ Volume mass of foamed material (mixture/matrix)
$D =$ Density of the solid material

Figure 2.15 illustrates a nomogram convenient for practical calculations plotted on the basis of Equation (2.2). The ratio $\varepsilon_1/\varepsilon_2$ of permittivities of the components is plotted on the left vertical axis (assuming $\varepsilon_1 > \varepsilon_2$). The volume content of the first component in the mixture, i.e. y_1 (in percentage) is plotted on the inclined scale. The dotted line exemplifies the plot for the solution of a particular problem with a given ratio of $\varepsilon_1/\varepsilon_2$ and the value of K.

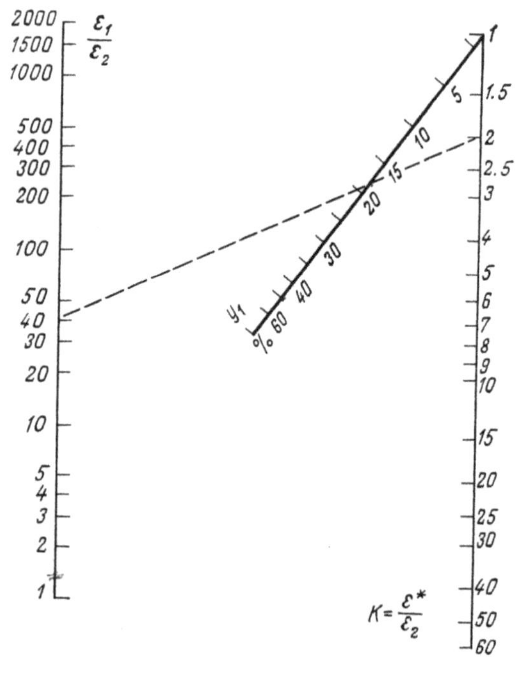

Fig. 2.15 Nomogram for calculations of dielectric constant of a mixture following Lichtenecker–Rother equation (Reprinted with permission, MIR Publishers Moscow)

After drilling through all these models a dielectric with permittivity ε_1 incorporating uniformly distributed (by volume) spherical inclusion of a material with permittivity ε_2 (volume concentration of inclusions y) can be written as [19]:

$$\varepsilon^* = \{\varepsilon_1(2\varepsilon_1 + \varepsilon_2 + 2y(\varepsilon_2 - \varepsilon_1))\}/2\varepsilon_1 + \varepsilon_2 - y\varepsilon_2 - \varepsilon_1)\} \tag{2.4}$$

Equation (2.4) can be applied to calculate the effective dielectric constant (K_{eff}) of the foam matrix (silesquioxane MSQ with pores). K_{S}, K_{p}, and K_{eff} are the dielectric constants of the bulk MSQ, the pores, and the mixture (foam), and y is percentage porosity. We can set up a mathematical equation to calculate the effective dielectric constant of the porous MSQ. The equation can be written as:

$$K_{\text{eff}} = K_s\{(1 + 2C)/(1 - C)\} \tag{2.5}$$

where $C = y\{(K_p - K_s)/(K_p + 2K_s)\}$. Several investigators have calculated the effective dielectric constants (K_{eff}) of porous dielectric materials applying different empirical formulas and have reported from time to time [20–23].

Figure 2.16 shows different sizes, distribution and structure of pores inside a low-K porous dielectric material. The porous dielectric materials can be (a) a constitutive porous dielectric or (b) a subtractive porous dielectric [24]. In a constitutive porous dielectric, the material is deposited either by various sol-gel reactions – usually a combination of hydrolysis and condensation processes – or by controlling the S/V ratio, where S represents the internal surface area of the reactor and V represents the reactor volume. Constitutive dielectric materials can also be obtained by doping silica with carbon or fluorine. On the other hand, subtractive porosity is introduced either by sol-gel processes or through the use of sacrificial nanoparticles, called *porogens*, that are desorbed during film cure.

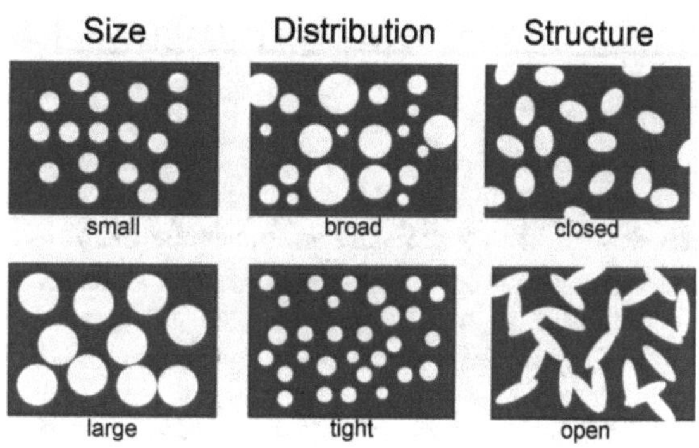

Fig. 2.16 Different sizes, distribution and structure of pores inside a low-K porous dielectric material (Reprinted with permission, AMD)

Commonly, the pores bearing low-K materials like xerogel and aerogel films have relatively broad pores, whereas pores in the porogen additive dielectric depend on the size of the chosen *sacrificial nano-particles* [25]. Care must taken in the time of introducing pores so that a monotonic increase in connectivity should not degrade the film properties.

Silica based dielectric materials can be non-porous or porous. Porosity in these materials can be either constitutive with pore size of 1 nm or subtractive where the pore size can reach as high as 10 nm.

2.2.4.1 Problems with Porous Films

One of the key problems with porous dielectric materials is the diffusion of metallic precursors through the pores and subsequent degradation of its electrical properties. Besides, thermal conductivity of silica based porous low-*K* materials is strongly dependent upon the porosity. As a matter of fact, with increasing porosity a strong decrease in thermal conductivity is observed (down to 0.3 W/m K at 50% porosity) [26–28]. The number of pores, pore size, and the distribution of the pores in the foam also affect the mechanical properties of the porous dielectric materials. For example, 45–50% porosity with subtractive techniques show an elastic modulus of around 2 Gpa or less. Besides, electrical properties and the adhesion of the porous dielectric materials are seen to decrease with percentage increase of porosity.

Swelling of low-*K* porous materials in the presence of liquids results in delamination. The degree of swelling during adsorption depends on the overall rigidity of the film skeleton. Further, the effect of different etch chemistries, e.g. non-polymerizing $CF_4/O_2/Ar$ (hard mask etch) versus polymerizing $C_4F_8/Ar/N_2$ (low-K etch), demonstrated undercut versus delay in the onset of damage. During reactive ion etching, porous low-*K* dielectric materials forms CF_4 compounds that decompose into volatile products, CO and CO_2. The depletion of carbon from these carbon containing compounds from the surface of the low-K leaves behind several dangling silicon bonds that ultimately change the physical, mechanical and electrical properties of the low-*K* dielectric materials.

Experimental data show that for $\leq$ 65 nm devices, the *K* value of the dielectric should be ≤ 2.2 which is primarily produced by introducing pores within the dielectric materials. Figure 2.14 shows different sizes, distribution, and structure of the pores in a porous film. One of the methods of introducing pores in dielectric films is by heating porogen. Unfortunately, the heating process leaves behind pores along with some carbon residue, which is conductive and dramatically reduces the breakdown voltage of the dielectric material. However, UV curing with the correct energy has been successful in minimizing carbon deposition compared to heating by electron beam or simple heating.

Once the pores are created, beam-positron annihilation lifetime spectroscopy (PALS) and transmission electron spectroscopy (TES) are used to study the void volumes and pore structures in the porous films [29–30]. The critical thickness of the porous films required in maintaining the continuity of the film is determined from the histograms of the peak height profile obtained from atomic force microscopy (AFM) [31–32].

Sometimes the pores can be interconnected (elongated, that is ~7 nm or longer in pore size) and can form a large open area within the dielectric film (Fig. 2.14). PALS measurements can detect them. The presence of moisture or oxygen inside these open areas can contaminate the barrier layer and can form oxide. It is not unlikely for the trapped oxygen to leak to the copper layer to form $Cu^{+}/^{++}$ions and to migrate into the silicon layer.

The degradation in mechanical and thermal properties of the porous dielectric material compared to the dense virgin material has become a reliability issue.

Moreover, if the pores are not *sealed* properly they can absorb moisture and can react with fluorine-containing material to form hydrofluoric acid (HF), which is detrimental to the ILD stack and to the associated layers. The atomic layer deposition (ALD) method has been reported to be successful in sealing the pores of the porous low-*K* material with silica [33–34].

2.2.4.2 Pore Sealing

The methods of pore sealing are dependent on the nature of the pore structures (mesoporous, microporous with micro- and meso-connections), morphology and chemical nature of the surface bearing the pores. In order to avoid the problems encountered with porous dielectric films, the pores are not generally sealed until the value of *K* reaches below 2.5 or the film porosity reaches more than 30%. The most common method practiced for pore sealing is plasma ashing followed by either physical vapor deposition (PVD) or plasma enhanced chemical vapor deposition (PECVD). However, care must be taken to ensure that the penetration of the sealing material should not act as a barrier but will give a good surface fitting to the copper.

Needless to say, deposition of a pore sealant in the pores should be compatible with the low-*K* dielectric with minimum thickness. Sealants like silicon oxycarbide, SiC:H, parylene, and benzocyclobutane (BCB) have been used so far and the results of their compatibility with low-*K* dielectric materials have been reported from time to time [35–39].

2.2.5 Achievement

2.2.5.1 Porous Silicon Dioxide (SiO_2) (the Xerogel)

In order to lower the effective dielectric constant (K_{eff}) of SiO_2, pores are introduced inside the SiO_2 matrix. The porous silica is termed *xerogel,* and is prepared by hydrolysis and condensation of either tetra-ethyl-orthosilicate (TEOS) or its derivatives [33]. Depending on the porosity and the processing technologies the *K* value of a porous *xerogel* can vary from 1.3 to 2.5. A highly porous xerogel is called *aerogel* [11, 40]. Final porosity during spinning can be controlled by using solvents, such as ethylene glycol (EG) or dimethylsulfoxide [41]. The drawback with the porous xerogel film is that it shows cracking patterns (Fig. 2.17) when the film is thin ($<1\ \mu m$) [42–43]. Experimental observations further show that all interfaces having energies below 5 J/m^2 lead to either film delamination or cracking and become a reliability threat during packaging and testing [44].

2.2.5.2 Carbon Doped Oxide

The synthesis of a nanopore carbon doped oxide (CDO) is reported with atomic composition of Si : C : O : H as 20.5 : 14.5 : 31 : 34. The porosity of the film is detected as $\sim$ 20% with pore size of 1–4 nm. A 200 nm commercial CDO film

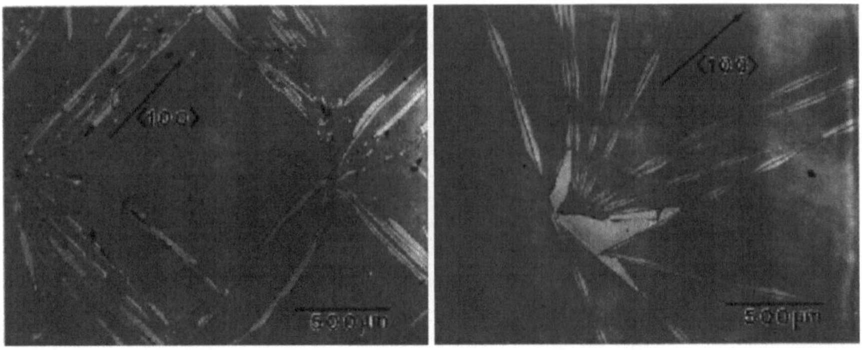

(a) (b)

Fig. 2.17 (**a**) Xerogel cracking on a <100> Si-wafer, and (**b**) cracking on a <111> silicon wafer (Photo courtesy, Prof. K.N. Tu, UCLA, CA)

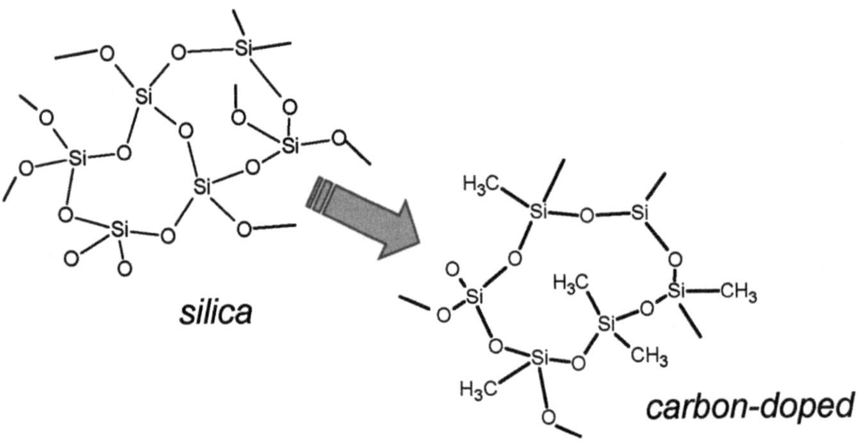

Fig. 2.18 The structure of silica and carbon doped silica (Reprinted with permission, Prof. R.H. Dauskardt et al., Stanford University, CA)

(Orion) deposited by PECVD shows a *K* value less than 2.2, with excellent thermal stability up to 600 °C and a leakage current of 1 nA/cm^2 at 2.5 MV/cm. Figure 2.18 shows the structural change of silica when it is doped with carbon.

2.2.5.3 Black Diamond

Black Diamond (BD) is a low-density silicon dioxide-like material formed generally by chemical vapor deposition (CVD). The precursors used include silicon-containing compound, typically an organosilane ($RSiH_{4n}$) and an oxidizer. The density and the *K* value (2.5–2.7) of the material can be controlled by appropriate choice of *R* (network terminating group). The CVD processed film is isotropic and amorphous in nature. The measured leakage current and breakdown field of the film have been reported as $\sim 5 \times 10^{-9}$ A/cm^2 at 1 V and >2.5 MV/cm, respectively. An eight-level Cu/black diamond integrated architecture is shown in Fig. 2.19, and Fig. 2.20 shows a nine-level metal with black diamond dielectric layer which has been fabricated by following the 130 nm process technology.

Fig. 2.19 Eight-level Cu/black diamond integration (Reprinted with permission, Applied Materials)

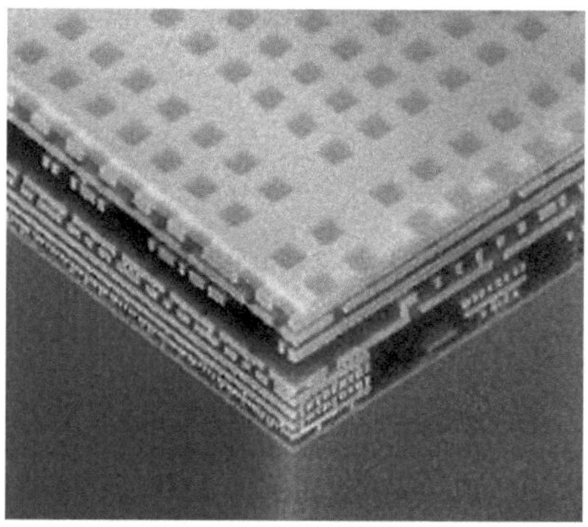

Fig. 2.20 AMD's 130 nm process, which uses nine levels of metal, black diamond low-*K* dielectric and BLOK dielectric barrier film (Photo courtesy Applied materials)

2.2.5.4 Bezocyclobutene (BCB) Based Polymer

In the late 1980s biscyclo (4,2,0) octa-1,3,5,-triene or 1,2-dihydrobenzenzocyclobutene, a low-*K* polymer, was formulated based on benzocyclobutene (BCB). Later on, a new formulation of BCB was reported as B-staged BCB which is a partially cross-linked 1,3,-divenyl-1,1,3,3 tetramethyldisiloxane-bisbenzocyclobutene (DVS-bis-BCB), a monomer in mesitylene solvent [33]. Figure 2.21a shows the structure of BCB and Fig. 2.21b represents DVS-bis-BCB.

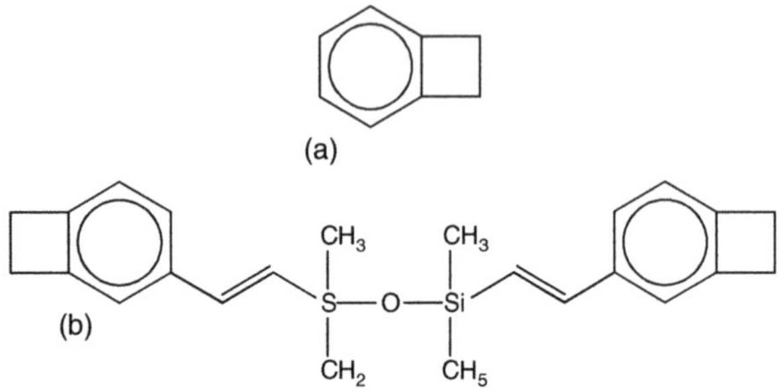

Fig. 2.21 Chemical structure of (**a**) BCB and (**b**) DVS-bis-BCB

Table 2.1 shows some of the physical characteristics of bezocyclobutane (BCB). The dielectric *K* can be reduced further by the addition of fluorine (F). However, fluorine (F) readily attacks tantalum based barriers, leading to volatile TaF_2 formation and ultimately loss of adhesion in the interface layer. The disadvantage of using this material is its higher moisture intake, which makes the film vulnerable to cracking.

Table 2.1 Characteristics of BCB dielectric film

Material	Dielectric constant (K)	Dielectric strength (V/cm)	Volume resistivity (Ω-cm)	CTE (ppm/ °C)	Thermal conductivity W/mK
BCB	2.49–2.65	3×10^6	1×10^{19}	52	0.29

[31,33,40, 43]

2.2.5.5 Aromatic (Hydrocarbon) Thermosetting Polymer (ATP)

Aromatic (hydrocarbon) thermosetting low-*K* polymer is receiving much attention for back-end-of-line (BEOL) applications in sub-100 nm node technology for its low moisture uptake and high fracture roughness [45]. The commercial product of ATP is *SiLK*– a cross-linked poly-aromatic matrix. It largely eliminates the nuclear polarizability component and shows high bond strength and excellent thermal resistance [46–47]. It is expected that this commercial product will serve the purpose of an ILD material in both Cu-damascene and Al-W systems. The polymer has an isotropic *K* value and some commercial houses are using this low-*K* dielectric material in a two-level Cu damascene process [44]. Figure 2.22 shows the structure of SiLK.

Figure 2.23 shows a scanning electron micrograph of the 130 nm Cu-damascene process architecture with SiLK as ILD [48].

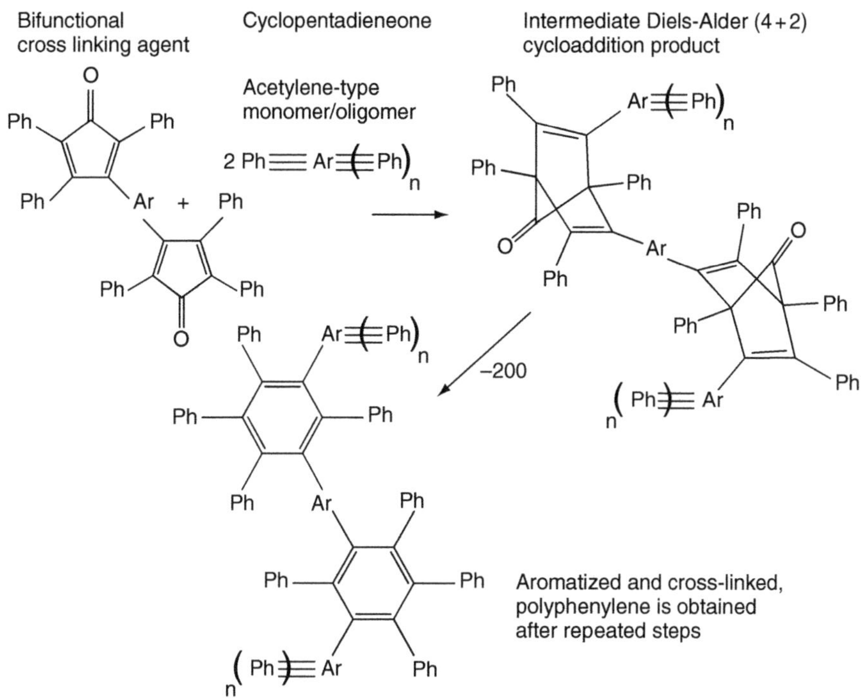

Fig. 2.22 Structure and synthesis of SiLK

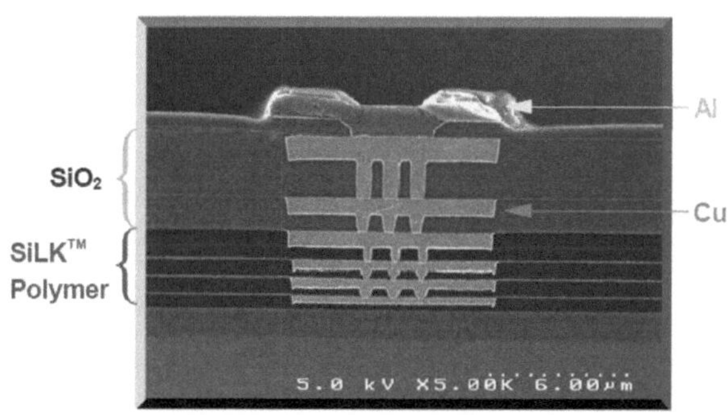

Fig. 2.23 Scanning electron micrograph of IBM's 130 nm Cu-damascene process with SiLK polymer (Reproduced with permission, IBM Research)

2.2.5.6 Hydrogen Silesquioxanes (HSQ, $HSiO_{1.5}$)

The chemical formula for HSQ can be written as (R Si $(OH)_xO_{(3-x)/2})_n$ where R is hydrogen or a methyl group, *n* is an integer whose value is greater than 8, and the value of *x* is between 0 and 2. The reactive mixture is composed of chlorosilane ($RSiCl_3$) and a polar organic solvent containing water or hydrogen. The solvent must be able to form hydrogen bonds with residual silanol (Si–OH) groups [49]. Ethyl acetate or methyl isobutyl ketone is the possible solvent which has been used in the experiment [50]. The dielectric constant (*K*) of HSQ is 2.9 [51], which can be further reduced either by introducing bonds with lower polarizability or by pores [52–53]. The nano-pore material forms an extra-low-*K* (XLK) material with reduced density.

FoX (flowable oxide), a commercial product of HSQ, condenses at elevated temperatures and forms a ladder network. During annealing care must be taken so that there is no exchange of hydrogen and/or oxygen atoms from moisture, which could raise the *K*-value of the material due to the presence of silonol (Si–OH) groups. In methyl silsesquioxanes (MSQ, $RSiO_{1.5}$, where, R stands for CH_3) this problem has been avoided by the substitution of the liable H atoms by hydrophobic organic R groups (Fig. 2.26). However, when an organic material is integrated into an inorganic resin matrix, they can segregate at the surface to lower their free energy, possibly leading to adhesion problems [48].

The synthesized HSQ can form two basic structures, namely a ladder structure (Fig. 2.24) and a cage structure (Fig. 2.25) and the film can be deposited by the spin-on-deposition (SOD) technique. The main drawback with HSQ is its thermal instability [54–57].

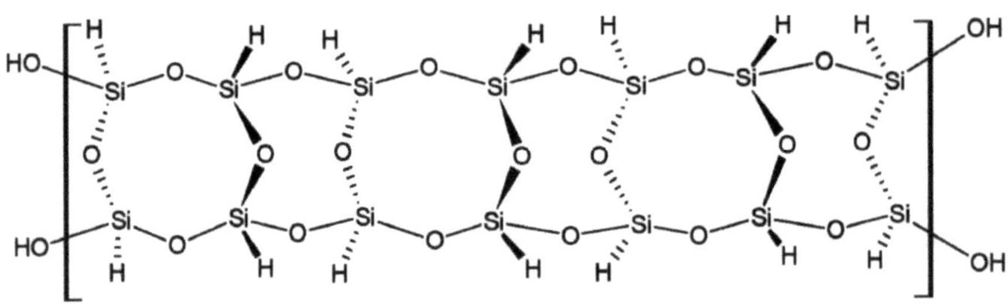

Fig. 2.24 Ladder structure of hydrogen silsesquioxane (HSQ)

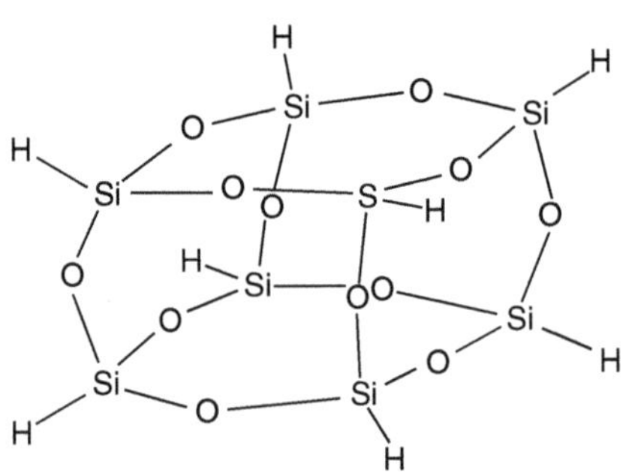

Fig. 2.25 Another form of hydrogen silsesquioxane (HSQ) which resembles a cage structure

2.2.5.7 Methyl Silsesquioxane (MSQ)

Methyl silsesquioxane (MSQ, $CH_3SiO_{1.5}$) is synthesized by the hydrolysis of methyl triethoxy silane [58–59]. Commercially available MSQ has a *K* value of ~2.7. Due to the presence of hydrophobic organic H atoms in MSQ, the hybrid polymer is protected from environmental oxygen to form any silanol groups that might increase its *K* value. Cross-linking (curing) takes place in MSQ through reactions of the alkoxyl silyl groups on the oligmers. The main difference between HSQ and MSQ is that MSQ is more stable at high temperatures (400 °C and above) [49]. Figure 2.26 shows the ladder structure of MSQ.

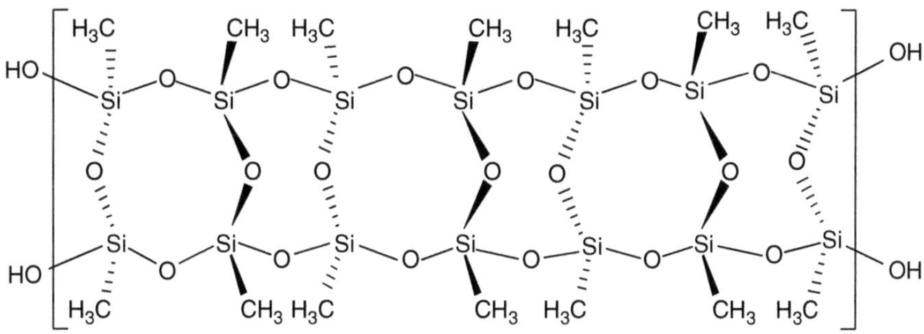

Fig. 2.26 Methyl silsesquioxane (MSQ) ladder structure

Processing of Low-K MSQ

Case I: By varying the percentage of *porogen* (organic polymer) from 1 to 30% in MSQ the porosity and the pore size inside the hybrid polymer can be changed from 12 to 34% and from 10 to 15%. As a result, the dielectric constant of the foam is found to change from 3.5 to 2.5 (Fig. 2.26). However, the wall density is found to be independent of the porogen content [60].

The inorganic–organic matrix gives rise to a hybrid system, which offers versatility and extendibility of the host matrix since the matrix material remains constant and only the loading of porogen varies [61]. Figure 2.27 shows the process sequence of blending thermosetting resin with dendritic polymer (porogen) and subsequent spin coating of the hybrid foam over a substrate. However, the main drawback of this hybrid matrix is that at elevated temperatures porogen decomposes and subsequently diffuses as a volatile product. Figure 2.28 shows the effect of porogen blending on the *K* (dielectric constant) value of the matrix.

Case II: Another way to change the *K* value of MSQ is to introduce an air gap by using a family of polynorborness [62–63]. It has been reported that the addition of 20% derivatives of norbornene (NB) polymer containing triethoxy silyl groups produces ~70 nm pores in MSQ and it reduces the *K* value of the foam from 2.7 to 2.3. The voids created in this method exhibit a closed pore structure, which is advantageous from the standpoint of moisture intake. In addition to that, NB changes the number of pores but not their sizes.

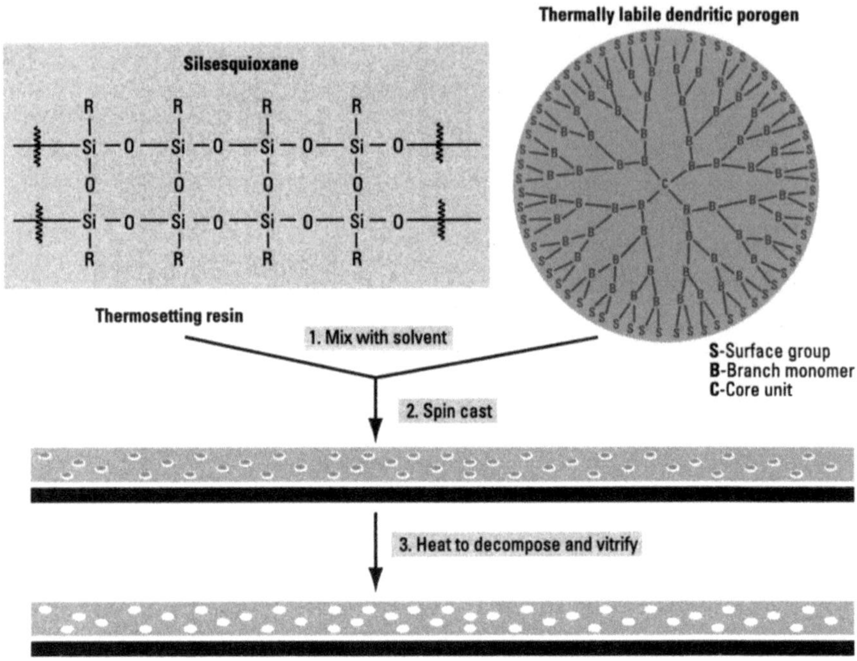

Fig. 2.27 The hybrid mix of inorganic silesquinone with organic porogen polymer solubilized in a common solvent (Reprinted with permission, IBM Research, [61])

Fig. 2.28 The initial porogen content (wt%) in MSQ versus change in K value of MSQ (Reprinted with permission, R.H. Dauskardt, Stanford University, CA)

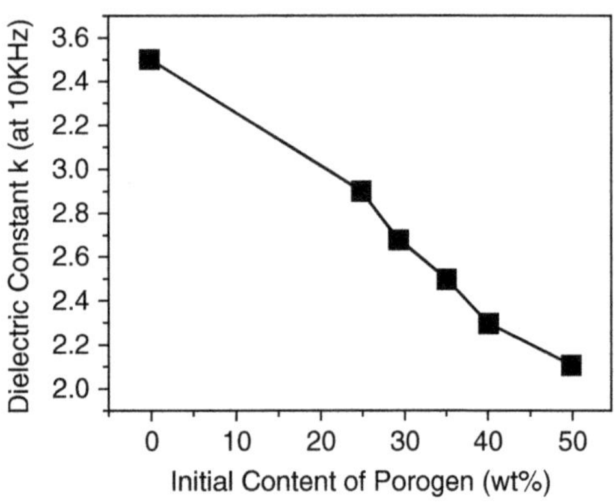

2.2.5.8 Poly (Arylene Ethers) (PAE)

Poly-arylene ethers (PAE) are synthesized by reducing bisphenol with activated defunctionalized aromatic precursor. The PAE family is a non-fluorinated polymer and shows a K value of ~2.8. The elimination of fluorine from the PAE has shown better results in the areas of low out-gassing after cure, thermal stability, low moisture uptake, good solvent resistance, and adhesion over the substrate [64]. The commercial products of the PAE family are FLARE and VELOX [65]. FLARE is a bridged PAE with a dielectric constant of 2.5 and has shown thermal stability up to 400 °C [66]. Figure 2.29 shows the structure of a PAE family.

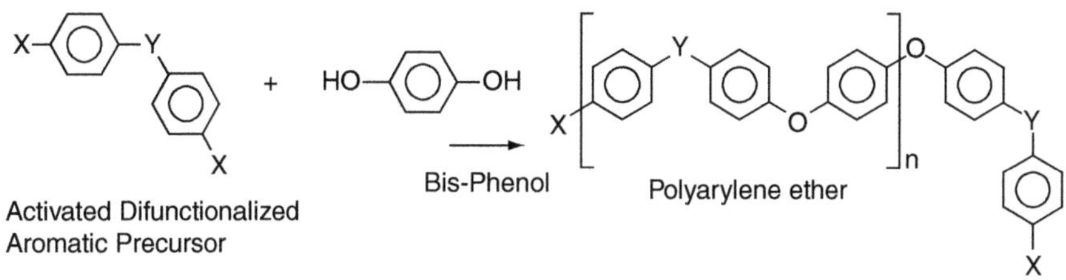

Fig. 2.29 The structure of poly-arylene ether low-*K* polymer

2.2.5.9 Parylene

Parylene is a common generic name for a unique series of polymers based on paraxylene. The thermoplastic polymer family is available in three trade names as: parylene N, parylene C, and parylene D [67]. The synthesis of the material is performed in a vacuum by pyrolysis of di-p-xylene (dimmer). The temperature during synthesis is very critical because it changes the morphology of the film from a homogeneous (non-porous) structure to a heterogeneous (porous) structure [68]. When fluorinated, parylene-N shows lower *K* (2.65–2.1) values but the adhesion of the co-polymer is impaired [70]. Experimental observations show that the CVD process produces films with better conformality and good gap filling compared to spin-on deposition (SOD) [69].

2.2.5.10 Teflon AF

This is a copolymer of tetrafluoroethylene and 2,2, -bis trifluoromethyl-4, 5-difluoro-1,3-dioxole [71], having optical and mechanical properties similar to other amorphous polymers including strength. The *K* value of the material is between 1.89 and 1.93, and it shows very good thermal and chemical stability at elevated temperatures [72].

The Teflon AF polymers are distinct from other fluoropolymers in that they are soluble in selected solvents, with high gas permeability, compressibility, creep resistance, and low thermal conductivity. The index of refraction of these polymers is between 1.29 and 1.34. Figure 2.30 shows the structure of the AF copolymer of Teflon and perfluoro-2,2-dimethyl-1,3-dioxole.

2.2.5.11 Diamond-Like Carbon (DLC)

DLC film shows high electrical resistivity and dielectric constant which can range from 10^2 to $10^{16}\Omega$-cm and 2.7–3.8, respectively, depending on the deposition conditions [73–74]. However, when doped with nitrogen, the film shows lower *K*

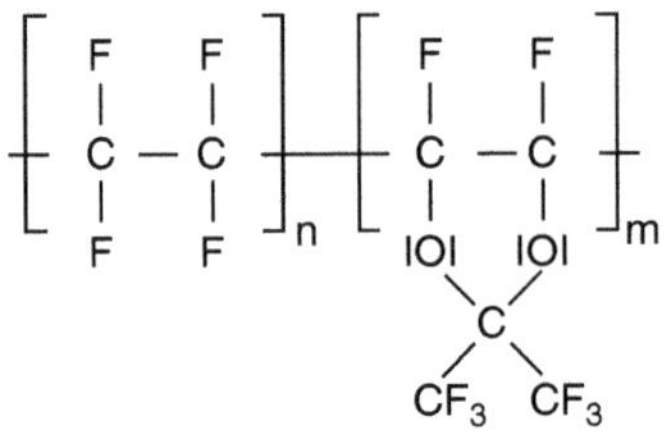

Fig. 2.30 The structure of Teflon AF®

(2.4–2.8) values and internal stress. A film of DLC can be deposited by the PECVD method using hydrocarbon as a precursor [75]. Besides the PECVD method, deposition of DLC has been reported by microwave plasmas [76], dc discharge, and ECR (electron cyclotron resonance) methods. DLC is an amorphous, metastable material and is grouped into four categories according to whether the material contains hydrogen or not [77].

2.2.6 Impact of Low-K ILD Materials on the Cu-Damascene Process

The choice of dielectric material has significant impacts on both lifetime and packaging reliability. Dual laid Cu-interconnect with low-*K* ILD has been introduced to address the *RC* delay reduction. However, porous low-*K* dielectric materials are generally less dense than SiO_2. As a result, integration of low-*K* materials in a Cu-damascene architecture has become a challenge (Fig. 2.31).

The pores in the dielectric medium are supposed to be spheroid and the porous dielectric medium can be considered as a random composite medium. In order to address the reliability issues of the porous dielectric, the dielectric layer is sandwiched between passivation layers and/or the pores are filled with some different material. Thus the total *K* value of the dielectric has to be calculated from the effective *K* value (K_{eff}) of the composite [78]. One way to determine the K_{eff} value is to apply the effective medium theory developed by Bruggeman, which is presented mathematically as [79–80]:

$$f_1\{(K_1 + K_{eff})/(K_1 + 2K_{eff})\} + f_2\{(K_2 - K_{eff})/(K_2 + 2K_{eff})\} = 0 \qquad (2.6)$$

where f_1 and f_2 represent the fraction of the two dielectric materials having K_1 and K_2 values. The value of K_{eff}, due to mixing of two dielectric materials, can be determined by substituting the values of f_1, f_2, K_1, and K_2 in the above equation. Bruggeman's model helps us to choose a barrier layer and an etch stop for a particular ILD material. From Fig. 2.32, we can see how the change of barrier layer from silicon nitride based material to silicon carbide based material significantly influences the K_{eff} value even before the low-*K* ILD is introduced.

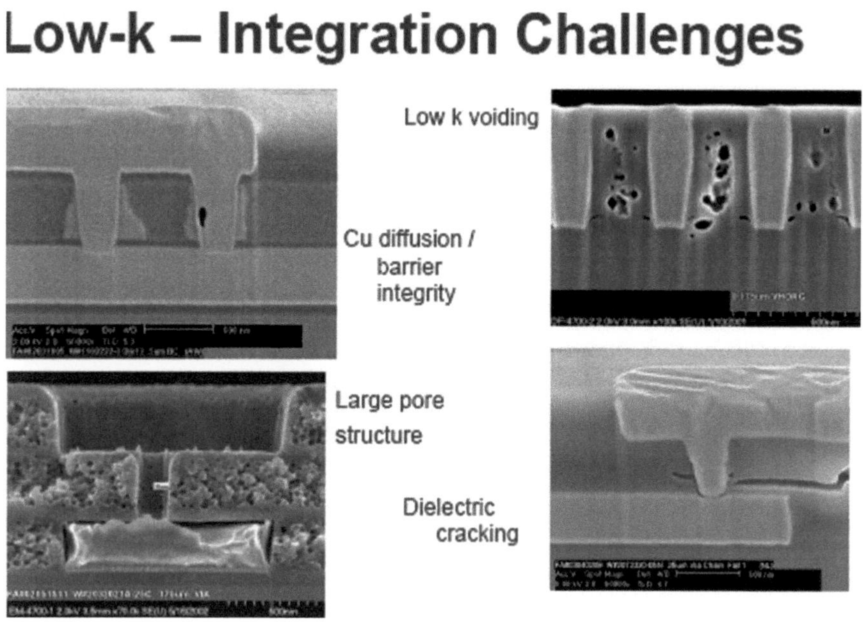

Fig. 2.31 Different challenges related to low-*K* integration in integrated circuits (ICs) (Courtesy, Peter J. Wolf)

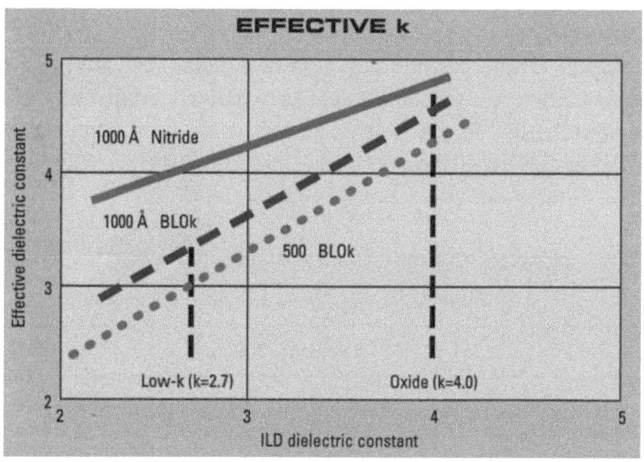

Fig. 2.32 The SiC based dielectric barrier impacts the effective *K* value much less dramatically than the traditional nitride film (Courtesy, Applied Materials)

The number of pores in a dielectric medium not only controls the *K* value but it also controls the mechanical strength and thermal conductivity of a dielectric film that ultimately controls the cracking, and delamination properties of the film [10]. The mechanical strength of a low-*K* dielectric material is important for chemical mechanical polishing (CMP) operations, where the dielectric strength impedes $Cu^{++}/^{+}$ diffusion at higher electric fields (approximately 1 MV/cm or above) [81–82].

To integrate the low-*K* dielectric material in Cu-interconnect structures it is helpful to sandwich them between capping stacks. The capping materials should also have low *K*, and should be compatible with photolithography processes [70]. Amorphous silicon carbide (SiC) [69] has low relative permittivity, good adherence properties on the substrate, high modulus and hardness [68], is chemically inert and moisture resistant [71–72], and has shown promising results as a capping material. Besides SiC, amorphous SiC:H has also been used as a cap layer and anti-reflecting coating (ARC) but dry etching of the material is hard. However, recent result shows that $Ar/CF_4/N_2/O_2$chemistry is able to dry-etch *a*-SiC: H [75].

Fluorinated oxide shares many integration similarities with undoped plasma oxides and is being extensively developed as the next generation ILD material with $K{\sim}3.5$. Fluorinating a dielectric is a common method of reducing *K* provided that the fluorine atoms are incorporated correctly. The external electric field cannot perturb the fluorine atom, because of its high electronegativity. Thus it is not possible to polarize it easily. But fluorine being a highly reactive element, excessive fluorination might corrode the metal and the dielectric.

Organic polymers, especially the polyimides, have good mechanical strength, thermal stability, and show high chemical resistance but their *K* values are typically more than 3.0. Moreover, they absorb moisture and show significant anisotropy in the dielectric constant. These shortcomings have stimulated the chemical industry to develop different families of low-*K* dielectric materials that are tailored to ILD integration. These include polyarylene ethers, derivatives of cyclobutane, polynobornenes, amorphous Teflon®, phase separated inorganic–organic hybrids, parylene-N, parylene-F, polynapthalene, and polytetrafluoroethylene [83].

Table 2.2 shows some of the promising materials for low-*K* application in Cu-interconnects and their suitable deposition methods.

Table 2.2 Low-*K* dielectric materials

Materials	Dielectric constant	Deposition process
Silicon dioxide (SiO_2)	3.8–3.9	PECVD
Carbon doped SiO_2	2.2–2.7	PECVD
Bezocyclobutane (BCB)	2.49–2.65	Spin-on
HSSQ	2.9	Spin-on
MSSQ	2.7	Spin-on
Polyarelene (PAE)	2.8	Spin-on
Parylene-N	2.8	CVD
Parylene-F	2.3–2.5	CVD
Teflon AF	1.89–1.93	Spin-on
Diamond like carbon (DLC)	2.7–3.4	PECVD
Fluorinated DLC	2.4–2.8	PECVD
Aromatic thermosets (SiLK)	2.6–2.8	Spin-on

2.2.7 Deposition Techniques

Deposition methods for different materials frequently used in Cu-damascene processes have been presented in a separate chapter. In order to keep the flow of the damascene architecture, a brief outline of the low-*K* deposition system is presented here. The chapters in the book have been arranged in sequence with the damascene process flow.

2.2.7.1 Flow chart for Damascene Architecture

(a) deposition of dielectric material;
(b) pattern generation (to create trenches);
(c) deposition of barrier layer;
(d) Cu-seed layer;
(e) electro-chemical deposition (ECD) of Cu;
(f) planarization; and
(g) via holes if necessary.

The deposition techniques used for the deposition of dielectric films include low-pressure processes such as physical vapor deposition (PVD), sputtering, plasma deposition, and chemical vapor deposition (CVD). The processes operating at atmospheric pressure may include thermal oxidation, conventional CVD, anodization, electrophoretic deposition, spin-on/spray-on technologies, screen printing and miscellaneous deposition processes, e.g. roller coating, offset printing, etc. [69–72]. Use of CVD processes offers an evolutionary approach for inter-metal low-*K* dielectric materials (IMD) with transition from plasma enhanced CVD (PECVD) SiO_2 to so-called organosilicate glass (OSG). However, the high plasma energy of PECVD is detrimental to the deposition of inorganic materials, especially the materials containing fluorinated amorphous carbon (FLAC).

Spin-on deposition (SOD), on the other hand, is very much like photoresist (PR) deposition, but the baking temperature of the SOD dielectrics is higher (400 °C and above) relative to the post-baking temperature of PR (130–140 °C). Many factors affect SOD processing, including track system cost, material cost (amount of material needed to coat), and track system throughput. Thus the process becomes very lengthy and expensive. Moreover, curing and pore generation at the same time are very difficult to achieve with SOD systems especially if the curing is done on a hot plate.

On the other hand, organic spin-on materials and inorganic CVD materials can be complementary, making hybrid integration a viable and attractive option. By eliminating process steps and allowing chip makers to use some of their current CVD equipment, the hybrid approach will allow the manufacturer to extend their knowledge in tackling the sub-100 nm node copper damascene process. It is most likely that organosilicate glass (OSG) will be the candidate for CVD technology while aromatic thermoset resin (ATR) could possibly be patterned with inorganic dielectric materials.

Thus there are two main strategies suitable for depositing low-*K* materials: (a) SOD and (b) CVD. For both SOD and CVD processes there are methods and materials that are unique combinations for a particular system. However, SOD is considered to be the most widely used method. CVD films, on the other hand, have given the biggest accomplishment in sub-100 nm processes. CVD grown OSG has the capability of producing film with $K<2.5$ which is not possible to grow with other methods. Although the CVD process has been the mainstream deposition process for silicon based inorganic dielectric materials, there are some shortcomings in vapor deposition methods when the dielectric material is a polymer. Therefore, one can say that the degree of molecular control one can exercise with the SOD process is not possible with a CVD system. Figure 2.33 shows a commercial low-*K* material (FLARE) which has been integrated in a Cu-damascene architecture.

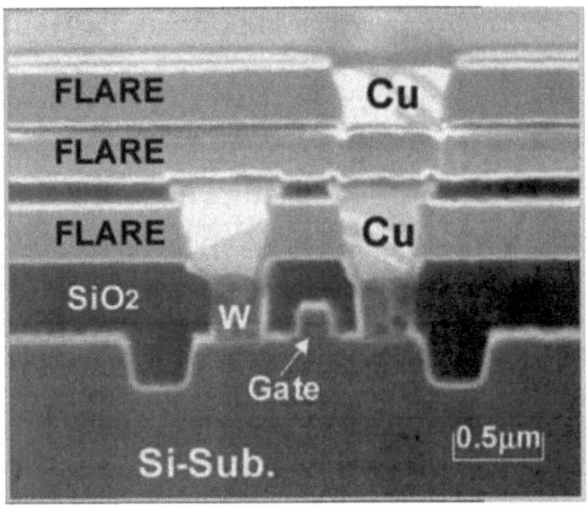

Fig. 2.33 Scanning electron micrograph (SEM) of FLARE integrated into the damascene structure (Courtesy of Honeywell Corp.)

However, there are controversies about SOD low-*K* materials and CVD carbon doped oxides. As a matter of fact, some of the spin-on organic varieties include polyarylene ethers, derivatives of cyclobutane, phase separated inorganic–organic hybrids, amorphous Teflon, xerogels, and polynorbornenes. On the other hand, the CVD alternative of organic polymers covers parylene-N and F, polytetrafluoroethylene (Teflon), and polynapthalene. But when the mismatch of the coefficient of thermal expansion (CTE) between two adjacent materials is an issue, the use of low-*K* SOD materials will be a matter of much concern. Moreover, delamination of SOD films at the point where interconnect via holes are attached to copper wire has been a troubling episode. Considering all of these facts, we can say that instead of using SOD or CVD alone, use of both in the 65 nm node technology might be a viable alternative.

Figure 2.34 shows a picture of a SOD coating system. Depending on the SOD material, the coat track design and processing requirements will differ. Generally, SOD track platforms are developed around resist tracks and are designed to continually remove evolved gases from the spin cup and hot plate areas.

Fig. 2.34 The spin coating process (SOD) for dielectric materials used in the copper damascene process (Courtesy of Tokyo Electron Limited)

2.3 High-K Dielectric Materials

2.3.1 Introduction

The semiconductor industry is feverishly ferreting after a dielectric material for a gate which will provide a very high-*K*, low leakage current and a low *equivalent oxide thickness (EOT)* compared to silicon dioxide (SiO_2) [84–85]. The search is spurred by the urgency of minimizing power consumption, particularly in battery-driven high-performance sub-100 nm devices [86–87]. As the thickness of silicon dioxide (SiO_2) approaches <1.5 nm, the leakage current becomes > 1 A/cm^2 and the tunnel current is seen to increase significantly. Therefore, it is expected that the future high-*K* materials for sub-100 nm node technology should provide excellent electrical characteristics such as dielectric constant (*K*) >30, interface density $< 1 \times 10^{11}$/cm^2-eV, tunneling current < 10 mA/cm^2, and negligible hysteresis [88–90]. An important issue preventing the implementation of high-*K* gate material is charge trapping in pre-existing traps inside the dielectric material, which affects the threshold voltage.

Scaling the thickness of the gate dielectric has long been recognized as one of the keys to scaling devices [91]. But when the oxide is made thinner, and the substrate doping is increased, the electric field applied to the oxide/silicon interface causes a significant quantization of the carriers perpendicular to the interface [92].

The International Technology Road Map for Semiconductors (ITRS) projected that for 45 nm node technology, the equivalent oxide thickness (EOT) should be <1.0 nm. The EOT scaling is very much dependent on high-*K* gate stacks, that include multilayer structures of SiO_2 interfaces with silicon (Si) substrates, including deposited high-*K* dielectrics. For efficient operation of the device, the interfacial layer created during surface preparation should have minimum interfacial charge to maximize channel mobility [93]. In addition to that, the high-*K* dielectric layer deposited by oxidants (ozone, O_3) during the CVD or PVD process should have sufficient barrier height, minimum charge-traps [94], and reduced tunneling leakage and the metal gate electrodes should be electrically and thermally stable with controllable work function [95].

During device fabrication with high-*K*, etching of Si wafer is generally done with hydrofluoric acid (HF) followed by an ammonium hydroxide (NH_4OH) surface treatment. The operation reduces EOT but degrades mobility due to surface nitridation. On the other hand, chemical oxidation by ozone (O_3) is seen to increase mobility at the expense of EOT. However, compared to chemical oxide, *in situ* steam generated interfaces exhibited better overall electrical performances.

Recently, oxides of the transition metals (TaO_2 and TiO_2) [96] Al_2O_3[97], ZrO_2, [98–99], HfO_2, [100–101], ferroelectric materials ($Pb(Zr,Ti)O_3$, $(Ba,Sr)TiO_3$) [102–104], and metal silicates ($ZrSi_xO_y$, $HfSi_xO_y$) [105–107] are showing promise for high-*K* gate materials in sub-100 nm node technology.

2.3.2 Impact on Scaling and Requirements

The gate dielectric has not been scaled as rapidly as feature size. It is expected that by the end of 2015, the equivalent oxide thickness (EOT $= (t_1$ (bottom) – t_2 (top)) (K_1/K_2)) will be a few ångstroms to minimize short channel effects and maximize device drive current. To meet all the stringent requirements of a sub-100 nm device, the gate dielectric must be physically thick, and it must have reasonably large band offsets to silicon (Si), so that it can minimize direct tunneling [88, 107–108].

On the basis of the requirement of ITRS [109] for a high-*K* dielectric material, a relation between the required *K* corresponding to the offset barrier of the material has been established and is shown in Fig. 2.35. However, the tradeoff between offset barrier height and *K* should meet the leakage current specification of a 22 nm device at low-stand-by power mode [88].

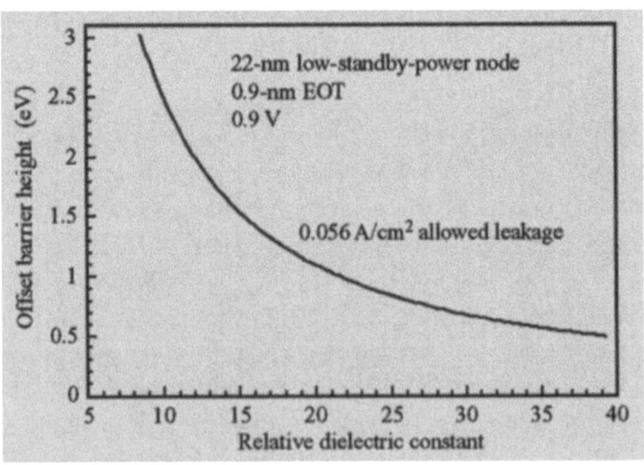

Fig. 2.35 The relation between relative dielectric constant of an insulator with its barrier height (Reproduced with permission, IBM Research, [88])

Table 2.3 Comparative analysis of band offsetts and dielectric constants (*K*) of different dielectric materials

Insulator	Bandgap (eV)	Relative dielectric constant	Conduction band offset (eV)
SiO_2	9	3.9	3.15
Si_3N_4	5.3	7.9	2.4
Al_2O_3	8.8	9.5–12	2.8
$ZrSiO_4$	~6	10–12	1.5
$ZrSiO_4$	4.5		0.7 (interfacial layer)
$HfSiO_4$	~6	~10	1.5
ZrO_2	5.7–5.8	12–16	1.4–1.5
HfO_2	4.5–6	16–30	1.5
La_2O_3	~6	20.8	1.5
Ta_2O_5	4.4	25	2.3
TiO_2	3.05	80–170	~0

Reprinted with permission IBM Research and Dev. [88]

Table 2.3 shows the comparative analysis of band offsetts and dielectric constants of different promising high-*K* materials [110–112]. From the table we can see that the conduction band offset decreases with increasing dielectric constant (*K*). The *K*, however, is related to the phase of the material and the first principles density functional approach confirms that the cubic phase has the highest *K* value [113]. However, the electronic density of states and the band gap do not have a profound effect on the electronic dielectric constant [114]. Thus it seems that novel structural modifications of a dielectric material might possibly provide a route to enhance the *K* value [115]. Above all, high-*K* gate dielectric materials should preserve high device channel mobility, low interfacial charge scattering centers, and should not degrade the interfacial roughness [116]. Furthermore, the conduction mechanism inside the dielectric should be electronic and not ionic because of the reliability issue.

2.3.3 Search for a Suitable High-K Dielectric Material

2.3.3.1 Nitrides and Oxynitrides

It is true that a single molecular layer of SiO_2, which contributes about 0.3 nm to the overall EOT, will not be able to meet all the requirements of a 65 nm device. Therefore, an alternative high-*K* dielectric material suitable for the sub-100 nm Cu-damascene process has to be identified.

It seems that oxynitrides and combinations of oxide and nitride of metals and semiconductors [117–118] including nano-crystalline alloys will serve as potential candidates as gate dielectric materials for advanced Si devices [119–120]. These materials have moderate to high *K* and can be optimized to a thickness of 1 nm to reduce the leakage current by about two orders of magnitude [121–122], which is not possible to achieve with SiO_2. However, it is very difficult to fabricate a reproducible gate material at the nanometer scale, which will be able to withstand the passage of a current density on the order of 10^5 A/cm^2(which means about 10^{11} C/cm^2 of charge over a ten-year life time) [111].

Amorphous silicon nitride (Si_3N_4) [123–124] is another potential candidate for a gate material. It can be deposited by any one of the following processes, such as: rf sputtering using an *a-Si* target with nitrogen discharge or by the reaction of SiH_4 and NH_3 in the presence of nitrogen (N_2) or with plasma enhanced cold wall systems. Si_3N_4 has a high dielectric constant (6–9 versus∼4.2 for CVD SiO_2), making it less attractive for inter-level insulation.

Oxynitride films of silicon (SiO_xN_y (H_z)), which has attracted attention as a gate material, have physical and chemical properties that are intermediate between silicon dioxide and silicon nitride [125]. The material can be deposited in the same way as Si_3N_4 with additional introduction of nitric oxide (NO) and ammonia (NH_3) gas. Properties of the material can be tailored for improved thermal stability, low stress, and crack resistance. However, the etch rate of the glassy mixture in buffered HF is about 35 times higher than Si_3N_4 films [126–127].

2.3.3.2 Tantalum Oxide (Ta_2O_5) and Titanium Oxide (TiO_2)

In recent years many investigators have renewed their interest in Ta_2O_5 and TiO_2 because these materials have high dielectric constant (*K*). Unfortunately, for a large number of high-*K* materials, the band gap is inversely proportional to *K*. To reduce leakage current tunneling, while maintaining the same gate capacitance, the material should have a large band gap and barrier height between the electrode and the conduction band of the dielectric [128–129].

Ta_2O_5 is the most thoroughly studied dielectric, used in thin film capacitors and as a gate dielectric [130–135]. Recently, atomic layer deposition (ALD) of Ta_2O_5 using TaI_5 and O_2 precursors is gaining importance as high-*K* material in dynamic memory circuits [136]. However, the oxides of Ta and Ti are thermally unstable when it is deposited over silicon, and an additional passivation layer raised concern about the scalability of EOT and process complexity [137].

2.3.3.3 Hafnium Oxide (HfO_2)

CVD and ALD deposited oxides (HfO_2), silicates ($HfSiO_x$ and HfSiON) and aluminates of hafnium (Hf) show higher *K* and lower leakage current than SiO_2 at the same EOT. The materials also offer a higher limiting offset barrier (0.5 eV) [138–139] besides being stable thermodynamically and electrically over silicon (Si) at high temperatures [140–144]. At the same time silicon dioxide and oxynitride, the traditional gate dielectric materials in field effect transistors (FETs), are rapidly approaching their ultimate thickness. Thus, the compounds of hafnium are under investigation.

Besides $HfCl_4$ and H_2O, Hf amides and Hf alkoxides are used as precursors to deposit HfO_2 during ALD [145]. On the other hand, in reactive dc magnetron sputtering [150], and CVD (MOCVD), Hf $(NO_3)_4$ is used to deposit HfO_2 films

[146–149]. However, when the temperature exceeds 700 °C, Hf $(NO_3)_4$ shows high oxygen diffusion, besides hysteresis, and crystallization. Moreover, the channel mobility of Hf $(NO_3)_4$ is inferior to the Si/SiO_2 interface [151].

2.3.3.4 Zirconium Oxide (ZrO_2)

Zirconium oxide (ZrO_2) has attracted much attention as a high-*K* material because of its high temperature stability and high mechanical strength [152]. Thin films of ZrO_2 can be prepared by many methods like sol-gel, spin coating, ALD, reactive sputtering, MOCVD, E-beam evaporation, pulse laser ablation, and rapid thermal CVD [153–155].

Unfortunately, oxides of hafnium (HfO_2) and zirconium (ZrO_2) tend to crystallize much more rapidly than SiO_2, which makes the interface surface rough. So the key issue will be to form amorphous oxides of the metal initially, which might transform into the crystalline state during thermal cycling. One current avenue of investigation is to incorporate silicon (Si), Al, or N_2 into the random network structure [115]. Indeed, adopting the above theory, zircon ($ZrSiO_4$), a compound of ZrO_2 and SiO_2, has shown better interface behavior with silicon (due to its dangling bonds) and has provided more symmetric band alignment with much higher conduction band offsets compared to ZrO_2 (2.10 eV compared to 0.64–1.2 eV of ZrO_2).

2.3.3.5 Lanthanum Aluminate ($LaAlO_3$)

Aluminate of lanthanum (La) is a promising material for a gate oxide. The localized d state in La oxide, as well as the energy relative to the silicon (Si) conduction band, play important roles in determining the magnitude of the direct tunneling current [156–157]. Molecular beam deposited 10 Å film shows a *K* value $\sim$24 with an optical band gap of 5.6 eV. The band offsets between $LaAlO_3$ and Si is 2.1 eV for electrons and 1.9 eV for holes [158].

2.3.3.6 Titanate Compounds of Barium ($BaTiO_3$), Barium Strontium ($BaSrTiO_3$), and Lead ($PbTiO_3$)

The *K* values of $BaTiO_3$, $BaSrTiO_3$, and $PbTiO_3$ are 1000, 1500 and 200–400, respectively [159]. Ferroelectric materials are ceramics and have nonlinear dielectric properties that show hysteresis under the action of an alternating voltage. Another interesting phenomenon observed with these dielectric materials is regions of spontaneous polarization, which is not caused by any external electric field. As a result, displacement of the titanium and oxygen layers results on a slightly distorted shape of the cell, and a special type of phase transition takes place when the temperature passes through the Curie point [149]. It is expected that these ceramics can be potential candidates for high-*K* materials, provided the conduction band offsets are comparable to alumina (Al_2O_3) (Table 2.3).

These titanate compounds can be deposited by direct and flash evaporation and CVD methods (by reacting metal lead (Pb) with ethyl titanate, Ti $(C_2H_5O)_4$).

Reactive sputtering from a lead titanium cathode in an argon–oxygen atmosphere can also be applied to fabricate thin films of titanium compounds.

The search for high-*K* materials to replace SiO_2 is still an ongoing process. There are several primary concerns about these high-*K* materials such as:

(a) *Threshold voltage pinning*: Gate stacks fabricated by polysilicon, HfO_2 and ZrO_2 show relatively high threshold voltage levels due to defects at the gate electrode/gate electrode interface. This reduces drive current and impairs device performance.
(b) *Phonon scattering*: Higher *K* dielectric materials show polarization effect and higher optical phonon vibration, that affect electron mobility and the performance of the device.

2.3.4 Deposition Technology for High-K Materials

Oxides of metals that are potential candidates for gate dielectrics can be deposited by reactive sputtering by introducing the second element (oxygen) in gaseous form. In order to increase the deposition rate, the second element can be introduced along with argon gas. However, other methods, such as atomic layer deposition (ALD) [160–163] and chemical vapor deposition (CVD) [164], can be applied to deposit the high-*K* materials with proper precursors. Physical vapor deposition (ionized physical vapor deposition (IPVD), a special type of sputtering system, for example) [165] and electron beam (EB) deposition methods [166] can also be used to deposit high-*K* gate materials. Details of the deposition methods have been presented in a separate chapter.

2.3.5 Summary

Scaling of device dimensions has historically made advances in silicon ULSI technology. As a result of this scaling, both the vertical and horizontal dimensions have been reduced to a point where silicon dioxide as a dielectric material cannot satisfy the requirements. The many benefits driving manufacturers to use low-*K* dielectric materials include increased device speed, reduced power, required heat dissipation, and reduced interline talk. However, the transition to low *K* has been delayed primarily because of challenges in their integration, including etching and stripping problems associated with the carbon content materials. At the same time, porous low-*K* materials are susceptible to absorbing moisture during planarization (CMP). The moisture inside the pores may react with fluorinated dielectric materials and can produce hydrofluoric acid, which is very corrosive. The moisture may also cause ionization of copper and may drive the ionized $Cu^{+/++}$ to silicon. Thus to reduce *RC* effects and to increase the speed of sub-100 nm devices the price is very high.

Low-*K* materials in general exhibit weak polarization when subjected to an external electric field. There are several guidelines in formulating these materials; the most obvious one is the choice of a non-polar dielectric system. In these materials polarity is weak because of symmetrical polar chemical groups that cancel the dipoles of chemical bonds between dissimilar atoms.

In the case of high-*K* materials, scaling the thickness of the gate dielectric has long been recognized as one of the keys to scaling devices. It is expected that for sub100 nm devices (CMOS) the gate oxide will be less than 2 nm. Thus it appears that SiO_2 will be unsuitable as a gate insulator in the sub-100 nm device. The vertical scaling requirements for gate stacks and for shallow extension junctions require 1 nm equivalent oxide thickness (EOT) to deal with leakage current tunneling. Some metal oxides, like HfO_2, ZrO_2 and their silicates, have been identified as promising materials that are supposed to work with 1 nm EOT. However, integration of high-*K* materials will be a challenging task because of the chemical bonding (ionic) and are likely to exhibit both ionic conduction and high charge levels.

The limited chemical and thermal stability are the two most important criteria associated with alternative high-*K* dielectrics and alternative gate electrode materials. In spite of the advantages of the polysilicon gate, the present process flow to integrate the future advanced gate stack employing these high-*K* materials is still questionable. Thus new integration schemes and device structures may be required to form source/drain junctions of MOS and all high temperature processes are to be done before the formation of the gate stack. At the same time, the chemical mechanical polishing (CMP) process used to planarize the surface has several advantages and challenges. The challenges include identifications of suitable materials for sacrificial gate stack and spacers and development of the consumables. Although several high-*K* materials and processes have been investigated, it still appears that the major concern is the cost.

References

1. S.P. Murarka, M. Eizenbergh, and A.K. Sinha (eds.), Interlayer dielectrics for semiconductor technologies, Elsevier/Academic press, Amsterdam, Boston, 2003 and M. Chudzik et al., IEEE VLSI Tech Dig., Issue 12–14, 194, (2007)
2. P.S. Ho, W.W. Lee, and J. Leu, Low dielectric constant materials for IC applications, Springer, New York, 2002 and E.P. Gusev, V. Narayan, and M.M. Frank, IBM J. Res. Dev., 50 (4/5), 387 (2006)
3. T. Taur et al., CMOS scaling into the 21st century, IBM J. Res. Dev., 39 (1/2), 245 (1995) and F. Fiorenza, R.L. Nigro, V. Raineri, and D. Slinas, Microelectron Eng., 84 (3), 441 (2007)
4. S. Momose, M. Ono, T. Yoshitomi, S. Nakamura, M. Saito, and H. Iwai, IEEE Trans. Electron. Dev. ED, 43, 1233 (1996) and M. Koh et al., Threshold voltage fluctuation induced by direct tunnel leakage current through 1.2–2.8 nm thick gate oxide for sealed MOSFETs, IEDM 98-919-34.2.1 (1998)
5. M.A. Alam, A critical examination of the mechanics of dynamic NBTI for PMOSFETs, IEEE Int. Electron. Dev. Mtg., 345, (8–10 Dec. 2003) and G. Chen et al., Dynamic NBTI of PMOS transistors and its impact on device life time, IEEE 41st Annual Int. Reliab. Phys. Symp., p. 196, April 2003
6. J.W. Mayer and S.S. Lau, Electronic materials science, Chapter 9, Macmillan Pub., New York, 1990 and G.C. Chen et al., Dynamic NBTI of PMOS transistor and its impact on device

lifetime, IEEE 41st Annual Int. Reliab. Phys. Symp., p. 196, April 2003 and A. Shiekova et al., NBTI Relib. Microelectron Reliab., 47 (4/5), 505 (2007)
7. M.T. Bohr and Y.A. El-Mansy, Technology for high performance microprocessors, IEEE Trans. Electron. Dev., 45 (3), 620 (1998) and T.K. Gupta, Hand -book of thick and thin film hybrid microelectronics, Chapter 6, Wiley, NJ, 2003 and R.H. Havemann and J.A. Hutchby, High performance Interconnects, IEEE Proc., 89 (5), 586 (2001) and also S. Sankaran et al., IEEE IEDM Tech. Dig., Issue 21 (2006)
8. N. Kawakami et al., Jpn. J. Appl. Phys. Part-II, 39, L182 (2000) and also R.P. Feynman, R.B. Leighton, and M. Sands, Lecture on physics, Wiley, New York, Chapter 10, p. 811 (1993)
9. M. Morgan, E.T. Ryan, J. Zaho, C. Hu, and P.S. Ho, Annu. Rev. Mater. Sci., 30, 645 (2000) and D.W. Hess, A century of dielectric science and technology, J. Electrochem. Soc., 150 (1), S-1 (2003)
10. L.C. Chen, Y.H. Xu, B. Dunn, K.N. Tu, Appl. Phys. Lett., 73, 2944 (1998) and H. Park and C.R. Helms, J. Electrochem. Soc., 139, 2042 (1992) and also S.C. Lee, A.S. Oates, and K.M. Chang, IEEE IITC Conf., SanFrancisco, CA (June 2008)
11. H.J. Lee, E.K. Lin, H. Wang, W.L. Wu, W. Chen, and E.S. Moyer, Chem. Mater., 14, 1845 (2002) and also D.J. Dumin, Int. J. High speed Electron. Syst., 11, 617 (2001)
12. A. Gill and V. Patel, Appl. Phys. Lett., 79, 803 (2001)
13. G.N. Taylor and T.M. Wolf, Polym. Eng. Sci., 20, 1086 (1980) and B. Kastenmeier, K. Pfeifer, and A. Knorr, Effective-K, Semicond. Int., 27 (8), 87 (July 2004)
14. H. Kitoh, M. Mroyama, M. Sasaki, M. Iwasawa, and H. Kimura, Jpn. J. Appl. Phys., 35, 1464 (1996)
15. A. Gill and V. Patel, Interaction of hydrogen plasma with extreme low-K SiCOH dielectrics, J. Electrochem. Soc., 151 (6), 133 (2004)
16. S.V. Nitta et al., J. Vac. Sci. Technol., B 17, 205 (1999) and also B. Peng, W.F. Yu, P.Lee, and M. Naik, A new CVD process for damascene low k application, Semicond. Fab. Tech. 10th ed., ICG Pub., UK, 285 (2000)
17. M. Bohr, Low dielectric constant material for ULSI interlayer dielectric applications, Proc. IEEE, Int. Electronic Device meeting, 10–13 Dec., Washington DC, pp. 241–244 (1995), and MRS Bull. Oct.1997 and also G.D. Wilk, R.M. Wallace, and J.M. Anthony, J. Appl. Phys., 89, 5243 (2001)
18. P. Nunan, The challenge of low-K, KLA-Tencor corp., Yield management solutions, Spring, San Jose, CA, p. 17 2000 and also S. Yang et al., Chem. Mater., 14, 368 (2002) and N. Nakamura et al., IEEE IITC June 4, SanFransisco, CA, (2008)
19. H.F. Wolf, Semiconductors, Wiley, New York, p. 336 (1971) and also B. Tareev, Physics of dielectric materials, Permittivity of mixtures, MIR Pub., Moscow, p. 116 (1975) and X. Zhao, D. Ceresoli, and D. Vanderbilt, Phys. Rev., B71, 085107 (2005)
20. S.M. Sze, VLSI technology, McGraw Hill, New York, p. 259 (1988) and also A.T. Kohl et al., Low-K porous methyli silsesquinoxane and spin on glass, J. Electrochem. Solid State Lett., 2 (2), 77 (1999)
21. S. Rogojevic et al., Interactions between silica xerogel and tantalum, J. Vac. Sci. Technol., B-19 (2), 354 March/April, p. 354 (2001) and also X. Ziang, K.S. Chen, R. Ghoddsi, A.A. Ayon, and S.M. Spearing, Residual stress and fracture in thick tetraethylorthosilicate (TEOS) and silane based PECVD oxide films, Sens. Actuators A, 91, 373 (2001)
22. B. Tareev, Physics of dielectric materials, Mir Pub., Moscow, p. 119 (1975) and also X. Gonze and C. Lee, Phys. Rev., B55 (10), 355 (1997) and P. Pulay, Chem. Phys. Lett., 73, 393 (1980)
23. U. Russow, Optical characterization of porous materials, Phys. Status Solidi. (a) 184 (1), March (2001) and also B. Shieh, K. Saraswat, M. Deal, and J. McVittie, Solid State Technol., 42, 51 (1999)
24. K. Maex, M.K. Baknalov, D. Shamiryan, F. Lacopi, S.H. Brongersma, and Z.S. Yanovitskaya, Low dielectric constant materials for microelectronics, J. Appl. Phys., 93 (11), 8793 (2003)

25. F. Iacopi, Z S. Tokei, M. Stucchi, S. Brongersma, D. Vanhaeren, and K. Maex, Microelectron. Eng., 65, 123 (2003)
26. S. Yang, P. Mirau, J. Sun, and D.W. Gidley, Characterization of nanoporous ultra low-K thin films templated by copolymers with different architectures, Radiation Phys. Chem., 68, 351 (2003) and H. Miyoshi et al., Jap. J. Appl. Phys., 43 (2), 498 (2004) and also R. Hoofman, R. Daamen, J. Michelon, and V. Nguyenhoang, SST, Alternatives to low-K nanoporous materials, 49 (8), 21 (2006)
27. C. Hu et al., Appl. Phys. Lett., 77, 145 (2000) and K. Maex et al., J. Appl. Phys., 93 (11), 8793 (2003)
28. K.W. Gerstenberg and M. Grischke, J. Appl. Phys., 69, 736 (1991) and F. Iacopi et al., J. Appl. Phys., 92, 1548 (2002) and also S. Narashimha et al., High performance 45 mn technology, IEEE IEDM Tech. Dig., 16 (1), 689 (2006)
29. E. Schaffer, Fracture mechanics of thin film dielectrics, The dow chemical co. sept. 2000 and also R.D. Miller, Science, 286, 421 (1999)
30. W.G.M. Van den, Solid State Technol., 48 (11), pp. 56–65 (Nov. 2005) and D.W. Gidley et al., Appl. Phys. Lett., 76, 1282 (2000)
31. J.G. Ryan et al., Copper and low-k dielectric integration challenges, Low-K dielectric materials Seminar, Semicon West 2000 and also T. Sui et al., Technology and reliability for advanced interconnects and low-K dielectrics, Mat. Res. Soc. Proc., 612, D1.2.1–D1.2.5 (2001)
32. C. Jin, S. Lia, and J.T. Wetzel, Evaluation of ultra low-K dielectric materials, J. Electron. Mater., 30 (4), 284–289 (2001) and P.D. Rouffignac, Z. Li, and R.G. Gordon, Sealing porous low-K dielectrics with silica, Electrochem. Solid-State Lett., 7 (12), p. G306 (2004)
33. D.W. Gidley, W.E. Frieze, T.L. Dull, A.F. Yee, E.T. Ruan, and H.M. Ho, Phys. Rev., B 60, 5157 (1999) and C. Gueds et al., Microelectron Reliab., 47 (4), 764 (2007)
34. M.E. Mills, P. Townsend, D. Castillo, S. Martin, and A. Achen, Benzocyclobutane (DVS-BCB) polymer as an interlayer dielectric (ILD) material, Microelectron. Eng., 33, 327 (1997)
35. C.M. Whelan et al., Sealing porous low-K dielectrics, Electrochem. Solid State Lett., 7 (2), F8–F10 (2004)
36. V. Jousseaume et al., Pore sealing of a porous dielectric by using a thin PECVD a-SiC:H conformal liner, J. Electrochem. Soc., 152 (10), F156 (2005)
37. J.J. Senkevich et al., Molecular caulk: A pore sealing technology for ultra low-K dielectrics, Mater. Res. Soc. Symp., 812, F1.2.1 (2004)
38. L. Peters, Is pore sealing key to ultra low-K adoption?, Semicond. Int., 28 (10), 49 Oct. (2005)
39. F. Iacopi, M.R. Baklanov, E. Sleeckx, T. Conard, H. Bender, H. Meynen, and K. Maex, J. Vac. Sci. Technol., 20, 109 (2002)
40. J.N. Sun, D.W. Gidley, W.E. Frieze, T.L. Dull, A.F. Yee, E.T. Ruan, S. Lin, and Z. Witzel, Probing diffusion barrier integrity on porous silica low-K thin films using positron anhilation spectroscopy, J. Appl. Phys., 89 (9), 5138 (2001)
41. J. Ning, Y. Hu, W.E. Frieze, W. Chen, And D. Gidley, How pore size and surface roughness affect diffusion barrier continuity on porous low-K films, J. Electrochem. Soc., 150 (5), F97 (2003)
42. A. Jain et al., Effects of processing history of modulus of xerogel films, J. Appl. Phys., 90 (11), 5832–5834 (2001)
43. J.B. Zhao et al., Reliability and electrical performance of low-K dielectric constant interlevel dielectric for high performance, Proc. IRPS, 156 (1996)
44. S.S. Prakash, T.J. Brinker, and A.J. Hurd, J. Non Cryst. Solids, 190, 264 (1995)
45. B.S. Martin, J.P. Godschalx, M.E. Mills, E.O. Shaffer II, and P.H. Townsend, Adv. Mater., 12, 1769 (2000)
46. C.J. Brinker, G.W. Scherer, Sol gel, Science Academic Pub., San Diego, CA , p. 507, (1999)

47. J. Hedrick et al., IEEE Int. Interconnect Tech. Conf. Proc., p. 261, (2000) and J.-P. Pascault, H. Sautereau, J. Verdu, and R.J.J. Williams, Thermosetting polymers, Marcell Dekker, New York, (Feb. 2002)
48. P.S. Foster, E. Ecker, E. Rutter Jr., and E.S. Moyer, US Patent 5,882,836 (1999) and also K. Mosig, T. Jacobs, K. Brenan, M. Rasco, J. Wolf, and R. Augur, Microelectron. Eng., 64, 11 (2002)
49. N. Aoi, Jpn. J. Appl. Phys., 36, 1355 (1997) and also G. Passemard, P. Fugier, P. Noel, F. Piresand, O. Demolliens, Microelectron. Eng., 33, 335 (1997)
50. US Patent, Dow Corning, #5,045592 (1975)
51. A. Modafe, N. Ghalichechian, B. Kleber, and R. Ghodssi, Electrical characterization of benzocyclobutene polymers for electrical micromachines, IEEE Trans. Dev. and Mater. Reliab., 4 (3), 495 (2004)
52. M. Morgan, E.T. Ryan, J.H. Zaho, C. Hu, T. Cho, and P.S. Ho, Annu. Rev. Mater. Sci., 30, 645 (2000)
53. C.V. Nguyen et al., Chem. Matter., 11, 3080 (1999)
54. M.E. Mills, P. Townsend, D. Castillo, S. Martin, and A. Achen, Benzocyclobutene (DVS-BCB) polymer as an interlayer dielectric (ILD) material, Microelectron. Eng., 33, 327 (1997)
55. M. Ikeda et al., Integration of organic low-K material with Cu-damascene employing novel process, IEEE Intl. Interconnect Tech. Conf. p. 131, (June 1998)
56. S.W. Chung, S.T. Kim, J.H. Sin, J.K. Kim, and J.W. Park, Comparative study of hydroorgano siloxane polymer and hydrogen silsesquioxane, Jap. J. Appl. Phys. Part I, 39, 5809–5815 (2000)
57. P.S. Ho, W.W. Lee, and J. Leu, Low dielectric constant materials for IC applications, Thermal properties, p. 43, Springer, New York (2002)
58. C.T. Chu, G. Sarkar, and X. Hu, J. Electrochem. Soc., 145, 4007 (1998)
59. M.J. Laboda, C.M. Grove, and R.F. Schneider, J. Electrochem. Soc., 145, 2861 (1998)
60. J.P. Godschalx et al., Polyphenylene oilgomers and polymers, US Patent 5965679, 1999
61. M. Padovani et al., Electrochem. Solid State Lett., 4, F25 (2001)
62. S.T. Martin et al., Development of low dielectric constant polymers for the fabrication of integrated interconnects, Adv. Mater., 12, 1769 (2000)
63. P.A. Kohl et al., Electrochem. Solid State Lett., 1, 49 (1998)
64. P.S. Ho, W.W. Lee, and J. Leu, Low dielectric constant materials for IC applications, Moisture uptake, Springer, New York, p. 46, 2002 and R.D. Miller, Science, 286, 421 (1999)
65. S.W. Chung, J.H. Shin, N.H. Park, and J.W. Park, Dielectric properties of hydrogen silsesquioxane films degraded by heat and plasma treatment, Jpn. J. Appl. Phys. Part-1, 38, 5214 (1999)
66. C.T. Coua, G. Sarkar, and X. Hu, J. Electrochem. Soc., 145, 4007 (2000)
67. K.G. Pruden, K. Sinclair, and S. Beaudoin, Characteristics of parylene N and parylene C photo-oxidation, J. Poly. Sci. Part-I, Poly-Chem., 41 (10), 1486–1496, Wiley, (2003)
68. Y.S. Yeh, W.J. James, and H. Yashuda, J. Poly. Sci. B, 28 (4), 545–568 (2003)
69. S.C. Selbrede and M.L. Zucker, Characterization of parylene-N thin films for low-K VLSI applications, MRS spring meeting, San Francisco, CA, spring 1997 and also A.J. Flewitt, A.P. Dyson, J. Robertson, and W.I. Milne, Thin Solid Films, 383, 172 (2001)
70. K. Taylor, M. Eissa, J. Gaynor, S.P. Jeng, and H. Nguyen, Parylene co-polymers, MRS spring meeting, San Francisco, CA (Spring 1997) and also R.L. Opila and D.W. Hess, A century of dielectric science and technology, J. Electrochem. Soc., 150 (1), S4 (2003)
71. G. Chen, Z. Xia, Y. Zang, and H. Zang, Preparing and polarizing stability of Teflon AF nonlinear optical polymer electret double layer thin film system, IEEE Trans. Dielect. Elect. Insulation, 6 (6), 929 (Dec. 1999)
72. P.T. Dao, D.J. Williams, and K.G. Berarduce, Constant current corona charging as a technique for poling organic non-linear optical thin film and the effect of ambient gas, J. Appl.

Phys., 73, 2043 (1993) and also A.F. Teflon, Amorphous Fluoropolymer, Technical Pub. DuPont Speciality Polymers Div., Wilmington, DE (1990)
73. P.W. May, Diamond thin films: A 21st century material, Phil. Trans. Royal Soc. Lond. A, 358, 473 (2000)
74. M. Ashman, J. Heberlin, and E. Pfender, Diamond Relat. Mater., 8, 1 (1999)
75. A. Gill, Plasma deposited diamond like carbon and related materials, IBM J. Res. Dev., 43 (1/2), 39 (1999)
76. E. Riedo, F. Comin, J. Chevier, F. Schmittisen, S. Decossas, and M. Sancrotti, Surface Coat. Technol., 125, 124 (2000)
77. E. Riedo, F. Comin, J. Chevior, and A.M. Bonnot, J. Appl. Phys., 88, 4365 (2000) and also A.Y. Liu and M.L. Cohen, Phys. Rev., B 41, 10727 (1990)
78. A.M. Campos, J. Torres, and J.J. Giraldo, Porous silicon dielectric function modeling from effective medium theories, Surface Rev. Lett., 9 (5/6), 1631 (2002)
79. D.A. Burggeman, Ann. Phys., 5, 636 (1935)
80. L. Lang, Y. Xia, M. Zhang, and W. Shi, Letter to the editor, Semi. Sci. Tech., 19 (3), L35, March (2003)
81. A.K. Sikdar, F. Giglio, J. Wood, A. Kumar, and J.M. Anthony, J. Electron. Mater., 30, 1522 (2002) and S. Gall et al., IEEE IITC, San Francisco, CA (June 2008)
82. K. Mosig, T. Jacobs, K. Brenan, M. Rasco, J. Wolf, and R. Augur, Micrelectron. Eng., 64, 11–24 (2002)
83. G.R. Yang, D. Mathur, X.M. Xu, S. Dabral, J.F. McDonald, and T.M. Liu, J. Electron. Matter., 25, 1778 (1996)
84. J.P. Chang, Y.S. Lin, S. Berger, A. Kepten, R. Bloom, and S. Levy, Ultra thin zirconium oxide films as alternative gate dielectric, J. Vac. Sci. Technol., B19 (6), 2571 (2001) and also H.J. Massoud, I. Baumvol, M. Hirose, and E.H. Pointdexter (eds.), The physics and chemistry of SiO_2 and the Si-SiO_2 interface, The electrochem. Soc. Pub., Pennington, NJ (2000), PV2000-2
85. D.A Buchanon, J.H. Sathis, E. Cartier, and D.J. Maria, Microelectron. Eng., 36, 329 (1997) and P.D. Krisch et al., Mobility enhancement of high-K gate stacks, IEEE IEDM Tech. Dig., pp. 1–4, Dec. (2006)
86. A.T. Kohl et al., Low K porous MSQ and spin on glass, Electrochem. Solid State Lett., 2 (2), 77 (1999)
87. B. Pang, W.F. Yu, P. Lee, and M. Naik, A new CVD process for damascene low k applications, Semicond. Fabr. 10th ed. ICG Pub., UK (2000)
88. J.J. Senkevich and S.B. Desu, Poly (tetra-fluoro-p-xylylene), a low dielectric constant chemical vapor polymerized polymer, Appl. Phys. Lett., 72, 258 (1998)
89. R. Leung et al., Porous and nonporous poly (arylene ether) thin films, Suitability as extra low-K dielectrics for microelectronics applications, Proc. 9th meeting of the symp. on polymers for microelectronics, Wilmington, DE (May 2000)
90. K. Postava, T. Yamaguchi, and T. Nakano, Characterization of organic low-dielectric-constant materials using optical spectroscopy, Opt. express, 9 (3), pp. 144–151 July (2001)
91. G.D. Wilk, R.M. Wallace, and J.M. Anthony, J. Appl. Phys., 89, 5243 (2001) and C.S. Park et al., Achieving low Vt and thin EOT , IEEE Proc. On VLSI-TSA, 12, p. 154, (2008)
92. R.M.A. Azzam and N.M. Bashara, Ellipsometry and polarized light, Elsevier Pub., Amsterdam (1977)
93. G. Bersuker et al., Interface induced mobility degradation in high-K transistors, Jap. J. Appl. Phys., 43, 7899 (2004) and J. Barnett et al., Cleaning role in high-K /metal gate substrate, Semicond. Int., 29 (2), 45 Feb. (2006)
94. B.H. Lee et al., Intrinsic characteristics high-K devices and implications of fast transient charging effects, Int. Electron. Dev. Meeting, (IEDM) Tech. Digest, 859–862 (2004)
95. D.Y. Cho et al., Control of silicidation in HfO_2/Si (100) interfaces, Appl. Phys. Letts., 86, 041913 (2005)

96. G.B. Alers et al., Appl. Phys. Lett., 72, 1308 (1998) and also D.J. Dumin, Int. High Speed Electron. Syst., 11, 617 (2001) and I. Kume et al., IEEE IITC, SanFrancisco, CA (June 2008)
97. L. Machenda et al., IEEE Tech. Dig. Int. Electron. Dev. Meet, (IEDM) 605 (1998)
98. J.P. Chang, Y.S. Lin, Y.M. Sun, J. Lozano, H. Ho, H.J. Park, S. Veldman, and J.M. White, Appl. Surf. Sci., 161, 115 (2000)
99. M. Copel, M. Gribelyuk, and E. Gusev, Appl. Phys. Lett., 76, 436 (2000)
100. L. Kang et al., Electrical characteristics of highly reliable ultrathin hafnium oxide gate dielectric, IEEE Dev. Lett., 21 (4), 181 (2000) and also M. Ritala, M. Leskelain, H.S. Nalwa (ed.), Hand book of thin film materials, Vol. 234, Academic press, Orlando, FL, p. 183 (2002)
101. M.M. Frank et al., Hafnium oxide gate dielectric grown from an alkoxide precursor, Mater. Sci. Eng., B 109, 6 (2004)
102. B. Van Dover, L.F. Schneemeyer, and R.M. Fleming, Nature, 392, 162 (1998)
103. R.A. Mckee, F.J. Walker, and M.A. Chrisholm, Phys. Rev. Lett., 81, 3014 (1998)
104. J.P. Han and T.P. Ma, Appl. Phys. Lett., 72, 1185 (1998)
105. W.J. Qi et al., Appl. Phys. Lett., 77, 1704 (2000) and E.P. Gusev, V. Narayanan, and M.M. Frank, IBM J. Res. Dev., 90 (4/5), 387 (2006)
106. B.H. Lee, L. Kang, W.J. Qi, and J.C. Lee, Appl. Phys. Lett., 76, 1926 (2000)
107. G.D. Wilk and R.M. Wallace, Appl. Phys. Lett., 74, 2854 (1999) and M.V. Fischetti, D.A. Neumayer, and E.A. Cartier, J. Appl. Phys., 90 (9), 4587 (2001)
108. R. Puthenkovilakam, E.A. Carter, and J.P. Chang, First principles exploration of alternative gate dielectrics: Electronic, structure of ZrO_2/Si/$ZrSiO_4$/Si interfaces, Phys. Rev., B-69, 155329 (2004)
109. J.L. Hedrick et al., Templating nanoporosity in thin film dielectric insulators, Adv. Mater., 10, 1049 (1988)
110. K.R. Carter et al., Process for manufacturing integrated circuit devices, US Patent 3895263 (1999)
111. C.J. Hawker, J.L. Hedrick, and R. Miller, Integrated circuit process for its manufacture, US Patent 5767014 (1998)
112. W.D. Gray and M.J. Laboda, New barrier layers can help Cu-low-K integration, Solid State Technol., 45 (3), 37 (2002)
113. X. Zhao and D. Venderbilt, Phys. Rev., B-65, 075105 (2002) and B-65, 233106 (2002)
114. X. Zhao and D. Vanderbilt, MRS. Fall Meeting, 745, N 7.2.1 (2003)
115. X. Zhao, D. Ceresoli, and D. Vanderbilt, Structural, electronic, and dielectric properties of amorphous ZrO_2 ab initio molecular dynamics, Phys. Rev., B-71, 085107 (2005)
116. P. Xu et al., BLOK-A Low-K dielectric barrier/etch stop film for copper damascene applications, Proc. IEEE Int. Conn. Technol. Conf. p. 109, (1999)
117. G. Lucovsky, J. Vac. Sci. Technol., A-16, 356 (1998)
118. T.P. Ma, IEEE Trans. Electron. Dev., 45, 680 (1998)
119. G. Lucovsky et al., Electronic structure of high-K transition metal oxides and their silicate and aluminate alloys, J. Vac. Sci. Technol., B-20 (4), 1739 (2002)
120. G. Wilk, R.W. Wallace, and J.M. Anthony, Micrelectron. Eng., 59, 329 (2001)
121. K.J. Wynne and R.W. Rice, Annu. Rev. Matter., 14, 297 (1984) and G.D. Wilk, R.M. Wallace, and J.M. Anthony, J. Appl. Phys., 87, 484 (2000)
122. M.J. Laboda, New solutions for interconnect dielectrics using trimethyl silane based PECVD processes, Pro Euro. Workshop on Mats. For Adv. Metallization, Oostende, Belgium (1999), Microelectron. Eng., 50, 15–23 (2000) and also G. Baccanari, M.R. Wordeman, and R.H. Dennard, IEEE Trans. Electron. Dev., 31, 452 (1984)
123. F.R. Ley, J. Am. Ceram. Soc., 83 (2), 245 (2000) and also W.A.P. Claassen, W.G.J.N. Valkenborg, M.F.C. Willemsen, and S.W. Haszko, J. Electrochem. Soc., 132, 893 (1995)
124. M.J. Hofmann, Si_3N_4 ceramics structure & properties of encyclopedia of materials science & technology, Elsevier Sci., NY, 4–5, 8469 (2001)

125. M.V. Fischeti and S.E. Laux, Long range Coulomb interactions in small Si-devices, J. Appl. Phys., 89, 1205 (2001)
126. J.R. Brews, W. Fichtner, E.H. Nicollian, and S.M. Sze, Generallized guide for MOSFET miniaturization, IEEE Electron Dev. Lett., 1, 2 (1980)
127. S. Wolf, Silicon processing for VLSI Era, Chapter 4, Vol. 4, Lattice press, Sunset beach, CA, pp. 145–180, 2004 and also A.C. Adams, Dielectric and polysilicon film deposition, in S.M.Sze (ed.), VLSI Technology, McGraw Hill, New York, p. 233 (1988)
128. C.M. Osburn et al., Vertically scaled MOSFET gate stacks and junctions, IBM J. Res. Dev., 46, 299–315 March/May (2002)
129. R. Sharma, A. Kumar, and J. Anthony, Advances in high-K dielectric gate materials for future ULSI Devices, JOL, Microelectron. Processing, 53, pp. 53–55 June (2001)
130. S.H. Lo, D.A. Buchanon, Y. Taur, and W. Wang, Quantum mechanical modeling of electron tunneling current from inversion layer of ultra thin oxide of n-MOSFET′s, IEEE Electron. Dev. Lett., 18, 209 (1997)
131. C.A. Ritcher, A.R. Hefner, and E.M. Vogel, A comparision of quantum mechanical capacitance voltage simulation, IEEE Electron. Dev. Lett., 22, 35 (2001)
132. J. Robertson, Band offsets of wide-band-gap oxides and implantation for future electronic devices, J. Vac. Sci. Technol., B-18, 1785 (2000)
133. G. Lucovsky, J. Vac. Sci. Technol., A-19, 1553 (2001)
134. International Technology Roadmap for Semiconductors (ITRS), Int. SEMATECH, Austin, TX, 2001
135. T. Yamaguchi, H. Satake, N. Fukushima, and A. Toriumi, Band diagram and carrier conduction mechanism in ZrO_2/Zr – silicate/Si MIS structure fabricated by pulsed laser ablation deposition, IEEE IEDM Tech. Digest, pp. 31–34 (2000)
136. L. Manchanda et al., Gate quality doped high-K films for CMOS beyond 100-nm, IEEE Int. Electron. Dev. Meeting (IEDM) Tech Digest, 9, 149 (2000)
137. L. Kang et al., Electrical characteristics of highly reliable ultrathin hafnium oxide gate dieelectric, IEEE Dev. Lett., 21 (4), 181–183 (2000) and E. Atnassova and A. Paskaleva, Microelectron. Reliab., 47 (6), 913 (2007)
138. J. Robertson, J. Vac. Sci. Technol., B-18, 1785 (2000) and also G. Lucovsky, J.L. Whiten, and Y. Zang, Microelectron. Eng., 59, 329 (2001) and K. Okada et al., IEEE Sym. On Int. Relib. Phys., April 27, Phoenix, AZ (2008)
139. M.V. Fischetti, Long range Coulomb interactions in small Si-devices, Part II: Effective electron mobility in thin-oxide structures, J. Appl. Phys., 89, 1232 (2001)
140. B.H. Lee et al., Ultrathin hafnium oxide with low leakage and excellent reliability for alternative gate dielectric applications, IEEE IEDM Tech. Dig., 133 (1999)
141. M.M. Frank et al., Hafnium oxide gate dielectrics grown from alkoxide precursor, Mater. Sci. Eng. 109, 6 (2004) and also G.D. Wilk, R.M. Wallace, and J.M. Anthony, J. Appl. Phys., 87, 484 (2000)
142. M. Ritla M. Leskela in H.S. Nalwa (ed.), Hand book of thin film materials, Vol. 234, Academic Press, Boca Raton, FL, p. 183 (2002)
143. M.R. Visokay et al., Appl. Phys. Lett., 80, 3183 (2002)
144. M. Kkoyama et al., Int. Electron. Dev. Meeting (IEDM), 849 (2002) and also J. Barnett et al., Cleaning's role in high-K/ metal gate success, Semicond. Int., 29 (2), 45 (Feb. 2006)
145. D.S. Hausmann, E. Kim, J. Becker, and R.G. Gordon, Chem. Mater., 14, 4350 (2002)
146. S.K. Ghandhi, VLSI fabrication principles, Wiley, New York (1983) and also P. O′Brien, N.L. Pickett, and D.J. Otway, Development of CVD delivery systems: Achemist′s perspective on chemical and physical interactions between precursors, Adv. Mater., Wiley-VCH, Weinheim, Germany, 14 (23), 237 (2002)
147. D.A. Buchanan, IBM J. Res. Dev., 43 (3), 245 (1999)
148. S.A. Campbell et al., IEEE Trans. Electron. Dev., 44, 104 (1977)
149. C. Chaneliere, J.L. Autran, R.A. Devine, B. Balland, Mater. Sci. Eng., R22, 269 (1998)

150. L. Kang et al., Electrical characteristics of highly reliable ultrathin hafnium oxide gate dielectric, IEEE Dev. Lett., 21 (4), 181–183 (2000)
151. H. Yang and G. Lucovsky, Integration of ultra-thin (1.6-2.0 nm) RPECVD oxynitride gate submicron C-MOSFETs, IEEE IEDM Tech. Digest, 245 (1999) and S. Kubicek et al., Low VT CMOS using doped Hf-based oxides, IEEE IEDM Tech Dig. (2007)
152. E.H. Kisi and C.J. Howard, Crystal structure of zirconia phases and interrelation, Key. Eng. Mater., 153, 1–36 (1998)
153. J.P. Chang, Y.S. Lin, S. Berger, A. Kepten, R. Bloom, and S. Levy, Ultrathin zirconium oxide films as alternative gate dielectric, J. Vac. Sci. Technol., B19 (6) 2212 (2001)
154. J.P. Chang, Y.S. Lin, Y.M. Sun, J. Lozano, H. Ho, H.J. Park, S. Veldman, and J.M. White, Appl. Surf. Sci., 161, 115 (2000)
155. M. Copel, M. Gribelyuk, and E. Gusev, Appl. Phys. Lett., 76, 436 (2000)
156. G. Lucovski, J.L. Whitten, and Y. Zhang, Microelectron. Eng., 59, 329 (2001)
157. W.A. Harrison, Elementary electronic structure, World Scientific, Singapore, Chapter 17 (1999)
158. P.W. Peacock and J. Robertson, J. Appl. Phys., 92, 4712 (2002) and also L.F. Edge et al., Measurements of band offset between amorphous $LaAlO_3$ and Si, MRS Symp., Dec. 1–4, 786, (2002)
159. Y. Xu, Ferroelectric materials and their applications, Amsterdam, North Holland (1991)
160. Y.S. Lin, R. Puthenkovilakam, and J. Chang, Appl. Phys. Lett., 81, 2041 (2002)
161. D.M. Hausmann, E. Kim, J. Becker, and R.G. Gordon, Chem. Mater., 14, 4350 (2002)
162. J.P. Chang, Y.S. Lin, and K. Chu, J. Vac. Sci. Technol., B-19, 1319–1327 (2001)
163. M. Putkone and L. Ninisto, J. Mater. Chem. Roy. Soc. Chem. London, 11, 3141 (2001)
164. J. Chang et al., J. Vac. Sci. Technol., B-19 (6), 2131 (2001)
165. J.P. Hopwood, (ed.), Ionized physical vapor deposition, Academic Press, New York (2000)
166. M. Hatmanova et al., J. Mater. Sci., 6 (11), 2387 (1996) and S. Pae et al., IEEE Symp. On Int. Reliab. Phys. April 27, Phoenix, AZ (2008)

Chapter 3
Diffusion and Barrier Layers

3.1 Diffusion

3.1.1 Introduction

Diffusion is a frequently used technique to incorporate impurities into a semiconductor. *Imperfection in a nearly perfect crystal* is the prime theme of intentional impurity diffusion in silicon crystals [1] for the formation of *p-n junctions*, *conduction channels*, and *source drain* regions. The performance of the devices depends critically on the impurity concentration and the impurity profile. For this reason the *diffusion of various impurities* in semiconductors has been studied extensively.

Figure 3.1 shows the *p and n junctions* inside a silicon (Si) transistor formed by diffusion of impurities inside Si. The surface of the diffused layer is metallized (granular structures) with aluminum (or Al-alloy) which is the conventional metal to make ohmic contacts before the Cu-damascene process.

In the Cu-damascene process, copper (Cu), the primary metal in the sub-100 nm silicon device, is considered a major insidious impurity that affects the minority carrier life time and device yield. However, for a long time, the detrimental effect of Cu on Si-device yield was uncertain. The extremely high diffusivity of copper in silicon, its small ionic radius, and its relatively weak interaction with the silicon lattice, make it highly mobile at room temperature and impacts on the stability of Cu-complexes [2].

Figure 3.2 shows an *in situ* X-ray diffraction pattern recorded during heating (200 °C) of a Cu-film deposited over silicon. As a matter of fact, the diffusion of copper at elevated temperatures and its precipitation in silicon depend upon various factors [2]. Thus there exists a need to study the diffusion mechanism of copper in silicon, because in the Cu-damascene process interconnecting lines are formed from electrodeposited copper film.

Figure 3.3 shows the Cu-interconnect damascene structure, which has replaced the conventional aluminum interconnect in modern sub-100 nm node technology.

As the state of 250 nm back-end processes changed to 130 nm, the propagation delays, signal dispersion, and/or cross-talk noise associated with Al/SiO_2 system limited the chip performance. Reduction of propagation delay due to *RC* effects of the interconnecting lines can minimize the capacitive cross-talk noise. Thus the obvious choice will be to replace the Al-interconnect with a better conductive

Tapan K. Gupta, *Copper Interconnect Technology*,
DOI 10.1007/978-1-4419-0076-0_3,

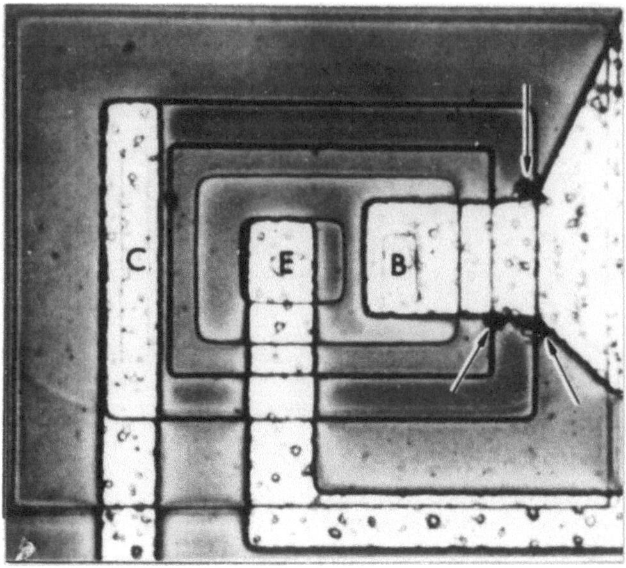

Fig. 3.1 Scanning electron microphotograph of diffused p and n junctions in silicon. E stands for emitter and B for base of a bipolar transistor (Photo courtesy Integrated Circuits Engineering, AZ)

Fig. 3.2 *In situ* X-ray diffraction patterns recorded during heating (4 K/min) of a Cu-layer deposited onto silicon (Photo courtesy M. Hecker et al., Highlights, 2001, TU Dresden, Germany)

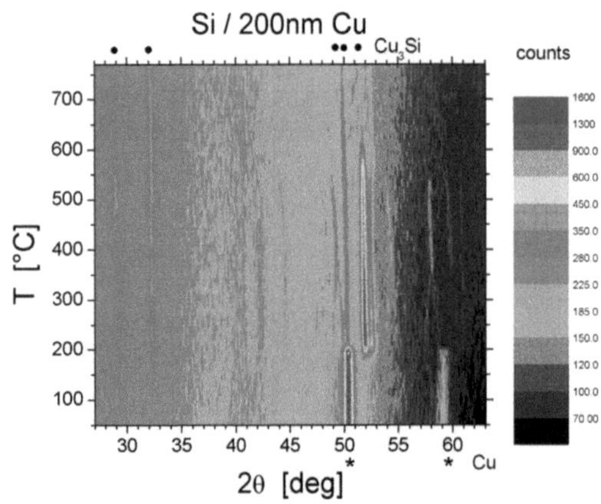

material and SiO_2 with a lower dielectric material. As a result, to reduce resistance (R) of the metal interconnect Cu has been evaluated as a better material than Al due to its higher electrical conductivity (Table 3.1). Not only this, to reduce the RC effect further, SiO_2 has been replaced by low-K dielectric material, and is in production level now.

Table 3.1 shows that the electrical conductivity of copper (Cu) is higher than that of Al and the electromigration property of Cu, which is related to the density of the metallic atoms, diffusion coefficient D, the effective valence Z^*, and the electrical resistivity ρ is lower than Al. Unfortunately copper (Cu) has high diffusivity in

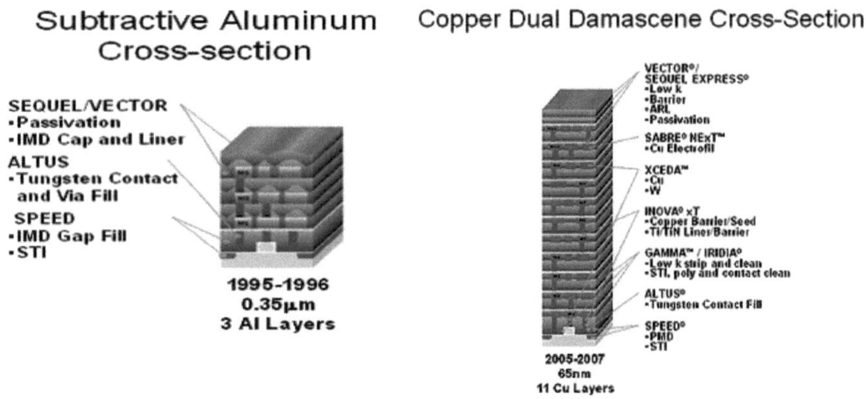

Fig. 3.3 Subtractive aluminum metallization and a modern dual damascene architecture where copper is used for metallization (Photo courtesy Novellus systems Inc.)

Table 3.1 Comparison of the bulk properties of prospective metals for forming interconnecting lines*

Metal	–Z*	ρ (μΩ-cm)	Diffusion coefficient D_0 (cm^2/s)	Z* ρD at 100 °C
Silver	9.3–23.3	1.57	1.1×10^{-26}	$2.8–7.0 \times 10^{-25}$
Aluminum	6.6–16.6	2.60	2.1×10^{-20}	$3.6–9.1 \times 10^{-19}$
Gold	5.7–7.5	2.36	2.2×10^{-27}	$3.0–3.8 \times 10^{-26}$
Copper	3.5–4.4	1.55	2.1×10^{-30}	$1.2–1.4 \times 10^{-29}$

*Adopted from different sources

silicon, and appears to have acceptor levels in Si at 0.24, 0.37 and 0.52 eV with respect to the valence edge, and forms copper silicide (Cu_3Si) at 200 °C (Fig. 3.2). When annealed at 1100 °C and air-cooled, large diameter Si-wafers show colonies of high-density copper precipitates [14].

3.1.2 Transitional Effects

The transition from aluminum (Al) interconnects to copper (Cu) interconnects requires several new and additional device processing. Not only that, the conventional back end process, which has been used in Al-metallization, including dry etching and reflow processes, has turned out to be inapplicable to copper Cu-metallization mainly for two reasons: (a) dry etching of Cu is difficult; and (b) higher reflow temperature. As a result, a new engineering process known as *damascene* has emerged into the arena to surmount the difficulties of dry etching of copper (Cu). Figure 3.4 shows the multilayer metallization scheme for filling via holes and trenches in the Cu-damascene process to surmount the difficulties of dry etching.

It is worth mentioning that Cu does not form a passivated stable oxide like Al [3] and the interface or the grain boundary (GB) triple-points become the active

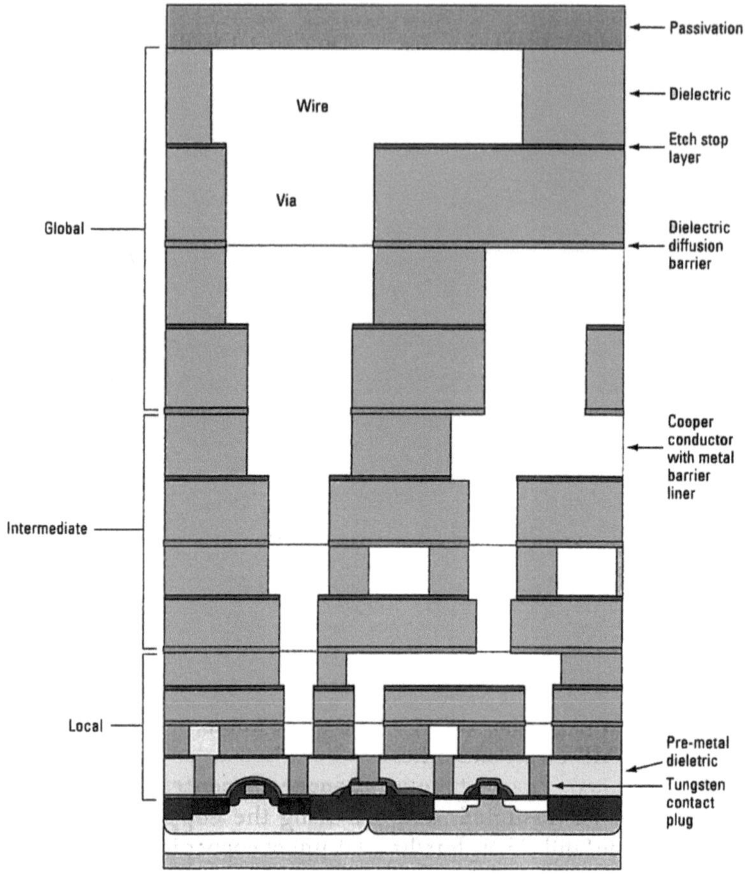

Fig. 3.4 Multilayer metallization architecture built by the copper damascene process (Courtesy ITRS, 1999)

flux divergence path in local Cu-interconnecting lines [4–5]. On the other hand, a very clean surface of a Cu-polycrystalline film shows very low activation energy ($\sim$0.47 eV), and favors interface diffusion in global interconnecting wires especially when GBs are not available [6–9]. As a matter of fact, GBs can be the fastest pathway for Cu-diffusion (because of a poor correlation of activation energy with grain size to film thickness), which does not support interfacial diffusion [10–13].

3.1.3 Mathematical Modeling of Diffusion in Cu-Interconnects

It is well established that applied bias and temperature enhance the diffusion of copper (Cu) from Cu-interconnecting lines sandwiched between the cap layer and the barrier layer (Ta/TaN) (Fig. 3.5).

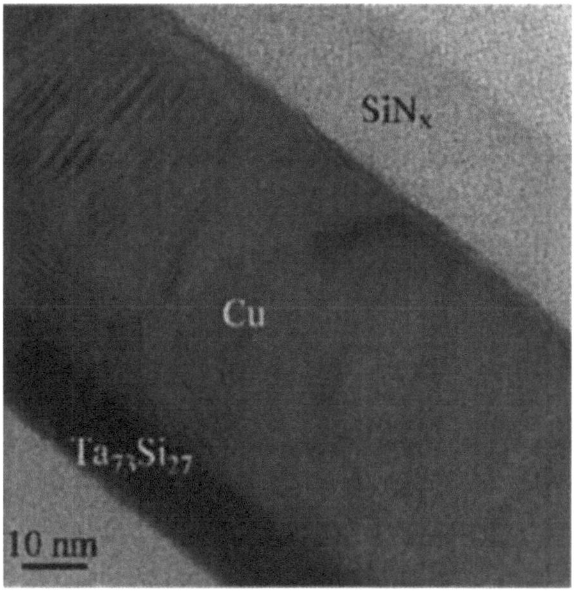

Fig. 3.5 The Cu-layer sandwiched between the cap layer and barrier layer (Reprinted with permission, Wiley-VCH, *Cryst. Res. Technol.* **40** (1/2,) 135 2005)

At room temperature, the grain growth of the electroplated copper (EP-Cu) deposited over tantalum (Ta) or tantalum nitride (TaN/Ta_2N) (barrier layer) is mostly along the <111> direction and has been detected both by atomic force microscopy (AFM) and focused ion beam (FIB). On the other hand, deposited Ta has a β-phase with either <001> or no preferred orientation depending on the process engineering [14], and it requires 1.4 eV energy for diffusion [15]. On the other hand, the reported enthalpies of Cu-diffusion in various barrier materials can vary from 1.3 to 4.5 eV [16–17].

The grain growth in nanostructured Cu-lines fabricated from EP-Cu film drastically affect the microstructure and the plasticity of the film [18–19] as the temperature and bias cause long-range internal stress. These are identified as diffusional creeps where diffusion occurs on the film surface and in the grain boundary (GB) (Fig. 3.6) [20–21].

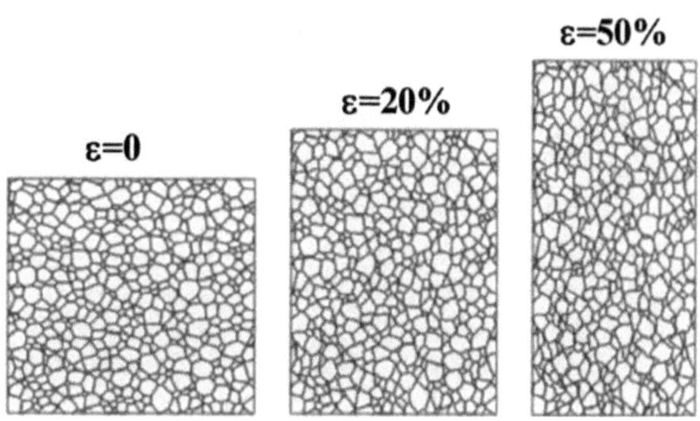

Fig. 3.6 Mesoscale simulation of grain boundary diffusional creep during grain growth in Cu-interconnects (Reprinted with permission, Dr. D. Wolf et al., Argonne National Lab)

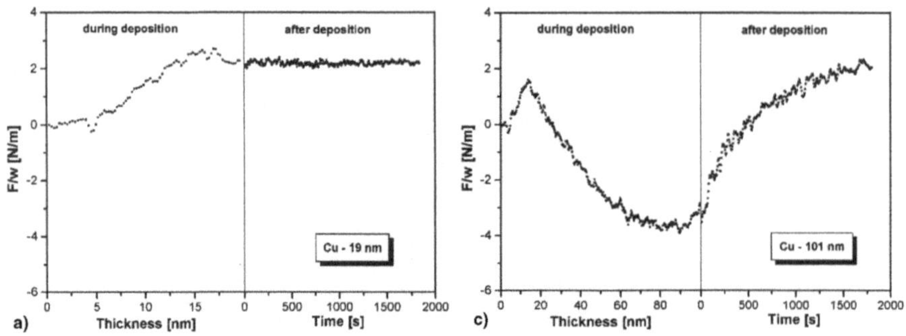

Fig. 3.7 Evolution of stress during and after deposition of copper films for different thickness: (**a**) 19 nm and (**c**) 101 nm (Reprinted with permission of Wiley-VCH Verlag)

Experimental observations show that the *stress* in thin films can exceed those corresponding to bulk materials and increases with decreasing film thickness. As a result, the physical properties of a film are very much influenced by the developed stress. Figure 3.7 shows the *evolution of stress* during and after deposition of copper films having 19 nm and 101 nm thickness.

To accommodate Cu-interconnects in place of aluminum (Al)/tungsten (W) systems, different exotic materials have been introduced. This has brought new failure modes and a shift in focus of interconnect reliability. For example, the drift of Cu ions in a dielectric under the influence of a high local electric field may cause time-dependent inter- and intra-level dielectric breakdown (TDDB). In order to testify the validity of the above statement, Cu-lines are tested under both thermal and electric field stress (BTS) conditions. From experimental observations, it has been accepted that diffused copper exists in an ionic state ($Cu^{+}/^{++}$) [22] in dielectric layers and the electric field distribution inside the dielectric will be modified with the Cu-concentration in it. For simulating the evolution of Cu-concentration in the dielectric layer, we have assumed that (a) Cu exists in the dielectric layer as a positive single ion (Cu^{+}, based on thermodynamic solid solubility of various copper species), and (b) Cu diffuses only normal to the surface, because diffusion is enhanced in parallel to the direction of the electric field. Therefore we can model the flux equation of Cu-diffusion as [23–24]:

$$J = -D\cosh(q\lambda E/2kT)(\partial C/\partial x) + (2D/\lambda)\sinh(q\lambda E/2kT)\,C \qquad (3.1)$$

where

D = Diffusivity of Cu in dielectric
λ = Jump distance of Cu (~2.7 Å)
E = Electric field
K = Boltzmann constant
T = Temperature in absolute
q = Electronic charge
$(\partial C/\partial x)$ = concentration gradient with respect to distance (x).

Now when $E << (2kT/\lambda)$ the above equation transforms to:

$$J = -D(\partial C/\partial x) + (DqE/kT)C. \tag{3.2}$$

Equation (3.1) is combined with Fick's second law ($-\partial J/\partial x$) and the diffusion equation is divided into two parts: (i) constant electric field mode; and (ii) variable electric field mode. The diffusion equation under constant electric field mode can be expressed as:

$$(\partial C/\partial t) = D\cosh\{(q\lambda E/2kT)\}\{(\partial^2 C/\partial x^2)\} - 2(D/\lambda)\{\sinh(q\lambda E/2kT)(\partial C/\partial x)\} \quad \textit{(Under constant electric field)}. \tag{3.3}$$

The diffused Cu^+ ion will remain in the dielectric in the ionized state, and with time the charge will change the electric potential profile ($\varphi(x)$) inside the dielectric layer. The change in the $\varphi(x)$ inside the dielectric can be expressed mathematically as:

$$(\partial^2\varphi(x)/\partial x^2) = -(\partial E(x)/\partial x) - qC(x)/\varepsilon_r\varepsilon_0 \tag{3.4}$$

where ε_r and ε_0 are the relative permittivity (of the dielectric), and the permittivity of the free space, respectively. We can set up the boundary conditions as: $\varphi(0) = V$ and $\varphi(L) = 0$ where V is the electric potential at the interface between Cu and dielectric, and L is the thickness of the dielectric layer. In our present case, as the electric field is a function of x, the diffusion equation under variable field can be expressed as:

$$\begin{aligned}(\partial C/\partial t) = D[&\cosh\{(q\lambda E/2kT)\}\{(\partial^2 C/\partial x^2)\} + \{(q\lambda E/2kT)(\partial E/\partial x)\\ &-(2/\lambda)\}\{\sinh(q\lambda E/2kT)(\partial C/\partial x)\}\\ &-(2/\lambda)\{\cosh(q\lambda E/2kT)\}\{(q\lambda E/2kT)(\partial E/\partial x)C\}]\\ &\textit{(Under variable electric field)}.\end{aligned} \tag{3.5}$$

By solving Equations (3.3) and (3.5) with boundary conditions as: $C(0,t) = C_0$, $C(L, t) = 0$, $C(x, 0) = 0$, where C_0 is the maximum solubility of Cu in a particular dielectric, the evolution of copper concentration inside the dielectric can be simulated.

The experimental evidence shows that the diffusivity of Cu in Ta is very small at 300 °C [25] and the simulated results confirm that diffusion of Cu in SiO_2 is very small at 200 °C. However, when the electric field is changed from 2.0 MV/cm to 0.1 MV/cm, there is an appreciable change in copper concentration inside SiO_2 [24]. As a matter of fact, at high electric field the drift of copper is appreciable and it accounts for device failure [26–27]. However, experimental results confirm that during high temperature annealing, (>500 °C) crystallization of the barrier layer favors *GB diffusion* [28]. Figure 3.8 shows an SEM of (a) as-deposited copper (Cu) on tantalum (Ta) and (b) crystallization of tantalum due to annealing at 400 °C.

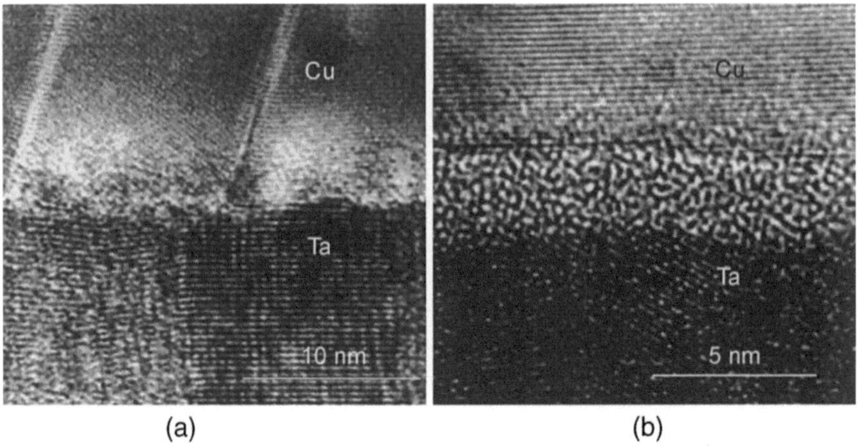

Fig. 3.8 SEM of (**a**) as-deposited copper (Cu) on tantalum (Ta) and (**b**) crystallization of tantalum after 400 °C annealing (Photo courtesy Professor S.S. Wong, Stanford University, CA)

3.1.4 Grain Boundary (GB) Diffusion

Qui, alors, peut calculer le cours d' une molécule? (Who then can calculate the course of a molecule?) [28]

The tensile stress in Cu-interconnects can vary from 230 MPa for a passivated line to 112 MPA for an unpassivated line and can produce a tensile strain ($\sim 10^{-3}$ eV) along the <111> direction [29–30]. The resulting strain depends on the direction of the applied stress. As a computational problem, diffusion near GBs in copper has been examined in the context of molecular dynamics [31] and has been found to be sufficient to cause *GB diffusion* [32–33].

The most frequently encountered defect in evaporated film is dislocation [34–37], which causes a misorientation between adjacent grains and ultimately increases the diffusion along grain boundary [38]. Figure 3.9 shows that when a tantalum (Ta)

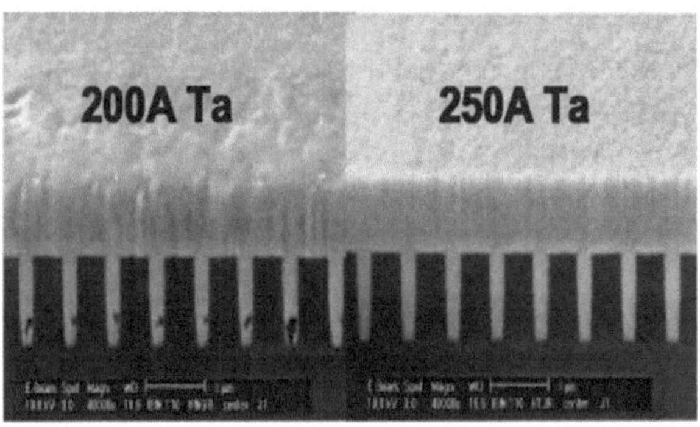

Fig. 3.9 SEM picture of vapor deposited (PVD) tantalum films (Photo courtesy Novellus System)

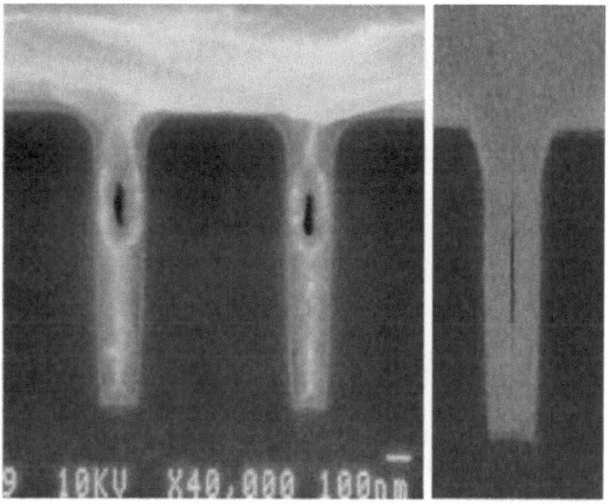

(a) (b)

Fig. 3.10 (**a**) The formation of voids and (**b**) formation of seams (Photo courtesy Novellus system)

film deposited over copper is 200 Å thick it shows voids due to dislocation and misorientation between adjacent grains. However, complete fill is observed when the Ta-film is 250 Å thick. Thus scaling of the Ta diffusion barrier is susceptible to forming voids that might be a diffusive path for $Cu^{+/++}$ ions.

Indeed, the candidates for fast diffusion paths in a damascene structure are (a) the grain boundaries (GBs) along the length of the interconnect line, (b) the Cu/barrier layer interface, (c) the Cu/over layer interface, and (d) gross *voids* or *seams* (Fig. 3.10) that may run down the length of the interconnect and form during the electroplating process [39]. As a matter of fact, the energies required for defect formation and migration of atoms in grain boundary (GB) are smaller than the energies required for defect formation and migration in the bulk lattice. As a result, GBs provide an easy path for mass transport in materials especially in polycrystalline thin films.

GB diffusion controls the kinetics of *structural changes, phase transformations,* and *solid state reactions* in engineering materials. It has been observed that diffusion in polycrystalline materials is six or more orders of magnitude higher than the diffusion in monocrystalline materials [40–41]. The diffusion model of Harrison deals with three types of regimes: A, B, and C (Fig. 3.11). In type A, lattice diffusion is the distinguishing feature with apparent diffusivity and lower activation energies. In type B, GBs are assumed to be isolated and the lattice fields and GBs are confined individually to themselves. In type C, atom migration is thought to occur only within the boundaries and lattice diffusion is considered to be negligible [42].

The study of the Harrison diffusion model shows that the diffusion of copper (Cu) will conform more closely to type C [43–44].

Ideally an amorphous dielectric layer has no grain boundary and should be ideal as a diffusion barrier for Cu. But when an amorphous layer of tantalum (Ta)

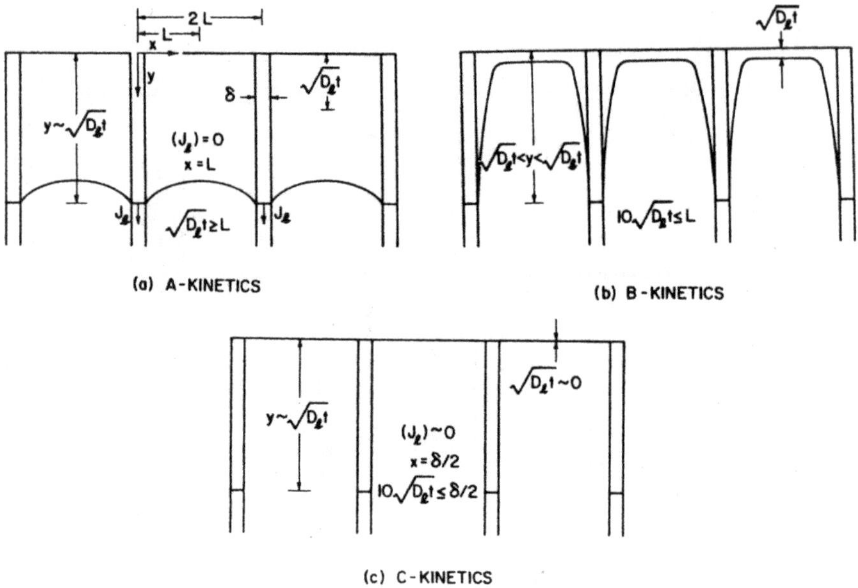

Fig. 3.11 Three types of grain boundary diffusion kinetics: (**A**, **B**), and (**C**). *Curved lines* are isoconcentration contours. The diffusion source coincides with the *top horizontal lines* (Reprinted with permission Noyes Pub. New York [40])

is present in front of a polycrystalline Cu-seed layer, the amorphous layer will not impose strict limitations on both spatial positions and the Burgers vectors of the grain boundary dislocations [45–46]. As a result, the spatially inhomogeneous growth of the amorphous phases forces the crystalline amorphous interface to be curved, which will increase its contribution to the total energy density and to the *anomalous diffusion* of Cu through the boundary of the amorphous-polycrystalline interface [47–49].

Another interesting thing, which has been observed in polycrystalline films, is the movement of the grain boundary due to *diffusion*, which is known as *induced grain boundary motion* (DIGM) [50–51]. In the *DIGM* model the interface moves due to disintegration of grains and the chemical composition of the moving interface and the bulk left behind will be different [51].

3.1.5 Vacancy Diffusion

The vacancy mediated diffusion of embedded atoms as proposed by Gastel et al., can be attributed to tantalum atoms forming a barrier layer formed for the Cu-interconnecting lines [52–54]. Indeed Ta/Si stacks when annealed at 550 °C show diffusion of Ta to the Cu-surface [55–56]. However, Ta diffusion due to the vacancies in Cu is seen to decrease when the Cu layer is provided with a silicon nitride (Si_3N_4) cap.

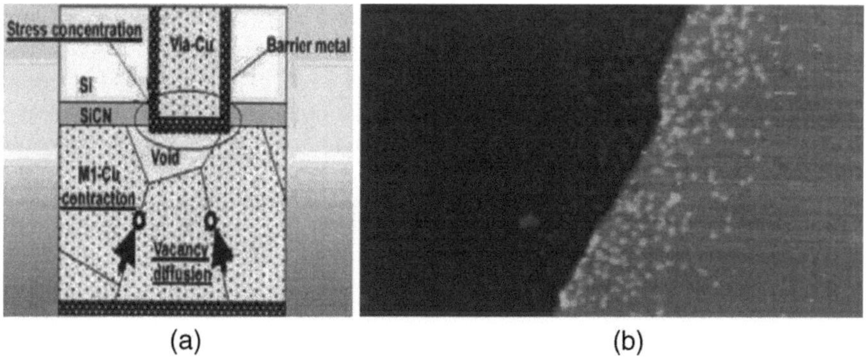

Fig. 3.12 (**a**) Vacancy diffusion in copper due to the migration of Cu atoms and (**b**) experimental verification of surface vacancies on low-index surface (Cu) by hopping of indium atoms (Photo courtesy R. van Gastel et al., Elsevier Pub.)

Figure 3.12 (a) shows a damascene structure with SiCN as low-K (~4.9) material and the deposited copper film serving as interconnect over the Ta/TaN bi-layer barrier. The figure further shows that due to atomic rearrangements on the low-index Cu-surface, adatoms from the surface have moved and created vacancies. These vacancies are identified as vacancy diffusion. Figure 3.12 (b) shows experimental verification of the hopping of indium (In) atoms into the vacancies created by Cu atoms due to its low index surface. This model can be clearly fitted into our system, where the barrier/cap layer is adjacent to the Cu-interconnect.

3.1.6 *Drift Diffusion*

The *drift diffusion* model can be set up in the spirit of the chemical oxidation of Cu via ambient gases like moisture and/or O_2 [57] from porous low-K dielectric materials or from SiO based dielectric films. Experimental evidence shows that after thermal cycling (annealing) at high temperatures (>500 °C), the oxygen species from the dielectric layer may diffuse through the grain boundary of Ta [58]. It further shows that under bias the $Cu^{+/++}$ diffuses into the neighboring layers (into the cap layer and/or underneath the Ta layer) [59–60]. As a matter of fact, the drift diffusion rate of $Cu^{+/++}$ is seen to increase as the percentage of hydrogen content in the a-SiC:H is increased.

We can set up a physical model of the drift of Cu^+ ions into the dielectric layer in order to explain the drift diffusion under time-dependent dielectric breakdown (TDDB) conditions. Let us consider two adjacent lattices, 1 and 2, a distance λ apart, where the concentration of Cu^+ in plane 1 is higher than that at plane 2. Then we can set up a flux equation for Cu^+ as:

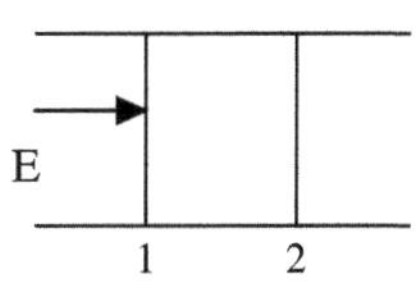

$$J(x,t) = [\partial C(x,t)/\partial x\lambda\partial\Gamma_+ + C(x+\lambda,t)(\Gamma_+ + \Gamma_-)]\lambda \tag{3.6}$$

where the number of atoms jumping out of plane 1 to 2 in a short period of time δt is $C(x,t)\ \Gamma_+\ \lambda\ \delta$ t. Similarly, the number of atoms jumping from plane 2 to plane 1 in time δt is $C(x+\lambda,t)\ \Gamma_-\ \lambda\ \delta t$. Now under TDDB conditions we will take it for granted that the electric field is very large and the influence of the diffusion flux is negligible (i.e. the first term of Equation 3.6). Therefore Equation (3.6) will transform into:

$$J(x,t) = J_{\text{drift}}(x,t) = C(x+\lambda,t)(\Gamma_+ + \Gamma_-)]\lambda. \tag{3.7}$$

The continuity equation is simplified to:

$$\{(\partial C(x,t))/(\partial t)\} = \{\lambda[\exp(-(E_a - q\lambda E/kT) - \exp(-E_a + q\lambda E/kT)\lambda]\}(\partial C(x,t)/\partial x) \tag{3.8}$$

where $\Gamma_+ = A\exp.-(E_a + q\lambda E/kT)$ is the the probability of a Cu^+ ion being available at position 1, and $\Gamma_- = -(E_a - q\lambda E/kT)$ is the the probability of a Cu^+ ion being available at position 2. The diffusion activation energy of Cu^+ is E_a. Now the TDDB (τ_{BD}) of the Cu^+ ion can be obtained by solving Equation (3.8) as:

$$\tau_{BD} = \{(B\exp E_a/kT)/(\exp(q\lambda E/kT) - \exp(-q\lambda E/kT)\}. \tag{3.9}$$

A model based on the one-dimensional solution of the drift-diffusion equation and aided by Poisson's equation with activation energy (1.8 eV), and electric field (ranges from 1.0×10^5 to 1.0×10^6 V/cm), gives rise to a diffusion depth of Cu^+ $\sim 5.6 \times 10^{-7}$ cm [57, 61]. However, the mathematical model of J.Y. Kwon et al., gives rise to a diffusion depth of Cu between 500 and 600 nm [23].

3.1.7 Interdiffusion

Interdiffusion of Cu into the cap layer creates voids in Cu-interconnecting lines in the micrometer range [62]. The interdiffusion is the result of thermal stress at high temperatures (>550 °C) and has been detected by glow discharge optical emission spectroscopy (GD-OES) and Auger electron spectroscopy (AES). The interdiffusion of Cu inside the cap layer (Si_3N_4) increases with increase of temperature and the distribution of Cu is noticed both along the vertical and lateral directions [63–64].

The diffusion of copper into the cap layer changes its electrical, thermal and mechanical properties [65–70]. We will discuss all of these in the course of our studies of the conductivity, electromigration and reliability of Cu-interconnects.

3.1.8 Diffusion of Copper and Its Consequences

Figure 3.13 shows the binary phase diagram of silicon. The copper in atomic percent (at. %) is plotted along the *X*-axis. The *Y*-axis represents the corresponding

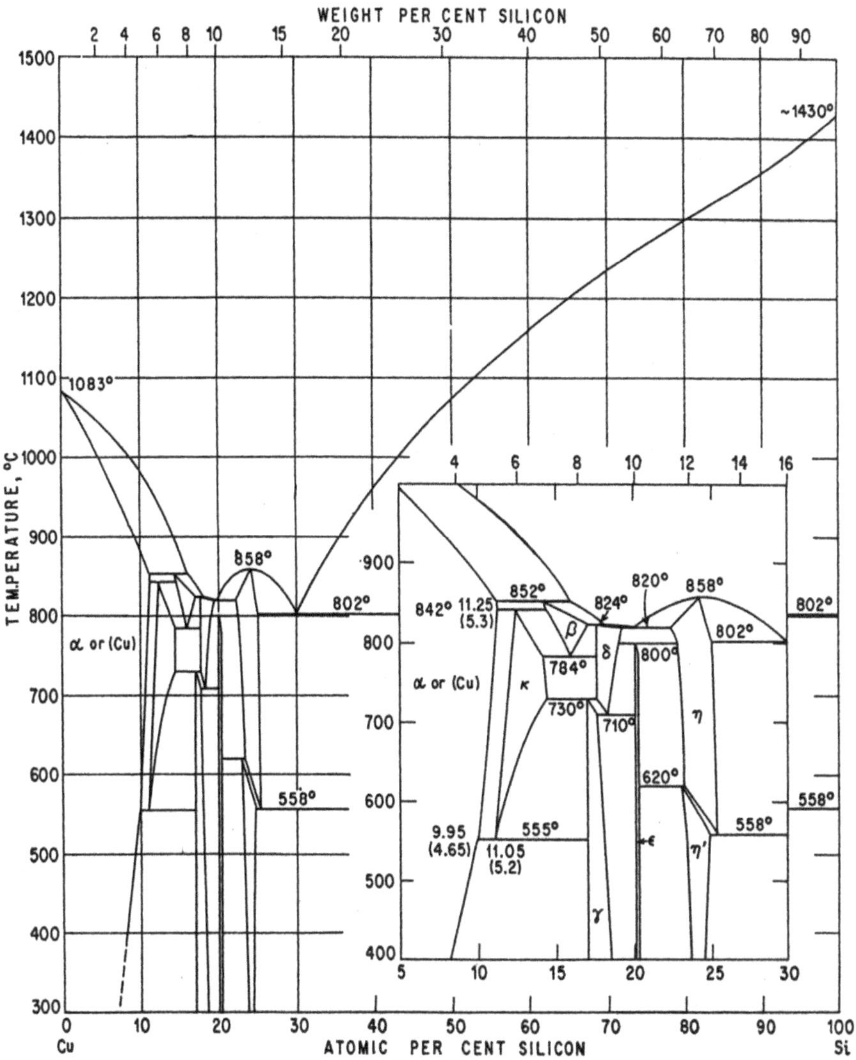

Fig. 3.13 The copper-silicon (Cu-Si) phase diagram (Reproduced with permission from McGraw Hill, New York)

temperature of the system. The liquidus consists of four branches corresponding to the crystallization of α, κ, β, δ, γ, ∈, η, η′ and silicon (Si) [71]. As the atomic percent of silicon (Si) is increased in copper the solidus curve changes from the α phase of copper to the β phase and finally to the η′ phase. The η′ phase (Cu_3 Si) of copper, which is shown in Fig. 3.14(a), has a slightly distorted γ-brass (trigonally distorted A_2 type) structure [72–73] and is isostructural with the high temperature Cu $_3$Ge phase [74].

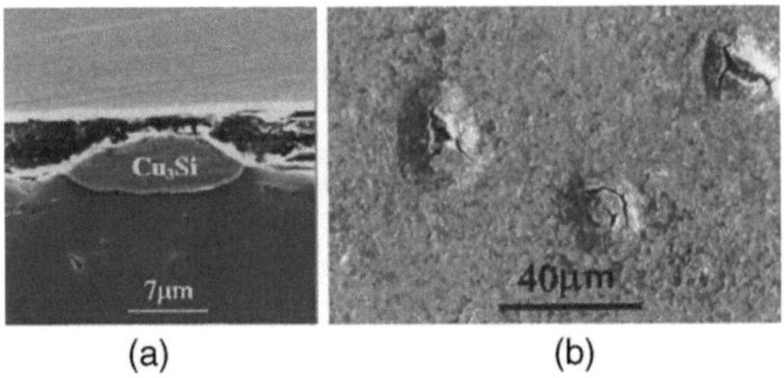

Fig. 3.14 Scanning electron micrograph of diffusion of copper inside silicon (Si): (**a**) the formation of Cu_3Si complex and (**b**) the swollen precipitates of copper in Si (Reproduced with permission, Elsevier Science, New York)

3.1.9 Precipitation

The use of copper as an interconnecting material in ultra-large-scale integrated (ULSI) circuits where Si is used as the substrate material has revived interest in the *diffusion* of copper in silicon and in the *precipitation* kinetics of copper in silicon (Fig. 3.14) [75]. The precipitation of copper as a pallet-like structure is possible in n- or in p-type silicon if the concentration of shallow dopants and copper donors becomes high ($\geq 10^{17}$ atoms/cm^3) [76]. Experimental observations show that these precipitates are unstable and form point defect centers that dissociate (dissociation energy is ~0.7 eV) within milliseconds [77–79]. These precipitates are distributed homogeneously through the bulk Si-wafer instead of colonies and the density can go up to 10^{13} atoms /cm^3. Energy dispersive spectrometry (EDS) reveals these swollen *precipitates* (Fig. 3.9b) as Cu $_3$ Si. Copper diffusion in silicon (Si) can lead to two types of failures: (i) deterioration of the dielectric layer leading to shorts or leaky paths between the conductors and the dielectric layer; and (ii) higher leakage currents in p-n junction devices due to the formation of deep level traps [80–82, 16].

Under moderate temperature and bias, ionic copper ($Cu^{+/++}$) diffuses very fast in silicon and in many dielectric materials (including SiO_2). Indeed, experimental observations have shown copper contamination during Cu-seed deposition, bulk fill of Cu by electroplating, chemical mechanical polishing (CMP), and cleaning processes. The diffusion of $Cu^{+/++}$ in the dielectric layer can degrade the properties of the dielectric [83–87].

Interstitial diffusion, a favorable diffusion mechanism of Cu in Si, can proceed via one of two mechanisms, namely, the direct *interstitial* or *interstitialcy* mechanism. At higher temperatures the movement rate of the atom is higher and direct interstitial jumps are not in favor of *grain boundary* (*GB*) diffusion. As a matter of fact, diffusion along {210} is dominated by the *vacancy* mechanism and diffusion along {310} is dominated by the interstitialcy mechanism [75, 88].

It has been estimated that copper diffuses along *grain boundaries* (GBs) more easily than through the the crystal lattice of silicon, since the energies required for defect formation and migration in GBs tend to be lower in polycrystalline film compared to the bulk lattice of a single crystal [89–90]. However, providing a barrier layer can restrict the diffusion of Cu; at high temperature, annealing the barrier layer may form crystallites that may result some diffusive paths for $Cu^{+/++}$ ions (Fig. 3.17). Figure 3.15 shows a transmission electron microphotograph (TEM) of an annealed TaN barrier layer with small grains and grain boundaries.

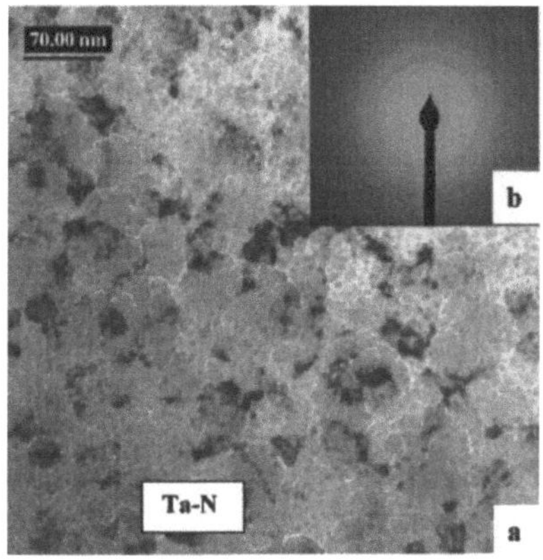

Fig. 3.15 (**a**) A dark field plane view TEM image of the TaN barrier layer with small grains and grain boundaries; and (**b**) electron diffraction pattern of a selected area (Reprinted with permission from Elsevier, New York, [160])

3.2 Barrier Layer for Cu-Interconnects

3.2.1 Theory

When a copper ion ($Cu^{+/++}$) migrates either to the cap layer or to the substrate it will change the intrinsic material properties of the layer (electrical, thermal, and mechanical properties).

Figure 3.16 shows a stack of films deposited following the damascene process where the barrier layer is sandwiched between a dielectric layer and a copper interconnect. The reliability of the barrier layer will depend upon the followings:

(a) the transport and loss rate of atoms from the respective layer;
(b) the thermodynamical stability of the layer;
(c) the adhesion of the layer to the substrate;
(d) the specific contact resistance of the layer;
(e) deposition of a void free conformal layer;
(f) mechanical and thermal stress resistance ability of the layer;
(g) the thermal and electrical conductivity of the layer.

Fig. 3.16 A stack of films deposited following the damascene process. A is copper, B is the barrier layer, C is the dielectric, and D is silicon

Regardless of the barrier layer material and the deposition method, control of barrier layer processing is the key to achieving high yield. For example, if the deposited layer becomes non-conformal, it will show a lack of adhesion and mass migration leading to void formation and Joule heating.

3.2.2 Ideal Barrier Layer

A single-crystal barrier layer is impractical, because of the presence of diffusion paths through the grain boundaries, the lattice and defect centers. So, an amorphous barrier layer will be ideal as a diffusion barrier. The self-diffusion or the transport property of the species in a solid is linked to viscosity, creep applied stress, electrical resistivity, internal friction, and the rate of crystallization. These parameters are again dependent upon the barrier film thickness, the process engineering and the processing temperatures. It has been observed that some of the amorphous barrier layers can retain their barrier efficiencies at higher temperatures even after crystallization had begun because the crystallites in the amorphous materials are discrete and separated by stretches of amorphous material. As a general rule, the higher the crystallization temperature [91–93], slower is the diffusion rate in the layer at a given temperature [91–93]. As a matter of fact, inter-diffusion of the overlay and the substrate material of the barrier layer show undesired phase transformation rather than simple crystallization.

Figure 3.17 shows the effect of thermal cycling (annealing) on the morphology of the barrier film [94]. The stability of a particular barrier layer is dependent on the annealing temperature and the deposition process [95, 16].

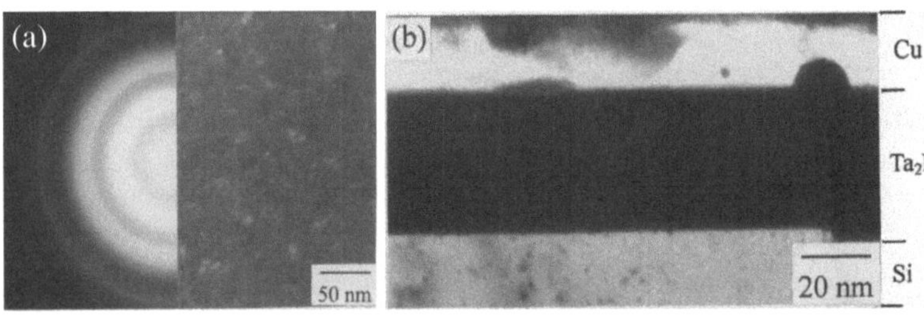

Fig. 3.17 TEM dark field image and diffuse diffraction pattern of a deposited TaN film and subsequent transformation to polycrystalline film (**a**) when it is annealed at 600 °C (Reprinted with permission Electrochemical Society, NJ, G.S. Chen and S.C. Haung)

3.2.3 Barrier Layer Architecture

In the Cu-damascene process, copper (Cu) interconnecting lines are created from the deposited Cu-film on a thin *barrier layer* (a thin film of tantalum, titanium, tungsten, or their stable compounds). The layer is sandwiched between copper interconnects and a dielectric layer (SiO_2 or low-K) to stop diffusion of copper into the neighboring area/s (Fig. 3.18).

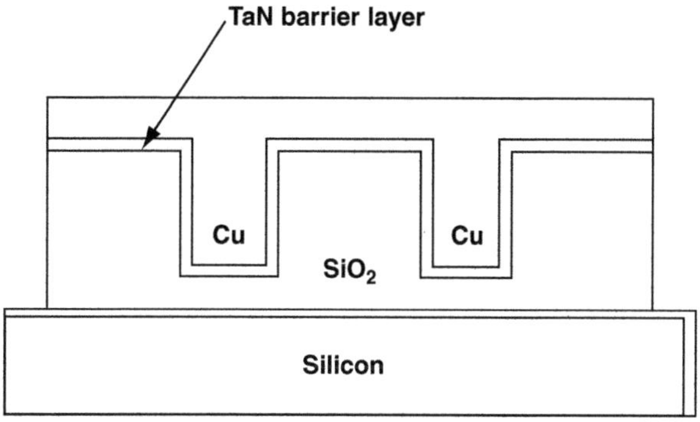

Fig. 3.18 Schematic of the damascene architecture and the deposited barrier layer over the dielectric

The diffusion of copper (Cu) from Cu-interconnecting lines into the dielectric layer and ultimately to silicon (Si) will depend upon the presence of a cap layer, the thickness of the barrier layer, the morphology of the barrier layer, the processing temperature, bias, and interfacial layer formation. The study of the *phase diagrams* will be helpful in understanding whether there will be any *interlayer reactions* and the formation of any *interfacial layer* between the two adjacent layers [96].

The thickness of the barrier layer should be kept as thin as possible (a nanometer thick) to leave as much cross-sectional area as possible for Cu. At the same time it should minimize the impact on the effective resistance of the Cu-interconnects. Figure 3.19 shows (a) a schematic of the via to reduce effective resistance of the Cu-lines in a multiple metal layer, and (b) resistivity versus different barrier layer thicknesses. Figure 3.19(b) also presents the expected requirements of the International Technology Roadmap for Semiconductors (ITRS).

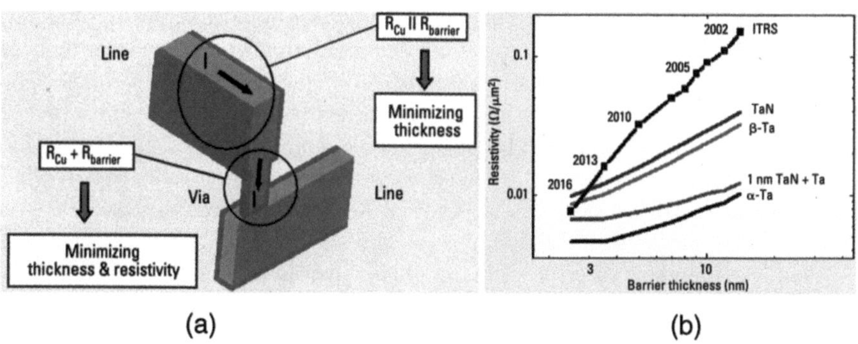

Fig. 3.19 (**a**) Schematic of the via to make ohmic contact to Cu-interconnect of a multilayer stack, and (**b**) specific resistivity versus barrier thickness and the ITRS requirements (Reprinted with permissions, Infineon Technologies AG, Munich, Germany)

3.2.4 Interlayer Reactions

Both thermodynamics and kinetics must be considered when investigating possible reactions between two elements or compounds. Equilibrium thermodynamic calculations provide a tool to predict whether a reaction is feasible or not under suitable conditions. For example, water (or moisture) can be used in conjunction with elevated temperatures to allow the barrier layer to react with fluorine (F) from a fluorinated *low-K* layer, to form fluoride/s of the barrier material and hydrofluoric acid (HF) [97]. However, a kinetic barrier may prevent the reaction from taking place.

The sign of the Gibb's free energy of reaction, ΔG_r, determines whether a chemical reaction is thermodynamically feasible at a temperature T, and pressure P. For example, if several possible reactions are thermodynamically feasible, the reaction with the most negative ΔG_r would be expected to dominate. For reactions occurring in the solid phase at relatively low temperatures, the T and P dependence of ΔG_r may be ignored and ΔG_r may be approximated as:

$$\Delta G_r = \Delta H_f - T\Delta S_f = \Delta G_f(\text{products}) - G_f(\text{reactants}) \tag{3.10}$$

where ΔG_r measures the Gibb's free energy, ΔH_f and ΔS_f are the changes in the enthalpy and entropy of a system, respectively, and T represents the temperature of the system. The analysis neglects the kinetics of the reaction and assumes equilibrium is attained, because equilibrium thermodynamic arguments have been very successful in accurately predicting the reaction products for various kinds of interfacial behavior [98–99]. It is, therefore, not infallible in predicting which reaction will dominate.

The Cu-damascene architecture with a barrier layer (Ta/TaN) in between copper and dielectric (SiO_2/low-*K*) can be visualized as three separate layers, such as Cu-Ta, Ta-TaN and the TaN-dielectric. The formation of interfacial layers between Cu-Ta and dielectric/TaN will ensure a strong *bonding* between the layers.

Bonding develops from physical or chemical interactions, interfacial frictional stress, and thermal stress due to mismatch between the coefficient of thermal expansion (CTE) of the intermediate (interface) layers and matrices. The formation of a strong interface layer is desirable from the standpoint of adhesion between the adjacent layers. The understanding and control of the underlying interfacial phenomena governing the transmission of thermal, electrical, and mechanical properties across the whole composite become of paramount importance in designing the barrier layer. A weak interface will lead to crack propagation following interface layer formation (if it forms). As we know that intermetallic phases (e.g. between Ta-TaN) are typically brittle and more susceptible to cracking [100].

Another important criterion of the interface layer is its ability to be wet by the liquid metallic matrix which will affect the surface roughness and crystallographic orientation of the layer thereby affecting the sheet resistance of the interface layer [101–102]. At the same time a new alloyed zone (interfacial layer) of sufficient width can form between Cu and the barrier layer or between the barrier layer and the dielectric layer. The initial stage of interface development constitutes the dilute limit, which will be characterized by the density of the metal adatoms. The dilute limit is very informative for a detailed understanding of the evolution of the chemistry of the interfacial layer. The structural aspects of the interface layer are strongly dependent on the deposition temperature and subsequent annealing history. Interfaces are often metastable, and inter-diffusion is accelerated at increased temperature and bias conditions.

Thermodynamically, a comparison of the free enthalpy of formation at various temperatures shows that many metals in the liquid state are reactive toward reinforcing materials in particular oxides, carbides, and fluorides (e.g. tantalum, Ta). However, in some cases though some reactive products are thermodynamically favored, some reactions are, however, not observed (e.g. Cu_xTa_{1-x}) and practically the kinetics of these reasons have to be considered in conjunction with thermodynamic data in order to evaluate the real potential of reactions.

The consequences of such interfacial layer formation may lead to chemical degradation of the interfacial layer associated with a decrease of the mechanical properties, the formation of brittle reactive products, as well as the release of highly reactive gases (e.g. the formation of unstable TaF, and release of fluorine gas). However, some experimental data show that a passive *cap* layer (SiCN) may sometimes prevent these reactions. Figure 3.20 shows a seamless low-*K* SiCOH stack technology. The MPS-SiCOH ($K = 2.5$) is the inter-metal dielectric, SiCN is the *cap* layer, and porous SiCOH (p-SiCOH) is the interlayer dielectric.

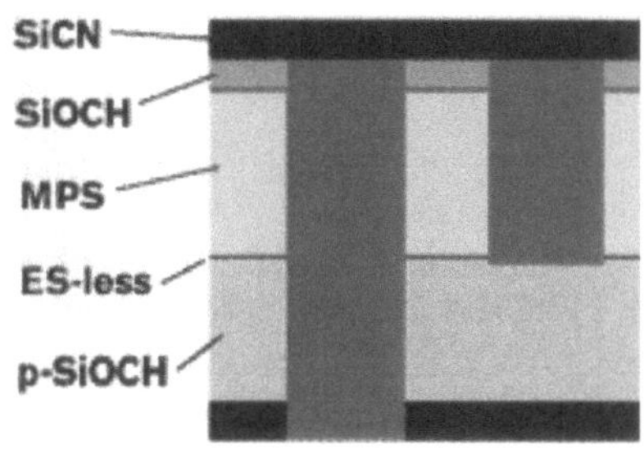

Fig. 3.20 Seamless low-*K* SiCOH stacks with SiCN as cap in 45 nm node technology (IITC preview, Semicond. Int'l, April, 2007)

3.2.4.1 Case I: Single Barrier Layer

(SiO_2)-(TaN or Ta): Generally, thin films of TaN, Ta TaN, and Ta and other transition metals and metal compounds are deposited by physical vapor deposition (PVD) for the barrier layer, which tends to be a line of sight process. However, PVD-deposited film coverage on the side walls, especially near the bottom of a high aspect ratio trench or via hole, can be several times thinner than that on the surface due to grazing phenomena [103–104]. One way to attend to conformal and continuous film inside the trench is to deposit a denser film, which requires high kinetic energies (100 eV for argon and 26 eV for copper), that may impinge the the sputtered atom deep into the substrate [105]. However, most recently developed ionized physical vapor deposition (IPVD) is expected to overcome these difficulties [106].

Experimental observations suggest that the constraints of the PVD deposition and the introduction of the low-*K* porous dielectric materials may necessitate a change over to chemical vapor deposition (CVD). Unfortunately, though the existing CVD methods provide conformal thin film, they have failed to deposit TaN because of the formation of the high resistivity Ta_3N_5 phase at relatively low temperatures. However, a CVD deposited TaSiN barrier layer has shown almost 85% side wall coverage and reasonably low resistivity. On the other hand, atomic layer deposited (ALD) TaN provides superior step coverage and uniformity at low temperatures but the deposition rate is very slow.

An ideal barrier layer should be thin, continuous, and conformal at the bottom and on the side wall of a high aspect ratio trench or via hole, and should offer low resistance. However, the formation of an interfacial layer (if it is formed) will change the resistivity of the layers. The formation of interface can be best understood from the phase diagram. Thus it is convenient to construct a ternary phase diagram on a Gibbs triangle (Fig. 3.21) [107].

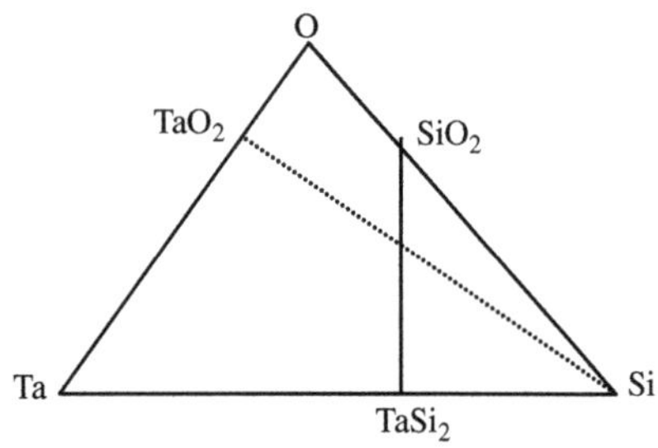

Fig. 3.21 Ternary phase diagram of Ta-Si-O system

Figure 3.21 shows the tie lines of Ta-Si-O ternary phase diagram [74]. The free energy of reaction between two possible crossing tie lines is considered. Along the O-Si line the stable compound SiO_2 is shown. Likewise along the Ta-Si line the possible compound is $TaSi_2$, which will form at a higher temperature (above 700 °C). The most stable compound of tantalum (at high annealing temperatures) with oxygen is TaO_5, and is shown along the Ta-O line. On the ternary phase diagram no tie line between Ta and SiO_2 is shown, because when Ta comes into contact with SiO_2 it does not form any stable compound. As a matter of fact, tantalum (Ta) does not have so much affinity to oxygen (O_2) as titanium (Ti), to form oxides of the metal, which can promote adhesion [108].

It is also interesting to study the effect of oxygen on such mixing or alloy formation. According to the surface potential model, the process of breaking a metal lattice and bonding to another metal lattice may be favored or disfavored according to the degree of electronegativity of the metals [109]. Copper has higher electronegativity (Cu = 1.75) compared to tantalum (Ta = 1.33). So according to the theory, the process of breaking the Cu atoms from Cu (interconnects) lattice, and their bonding to the Ta lattice, will be reduced in the presence of oxygen whereas the release of Ta atoms and bonding to Cu lattice will be favored [110–111].

Cu-Ta_xN : Tantalum nitride (Ta_xN) is deposited in between the dielectric layer and the copper. Figure 3.22 shows a high-resolution transmission electron micrograph (HRTEM) and the electron diffraction pattern (EDP) of an amorphous Ta_xN film. When annealed at 650 °C, the Rutherford back-scattering (RBS) spectrum shows a broad peak, due to chemical reactions at the interface and the formation of voids in the Cu-film. The energy of the Ta signal changes to the higher energy side after annealing, which suggests out-diffusion of Ta into the Cu film (as we have noticed with In, in Fig. 3.12b). Thus it will be more realistic to construct a ternary phase diagram of the Cu-Ta-N system (shown in Fig. 3.23) to see the probable reactions that may occur between the two layers.

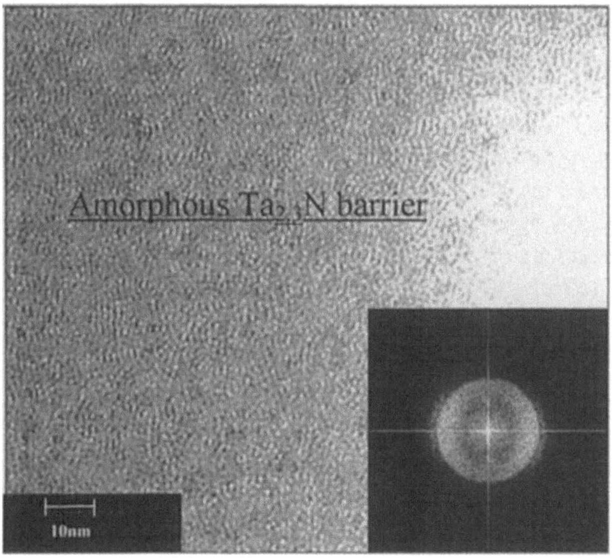

Fig. 3.22 High resolution TEM picture and electron diffraction pattern of $Ta_{2.3}N$ film (Reprinted with permission [149])

Fig. 3.23 The phase diagram of TaN and Cu

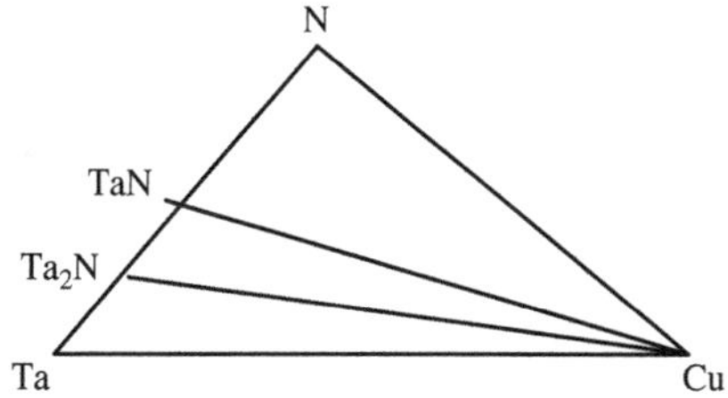

Figure 3.23 presents the ternary phase diagram of tantalum (Ta), copper (Cu), and nitrogen (N_2). The vertices of the triangle represent the pure state of the elements. Solid tie lines are constructed on the basis of the formation of stable compounds.

It is seen that Cu is in equilibrium with both Ta_2N and TaN. Thus the reactions at the interface of Cu/TaN do not have a driving force because the interface is already in local thermodynamic equilibrium. But at high temperatures (above 500 °C) tantalum nitride will crystallize to form hexagonal Ta_2N [98]. When the sample is annealed at 775 °C an amorphous layer of a noble refractory metal such as Cu_xTa_{1-x} will be formed with x centering around 0.5 [112–113]. This indicates that at high temperatures Ta can diffuse inside Cu and can form an alloy. However, according to some observers the formation of Cu_xTa_{1-x} is not noticeable even when the processing temperature is 1000 °C [114].

3.2.4.2 Case II: Double Layer Barrier Cu-Ta-TaN

TaN adheres very well to SiO_2 and most of the low-K dielectric materials, whereas Ta adheres well to Cu. For this reason a bi-layer of Ta/TaN is deposited between Cu and the dielectric layer. Moreover, the microstructure of the EP-Cu layer remains unchanged even if the barrier layer changes from Ta to TaN. However, the presence of <111> texture in the Cu-layer is observed more with a single-layer barrier than with a bi-layer barrier. A bi-layer barrier also increases the twin fraction associated with weakening of the texture of the copper film. Twining in grains (which is due to stacking faults) results in randomization of the texture and is a common phenomenon in fcc metal-like copper [115]. According to some observers, the texture of Cu-film is more dependent upon the thickness of the Cu-interconnecting line [116].

3.2.5 Influence of the Barrier Layer Properties on the Reliability of Cu-Interconnects

Experimental evidence shows that a minimum thickness of 25 nm of the barrier layer is required to deposit a continuous film of Cu [117]. Experimental evidence also shows that the formation of an interface layer between Cu and the barrier layer may change the overall volume of the adjacent layers and may induce stress, which ultimately can produce *voids*/or *buckling* due to relief of *biaxial stress* (formation of oxides of metals) [118].

3.2.5.1 Morphology of the Barrier Layer

In thin polycrystalline films *surface roughness* and *surface morphology* are two important structural parameters that affect grain growth at room temperature in Cu-interconnecting lines [119–121]. Both the *grain structure* and the *surface roughness* are related to one another via *triple lines* [122], where grain boundaries (GBs) intersect the surface. In turn, the surface roughness is known to influence the migration of GBs [123–124], which may evolve into low angle with low mobility, and can affect the grain growth [125]. On the other hand, annealing at higher temperatures makes

GBs mobile and they quickly re-orient themselves in lower energy configurations (cusps in the energy versus orientation of GBs).

Film *roughness* is a function of grain size, and the larger grain size in Cu-interconnects shows lower resistivity [126]. The minimum feature dimensions of width and height of the damascene trenches also limits average grain size [127].

An analytical model suggests that the rate of in-plane *texturing* is related in part to the allowed angular range of incident adatom flux and a *rough* surface can alter the amount of incoming adatom flux. As a result, the *texture* of a PVD deposited Cu-seed is seen to depend upon the surface roughness of the barrierlayer [128, 116]. On the other hand, in electroplated Cu-interconnects a small weakness of <111> *texture* is observed with increasing number of sublayers in the barrier film. However, the thickness of the barrier does not have a major influence on the microstructure of the Cu-film.

There are some conflicting reports about the *textural* growth of the Cu-film over barrier layer. According to one group, texture <111> of copper film is found to weaken with the deposition temperature of the barrier layer. Another group's findings are that the microstructure of Cu-interconnects is independent of the barrier layer and is dependent on the Cu-line width.

The merits of an amorphous layer as an effective diffusion barrier are well established (*because of the linear, isotropic, and homogeneous layer with no extended defects, such as grain boundaries, dislocations, lower diffusion of dopants or metals in the case of metal gates*). Amorphous films with short range order have been found to possess various unique and attractive properties compared to crystalline films that have voids and grain boundaries. *Voids* and *grain boundaries* present in crystalline films act as defect centers and serve as diffusion paths. Thus a polycrystalline Ta barrier can be a source of diffusion paths for Cu during thermal cycling or annealing [129–130]. There are reports of deposition of amorphous layers of $Ta_{73}Si_{27}$ by *radio frequency magnetron sputtering* to avoid *grain boundary* diffusion [131]. However, during *annealing* at high temperatures (>600 °C) this metastable amorphous structure transforms into a crystalline structure [132–134].

3.2.5.2 Stress

Microstructure and texture are known to influence *mass migration* and *stress voiding* performance of a thin film [135, 39]. The *stress* may be *compressive* and in extreme cases it may *buckle up* on the substrate, as has been observed in sputtered Ta film on *xerogel* (Fig. 3.25) [136]. Alternatively, the film may be under *tensile stress* (as is observed for the films of copper (Cu), silver (Ag), and aluminum (Al)) under normal deposition conditions and might cause *cracking* [137]. The stress condition of the substrate affects the microstructure of the deposited film. It has been observed that a film deposited on the wall of a high aspect ratio trench undergoes more stress than when the film is deposited on a conformal blanket [138]. The predominant mechanism of microstructure evolution is surface energy minimization for thin films with high initial stress. Surface energy calculations from experimental data show that usually <111> textured films have minimum surface energy.

In certain cases stress may depend upon the way films are prepared, e.g. the stress in electroplated Cu (EP-Cu) can change with the additives in the electrolyte. Likewise, stress inside PVD films can show different stress conditions according to the reactive pressure, gas used, and gas pressure. It has been observed that stress in a Cu-seed layer is higher when it is deposited over a TaN barrier layer than when is deposited over a Ta layer. However, the stress has been observed to be a minimum when it is deposited over TaSiN [102].

Figure 3.24 shows the residual stress in as-deposited and annealed Cu-film on a Ta barrier layer. It is interesting to see that the residual stress increased after annealing.

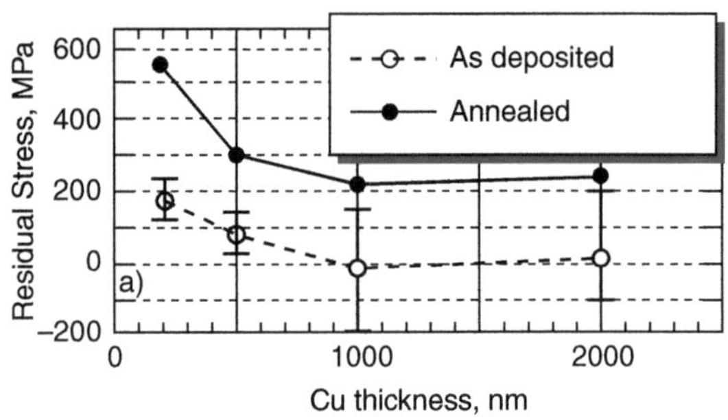

Fig. 3.24 The residual stress in as-deposited and annealed Cu-film on Ta layer (Reprinted with permission Materials Research Society, PA)

There is an another form of stress known as *thermal stress*, which is noticeable in multilayer damascene architecture with layers having different coefficient of thermal expansion (CTE). It is also true that t*hermal stress* related *adhesion* failure in thin films is not an uncommon phenomenon [139–140].

3.2.5.3 Adhesion

Adhesion of thin films has been of interest for many reasons. One of the main reasons is that the durability of a deposited thin film is governed by the adherence of the film to the substrate. The durability is particularly noticeable if the film or the substrate is subject to corrosion due to a humid atmosphere [141].

Adhesion, stress and defects are on-going problems in thin films whose magnitudes vary with material, substrate, deposition technology, ambient and the environmental conditions during deposition such as temperature, pressure and residual gases inside the chamber. Nearly all thin films, no matter how they are deposited, are subject to internal stress and the magnitude of the stress increases as the feature size of the film decreases.

Adhesion of copper on the barrier layer should be high, otherwise void formation in Cu-film may result in delamination and discontinuity in the film. Titanium and its compound adheres well with almost any substrate and acts as an adhesion promoter (glue layer) [142]. But Ti and its compounds do not provide an effective barrier layer to Cu-interconnects, especially at moderate temperatures (above 500 °C) [143]. On the other hand, Ta and its compounds are better barrier materials compared to Ti and its compounds, but TaN does not adhere well to copper. This has led to the deposition of a bi-layer barrier of Ta/Ta_xN where the Ta layer promotes the adhesion of Cu and the Ta_xN layer retards fast diffusion of Cu at high temperatures [144–145]. Most recently the adhesion property of copper (Cu) has been estimated from experimental data on the lattice misfit concept. The data show that *hcp* elemental metals like ruthenium (Ru) and osmium (Os) have better matching interfaces with fcc copper than with *bcc* elemental metals like molybdenum (Mo), tantalum (Ta) and tungsten (W) [145].

3.2.6 Low-*K* Dielectric-Barrier Layer

The porous low-*K* dielectric materials like xerogel, aerogel, or any porous silica network (e.g. p-SiCOH) may contain moisture and can be a potential source of oxygen and/or hydrogen. During sputtering, high energetic metal (present in the barrier layer) atoms can diffuse inside the pores and the presence of large concentrations of oxygen in the annealing environment can form oxides of the metal and can lead to *buckling* or may introduce *stress induced voids* [136]. Sometimes porous dielectric materials like polyarylene ethers (PAEs), or bisbenzocyclobutane, (BCB) are fluorinated to lower their *K* values. It is not unlikely that the free fluorine ions available from the fluorinated dielectric will form fluorides of the metal (present in the barrier layer) and the presence of any moisture can form hydrofluoric acid (HF) which is very corrosive.

Inorganic polymers like hydrogenated silsesquioxane (HSQ) built on an O-Si-O backbone can be a source of oxygen and free silicon atoms to form a very brittle interface layer during thermal cycling. On the other hand, some porous ultra-low-*K* materials (Si-C-O-H) are hydrophobic and after plasma etching form Si-OH, Si-O-Si bonds through condensation reactions and dangling bonds. The available oxygen from the chain can form metal oxides and ultimately can introduce stress in the film. However, although thermodynamic formation of oxides and silicides of the metals is possible, under certain conditions some metals do not form metal silicides [145–146]. On the other hand, polymers like aromatic hydrocarbon can form trap centers under bias conditions (due to Poole–Frenkel conduction), that can seriously degrade their dielectric characteristics [147–148].

3.2.7 Reaction Rates

The integrity of a barrier layer can be preserved so long as no interface metal alloy is formed during temperature cycling. An amorphous barrier layer can form a crystalline alloy, which can be a source of fast diffusion path [149]. Thus reaction rates between the layers above and below the barrier layer should be either slow or negligible. The utility of thermodynamic analysis will be demonstrated by examining which chemical reactions would be predicted to occur during the formation of interfaces between the dielectric layer (SiO_2 or low-K material) and the barrier layer (TaN/Ta)

First we will consider the deposition via homogeneous nucleation in the gas phase of Ta on to a SiO_2 substrate. Thermodynamic analysis predicts that SiO_2 in intimate contact with Ta is stable and no reaction will occur. However, for many deposition techniques, the formation of TaO_2 proceeds via heterogeneous nucleation in which Ta reacts with oxygen at the semiconductor surface (the accuracy of ΔG_r is not sufficient to predict this unequivocally). But SiO_2 will not react with Ta, to form Ta_2O_5 or TaO_2 because thermodynamically that is not feasible. However, if free oxygen were available in SiO_2 to react with Ta and the SiO_2 substrate, the most probable reaction products would be TaO_2 and Si. In other words, if oxygen is available, the substrate will tend to oxidize Ta and stoichiometric SiO_2 will tend to form. It is very unlikely that the dielectric substrate (SiO_2) will be clean and free from any oxygen upon the initiation of Ta deposition. As a result a SiO_2 surface will tend to oxidize Ta to form Ta_2O_5 and ultimately to a stable TaO_2.

Table 3.2 is a summery of the thermodynamic predictions when Ta metal is deposited upon either one of the dielectric substrates such as SiO_2, Si_3N_4 (cap layer), or low-K. When Ta is deposited on low-K dielectric material, various reactions may occur depending on the constituents of the dielectric. Ta is predicted to react with fluorine (F) atoms present in fluoro-silicon glass (FSG) regardless of whether they are homogeneously or heterogeneously nucleated. On the contrary, reactions are predicted to occur for the Si-glass compound only when deposition occurs via heterogeneous nucleation at the surface.

In the presence of fluorinated low-K, deposited Ta species of the barrier layer may form an interfacial TaF_5 layer and the presence of moisture inside the pores of a porous low-K silica can ultimately form TaF_5 and HF. Similarly, if the low-K material is chlorinated, it will form $TaOCl_3$ and HCl (in the presence of moisture). On the other hand, when a low-K material contains carbon (which is used to lower the K value of the dielectric material) it can form thin tantalum carbide (TaC) at the interface. Thermodynamic analysis indicates (Table 3.2) that TaC formation is feasible for homogeneous nucleation since the formation of TaC is energetically favorable for low-K porous dielectric materials [150]. This will be followed by the formation of $TaSi_2$, which might lead to the dissociation of TaC, which will be completely used in the reaction. Figure 3.25 shows the isothermal section of the Si-Ta-C ternary phase diagram at 800 °C.

Tantalum (Ta) atoms emerging from the gas phase can react with oxygen (O_2) and can form tantalum oxide, TaO_2 or Ta_2O_5. Studies of X-ray diffraction (XRD) spectra reveal that transitional sub-oxides like Ta_6O, Ta_4O, and Ta_2O can be formed in between [136]. Sometimes the possibility of formation of a complex ternary oxide Si-Ta_2O_x with a dielectric constant at least twice that of SiO_2 ($\sim$3.9) cannot be ignored [151]. From the table of thermodynamic quantities, one can gather information for the heats of formation (H_f) of each component in the reactions and can infer that there is a much larger thermodynamic driving force favoring oxidation over carbonation [152]. On the other hand, solubility of O_2 in Ta at 700 °C is 1.5 at% whereas the solubility of N_2 in Ta is 1.8 at% at 500 °C. However, the extent of the

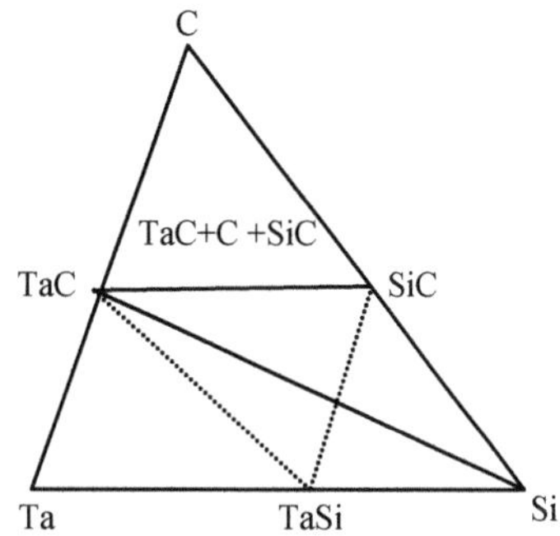

Fig. 3.25 An isothermal section of the calculated ternary Si-Ta-C phase diagram at 800 °C

reaction of Ta with N_2 is significantly lower than with O_2 at 450 °C [153]. Gallis et al [145] have found that a Ta layer deposited over SiO_2 can form oxides and silicides of tantalum. However, some kinetic limitations and surface impurities may prevent the reaction between SiO_2 and Ta (Table 3.2) [146,154]. In order to avoid

Table 3.2 A summary of thermodynamic predictions for Ta deposited on a dielectric layer (SiO_2 and Low-*K*)

Case I: Homogeneous nucleation		
Deposition species	Substrate	Reaction Products
Ta (g)	SiO_2	No reaction (assuming no free oxygen available)
Si_3N_4 (g)	SiO_2	No reaction (assuming no free oxygen available)
Ta (g)	Low-*K* (FSG)	No reaction (no free F^- available)
Ta (g)	Xerogel type (porous silica)	No reaction (assuming no free oxygen available)
Ta (g)	Carbonated low-*K*	No reaction (assuming no free carbon available)
Case II: Heterogeneous nucleation		
Deposition species	Substrate	Reaction Products
Ta (g)	SiO_2	$TaO_2 + Si$
Ta (g)	SiO_2	$3TaSi_2 + 2Ta_2O_5 + TaO_2$
Ta	SiO_2	No reaction (assuming no free oxygen available)
Ta (g)	Low-*K* (FSG), Fluorinated silica glass, fluorine doped SiO_2	TaF
Ta (g)	Xerogel type (porous silica)	TaO_2
Ta (g)	MSQ (carbon doped)	TaC
Ta (g)	Chlorinated SiO_2	$TaOCl_3$, $TaOCl_5$
Ta (g)	HSQ (H_2 doped oxide)	TaO_2
Ta (g)	SiCOH type low-*K*	TaO_2

out-diffusion of Cu into the barrier layer during the crystallization process, the Cu-layer is capped with Si_3N_4. As a result Cu diffusion intensifies along the Si_xN_y layer forming voids within the Cu-film [155]. However, experimental evidence also shows that the interface energy release rate between Cu/Si_3N_4 is less than Cu/TaN [156–157].

3.2.7.1 Effect of Ambient

The interdiffusion or alloy formation between two thin films in contact with each other is very much affected by ambient conditions. According to the surface potential model, the interface is reduced when over coating polycrystalline metal is more electronegative and an increased work function is induced on the metal by the ambient (e.g. oxygen) [158–159]. Optical micrographs of the surface of Ta films sputtered on ~70% porous xerogel at argon (Ar) pressure of 5 mtorr, show evidence of buckling due to the oxidation of tantalum [136]. The extent of buckling is increased upon annealing (Fig. 3.26).

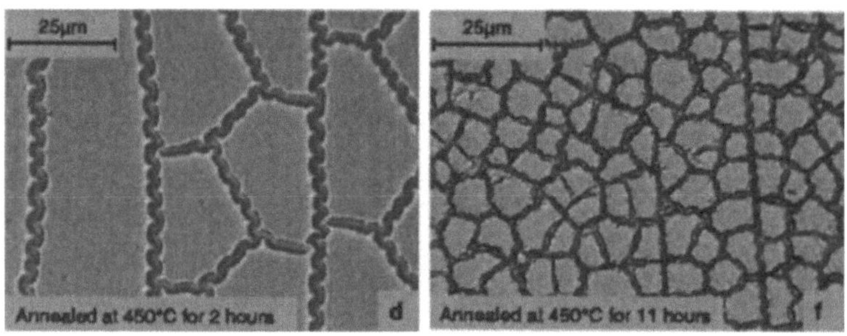

Fig. 3.26 An optical micrograph of the buckling of the Ta films on xerogel (Reprinted with permission American Vacuum Society AIP, [136])

Another important interaction of the ambient (N_2) has been observed when the partial pressure of N_2 changes from 0 to 30%, metallic Ta (bcc) changes to a mixture of Ta(N) (bcc), and hexagonal Ta_2N, then sequentially to fcc and a mixture of TaN with N-rich phases [160]. The formation of Ta_2N from TaN will change the electrical resistivity (Fig. 3.27) as well as the mechanical properties besides the texture, which will ultimately affect the effective resistivity of the Cu-interconnecting lines [161, 160].

3.2.7.2 Processing Temperature

The reaction rates are increased as the annealing temperature is increased. However, the initiation temperature for the reaction is different with different materials. For example, the Cu/Ta/SiO_2 system shows Cu-diffusion between Ta and SiO_2 after annealing at 800 °C. However, implanted (implantation of N_2 or O_2) Ta barriers are seen to show better results at high temperatures [162–163]. Therefore, it is expected that processing temperatures during deposition of a thin film should be

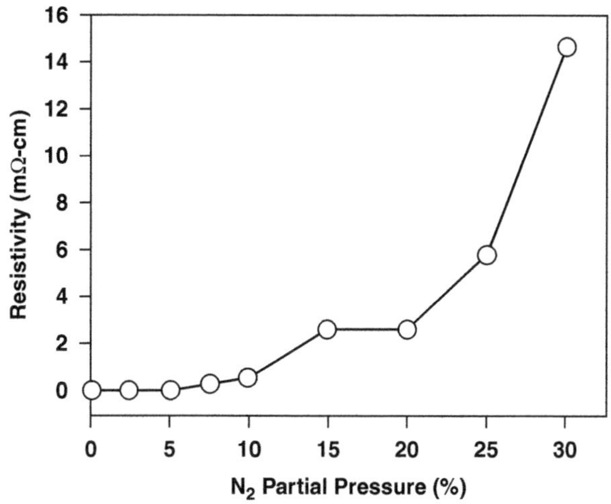

Fig. 3.27 The change in electrical resistivity of TaN films with partial pressure of N_2 during deposition (Reprinted with permission, Springer-Verlag 2000 [65])

as low as possible. This is one of the reasons why low-temperature CVD and ALD have become so much popular for the deposition of sub-micron films. However, the choice of precursors in CVD and ALD processes are very critical, because some of these precursors are seen to leave almost 2–10 at% of carbon during deposition which can bring a behavioral change in the film.

3.2.8 Influence of the Barrier Layer on the Electrical Conductivity of Cu-Lines

Figure 3.28 presents a schematic of an interconnecting line having resistance R. The resistance R (inverse of conductance) of a conductive material can be written mathematically as: $R = \rho\ (L/A)$, where ρ is the resistivity of the material, L is the length of the conductive material, and A is the area, which is equal to the width times the thickness of the conductive material.

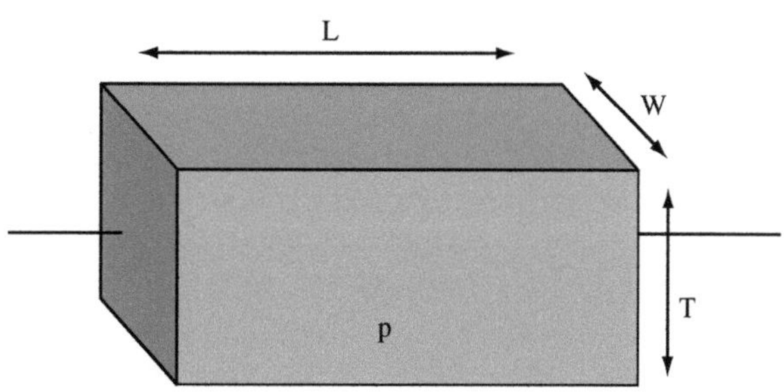

Fig. 3.28 The basic structure of the resistance

In general, for the measurement of resistance of a metal line, the resistivity of a bulk material is taken as reference. However, in the Cu-damascene process copper lines are fabricated over a barrier metal or metal compound having different bulk resistivity. As a result the Cu-line or the interconnecting line is confined within the walls of the barrier layer. The electrical conductivity of the barrier layer is important, because it will determine the equivalent resistivity of the Cu-lines. In our present model, we are considering a high performance liner in the Cu-damascene multilevel process consisting of a bi-layer (hcp/bcc-TaN followed by a bcc-Ta (α-Ta)) deposited sequentially, in a single PVD chamber. It has been observed that at least a 2 nm TaN layer is required for the formation of α-Ta (15 μΩ-cm) which is much more conductive than β-Ta (150 μΩ-cm). However, the β phase columnar structure of tantalum is more resistant to copper than the mixed α (bcc)-β phase [164–165, 130].

Figure 29 shows (a) the confined Cu-line inside the barrier layer, (b) the top view of the damascene trench structure, and (c) represents deposited Cu over the barrier layer.

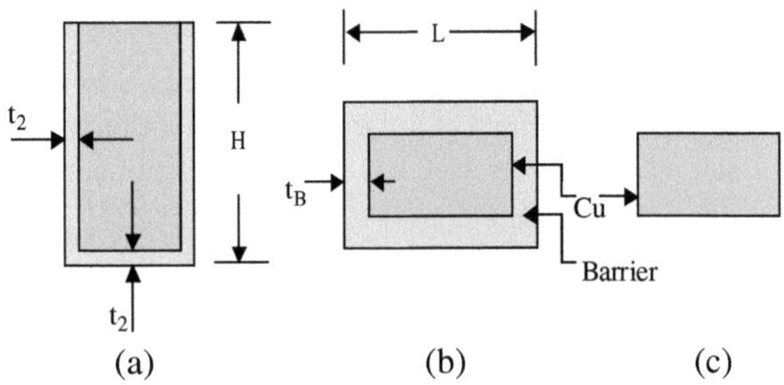

Fig. 3.29 (**a**) The confined Cu-line inside the barrier layer; (**b**) the *top* view of the damascene trench structure; and (**c**) deposited Cu

The impact of the side wall of the barrier layer on very narrow (reduced grain size) metal lines of copper changes the equivalent resistance (R_{eq}) of the central copper wire over the barrier line. However, we can make an approximation of the impact of these parameters as long as the average grain size for wide lines is greater than about six times the barrier thickness over a flat surface. The approximation can be made by modeling the adjacent layers of copper (Cu) and barrier as two parallel resistances R_{Cu} and R_B connected together. As the Cu-interconnecting lines are scaled (the line width becomes <2× the nominal Cu-grain size, and the thickness of the barrier layer is >6× the grain size of nominal Cu-line) we can approximate the equivalent resistance (R_{eq}) as:

$$(1/R_{eq}) = (1/R_{cu} + 1/R_B) = \{(W_{cu}t_{cu}/(\rho_{Cu}L)\} + \{(W_B t_B)/(2\rho_B L)\}. \quad (3.11)$$

Now, the Cu-line is embedded on the barrier layer and both are inside the trench (shown in Fig. 3.27(a)). As a result, Equation (3.11) is transformed into: [166]:

$$1/R_{eq} = [\rho_1.\rho_2 L] \ / \ [\{\rho_2 \{(W-2t_2)(H-2t_2)\} + \{\rho_1 \{(2t_2)(H+W-2t_2)\}], \quad (3.12)$$

where,

R_{eq} = Equivalent resistance of the Cu-interconnect
ρ_1 = Resistivity of copper
ρ_2 = Resistivity of the barrier metal /metal compound
L =The length of the trench
W = Width of the trench
t_2 = Thickness of the barrier layer (t_B)
H = Height of the trench.

According to the ITRS road map, the thickness of the barrier layer should not take more than 15% of the width of the Cu-interconnect. However, it has been argued elsewhere that the minimum thickness of the barrier layer should be at least the thickness necessary to close the substrate surface [167]. It is also likely that the electrical resistance of the cap layer can change due to the phase transformation of the substrate material [168–169]. Following Equation (3.12) the generic *RC* time constant (time delay) of the Cu-interconnect when deposited over a barrier layer having higher resistivity can be modeled as:

$$R_{eq}\,C_{eq} = 2\,R_{eq}\,(C_V + C_L) = 2R_{eq}\left[\{(\varepsilon_0\varepsilon_r\,WL)/(H)\} + \{(\varepsilon_0\varepsilon_r\,TL)/(X)\}\right] \tag{3.13}$$

where,

R_{eq} = Equivalent resistance from Equation (3.12)
C_V = Inter-layer capacitance
C_L = Intra-layer capacitance
ε_0 = The permittivity of free space
ε_r = The relative permittivity of the dielectric
H = Height of the dielectric layer
T = The thickness of the Cu/barrier layer
W = width of the Cu/barrier layer
X = spacing between adjacent interconnects
P = Pitch of the line.

From Equation (3.13) we can see that the *RC* time constant is directly proportional to the resistivity. Therefore a 40% reduction in time constant can be achieved by using copper (Cu) rather than aluminum (Al). However, when the lines are scaled the 40% the gain is very much in doubt.

3.2.9 Influence of Barrier Layer Thermal Conductivity on Cu-Line

The ITRS road map predicts that the number of transistors in high performance microprocessors will exceed a billion in the year 2007. As a result, the number of interconnects connecting these devices will also increase and the temperature rise due to scaled devices will increase enormously. Due to this fact, the temperature dependence of clock delay and clock skew have received considerable attention. Signal delay is directly proportional to interconnect resistance and is affected by the temperature dependence of the interconnect's electrical resistivity. On the other hand, typical low-*K* dielectric materials have up to an order of magnitude lower thermal conductivity than conventional SiO_2. At the same time, multistack interconnect architecture is characterized by a wide range of length scale and significant material inhomogeneity (Fig. 3.30). The current density (*j*) through the Cu-line and the barrier Ta (Tantalum or its compound) can be written as:

$$j_{\mathrm{cu}} = \left[\{(I/A_{\mathrm{cu}})\}\right]\left[\{(\rho_{\mathrm{Ta}}/A_{\mathrm{Ta}})\} / \{(\rho_{\mathrm{Cu}}/A_{\mathrm{cu}}) + (\rho_{\mathrm{Ta}}/A_{\mathrm{Ta}})\}\right] \quad (3.14)$$

$$j_{\mathrm{Ta}} = \left[\{(I/A_{\mathrm{cu}})\}\right]\left[\{(\rho_{\mathrm{Cu}}/A_{\mathrm{Ca}})\} / \{(\rho_{\mathrm{Cu}}/A_{\mathrm{cu}}) + (\rho_{\mathrm{Ta}}/A_{\mathrm{Ta}})\}\right] \quad (3.15)$$

where ρ is the electrical resistivity, *A* is the cross-sectional area, and *I* is the total current and the subscripts represent the corresponding metal layer. The heat dissipated in the line of length *L* can be written as:

$$Q = L\,(j2_{\mathrm{Cu}}\rho_{\mathrm{Cu}}A_{\mathrm{Cu}} + j2_{\mathrm{Ta}}\rho_{\mathrm{Ta}}A_{\mathrm{Ta}})\,. \quad (3.16)$$

In order to reduce the number of nodes, the metal line is assumed to be homogeneous material with its effective thermal conductivity derived from two parallel resistances for copper line and barrier layer. This can be justified since the thermal conductivity of metal is much higher than the surrounding dielectric and the thickness of Ta is much smaller in comparison to the line cross-section [168]. The unit

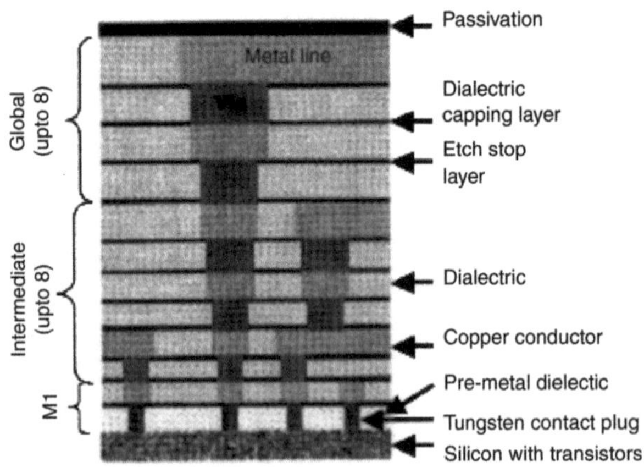

Fig. 3.30 Multilevel interconnect stack used in Cu-damascene technology (Courtesy ITRS 2004)

cell model from Ramkrishna et al. shows an exponential rise of temperature with increase of current density [169].

Figure 3.30 shows the multilevel stack typically used in the Cu-damascene structure. There might be more that seven metal lines at the intermediate level and five or more in the global interconnecting system.

Figure 3.31 shows the conduction of heat from Ta-layer to Cu-layer. The heat conduction can be modeled by considering the layers as slabs 1 (Ta), and 2 (Cu) having conductivity K_1 and K_2, thickness L_1 and L_2 and A is the cross-sectional area. Let us assume that the temperatures of the outer surfaces are T_H and T_C, and let T_X be the temperature at the interface. In a steady state the rate of heat transfer will be

$$Q = \{A\,(T_H - T_C)\} / \{(L_1 K_1) + (L_2/K_2)\}\,. \tag{3.17}$$

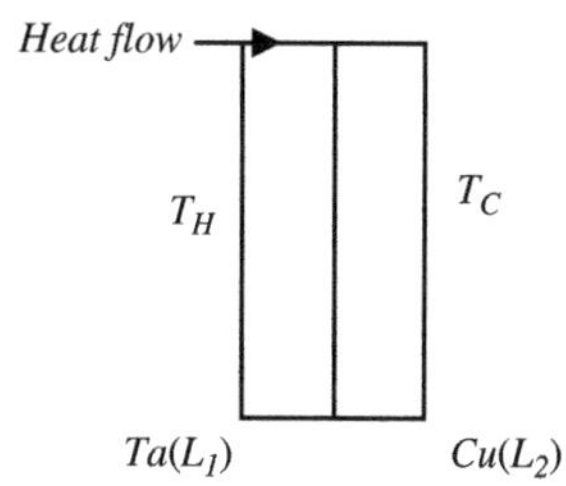

Fig. 3.31 The conduction of heat through tantalum (or its compound) and copper

The thermal conductivity of two adjoining layers can be modeled as: [170–171]

$$K_{bi-layer} = 2\,\{(K_1K_2)/(K_1 + K_2)\}\,, \tag{3.18}$$

where K_1 is the thermal conductivity of layer 1 (either Cu when the bi-layer is considered Cu/Barrier layer, or barrier when the bi-layer is barrier/dielectric) and K_2 is the thermal conductivity of layer 2 (either barrier or dielectric). Equation (3.18) will give the equivalent thermal conductivity of the bi-layer and we will be able to determine that value if we know the values of K_1 and K_2.

Interfacial thermal resistance (Θ) normal to the interface ($K_\perp$) is due to the commutative impact of scattering phenomena and can be represented by [172]:

$$K_\perp = \left[\{(t_{Cu} + t_{Ta})\}\right] / \left[\{(t_{Cu} - \delta) / (K_{cu})\} + \{(t_{Ta} - \delta) / (K_{Ta})\} + \left\{(2\delta) / \left(K_{bi-layer}\right)\right\}\right]. \tag{3.19}$$

where t_{Cu} and t_{Ta} are the thickness of the Cu and Ta films, δ the effective thickness consumed from the individual layer due to interface formation, K_{int} the thermal conductivity of the interface layer, and K_{cu} and K_{Ta} are the thermal conductivities of Cu and Ta. We can clarify further, from some experimental evidence, that heat must flow transversely through grain boundaries and interior grains, when the grain size is equal to the thickness of the film (0.3–3 μm) [173].

Following Equation (3.19) the thermal resistance for heat flowing through (normal to) a bi-layer (Cu/barrier (B) or (B/dielectric layer) can be obtained by summing the thermal resistance, thickness, and thermal conductivity of the bi-layer [174] and can be modeled as: [175]

$$\Theta_{(\text{interface})} = \partial \{(2/k_{\text{int}}) - (1/K_{\text{cu}}) - (1/K_{\text{Ta}})\} \tag{3.20}$$

Table 3.3 shows some of the structural, thermal and electrical properties of the transition metals and their compounds.

Table 3.3 Physical properties of some of the probable interactive materials

Material	Structure	TCR (ppm./°C)	Conductivity (Electrical) $(\mu\Omega\text{-cm})^{-1}$	Conductivity (Thermal) Watts/cm-°C
Ta	bcc (α)	500–1800	0.04–0.02	0.54
	tetragonal (β)	–100 to + 100	0.0055–0.0045	
TaO_2				
TaN	Amorphous	–60	0.0004	
	Cubic <111>		0.0066–0.005	
$TaSi_2$	Hexagonal	12.5	0.066–0.02	
TaC	Cubic	6.3	0.01–0.0066	0.22
TaF_5	tetrameric			
$TaOCl_3$				
Cu	Cubic	0.393	0.64	0.053
Cu_3Si				

3.2.10 Classification of Barrier Layer

The diffusion of copper in silicon along the grain boundaries (GBs) can be 100 times larger than the diffusion along the lattice of single crystal silicon. A large number of transition metals and their compounds have been tried as barrier layer and have been reported from time to time. The barrier layer extends the lifetime of the resultant contact structure by retarding interdiffusion or resisting the chemical reactions between the two layers to form a new phase [176–178]. The diffusion barriers can be classified into three categories namely: (a) *passive* (b) *sacrificial,* and (c) *stuffed* [139].

A *passive barrier* is an inactive and protective layer without responding to or initiating any interactions (chemical, physical, or metallurgical) whatsoever and it has high negative heat of formation. A *sacrificial layer,* on the other hand, reacts between layers very slowly and within the lifetime of the device there is no diffusion of the metal. The third one is the *stuffed layer,* which provides some kind of barrier species to block the grain boundaries so that no diffusion occurs. Nitrides, silicides, borides, and carbides of transition metals are possible materials that have been used as barrier layers in advanced semiconductor device technology [146,179–180]. Figure 3.32 shows a schematic of three major diffusion barriers.

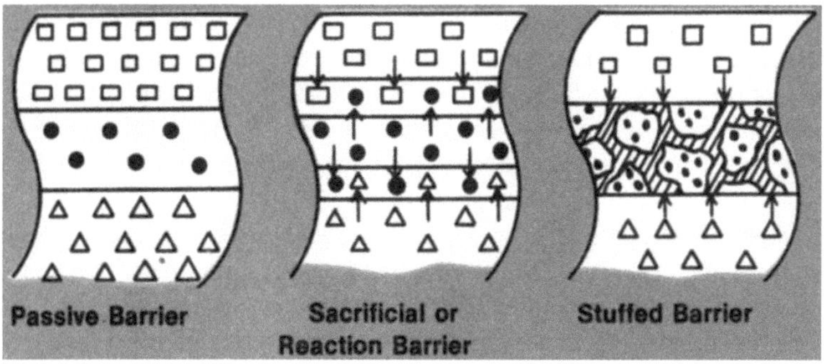

Fig. 3.32 Schematic of the three major diffusion barrier types (Reprinted with permission, Solid State Technology, **28**, 10, 1985)

3.2.11 Properties of Different Barrier Layer Materials

3.2.11.1 Tantalum (Ta) and Its Compounds

Tantalum (Ta) is a refractory metal and belongs to the class of metals known as valve metals, which form tough self-protecting oxides [181–183]. Ta has a high melting point (mp~3000 °C) and PVD of Ta has been found to be the most suitable method for deposition of the film. The difficulty with PVD film is that the structure of the deposited film depends on the processing conditions [184–185]. Sputtered Ta is known to produce bcc-Ta (α), β-Ta, or even fcc-Ta, and the film roughness increases as the density decreases with orientation angle of the film [186].

As-deposited Ta-film has a β-phase with polycrystalline columnar structure [187] and it grows along the <002> direction with an additional shoulder along <202>. The Ta layer texture over Si remains unchanged (no out-diffusion of silicon) at 600 °C. However, annealing at higher temperatures shows the formation of $TaSi_2$, which grows along <111> with secondary growths along <200> and <112>. Some observers have reported the formation of amorphous Ta_5 Si_3 at the interface of Ta/Si, which transforms to $TaSi_2$ crystallites when annealed at 700 °C.

Tantalum nitride: Ta and its nitrides have been found to be promising materials for barrier layers in copper dual damascene processes. Tantalum nitride can be deposited by (i) reactive sputtering [188–189], (ii) MOCVD [190], (iii) ion beam assisted deposition (IBAD) [191], and (iv) implantation of N_2 (>25 at %) [192].

Low-pressure (5×10^{-5} torr or less) deposition results in Ta_2N composition whereas the formation of TaN is favorable above 10^{-4} torr. Hcp TaN shows lower resistivity, and higher reliability at 800 °C compared to bcc TaN [193]. TaN is a chemically inert refractory compound and can be present in various phases, such as β-Ta_2N, θ- TaN, η-TaN, δ-TaN_{1-x}, Ta_2N_6, Ta_4N_5 and Ta_3N_5 [194]. Moreover, tantalum and its nitrides (Ta_2N and TaN) a do not form Cu-Ta or Cu-N alloys (e.g. Cu_3N is very unstable and decomposes at ~300 °C) [193].

Sputter deposited tantalum nitride (TaN) is cubic and growth is seen along <200> with strong secondary growth along <111>. The structure remains unaltered below 800 °C, and shows better integrity over Ta film. [194–198]. The sputter deposited Ta_2N film on the other hand is amorphous and forms crystallites above 600 °C. However, above 600 °C sputter deposited amorphous layers of TaN over Ta_2N show better reliability and minimum intrinsic stress [199].

The resistivity of TaN layers is very much influenced by the amount of N_2 present in the system. As the ratio of (N/Ta) increases the activation energy related to lattice out-diffusion of copper decreases. According to some observers [200], annealing between 500 and 900 °C induces more strain in the single layered TaN film, compared to the double layered Ta_2N/TaN film.

Tantalum forms a carbide of the metal (TaC) which is similar to tantalum nitride (Ta_xN). However, in the case of TaC there is no equilibrium ternary phase between TaC and Si. When deposited over SiO_2 the oxygen from SiO_2 can form an amorphous layer of TaO_x. It has been observed that tantalum carbide (TaC_x) film deposited by magnetron sputtering over silicon dioxide grown on n-type silicon is almost amorphous in nature [201].

3.2.11.2 Titanium(Ti) and Its Compounds

Titanium metal has a hexagonal closed packed (hcp) lattice and resembles other transition metals such as iron (Fe) and nickel (Ni). It is hard, refractory (mp = 1680 °C) and is a good conductor of heat and electricity [201–202]. Titanium reduces silicon dioxide (SiO_2) at relatively low temperatures and forms a strong bond with SiO_2. It is quite light in comparison to other metals of similar mechanical properties and usually resistant to certain kinds of corrosion. At elevated temperatures it reacts with almost all non-metals, like hydrogen (H_2), oxygen (O_2), nitrogen (N_2), carbon (C), boron (B), silicon (Si), and sulfur (S),. The possible metal compounds that are formed with N_2, C, and B are nitride of Ti, (TiN), carbide of Ti (TiC) and borides of Ti (TiB, and TiB_2) that are very stable, hard and refractory and they are potential candidates for a barrier layer. Titanium compounds are also considered as adhesion promoters.

Titanium borides (TiB and TiB_2) deposited by magnetron sputtering have been used as barrier layers and have been found to be reliable below 700 °C [202]. There has been an attempt to use CVD deposited titanium silicon nitride (TiSiN) as a barrier layer, but it has not been successful due to poor copper wetting and poor adhesion of the film to low-*K* dielectrics.

TiN film is most commonly deposited by reactive sputtering in the presence of N_2 and Ar [203] with good stoichiometry and purity. When deposited over small via holes, the film shows lack of conformality and uniformity. However, ionized metal plasma or MOCVD deposited films are conformal [204–205]. Conformal, and void-free TiN layers have also been deposited successfully at low temperatures by atomic layer deposition (ALD) [206].

Sputter deposited titanium carbide (TiCx) has been reported to be a stable barrier for copper below 600 °C. However, the formation of localized defects or grain boundaries above 600 °C becomes the primary source of Cu-diffusion [207–208]. Moreover, the adherence of Cu is poor on titanium carbide (TiC) film due to the presence of carbon.

3.2.11.3 Molybdenum (Mo) and Its Compound

Molybdenum and tungsten are both similar in nature, but they form different compounds. Both elements are inert to oxygen at ordinary temperatures, but at high temperatures they can form higher oxides of the elements (MoO_3, WO_3, Mo_2O_5 and WO_2). The resistivity of MoO_3 increases monotonically with the percentage of oxygen content in the film regardless of sputtering conditions. A similar trend has also been reported for tungsten nitride (WN) films [209–210].

Pulsed layer deposited (PLD) and magnetron sputter deposited (MSD) molybdenum (Mo), carbon (C), and carbon boride (B_4C) layers have been reported by S. Braun et al. [181]. Studies show that the MSD method grows more amorphous inter-layers compared to the PLD method. It is expected that thin films of amorphous silicon molybdenum nitride grown by reactive co-sputtering of silicon (Si) and molybdenum (Mo) targets in presence of nitrogen can be a potential candidate for barrier layers [182–183].

3.2.11.4 Tungsten (W) and Its Compounds

Tungsten nitride (W_xN) is deposited either by radio frequency plasma or CVD methods by the reduction of hexafluoride of tungsten (WF_6) and is considered as a good barrier layer against Cu-diffusion. The tungsten to nitrogen ratio determines the morphology of W_xN and its effectiveness as a barrier layer [184]. There is no structural change noticed at 600 °C; however, above 600 °C, the film forms crystallites [85, 211]. Compounds of tungsten with carbon and nitrogen, WCN, have been deposited by atomic layer deposition (ALD) and has been integrated successfully as a barrier layer with low-K dielectric materials [212, 206]. Another potential candidate for barrier layer is tungsten carbide, which has been prepared by plasma assisted ALD, using *bis*(tert-butylimido) and *bis*(dimethylamido) at 250 °C.

3.2.11.5 Cobalt

Cobalt is a hard, bluish white metal, which melts at 1493 °C and is found in nature in association with nickel and arsenic. It dissolves in mineral acids, and does not combine directly with hydrogen or nitrogen to form hydrides or nitrides of the metal. At elevated temperatures, it reacts with atmospheric oxygen and in moist atmosphere it forms oxides of the metal (CoO and Co_3O_4). There is no evidence of the existence of a Co_2O_3 compound. Co has been used in the Cu-damascene process as an adhesion promoter [213]. The thin epitaxial barrier layer of $CoSi_2$ has shown promise because of its morphological integrity above 550 °C [214]. Ultra-thin (10 Å) cobalt film deposited by e-beam evaporation and rapid thermal processing shows low resistivity (14 μΩ.cm) and the surface roughness of the film remains unchanged above

500 ºC [215–216]. Low-resistivity Co has also been used as a contact metal in semiconductor devices (Fig. 3.33).

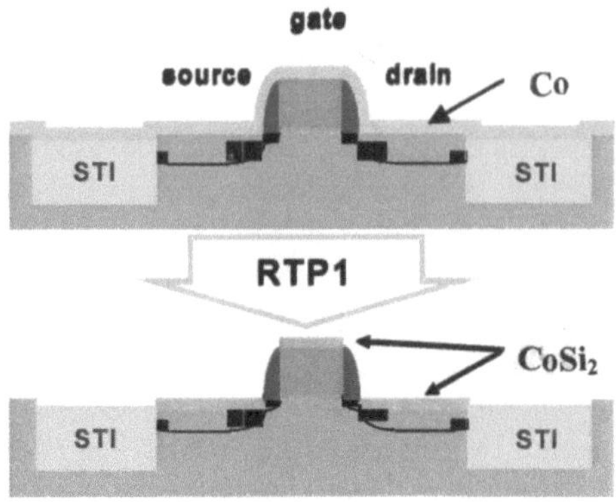

Fig. 3.33 The fabrication of a MOS device where cobalt (Co) has been used as low-resistivity contact over the source and the drain regions (Reprinted with permission of AIP)

Recently cobalt tungsten phosfide (CoW P) deposited by the electroless method [217] has been used as *cap* material in copper interconnects. *Capping* the copper in the Cu-damascene process with CoWP has improved device reliability [218].

3.2.12 Cap-Layer, Its Properties and Functions

In order to reduce copper oxidation, corrosion, interface diffusion and inelastic deformation due to diffusional creep, a cap layer has been used with Cu-interconnects (Fig. 3.34) [219–220].

The cap layer has the ability to suppress vacancy generation [221] in Cu; besides, the adhesion energy of the cap layer is very high compared to the dielectric surface

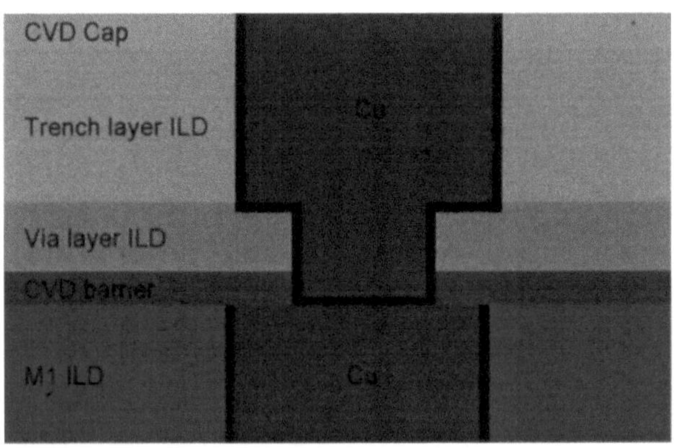

Fig. 3.34 Schematic of the cap layer (Reprinted with permission AMD)

(40 J/m^2 : 10^{-20} J/m^2). The adhesion energy of the cap layer is proportional to the electromigration (EM) activation energy. Cobalt tungsten phosphide (CoWP) has the highest activation energy compared to other cap layers (SiC, SiN, and NSiC) that are used in the Cu-damascene process (Fig. 3.34). As a result, CoWP is believed to be the most effective cap layer to minimize EM in Cu-interconnects. Figure 3.35 shows two different approaches (electroless) to form an effective (CoWP) cap-layer. However, a palladium (Pd) activated cap is becoming unpopular because of its detrimental effects, such as: increased line resistance due to Pd diffusion into copper (Cu), Cu-etch loss, and corrosion of Cu-barrier layer interface, besides processing and consumable costs.

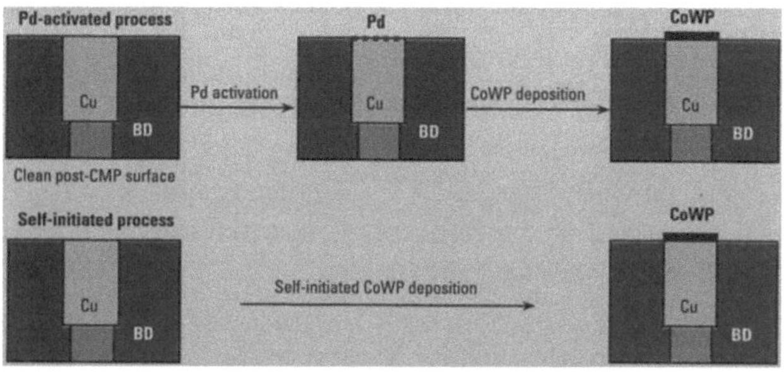

Fig. 3.35 Two different approaches to deposit cobalt based capping layers (Reprinted with permission from Semiconductor Internationl, Oct. 2005, p. 45)

The X-ray diffraction (XRD) pattern of high temperature an annealed CoWP cap layer material shows a broad peak which is supposed to be due to chemical reactions at the interface and formation of voids (Fig. 3.37). Figure 3.36 shows different interface-debond energy versus void growth rate of different cap materials.

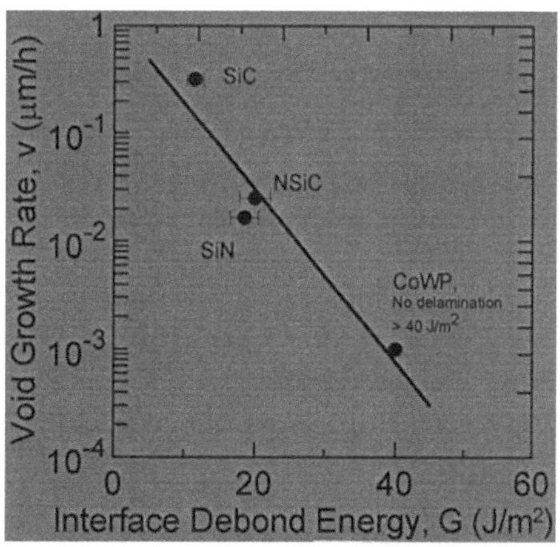

Fig. 3.36 The comparative analysis of different cap materials and their ability to suppress voids (Reprinted with permission, IBM Research)

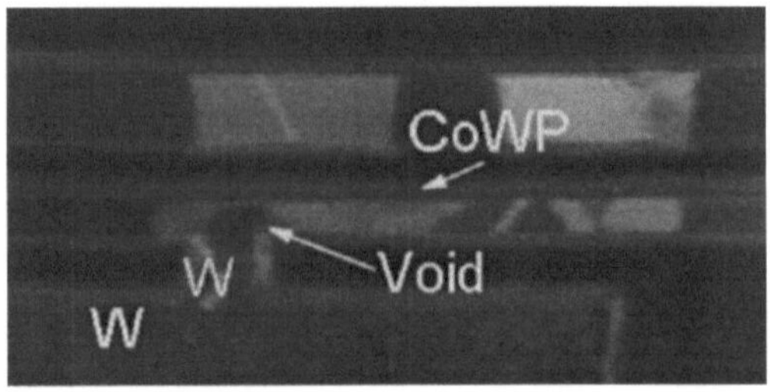

Fig. 3.37 Void formation at high temperatures (Reprinted with permission, Mat. Res. Soc. Adv. Met. Conf. 2003)

Autocatalytic chemical reduction (electroless deposition) of Co-alloy with W, and phosphide is a very popular method to deposit a CoWP cap-layer [222]. Electroless deposition of the cap layer when self-aligned to Cu forms a smooth conformal film (Fig. 3.37). The CoWP capping layer not only suppresses electromigration (EM), but it also increases the current carrying capacity in the Cu-interconnect. According to some observers, the use of CoWP could enable a change in the choice of material for the etch stop dielectric layer, now typically SiN ($K \sim 7.0$), or SiC ($K \sim 3.5$), or SiCN ($K \sim 4.9$) is used as cap layer.

3.3 Summary

There has been considerable progress in the use of copper (Cu) as interconnecting metal in sub-100 nm devices due to its higher conductivity and electromigartion resistance than aluminum. However, the high diffusivity of Cu in silicon needs an effective barrier layer. The barrier layer is usually laid on between the Cu-interconnecting lines and the dielectric layer. The barrier layer must be able to prevent diffusion, should exhibit low equivalent resistance with Cu-lines, should have good adherence properties over the substrate and the over layer, and be compatible to chemical mechanical polishing (CMP). In addition to all of these, the barrier layer must be thin, conformal, and continuous to fully encapsulate the Cu-lines. Transitional metals and their compounds, especially Ta/TaN, have high melting points (therefore high activation energies are required for both lattice and grain boundary self-diffusion) and have shown promising results. At moderate temperatures, these barrier layers keep their morphological identities, do not form intermetallic compounds, and provide a relatively stable interface with Cu. However, the formation of a conformal, thin continuous barrier layer will depend upon the deposition technology and several processing conditions. For example, ionized sputtering technology has replaced the conventional PVD method, to produce thin uniform coverage in deep via holes or trenches. On the other hand, ALD processes have successfully demonstrated the deposition of highly uniform and conformal thin films by alternating exposures of a surface to vapors of two chemical reactants. Likewise, the innovative low-temperature processing of MOCVD has been successfully applied to deposit good barrier films with fine-tuning of the precursors.

References

1. H.J. Queisser and E.E. Haller, Defects in semiconductors: Some fatal, some vital, Science, 281, 945 (1998) and also P. Stiffer and E. Krimmel (eds.), Silicon evolution and future of a technology, p. 249, Springer, New York, (2004) and B.I. Boltaks, A treatise on diffusion in semiconductors, Academic, San Diego, CA., (1963) and O. Muscato, Physica, 365 (2), 409 (2006) and IEEE IEDM News, Dec 16, (2007)
2. A.A. Istratov and E.R. Weber, Physics of copper in silicon, J. Electrochem. Soc., 149 (1), G21 (2002) and also T. Heiser, S. McHugo, H. Heislmair, and E.R. Weber, Transient ion drift direction of low level copper contamination in silicon, Appl. Phys. Lett., 70, 3576 (1997) and Z. Xi, J. Chen, D. Yang, A. Lawerenz, and H.J. Moeller, Copper precipitation in large diameter Czocharlski silicon, J. Appl. Phys., 97, 094909 (2005) and also S-H Kim et al., Electrochem. Solid State Lett., 11 (5), H-127 (2008)
3. J. Li, Y.S. Diamond, and J.W. Mayer, Copper deposition and thermal stability issues in Cu-based metallization for ULSI technology, Mater. Sci. Rep., 9, 1 (1992) and also A. Correia, D. Ballutaud, A. Bountry-Forveille, and J-L. Maurice, Effects of copper and oxygen precipitation during thermal oxidation of silicon, J. Appl. Phys., 78, 6543 (1995)
4. K.M. Crosby, Grain boundary diffusion in copper under tensile stress, Cond. Matter., 1, 65 (2003)
5. K. Masu et al., Jpn. J. Appl. Phys., 30 (1), 3642 (1991) and also I. Kaur, Y. Mishin, and W. Gust, Fundamentals of grain and interphase boundary diffusion, Wiley, Chichester, (1995) and also K.F. Vaz et al., Microelectron. J., 34 (5/8), 599 (2003)
6. J.H. Choy and K.L. Kavanagh, Appl. Phys. Lett., 84 (25), 5201 (2004)
7. E.T. Ogawa, K.D. Lee, V.A. Blaschke, and P.S. Ho, Electromigration reliability issues in dual damascene Cu interconnections, IEEE Trans. Reliab., 51 (4), 403 (2002)
8. J.R. Lyod and J.J. Clement, Thin Solid Films, 262, 135 (1995) and also M.J. Kerr and A. Cuevas, Recombination at the interface between silicon and stoichiometric plasma silicon nitride, Semicond. Sci. Technol., 17, 166 (2002)
9. C.S. Hau-Riege and C.V. Thompson, Appl. Phys. Lett., 78, 3451 (2001)
10. C.K. Hu, R. Rosenberg, and K.Y. Lee, Appl. Phys. Lett., 74, 2945 (1999)
11. M. Nathan et al., Appl. Phys. Lett., 77, 3355 (2000) and also M.R. Sorensen and A.F. Voter, J. Chem. Phys., 112, 9599 (2000)
12. J.H. Choy and K.L. Kavanagh, Appl. Phys. Lett., 84 (25), 5201 (2004)
13. T. Laurila, K. Zeng, J.K. Kivilathi, J. Molarius, and I. Suni, Chemical stability of Ta diffusion barrier between Cu and Si, Superfices y vacio, 9, 199 (Dec. 1999) and also E.T. Ogawa, K.D. Lee, V.A. Blaschke, and P.S. Ho, Electromigration reliability issues in dual damascene Cu interconnections, IEEE Trans. Reliab., 51 (4), 403 (2002)
14. A.E. Kaloyeros, X. Chen, S. Lane, and H.I. Frisch, Tantalum diffusion barrier grown by inorganic plasma promoted chemical vapor deposition, JMR, 15 (12), 2800 (2000) and M.A. Meyer, Ph.D. Thesis, Brandenburg Univ. of Technol., Cottbus, Germany, (2007)
15. J.D. Torre et al., Microstructure of thin tantalum films sputtered onto inclined substrate: Experiments and atomistic simulations, J. Appl. Phys., 94 (1), 263 (2003) and also P. Klaver and B. Thijsse, Thin Solid Films, 2, 467 (1968)
16. H. Mehr and L. Bornstein (eds.), Diffusion in solid metals and alloys, Gr. III, Vol. 26, Springer, Berlin, (1990)
17. K. Holloway and P.M. Fryer, Appl. Phys. Lett., 57, 1736 (1990) and T. Oku et al., Appl. Surf. Sci., 99, 265 (1996) and C.K. Hu, L. Gignac, and R. Rosenberg, Microelectron. Relib., 46, 213 (2006)
18. D.E. Kramer, A.A. Volinsky, N.R. Moody, and W.W. Gerberich, JMR, 16 (11), 3150 (2001) and D. Chocyk et al., Evolution of stress and structure in Cu thin films, Cryst. Res. Technol., 40 (4/5), 509 (2005)
19. W.W. Gerberich, A.A. Volinsky, N.I. Tymiak, N.R. Moody, Mater. Res. Soc. Symp., 594, 351 (2000) and L. Lu, N.R. Tao, L.B. Wang, B.J. Ding, and K. Lu, Grain growth and strain

release in nanocrystalline copper, J. Appl. Phys., 89 (11) 6408 (2001) and also M.A. Meyer and E. Zschek, Proc. 9th Int. Workshop on stress induced phenomena and metallization, AIP Conf. Kyoto, Japan, 4–7 April (2007)
20. M.J. Buehler, A.Hartmaier, and H. Gao, Hierachical multi-scale modeling of plasticity of sub-micron thin metal films, Model. Simul. Mater. Sci. Eng., 12, S391 (2004) and also K. Chen, A. Fan, and R. Reif, J. Electron. Mat., 30, 331 (2001)
21. H. Gao, L. Zhang, W.D. Nix, C.V. Thompson, and E. Artz, Crack like grain boundary (GB) diffusion wedges in thin metal films, Acta. Mater., 47, 2865 (1999) and K. Mohseni, A. Shakouri, R.J. Ram, and M.C. Abraham, Phys. Fluids, 17, 100602 (2005)
22. B.G. Wills and D.V. Lang, Oxidation mechanism of ionic transport of copper in SiO_2 dielectrics, Thin Solid Films, 467, 284 (2004) and also V.S.C. Len, R.E. Hurley, N. McCuster, D.W. McNeil, B.M. Armstrong, and H.S Gamble, Solid State Electron., 43, 1045 (1999)
23. J.Y. Kwon, K.S. Kim, Y.C. Joo, and K.B. Kim, Simulation of the copper diffusion profile in SiO_2 during bias temperature stress (BTS) test, Jpn. J. Appl. Phys., 41, L99 (2002) and A. Mineji et al., IEEE Int. Workshop Jn. Technol. Paper, S 4–8 (2007)
24. J.D. McBrayer, Ph.D. dissertation, Stanford University, 1983 and P.G. Sverdrup et al., Int. conf. on simulation and semicond. process & devices (SISPAD), p. 54, Sept 6–8, Seattle, WA, (2000)
25. J.C. Lin and C. Leez, Electrochem. solid state Lett., 2 (4), 1812 (1999)
26. A.L.S. Loke, Ph.D. Dissertation, Stanford university, (1999)
27. H. Miyazaki, K. Hinode, Y. Homma, and N. Kobayshi, Jpn. J. Appl. Phys., 35, 1685 (1996)
28. V. Hugo, The house in the RUE plumet, Les miserables, Vol. 4, Chapter 235, Book 3rd ed., Signet, New York, (1987)
29. D. Gan, P.S. Ho, R. Haung, J. Len, J. Maiz, and T. Scherban, Isothermal stress relaxation in electroplated Cu-films, J. Appl. Phys., 97, 103531 (2005)
30. A.A. Volinski et al., Residual stress and microstructure of electroplated Cu fil on different barrier layers, MRS. Symp., 695, L1.11.1 (2002) and also J.J. Tommy, S. Hymes, and S.P. Murarka, Stress effects in thermal cycling copper thin films, Appl. Phys. Lett., 66, 2074 (1995)
31. M. Sorensen, Y. Mishin, and A. Voter, Phys. Rev., B-62, 3658 (2000)
32. H. Van Swygenhoven, D. Farakas, and A. Caro, Phys. Rev., B-62, 831 (2000)
33. K.M. Crosby, Grain boundary diffusion in copper under tensile stress, Cond. Mater., 1, 065 (July 2, 2003)and also Y. Mishin, Philos. Mag., A72, 1589 (1995) and M. Hauschudtet al., J. Appl. Phys., 101 (4), 043523 (2007)
34. L.I. Maissel and R. Glang, Handbook of thin film technology, McGraw Hill, New York, (1983)
35. J. Thibault, J.L. Rouviere, and A. Bourret, Grain boundaries in semiconductors, In Hand book of semiconductor technology, K.A. Jackson and W. Schroter, Vol. I, Wiley and VCH, New York, (2000)
36. W.T. Read, Dislocations in crystals, McGraw Hill, New York, (1953)
37. J.W. Mayer and S.S. Lau, Electronic materials science for integrated circuits in Si and GaAs, p. 172, Mcmillan Pub., New York, (1990)
38. J. Nucci, R. Keller, D.P. Field, and Y.S. Diamond, Appl. Phys. Lett., 7 (10), 1242 (1999)
39. L. Vanasupa, Y.C. Joo, P.R. Besser, and S. Pramanick, J. Appl. Phys., 85 (5), 2583 (1999)
40. B.L. Sharma, Diffusion in semiconductors, pp. 87–126, Trans. Tech. Pub., Germany, (1970)
41. B.I. Boltaks, Diffusion in semiconductors, D. Shaw (ed.), Atomic diffusion in semiconductors, Academic Press, New York, (1973)
42. J.C. Tsai, Diffusion, In VLSI Technology, (ed.) S.M. Sze, McGraw Hill, New York, Chapter 5, (1983)
43. M. Nathan et al., Appl. Phys. Lett., 77, 3355 (2000) and also D. Gupta and P.S. Ho, Some fundamental aspect of diffusion in diffusion phenomena in thin films and microelectronic materials, Noyes Pub., Park Ridge, (1988)

44. J.H. Choy and K.L. Kavanagh, Appl. Phys. Lett., 84 (25), 5201 (2004)
45. A.P. Sutton and V. Viter, Phil. Trans. Roy. Soc. London, A-301, 1–68 (1983)
46. A.P. Sutton and R.W. Ballufi, Interfaces in crystalline Mats., Clarendon Press, London, (1995)
47. A.L. Kolesnikova, I.A. Ovid'ko, and A.B. Reizs, J. Mater. Proc. Manuf. Sci., 7, 5 (1999) and also L. Kuipers, M.S. Hoogeman, and J.W.M. Frenken, Phys. Rev. Lett., 71, 3517 (1993)
48. R.A. Masumura and I.A. Ovid'ko, Enhanced diffusion near amorphous grain boundaries in nano crystalline and polycrystalline solids, Mater. Phys. Mech., 1, 31–38 (2000)
49. N.F. Mott and E.A. Davis, Electronic processes in non-crystalline materials, Clarendon Press, Oxford, London, (1979)
50. U.F. Mayer and G. Simonett, On diffusion-induced grain boundary motion, Contemporary mathematics Am. Math. Soc., 238, 231–240 (1999) and A. Gouldstone et al., Acta Mater., 55, 4015 (2007)
51. J.W. Chan, P.C. Fife, and O. Penrose, A phase field model for diffusion induced grain boundary motion, Acta Mater., 45.(10), 4397–4413 (1997)
52. R. Gastel, E. Somfai, S.B. Albada, W. Saarloos, and J.W. Frenken, Nothing moves a surface: Vacancy mediated surface diffusion, Phys. Rev. Lett., 86 (8), 1562 (2001)
53. L. Hansen, P. Stoltze, K.W. Jacobsen, and J.K. Norskov, Phys. Rev., B-44, 6523 (1991) and A. Gouldstone, K.J. Vanviet, and S. Suresh, Nature, 411, 656 (2001)
54. T.Flores, S. Junghans, and M. Wuttig, Atomic mechanism for the diffusion of Mn atoms incorporated in the Cu(100) surface: an STM study, Surf. Sci., 371, 1 (1997)
55. J.W. Mayer and S.S. Lau, Electronic materials science, Chapter 7, p. 208, Mcmillan Pub., New York, (1990) and also A.E. Kaloyeros, X. Chen, S. Lane, H.L. Frisch, and B. Arkles, JMR, 15 (12), 2800 (2000) and J. Boreland et al., Int. Workshop on Jn. Transistors, S4–7, 69 (2007)
56. K. Holloway, P.M. Freyer, and C. Cabral, Jr., J. Appl. Phys., 71, 5433 (1992) and C. Nieh et al., IEEE EDL, 27 (12), 969 (2006)
57. K. Nagas, J.B. Neaton, and N.W. Asheroft, Phys. Rev., B68, 125403 (2003) and also S.M. Sze, Physics of Semiconductor Devices, 2nd ed., Wiley, NY (1981) and B.G. wills and D.V. Long, Oxidation mechanism of ionic transport of copper in SiO_2 dielectric, Thin Solid Films, 467, 284 (2004) and O. Muscato, Physica A, 365 (2), 409 (2006)
58. F. Lanckmans, B, Brijs, and K. Maex, The role of H in Cu^+ drift diffusion in plasma deposited a-SiC: H, J. Phys. Cond. Matter., 14, 3565 (2002) and also A.E. Kaloyeros, X. Chen, S. Lane, H.L. Frisch, and B. Arkles, JMR, 15 (12), 2800 (2000)
59. Y. Li, J. Hunter, and T.J. Tate, SIMS study of Cu trapping and migration in low-K dielectric film, Appl. Surf. Sci., 231, 791 (2004)
60. W. Li. Sung and Bi-S. Chiou, J. Electron. Mater., 31 (5), 472 (2002)
61. K. Nago, J.B. Neaton, and N.W. Ashcroft, Phys. Rev. B, 68, 125403 (2003)
62. K.N. Tu, In Thin film interdiffusion &reactions, J.M. Potate, K.N. Tu, and J.W. Mayer (eds.), pp. 360–403, Wiley, New York, (1978), and also E.T. Ogawa, K.D. Lee, V.A. Blaschke, and P.S. Ho, Electromigration reliability issues in dual damascene Cu interconnections, IEEE Trans. Reliab., 51 (4), 403 (2002)
63. R. Hubner et al., Void formation in the Cu layer during thermal treatment of $SiN_x/Cu/Ta_{73}Si_{27}/SiO_2/Si$ systems, Cryst. Res. Technol., 40 (1/2), 135 (2005)
64. K.N. Tu, J. Appl. Phys., 94, 5451 (2003)
65. H.B. Nie et al., Appl. Phys., A-73, 229 (2002)
66. S.P. Hau-Riege et al., Mat. Res. Symp. Proc., 612, D2.2.1 (2000) and also A.A. Istratov and E.R. Weber, Electrical properties and recombination activity of copper, nickel and cobalt in silicon, Appl. Phys., A66, 123 (1998)
67. J. Gambino et al., Effect of CoWP cap thickness on via yield and reliability for Cu-interconnect with CoWP-only cap process, IITC, IEEE, (2005) and also A.E. Kaloyeros et al., J. Mater. Res. Soc., 15 (12), 2800 (2000)
68. A.A. Volinsky, J.B. Vella, and W.W. Gerberich, Thin Solid Films, 429 (1/2), 201 (2002)

69. D. Cramer et al., Acta Mater., 47, 333 (1999)
70. Y. Liu and A.H.W. Ngan, Scripta Mater., 44, 237 (2001)
71. C.S. Smith, J. Inst. Metals, 40, 359–371 (1928)
72. S. Arrhenius and A. Westgren, Z. Physik. Chem., B-14, 66–79 (1931)
73. K. Schubert and G. Brandauer, Z. Metalkunde, 43, 267–268 (1952) and also M. Seibt, H. Hedeman, A.A. Istratov, F. Riedel, A. Sattler, and W. Schroter, Structural and electrical properties of metal silicide precipitates in silicon, Phys. Status Solidi A 171, 301 (1999)
74. P.M. Hansen, Constitution of binary alloys, 2nd ed., p. 632, McGraw Hill, New York, (1958)
75. A. Loke, C. Ryu, C.P. Yue, J.S. Cho, and S.S. Wong, Kinetics of copper drift in PECVD dielectrics, IEEE Trans. Electron. Device Lett., 17 (2), D549–551 (1996) and also S. Kim et al., Electrochem. Solid State Lett., 10 (6), 193 (2007)
76. R.S. Muller and T.I. Kamins, Device electronics for integrated circuits, pp.11–14, Wiley, New York, (1977)
77. K.H.J. Buschow, R. Cahn, M. Flemings, B.I. Ilschner, E.J. Kramer, and S. Mahajan, Copper alloying, Encyclopedia of materials: Science and technology, Elsevier Science Ltd., Oxford, UK (2000)
78. M. Hansen, Constitution of binary alloys, p. 2000, McGraw Hill, New York, (1958)
79. F. Pawlek and K. Reichel, The effect of impurities on the electrical conductivity of copper, Z. Metallk, 47, 347–356 (1956) and also R. Hoelzl, K.J. Range, and L. Fabry, Modeling of copper gettering in p-and n-type silicon and in polysilicon, Appl. Phys., A75, 525 (2002)
80. A.G. Milnes, Deep level impurities in semiconductors, Wiley, New York, (1973)
81. R.N. Hall and J.H. Racette, Diffusion and solubility of copper in extrinsic and intrinsic germanium, Silicon and gallium arsenide, J. Appl. Phys., 35 (3), 379–385 (1964)
82. A.L.P. Rotondaro et al., J. Electrochem. Soc., 143 (9), 3014–3018 (1996)
83. K. Takeda, K. Hinode, I. Oodake, N. Oohashi, and H. Yamaguchi, Enhanced dielectric breakdown life time of the copper silicon nitride/ silicon dioxide structures, IEEE RPS, Proc. p. 36, (1998)
84. M. Vogt, M. Kachel, M. Plotner, and K. Drescher, Dielectric barriers for Cu-metallization systems, Microelectron. Eng., 37/38, 181 (1997)
85. S.K. Rha, S.Y. Lee, W.J. Lee, Y.S. Hwang, C.O. Park, D.W. Kim, Y.S. Lee, and C.N. Whang, J. Vac. Sci. Technol., B16, 2019 (1998) and also A.A. Istratov and E.R. Weber, Physics of copper in silicon, J. Electrochem. Soc., 149 (1), G-21–G-30 (2002)
86. C.P. Flynn, Point defects and diffusion, Clarendon Press, Oxford, London, (1972)
87. H.F. Wolf, Semiconductors, Chapter 2, p. 147, Wiley, New York, (1971)
88. I. Kaur, Y. Mishin, and G. Ggust, Fundamentals of grain and interphase boundary diffusion, Wiley, Chichester, (1995)
89. M.R. Sorensen, Y. Mishin, and A.F. Voter, Diffusion mechanism in Cu-grain boundaries, Phys. Rev. B, 62 (6), 3658–3673 (2000)
90. K.N. Tu, Recent advances on electromigration in very large scale integration of interconnects, J. Appl. Phys., 94 (9), 5451 (2003)
91. J.D. Wiley, J.H. Perepezko, J.E. Nordman, and K.J. Guo, IEEE Trans. Ind. Electron., 29, 154 (1982) and A. Horsfall, A. Curthbertsen, S. Bull, and A.O. Neill, Microelectron. Eng., 84 (11), 2486 (2007)
92. A.L. Greer, In Diffusion phenomena in thin films, (ed.) D. Gupta and H.S. Ho, Noyes Pub. Park Ridge, N.J., (1987)
93. M.A. Nicolet and M. Bartur, J. Vac. Sci. Technol., 19, 786 (1981)
94. G.S. Chen and S.C. Huang, Intrinsic properties and barrier behaviors of thin films of sputtered deposited single layered and alternate layered tantalum nitrides, J. Electrochem. Soc., 148 (8), G-424 (2001)
95. A.A. Istratov and E.R Weber, In Copper interconnects, New contact metallurgies /structures, and low-K dielectrics, G.S. Mathad et al. (eds.), p. 90, The Electrochem. Soc. Pub. Pennington, NJ, (2001)

96. J. Stake and W. Brenig (eds.), Amorphous and liquid semiconductors, Vol. 1 and 2, Taylor and Francis, London, (1974) and H.Y. Wong, N.F.M. Shukor, and N. Amin, Microelectron. J., 38 (6/7), 777 (June 2007)
97. J.F. Wagner and C.W. Wilmsen, The deposited insulator/III-V semiconductor interface, In C.W. Wilmsen (ed.), Physics and Chemistry of II-V semiconductor interface, Plenum press, New York, (1985)
98. L.J. Brillson, Advances in understanding metal semiconductor interfaces by surface science technique, J. Phys. Chem. Solids, 44, 703 (1983)
99. S.P. Kowalcxyk, J.R. Waldrop and R.W. Grant, Interfacial chemical reactivity of metal contacts with thin native oxides of GaAs, J. Vac. Sci. Technol., 19, 611 (1981)
100. National Center for Manufacturing Sciences (NCMS), The lead free solder project, NCMS Rept. 0401 RE 96 (1997)
101. M.J. Rost, D.A. Quist, and J.W.M. Frenken, Phys. Rev. Lett., 91 (2), 026101-1 (2003)
102. S. Balakumar et al., Effect of stress on the properties of Cu-lines in Cu-interconnects, J. Electrochem. Soc., 7 (4), G-68 (2004)
103. J.D. Torre et al., Microstructure of thin films sputtered into inclined substrates, experiments and atomistic simulation, J. Appl. Phys., 94, 263 (2003)
104. A.G. Dirks and H.J. Leamy, Thin Solid Films, 47, 219 (1997)
105. D.E. Hanson, J.D. Kress, A.F. Voter, and X.Y. Liu, Phys. Rev. B, 60 (11), 723 (1999)
106. J. Hopwood (ed.), Ionized Physical Vapor Deposition, Academic Press, San Diego, CA, (2000) and Y Yang, M. Wong, and M. Kushner, IEEE IITC, Burlingame, CA, (April 2008)
107. D.R.F. West and N. Saunders, Ternary phase diagrams in material science, 3rd. ed., Maney Pub., London, (2002)
108. S.P. Murarka and D. Fraser, J. Appl. Phys., 51, 1593 (1980)
109. C.A. Chang, J. Electrochem. Soc., 127, 1331 (1980)
110. C.A. Chang, Appl. Phys. Letts., 38, 860 (1981)
111. C.A. Chang, Reduced Cu-Cr mixing and reduced Pt-Cu interdiffusion by oxygen in Cu/Cr and Pt/Cu/Cr thin films, J. Appl. Phys., 53 (10), 7092 (1982)
112. J.W. Mayer and S.S. Lau (eds.), Electronic materials for integrated circuits, p. 333, Mcmillan Pub., New York, (1990)
113. M.A. Nicolet, I. Suni, and M. Finetti, Solid State Technol., 26, 129 (Dec. 1983) and also F.W. Saris et al., Proc. MRS., 54, 81 (1986)
114. P.M. Hansen, Constitution of binary alloys, McGraw Hill, New York, (1958)
115. F.J. Humphreys, Y. Haung, I. Brough, and C. Harris, J. Micrelectron. Mater., 195, 212 (1999)
116. T. Muppidi, D.P. Field, J.E. Sanchez, and C. Woo, Barrier layer geometry and alloying effects on the microstructure and texture of electroplated copper thin films and damascene films, Thin Solid Films, 471, 63 (2005)
117. Z.C. Wu et al., J. Electrochem. Soc., 146, 11 (1999)
118. K.M. Crossby and R.M. Bradley, Phys. Rev., E-59, R-2542 (1999)
119. C.V. Thompson et al., J. Appl. Phys., 67, 4099 (1990)
120. J.A. Floro et al., MRS. Bull., 27, 19 (2002) and also J.W. Christian, The theory of transformations in metals and alloys, Part 1, 2nd ed. Pargamon Press, Oxford, London, (1975)
121. C.V. Thompson, Annu. Rev. Mater. Sci., 20, 245 (1990)
122. W.W. Mulins, J. App. Phys., 28, 333 (1957)
123. W.W. Mulins, Acta Metall., 6, 414 (1958)
124. R. Dannenberg et al., Thin Solid Films, 370, 54 (2000)
125. R.D. Doherty et al., Mater. Sci. Eng., A 238, 219 (1997)
126. M.J. Rost, D.A. Quist, and J.W.M. Frenken, Phys. Rev. Letts., 91 (2), 02601-1 (2003)
127. Q.T. Jiang et al., Analysis of copper grains in damascene trenches after rapid thermal processing, J. Electron. Mater., 31 (2), (2002)
128. J.F. Whitacre, Z.V. Rek, J.C. Bilello, and S.M. Yalisove, Surface roughness and in plane texturing in sputtered films, J. Appl. Phys., 84 (3), 1346 (1998)

129. L. Liu, Y. Wang, and H. Gong, J. Appl. Phys., 90, 416 (2001) and J.H. An and P.J. Ferreira, Appl. Phys. Lett., 80, 151919 (2006)
130. M. Stavrev, D. Fisher, F. Praessler, C. Wenzel, and K. Drescher, J. Vac. Sci. Technol., A 17, 993 (1999)
131. F.A. Baicchi, N. Lifshitz, T. Sheng, and S.P. Murarka, J. Appl. Phys., 64, 6490 (1988)
132. A.A. Istratov and E.R. Weber, Appl. Phys. A, 66, 123 (1998), and (iii) degradation of the lifetime W.B. Henley, D.A. Ramappa, and L. Jastreszbski, Detection of copper contamination in silicon by surface photovoltage diffusion length measurements, Appl. Phys. Letts., 74, 278–280 (1999)
133. T. Heiser, A.A. Istratov, C. Flink, and E.R. Weber, Electrical characterization of copper related defect reactions in silicon, Mater. Sci. Eng., B58, 149 (1999)
134. A.A. Istratov, C. Flink, H. Hieslmair, E.R. Weber, and T. Heiser, Intrinsic diffusion coefficient of copper in silicon, Phys. Rev. Lett., 81, 1243 (1998)
135. L. Arnaud et al., Microelectron. Reliab., 40, 77 (2000)
136. S. Ragojevic et al., Interactions between silica xerogel and tantalum, J. Vac. Sci. Technol., B 19 (2), 354 (2001) and G.X. Cao and X. Chen, Phys. Rev. B, 73 (15), 155435 (2006)
137. M.D. Thouless, J. Vac. Sci. Technol., A 9 (4), 2570 (1991) and A. Gouldstone et al., Acta Mater., 55, 4015 (2007)
138. J.M.E. Harper and K.P. Rodbell, J. Vac. Sci. Technol., B15, 763 (1997)
139. M.A. Nicolet, Thin Solid Films, 52, 415 (1978) and S. Donovan and M.J. Moll, IEEE workshop on Microelectron. Dev. Boise, Idaho, (2006)
140. L.P. Buchwalter and J. Adhes, Sci. Technol., 9, 97 (1995)
141. D.S. Campbell, Mechanical properties of thin films, In L.I. Maissel and R. Glang (eds.), Handbook of thin film technology, pp. 12–16, McGraw Hill, New York, (1983)
142. B.Y. Tsui, J. Yang, and T.K. Ku, Micoelctron. Reliab., 41, 1889 (1999) and A. Sakata et al., IEEE IITC, (2006)
143. W.F. Wu et al., Electrochem. Solid State Lett., 2, 342 (1999)
144. W.F. Wu, K.L. Ou, C.P. Chou, and J.J. Hsu, PECVD-Ti/TiNx barrier with multilayered amorphous structure and high thermal stability for copper metallization, Electrochem. Solid state Lett., 6 (2), G27–G29 (2003)
145. P. Gallais, J.J. Hantzpergue, and J.C. Remy, Thin Solid Films, 165, 227 (1988) and also H. Kim et al., Material consideration on Ta, Mo, Ru, and Os as glue layer for ultra large scale integration Cu interconnects, Jap. J. Appl. Phys., 45 (4A), 2497 (2006)
146. S.P. Murarka, Silicides for VLSI applications, Academic, New York, (1983) and J. gambino, IEEE IITC, Proc., p. 22, (2007)
147. P.T. Liu, T.C. Chang, S.T. Yan, C.H. Li, and S.M. Sze, Electrical transport phenomena in aromatic hydrocarbon polymers, J. Electrochem. Soc., 150 (2), F-7 (2003) and C-C Huang, J-L Huang, Y.L Wang, and J.J. Chang, J. Vac. Sci. Technol., B, 24 (6), 2621 (2006)
148. A.S. Loke et al., IEEE Trans. Electron Dev., 46, 2178 (1999) and A. Sakata, 9th Int. Workshop on Stress induced phenomena in metallization, 4th April, Kyoto, Japan, (2007)
149. K.P. Yap et al., Integrity of copper tantalum nitride metallization under different ambient conditions, J. Electrochem. Soc., 147 (6), 2312–2318 (2000)
150. T. Laurila et al., TaC as a diffusion barrier between Si and Cu, J. Appl. Phys., 91, 5391 (2002)
151. G.B. Alers et al., J. Appl. Phys. Letts., 73 (11), 1517 (1998)
152. M. Yamaguchi, Thermal nitridation of InP, Jpn. J. Appl. Phys., 19, L401 (1980)
153. G.E. Miller, Tantalum and Niobium, Academic press, London, (1959)
154. R. Pretorius, J. Harris, and M.A. Nicolett, Solid State Electron., 21, 667 (1978)
155. R. Hubner et al., Void formation in the Cu-layer during thermal treatment of SiNx/Cu/$Ta_{73}N_{27}$/SiO_2/Si systems, Cryst. Res. Technol., 40 (1/2), 135 (2005)
156. D.Y. Kim, Ph D. Thesis on study on reliability of VLSI interconnection structures, (Dec. 2003) and also S.R. Gess, H. Donohue, K. Buchanan, N. Rimmer, and P. Rich II,

Microelectron. Eng., 64, 307 (2002) and H. Ono, T. Nakano and T. Ohta II, Appl. Phys. Lett., 64, 1511 (1994)

157. W.C. Johnson, P.W. Voorhees, and D.F. Zupon, The effect of elastic pressure on the kinetics Oswald ripening: Two particle problem, Met. Trans. A (Physical Met. And Met. Sci.), 20A (7), 1175 (1989)
158. L. Pauling, The nature of chemical bond, 3rd ed. Cornel Univ. Pub., New York, (1960)
159. C.A. Chang, Reduced Cu-Cr mixing and reduced Pt-Cu interdiffusion by oxygen in Cu/Cr and Pt/Cu/Cr thin films, J. Appl. Phys., 53 (10), 7092 (1982) and also F. Lanckmans, B. Brijis, and K. Maex, The role of H in Cu^{+} drift diffusion in plasma deposited a-SiC:H, J. Phys. Cond. Mater., 14, 3665 (2002)
160. H.B. Nie et al., Structural and electrical properties of TaN thin films fabricated by using RF magnetron sputtering, J. App. Phys., A-73, 229 (2001) and B. Predel, Phase equilibria, Crystallographic, and thermodynamic data, electronic Mat. and Semicond., O. Madelung (ed.), Springer Verlag, AG, Germany (2006)
161. M. Takeyma, A. Noya, T. Sasse, and A. Ohta, J. Vac. Sci. Technol., B14, 674 (1996) and also A.E. Kaloyeros, X. Chen, S. Lane, and H.L. Frische, Tantalum diffusion barrier grown by inorganic plasma promoted chemical vapor deposition, J. Mater. Res., 15 (12), 2800 (2000) and M. Stavrev et al., J. Vac. Sci. Technol., A17 (3), 993 (1999)
162. 9th Europhysical Conf. On Defects in insulating materials, Wroclan, Poland, (June30–July5, 2002) and J. Gambine et al., IEEE Proc. On IITC, p. 22, (2007)
163. E. Weiser et al., Thin Solid Films, 410, 121 (2002) and T. Murata et al., Jap. J. Appl. Phys., 47, 2488 (2008)
164. D. Edelstein et al., Int. Interconnect. Tech. Conf., San Fransisco, CA, (June 4–6, 2001) and also L.I. Maissel and R. Glang, Hand book of thin film technology, Chapter 2, 4, 5 and 19, McGraw Hill, New York, (1983)
165. D. Fischer et al., Surf. Interface Anal., 25, 522 (1997) and H. Sakai et al., Adv. Metal. Conf., Univ. of Tokyo, Japan, (Sept. 26, 2006)
166. Y.S. Diamond, (Tutorial), On advanced matellization conference (AMC) San Diego, CA, (Oct. 3–5, 2000)
167. G. Beyer et al., Development of sub-100 nm ALD barrier for Cu-low-K interconnects, Microelectron. Eng., 64, 233 (2000)
168. C.W. Chang, B.C. Regan, W. Mickelson, R.O. Ritche and A. Zettl, Probing structural phase transitions of crystalline C60 resistivity measurements of metal film over layers, Solid State Commun., 128, 359 (2003) and S.O. Gurrum, Ph.D. Thesis, G.W. Woodruff, School of mechanical engineering, Gerogia Tech., (May 2006)
169. C. Wen et al., App. Phys. Lett., 61, 2162 (1992) and also K. Ramkrishna, M. Gall, P. Justison, and H. Kawasaki, Prediction of maximum allowed rms currents for electromigration design guidelines, Proc 7th Int. Workshop on stress induced phenomena in metallization, Vol. 741, pp. 156–164, Melville, New York, AIP, (2004)
170. M.M. Yovanovich, Theory and applications of construction and spreading resistance concepts for microelectronic thermal management, cooling techniques for computers, W. Aung (ed.), p. 277, Hemisphere Pub., Boulder, CO (1991)
171. M.M. Yovanovich and V.W. Antonetti, Application of thermal contact resistance theory to electronic packages, Advance thermal modeling of electronic components and systems, Vol. 1, A. Barcohen and A. Kraus (eds.) , p. 79, Hemisphere Pub., Boulder, CO (1988)
172. J. Bonevich, D. Van Heerden, and D. Rossell, J. Mater. Res., 14, 1977 (1999)
173. D. Josell, A. Cezairliyan, and J.E. Bonevich, Int. J. Thermophys., 19, 525 (1998) and also B. Clemens, G.L. Eesly, and C.A. Paddock, Phys. Rev., B-37, 1085 (1988)
174. J.Bonevich, D. Van Heerden, and D. Rosell, J. Mater. Res., 14, 1977 (1999)
175. T.K. Gupta, Hand book of thick and thin film hybrid microelectronics, Wiley, NJ, (2003)
176. M. Wittmer, J. Vac. Sci. Technol., A2 (2), 273 (Aprl./June 1983)
177. J.M. Potate, K.N. Tu, J.W. Mayer (eds.), Thin Film—inter diffusion and reactions, Wiley, New York, (1978)

178. L. Loth, Transition metal carbide and nitrides, Academic Press, New York, (1971) and also H.L. Goldschmidt, Interstitial alloys, Plenum Press, New York, (1967)
179. P. Schwartzkopf and R. Kieffer, Refractory hard metals, MacMillan, New York, (1967)
180. M.A. Nicolett, Solid State Technol., 26 (2), 129 (Dec. 1983) and also T. Laurila, Tantalum-based diffusion barriers for copper metallization, Ph.D. Thesis dissertation at Helsinki University of Tech., Helsinki, Finland, (Dec. 2001)
181. S. Braun, H. Mai, M. Ross, R. Scholz, and A., Leson, Mo/Si multilayers with different barrier layers for applications as extreme ultra violet mirrors, Jpn. J. Appl. Phys., 41 (6B), 4074 (2002)
182. H. Kattelus et al., Microelectronics Eng., 60, 97 (2002)
183. M.A. Nicolet and P.H. Giauque, Microelectron. Eng., 55, 357 (2001)
184. S. Ganguly, L. Chen, T. Levine, B. Zeng, and M. Chang, Development of tungsten nitride film as a barrier layer for copper metallization, J. Vac. Sci. Technol., B 18, 237 (2000)
185. J.S. Becker and R. Gordon, Diffusion barrier properties of tungsten nitride films grown by atomic layer deposition (ALD) from bis (tert-butylimido) bis (dimethylamido) tungsten and ammonia, J. Appl. Phys. Lett., 82 (4), 2239 (2003)
186. J.D. Torre et al., Microstructure of thin Ta-film sputtered onto inclined substrate, J. Appl. Phys., 94 (1), 263 (July 2003)
187. A.G. Dirks and H.J. Lemy, Thin Solid Films, 47, 219 (1997)
188. K. Radhakrishanan, Ng. Geok, R. Gopalkrishnan, Mat. Sci. Eng., B-57, 224 (1999)
189. J.C. Lin, G. Len, and C.Lee, J. Electrochem. Soc., 146, 1835 (1999)
190. M.H. Tai, S.C. Sun, H.T. Chiu, C.E. Tsai, and S.H. Chung, Appl. Phys. Lett., 67, 1128 (1995)
191. K. Baba, H. Hatada, K. Udoh, K. Yasuda, Nucl. Inst. Method. Phys. Res., B127/28, 841 (1997)
192. F. Weiser et al., Improvement of Ta-based thin film barrier for Cu-metallization by ion implantation of nitrogen and oxygen, Adv. Met. Conf. (AMC), Orlando, FL, (Sept. 28–30 1999)
193. J.J. Boland and J.H. Weaver, A surface view of etching, Phys. Today, 51, 34 (1998) and T. Oku, E. Kwakami, M. Uekebo, K. Takahiro, S. Yamaguchi, and M. Murakami, Diffusion barrier property of TaN between Si and Cu, Appl. Surf. Sci., 99, 265 (1995)
194. H.O. Pierson, Hand book of refractory carbides and nitrides, Properties characteristics, processing and applications, Noyace, NJ., (1996) and N-H. Kim, S-Y. Kim, W-S. Lee, and E. Chang, Microelectron. Eng., 84 (11), 2663 (2007)
195. K. Holloway and P.M. Fryer, Appl. Phys. Lett., 57, 1736 (1990)
196. K.H. Min, K.C. Chun, and K.B. Kim, J. Vac. Sci. Technol., B14, 3263 (1996), and also J.C. Lin and C. Lee, Electrochem. Soild State Lett., 2, 181 (1999)
197. J.O. Olowilfe, J. Li, and J.W. Mayer, Appl. Phys. Lett., 58, 469 (1991)
198. T. Laurila et al., Tantalum carbide and nitride diffusion barriers for Cu metallization, Microelectron. Eng., 60, 71 (2001)
199. M. Hansen, Constitution of binary alloys, McGraw Hill, New York, (1958)
200. J. Chen and J.L. Wang, Diffusion barrier properties of sputtered TiB_2 between Copper and Silicon, J. Electrochem. Soc., 147 (5), 1940–1944 (2000)
201. H.Y. Tsai, S.C. Shun, and S.J. Wang, Characterization of sputtered tantalum carbide barrier layer for copper metallization, J. Electrochem. Soc., 147 (7), 2766–2772 (2000) and A. Sakata et al., IEEE Proc. On IITC, (2006)
202. J. Lu, H. Hsu, Q.H. Dixit, J. Luttmer, R. Havemann, and L. Magel, A novel process for fabricating conformal and stable TiN-based barrier films, J. Electrochem. Soc., 143, L279 (1996) and also F.A. Cotton and G. Wilkinson, Advanced Inorganic chemistry, p. 807, Wiley, New York, (1972)
203. D.S. Yoon, J.S. Roh, S.M. Lee, and H.K. Baik, Acta Matter. 51 (9), 2531–2538 (2003)
204. A. Westerheim et al., Integration of CVD titanium nitride for 0.25 μm contacts and vias, J. Vac. Sci. Technol., B-16 (5), 2729 (1998)

205. B. Weiller and S. Adamson, Effect of dimethylamine on CVD of TiN from Tetrakis (dimethylamido), J. Electrochem. Soc., 144, L40 (1997)
206. K.E. Elers et al., Diffusion barrier deposition on a Cu-surface by ALD, Wiley Intersci., NJ (Jan 2002)
207. K. Tao, D. Mao, and J.P. Hopwood, J. Appl. Phys., 91 (7), 4040 (2002)
208. S.J. Wang, H.Y. Tsai, and S.C. Sun, Characterization of sputtered titanium carbide as diffusion barrier for copper metallization, J. Electrochem. Soc., 148 (8), C563 (2001)
209. B. Borovski, M. Krueger, E. Ganz, Phys. Rev., B59, 1598 (1999) and T. Zhong, V. Dubin, and M. Fang, US Patent, 7416980, (Aug. 2008)
210. B.S. Swartzentruber, Phys. Rev. Lett., 76, 459 (1996)
211. D.T. Shaw, DOE 2003 Wire Development Workshop, St. Petersburg, FL, E.K. Broadbent, Tungsten and other refractory metals for VLSI applications, (ed.) R.S. Blewer, Materials Res. Soc. Pub. Pittsburgh, PA, (1986)
212. G. Beyer and M. Bavel, Using atomic layer deposition to prepare future generation copper diffusion barrier, Micro. 20 (9), 51–58 (2002)
213. Z. Li, R.G. Gordon, D.B. Farmer, Y. Lin, and J. Vlassak, Nucleation and adhesion of ALD copper on cobalt adhesion layers and tungsten nitride diffusion barriers, Electrochem. Solid-State lett., 8 (7), G182 (2005)
214. J.R. Creighton and J.W. Rogers, Jr., Tungsten and other refractory metals for VLSI applications III, p. 63, Proc. Mater. Res. Soc., V.A. Wells (ed.), MRS Pub., Warrendale, PA (1988), and also K.N. Tu, Surface and interfacial energies of CoSi, and Si films, IBM. J. Res. Dev., 34 (6), 868 (1990)
215. M. Kleinschmit, M. Yeadon, and J.M. Gibson, Appl. Phys. Lett., 75 (21), 3288 (1999)
216. S. Kal, I. Kasko, and H. Ryssell, J. Electron. Mater., 24 (10), 1349 (Oct. 1995)
217. N. Petrov, Y. Sverdlov, and Y.S. Diamond, Electrochemical study of the electroless of Co(P) and Co(W,P) alloys, J. Electrochem. Soc., 149 (4), C-187 (2002)
218. A. Khon, M. Eizenberg, Y. Diamond, B. Israel, and Y. Sverdlov, J. Micrelectron. Eng., 55, 297 (2001)
219. C.K. Hu et al., Reduced Cu-interface diffusion by CoWP surface coating, Microelectron. Eng. 70 (2/4), 506–511 (Nov. 2003) and J. Gambino et al., IPFA, p. 59, (2007)
220. M.J. Kobrinski, C.V. Thompson, and M.E. Gross, J. Appl. Phys., 89 (1), 91 (2001) and G.C. Schwartz et al. (eds.), Hand book of Semicond. Interconn., p. 326, CRC Press, Boca Raton, FL, (2006)
221. V.G. Weizer and N.S. Fatemi, Metal-silicon reaction rates—the effects of capping, J. Electron. Mater., 18 (1), 7 (1989)
222. N. Petrov, Y. Sverdlov, and Y.S. Diamond, Electrochemical study of electroless deposition of Co (P) and Co (WP) alloys, J. Electrochem. Soc., 149 (4), C187 (2002) and K. Chattopadhya et al., IEEE Int. Reliab. Phys. Symp., p. 128, (2006)

Chapter 4
Pattern Generation

4.1 Photolithography

4.1.1 Introduction

The desire for small feature size continues to make tremendous gains in device performance but with added circuit complexity. Fine pattern generation on the deposited dielectric layer is the usual practice in damascene (single/dual) architecture. The main driver behind this is advanced *photolithography (or lithography which uses a high polymer, called a resist, and a proper radiative exposure with proper optical systems)*. In an effort to keep up the trend of reduction in feature size, semiconductor industries have switched over from conventional ultraviolet (UV) to deep ultraviolet (DUV) lithography. Figure 4.1 shows a picture of the trenches produced by using a DUV resist with resolution factor k_1 as small as 0.25 (the 2004 ITRS requirements for technology node is *hp90* and is expected to be *hp65* by the year 2007, where hp 90 means 90 nm node technology). The images show (a) 80 nm lines with 120 nm spacing and (b) 90 nm lines with 135 nm spacing.

Photolithography is a prime example of how the International Technology Roadmap for Semiconductors (ITRS) drills down to its equipment and materials underpinnings. The growth of the industry has been the result of improved lithographic resolution and overlay across increasingly larger fields. Tool technology has progressed to successively shorter wavelengths and high *numerical aperture* (NA) lenses. NA has been a factor in extending optical lithography and its various iterations for some time. However, the approach of high NA and shorter wavelength has increased the resolution capability of the exposure tools at the cost of reduced the intrinsic depth of focus (DOF) of the aerial image. Figure 4.2 shows the future requirements for lithography including exposure tools. The legend indicates the approximate time that will be required for the development and qualification/pre-production.

According to the International Technology Roadmap for semiconductors (ITRS) linear device features have decreased at the rate of 70% every three years, and *photolithography* has evolved to make smaller and smaller devices (Fig. 4.3).

Tapan K. Gupta, *Copper Interconnect Technology*,
DOI 10.1007/978-1-4419-0076-0_4, © Springer Science+Business Media, LLC 2009

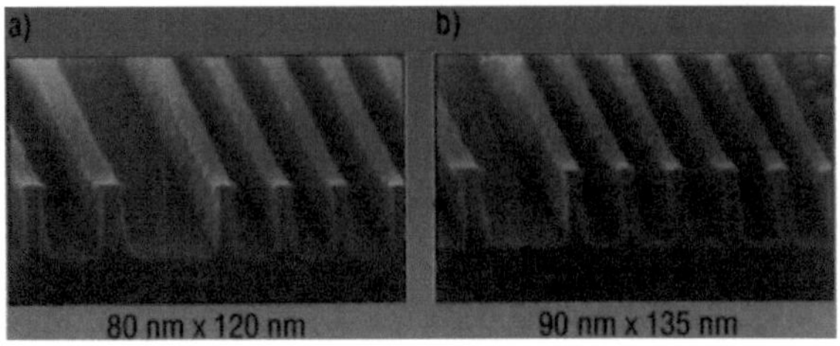

Fig. 4.1 A picture of the trenches produced by photolithography (**a**) 80 nm lines and 120 nm spaces, (**b**) 90 nm lines and 135 nm spaces (Reprinted with permission, *Solid State Technology*, Nov. 1998 p. 62)

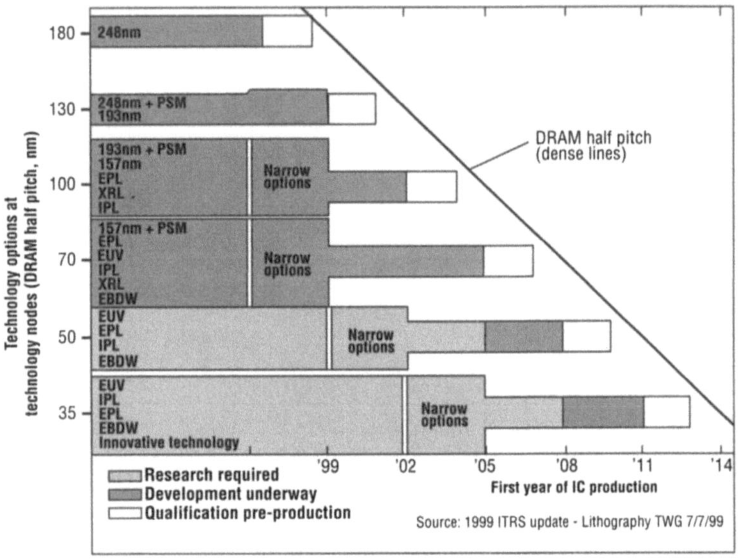

Fig. 4.2 The detail future photolithography requirements and exposure tool potential solutions (Reprinted with permission ITRS update, 1999, SEMATECH, Austin, TX)

Figure 4.3 shows one of the 58 millions tiny transistors fabricated on a silicon on insulator (SOI) substrate 118 mm square, following 130 nm node technology. Patterned interconnecting Cu-lines 400 meters long and 1/800th the width of a human hair have been used to connect the sub-micron transistors (Fig. 4.4) [1–2]. Advanced *photolithography* has demonstrated that printing and *pattern* transfer below 10 nm is possible as predicted by the simulation results [3–5].

The photolithography starts with a circuit pattern etched in chrome metal on glass [6], and illumination of the patterned photomask in the optical stepper. An image of a circuit pattern is captured on the silicon wafer as a latent image by photosensitive polymer, known as *photoresist (resist)*.

In the early 1990s, there was significant concern that the industry could not make the transition from i-g line lithography to 248 nm deep ultraviolet (DUV) lithography, because of too many challenges to overcome. Now DUV lithography has matured and has been very successful in high volume manufacturing [7]. As a matter

Fig. 4.3 Scanning electron microphotograph of one of the 58 million transistors fabricated on a single chip following 130 nm-node technology. The interconnections are made out of copper (Photo courtesy, IBM, Research, G-5 processor)

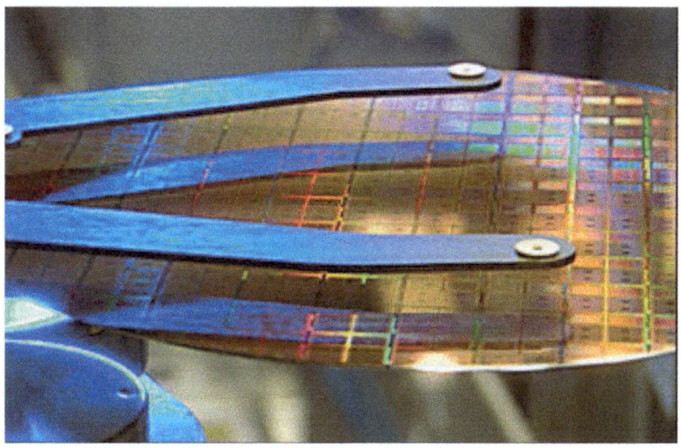

Fig. 4.4 Patterned copper wires less than 1/800th the width of a strand of human hair (Reprinted with permission, IBM research)

of fact, starting with i-line (365 nm) optical lithography has progressed steadily to DUV regions (248, 193, and 157 nm) and finally has landed int the extreme ultraviolet lithography (EUVL) region. Now DUV methodology and the exposure tools with numerical aperture (NA) >0.80 are readily available. Lithography equipment suppliers are ready to deliver production quantities of 0.75 NA for 248 nm and 0.68 NA for 193 nm exposure tools to support 90 nm node technology on 300 mm wafers. Figure 4.5 shows predictions about the evolution of lithography from 1996 to 2002.

It is expected that 65 nm node technology will require 157 or 193 nm lithography with *alternating phase shift mask* (Alt. PSM) [8]. Alternately, *optical proximity* correction can extend KrF (248 nm) technology down to 180 nm [9] and ArF laser and associated technology to 100 nm. However, there are many challenges for 157 nm lithography, like the development of large diameter high purity CaF_2 crystals for optics, the development of pellicles with high transparency to protect masks, and large scale manufacturing capability of resist materials [7]. Because the resist materials that are used in 193 nm system are highly absorptive, they cannot be used for 157 nm exposure. However, some resist materials have been reported recently that have sufficient transparency and are expected to work with 157 nm [10].

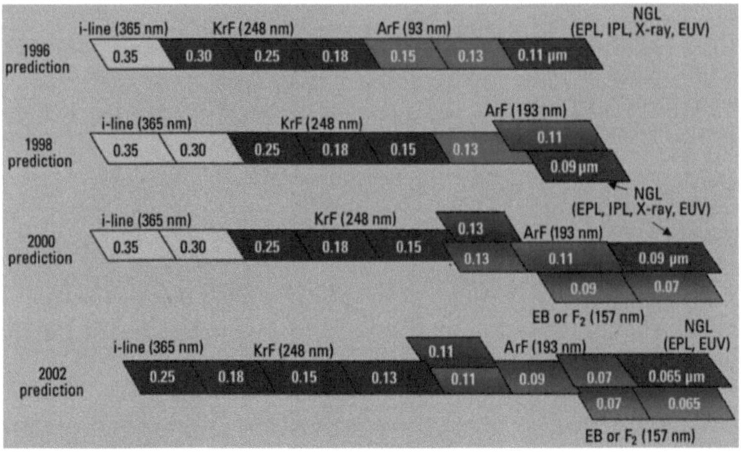

Fig. 4.5 Predictions about the evolution of lithography technology from 1996 to 2002 (Reprinted with permission, *Semiconductor International,* Feb. 2003, p. 58)

Nevertheless, the transition to 157 nm wavelength is by no means straightforward. In general, material development is the most critical aspect of 157 nm lithography, and it relies on the simultaneous development of exposure and *resist material.* For the purpose of sharp pattern development, experience suggests that at least 40% of incident light should make it to the bottom of the resist layer. The first polymer to demonstrate such transparency was Teflon AF [11].

Table 4.1 shows how the miniaturization of the feature sizes has changed the exposure wavelengths and the chemistry of the resist materials. We will witness all these in the following paragraphs.

Table 4.1 Chronology of lithography

Year	Line width (nm)	Wavelength
1990–1991	500	365
1994–1995	350	365/248
1997–1998	250	248
1999–2000	180	248
2001–2002	130	248
2002–2003	90	193
2003–2005	65/45	193/157
2005–2010	45/32	193/157

4.1.2 Resolution Limits of Optical Lithography

The minimum feature size that can be printed with optical lithography is determined by Rayleigh's equation:

$$W_{\min} = \{(k_1\lambda)/(\mathrm{NA}) = \{(k_1\lambda)/(n\ \sin\alpha) = \{(k_1\lambda)\}/\{(0.61\lambda)/(2\mathrm{b})\} \qquad (4.1)$$

where k_1 is the resolution factor and is dependent on the optical system, λ is the wavelength of the exposure, NA is the numerical aperture, n is the refractive index of the medium, α is the angle of acceptance, and $2b$ the separation between two images. k_1 is a complex factor of several variables such as resist material, resolution, enhancement techniques, off-axis illumination, and optical proximity correction.

Figure 4.6 shows how the values of k_1 have changed with time (year). With time, the exposure wavelengths, the chemistry and the processing technology of the resists have changed. However, the practical lower limit of k_1 is still higher than 0.25.

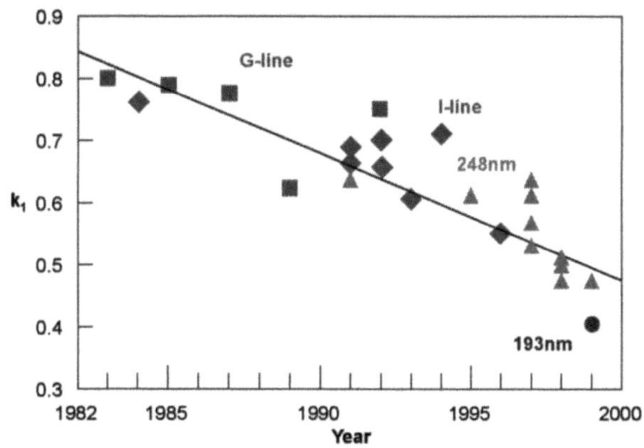

Fig. 4.6 The trends of k_1 values with the development of the lithography system (Reprinted with permission, Scotten W. Jones IC Technology Report, IC Knowledge, 2006)

4.1.2.1 Effect of the Media on Resolution

The numerical aperture (NA) in Equation (4.1) is equal to (n sinα), where n is the refractive index of the medium, and α is the angle of acceptance. Theoretically, if the value of n is changed from air ($n = 1$) to a medium of higher value, it is possible to make the minimum feature size smaller. Recently, interest has grown dramatically in exploring the use of a high index liquid immersion medium between the exit face of the lens and the photoresist surface to further extend optical lithography. The process is known as *immersion lithography*.

Thus, following the above theory, the effective wavelength (λ_{eff}) can be affected by λ_0 / n (λ_0 is the wavelength of the exposure light, and n is the refractive index of the medium) through the use of deionized (DI) water ($n \approx 1.44$) as the immersion medium. Theoretical calculations show that it will potentially extend 193 nm lithography by 44%, twice the resolution enhancement offered by 157 nm lithography. However, transparency is a key requirement for immersion fluids. It has been found that many saturated hydrocarbons such as linear alkaline groups show transparency, required viscosity, and promising results with 193 nm exposure. Therefore, it is hoped that 193 nm-*immersion technology* will be the choice in nano-lithography for several generations [12]. Figure 4.7 shows an experimental set up for 193 nm-immersion lithography with water as the immersion liquid.

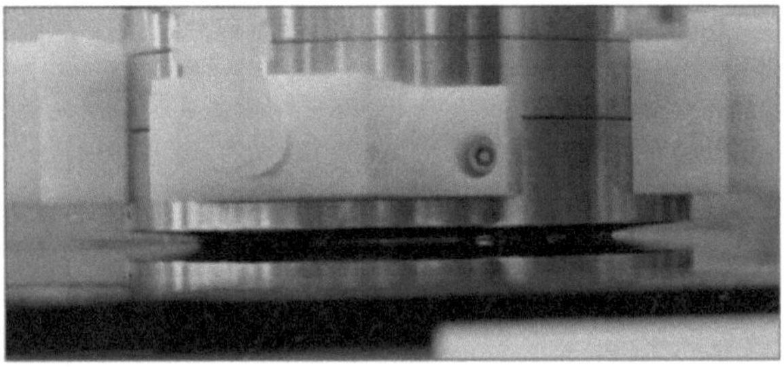

Fig. 4.7 Experimental set up of an immersion technology (Courtesy of Tokyo Electron Limited)

4.1.2.2 Effect of Wavelength Exposure on Resolution

The shorter the wavelength (λ) of the exposure used, the smaller the circuit features that can be copied. Most of the commercial production exposures rely on excimer lasers that will cover 248 nm (KrF), 193 nm (ArF) and 157 nm (F_2) wavelengths [13]. The use of commercial F_2 excimer lasers including suitable *masks*, transparent pellicles, and reticles is still under development [14–15].

4.1.2.3 Effect of Numerical Aperture (NA) on Resolution

Besides the wavelength (λ), the resolution of the exposure light will also depend on the degree of coherence, and the numerical aperture (NA). The NA, on the other hand, is related to the depth of focus as $DOF = \lambda/NA^2$. It is derived from the usable diameter of the objective lens, the distance from the reticle to the objective lens, and the refractive index of the media. Now, manufacturers have the capability of making NA greater than 1, – the so-called hyper-NA. It is true that building a larger usable objective lens can increase NA, but it will require exceptionally high quality lens material, which has proven to be a technological hurdle. It is also true that increased NA leads to increase unregulated and undesirable polarization effects throughout the optical path of an exposure system (Fig. 4.8). Therefore, an appropriate compromise between high resolution, NA and depth of focus must be established in designing *aligners*.

4.1.2.4 Modulation Transfer Function and Critical Dimension (CD)

Photolithography patterns are imaged from a mask plate, which is illuminated by an external light source with its own optical characteristics. Therefore, the performance of photolithography equipment can be measured by using criterion other than the *Rayleigh limit*, which is known as the *modulation transfer function* (*MTF*). The

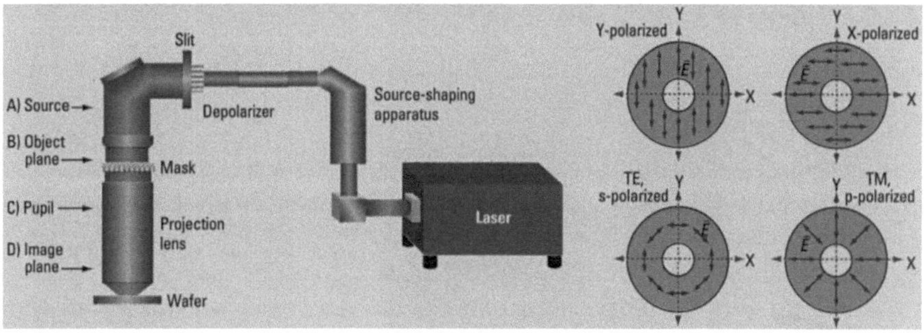

Fig. 4.8 Potential influences in the optical path when hyper-NA (>1) is used in the Illumination path (Reproduced with permission *Semiconductor International*, Feb. 2006, p. 32)

MTF depends on (i) NA, (ii) λ, (iii) the mask feature size, and (iv) the degree of *spatial coherency* of the exposure light. Dimensions are most often measured on line-type patterns, and the dimensional control is frequently referred to as the line width control or better known as the critical dimension (CD). If CD *pushing* is a common part of a process, masks should be ordered with bright field dimensions at their minimum expected values. Too much CD *pushing* often results to an irreproducible process [16].

When analyzing imaging requirements for future lithographic tools, the relation between critical dimension (CD), depth of focus (DOF), and the numerical aperture will be helpful. Figure 4.9 shows the relation between lens NA versus DOF and CD.

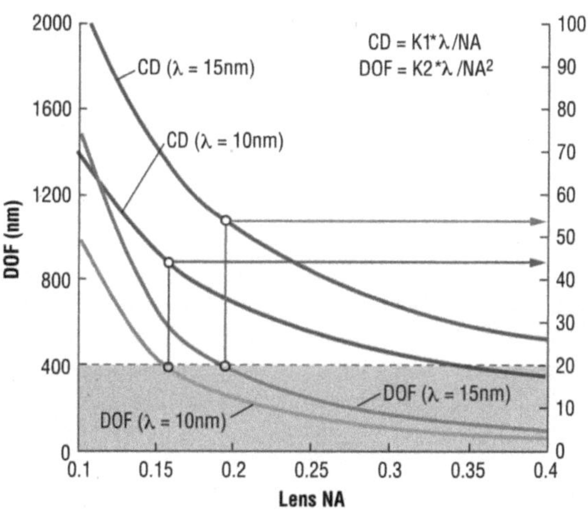

Fig. 4.9 Relation between lens NA and DOF and CD (Reprinted with permission, *Solid State Technology*, Feb. 2002)

4.1.2.5 Effects of Antireflective Coating (ARC) on Resolution

Photoresist line width variations are difficult to control in the vicinity of steps because of: (a) differences in energies due to resist thickness; (b) scattering of light; and (c) standing wave effects. A variety of procedures has been suggested including use of *high γ (contrast) resist*, dyes to make the resist more optically absorptive, and *anti-reflecting coating* (ARC). Experimental observations show that the formulation of un-bleachable dyes is simpler than ARCs but require increased exposure time (i.e., reduced throughput).

ARCs are polymer films (e.g. an anthracene compound with stabilized chromophore) that are highly absorbing and non-bleaching at the exposure wavelength. Recently, a spin on SiO based back anti-reflecting coating (BARC) layer, referred to as a sacrificial light absorbing material (SLAM), has shown promising results for via first patterning, as a result of minimized standing wave effects and reflections from the substrate surface (Fig. 4.10). In addition to that, the ARC layer can be utilized as a planarization layer. However, the price paid for all these benefits is process complexity and possible loss of dimension control [13].

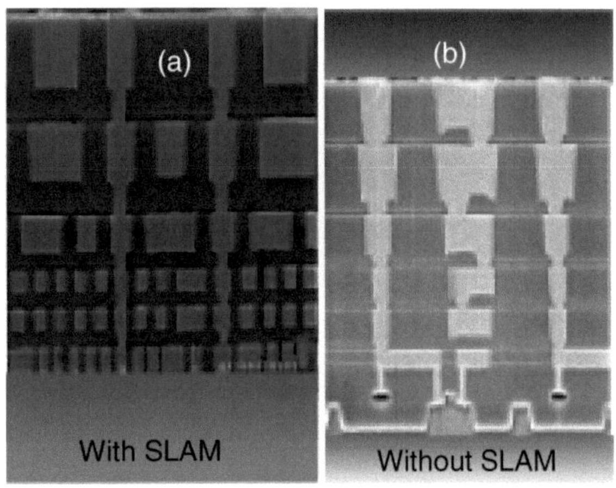

Fig. 4.10 Scanning electron micrograph comparing the trench structures (**a**) with a sacrificial light absorbing material and (**b**) without a sacrificial light absorbing material (Photo courtesy Honeywell, CA)

4.1.3 Deep Ultraviolet (DUV) Lithography

The journey through micro-lithography, which started with 365 nm exposure and with o-quinonediazides formulation of the resist materials to resolve 0.5 μm feature, came to a halt in the early 1990 s because of its inability to resolve lines below 250 nm. Experimental observations showed that one of the methods of achieving an improvement in resolution is through the use of high-energy shorter wavelength radiation exposure [17].

DUV lithography, which needs shorter exposure wavelength than UV and mid-UV lithography, should be supported by proper resist material and an optical hardware system. Optical hardware limitations in the deep-UV region revolve around lens design and power increase of the radiation source. As a result, the industry introduced tools with lenses >0.80 NA and powerful excimer lasers [18]. However, there is a large number of issues concerned with the use of 193 nm lasers in production steppers that need to be addressed.

DUV resist materials formulated on various chemistries are mostly poly-methylmethacrylate (PMMA), methacrylates with large alicylic group, poly (p-hydroxystyrene), and certain fluoropolymers and polysiloxanes [19–20]. The spectral absorption of methacrylates shows absorption below 260 nm with a peak between 225 and 220 nm. The absorption is due to the n–π^* transition of the carbonyl group, with maximum absorption coefficient between 0.27 and 0.47 μm^{-1} at 215 nm. Several methacrylate based DUV resists have been formulated but notable among these are copolymers of methyl methacrylate and indenone (3%) and those require an exposure of only 60 mJ/cm^2 at 248 nm. However, PMMA and some of its copolymers exhibit poor resistance to the dry etching process. There has been a negative resist for DUV application too, which is based on cyclized poly-isoperene and several bisazides with a sensitizer, which is not absorbed in the visible region.

The modern DUV-resist contains multiple components including a polymeric binder [21]. For example, the 248 nm resist system is based on a chemically amplified photoacid-catalyzed process [22], and it contains a para-hydrostyrene binder, and a high etch resistance material with aromatic rings. Some of the hydroxyls in the binder are protected as acid-labile tert-butyl carbonates [23]. However, the presence of aromatic groups in the binder greatly reduces their radiation sensitivity at shorter wavelengths (below 254 nm) [24]. As a result, new formulation of resist materials for 193 nm photolithography is based on norbornene/maleic anhydride copolymers or poly(meth)acrylates having a polycylic (e.g. adamantyl) group as their side chains [25].

Resist out-gassing (because of vapor pressure, boiling point, etc.) during exposure is a problem because it degrades lens transmission. However, experimental evidence shows that the use of photo-decomposable aminosulfonate moieties (capable of photogenerating free aminosulfonic acid) can alleviate the problem.

4.1.3.1 Technology for 193 nm Lithography

The transition from 248 to 193 nm brought several issues in the formulation of resist chemistry. For example, the photoacid generator (PAG) present in 248 nm resist has to face higher energy photon (4.89–6.42 eV) at 193 nm. At the same time, the resist is highly reflective at 193 nm, and it requires a methacrylate based antireflective coating (ARC), because phenyl-containing ARC is not sufficient. Besides, the photo-speed of the resists to be used in 193 nm lithography has to be addressed properly.

The para (p)-hydroxystyrene (PHOST) based 248 nm resist is too opaque for 193 nm and it requires the development of resist materials and chemistries such as hydrophilicity, glass transition temperature (Tg), and etch resistance (presence of aromatic ring in high concentration). Besides, it requires the addition of some acid-cleavable functionality such as carboxylic acid to promote the solubility in tetramethyl ammonium hydroxide (TMAH) [26]. The recently formulated *PHOST* contains a hydroxyl group which is acidic, and is shown in Fig. 4.11 with the distribution of negative charges.

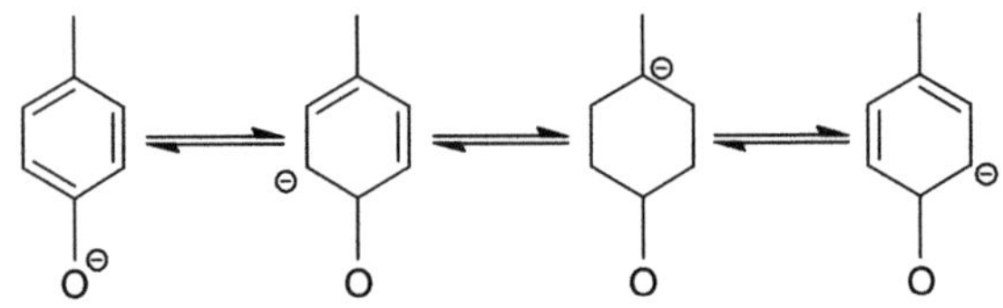

Fig. 4.11 The distribution of negative charge on the anion due to acidity of PHOST (Reprinted with permission, *Solid State Technology* (SST), March 2000)

The stabilization in *PHOST* is achieved by the distribution of the negative charges on several π electron systems of the adjacent aromatic ring of the anion (Fig. 4.12) [27].

Fig. 4.12 The negative charge on the carboxylate anion (Reprinted with permission, SST, March 2000)

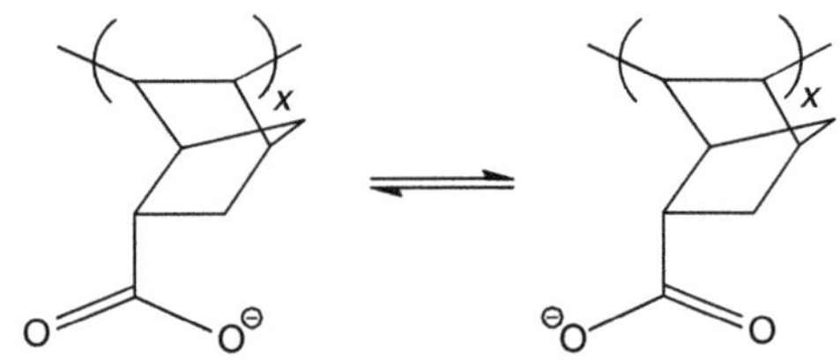

Recently, attention has been concentrated on the derivatives of acrylate and methacrylate copolymers with large alicylic groups such as menthyl, adamantyl, isobornyl, and tricyclodecyl.

Fig. 4.13 shows free radical alternating co-polymerization of maleic anhydride and norbornene.

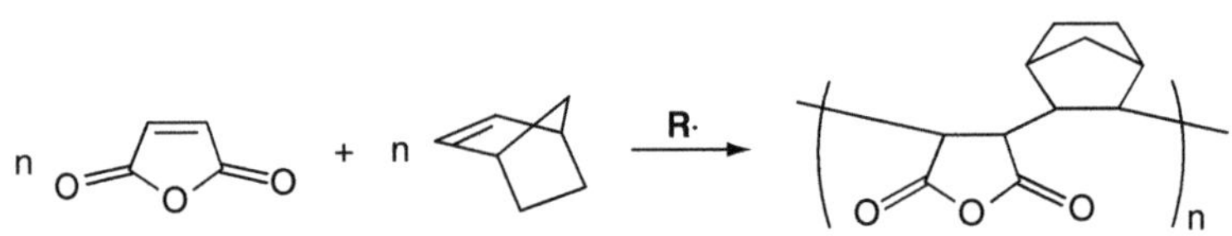

Fig. 4.13 Free radical alternating co-polymerization of maleic anhydride and norbornene (Reprinted with permission, SST, March 2000)

As cyclic olefin hydropolymerizes poorly via free radical methods, a different formulation of norborne has been developed. It contains maleic anhydride, acrylic acid, and t-butyl acrylate monomer polymerized by free radicals and it works well with 193 nm ArF (argon fluoride laser) exposure [28–29]. The incorporation of acrylic acid provides aqueous base solubility and tert-butyl acrylate is similar to poly (norbornene-alt-maleic) anhydride [30].

It has been found that incorporation of maleic anhydride, via free radical alternating co-polymerization with cycloaliphatic olefin monomers, dramatically improves cracking and adhesion problems [31]. Recently, a co-polymer with 2-wt% of triphenyl sulfonium trisulfate (TPSOTf) and propylene glycol ether-2 acetate has been used as a 193 nm resist system to create 100 nm structures [32]. However, synthesized cycloaliphatic co-and terpolymers using Pd (II) metal catalyzed polymerization are seen to work well with 193 nm exposure.

The performance criteria for 193 nm resists are continually being improved by reducing line edge roughness, improving *etch resistance*, and preventing pattern collapse due to capillary effects together with the mechanical rigidity of the resist. However, some of the resist materials used for 193 nm show shrinkage in dimension during processing with low-K. The probable cause might be due to their lower glass transition temperature (T_g) and higher thermal expansion coefficient.

193 nm resists are mostly used in the manufacture of dynamic random access memories (DRAMs), or high volume microprocessors. Today, 193 nm resist is matured and is capable of printing 70 nm isolated lines. Though out-gassing is a problem with most of these DUV resists, it is considered to be a manageable problem [33].

4.1.3.2 193 nm Exposure System

Argon fluoride lasers (ArF) provide greater depth of focus because of their shorter wavelength. The pulse energy that is used as a radiation source for 193 nm wavelength is 100 mJ. However, increasing the pulse rate to the multi-kHz level the efficiency of the exposure can be increased with a number of risks [34].

Stepper

The ultimate advantage of *steppers* over scanner-type aligners is higher image resolution. The projection system in steppers uses *refractive* instead of *reflective* optics to project the mask image on to the wafer.

Figure 4.14 shows a schematic of (a) a stepper and (b) a scanner. The use of an all-refractive lens in a stepper requires a line-narrowed laser source for chromatic correction. Commercial steppers use an argon fluoride laser (at 193 nm) as a radiation source. The line narrowing system utilizes gratings instead of using transmissive etalons to avoid optical damage of the bulk silica and etalon coatings. In order to expand the beam size on the grating without reducing the line width per pass, one or more prism is used in the optical path. A small field, 193 nm micro-stepper schematic diagram is shown in Fig. 4.15. It has a 0.5 NA $\times$ 10 imaging lens, and a table to mount the wafer. The table can be adjusted through X, Y, Z axes and can be rotated through an angle θ.

Step-and-Scan System

Unlike the stepper, which instantly exposes a full image field, a scanner exposes through a small slit of the lens image field and travels across the entire field. The fundamental difference allows a scanner to average the lens wave-front aberration along

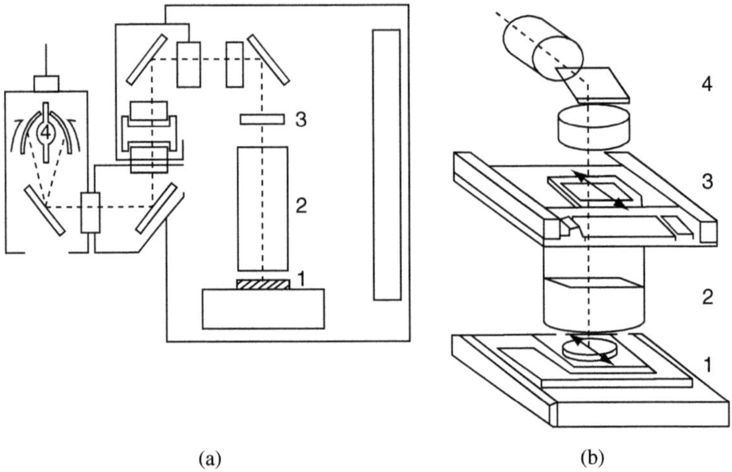

Fig. 4.14 The schematic of (**a**) stepper: (1) wafer, (2) projection lens, (3) reticle, and (4) light source, and (**b**) scanner: (1) wafer scanning stage, (2) projection lens slit, (3) reticle scanning stage, (4) illumination system

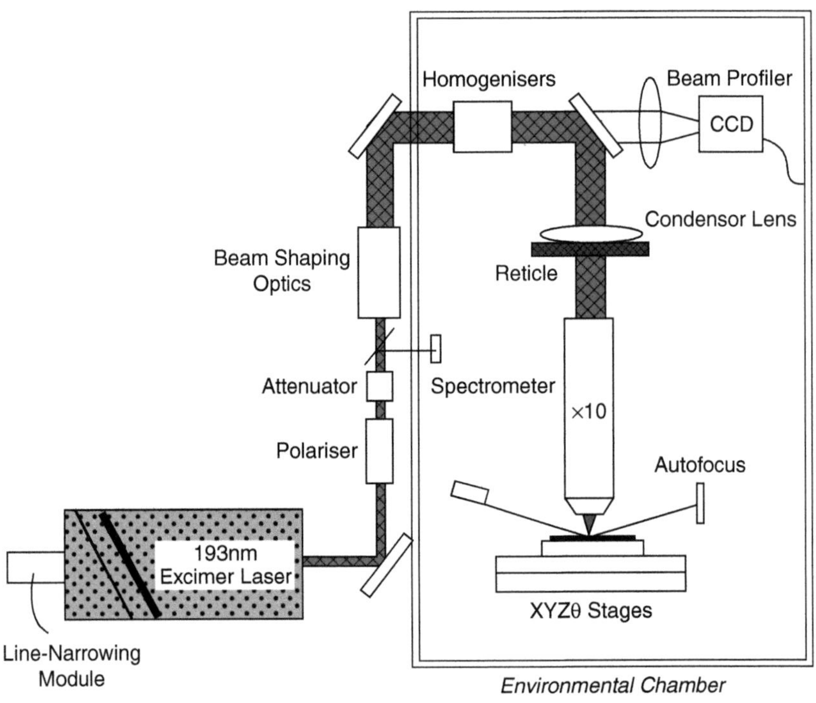

Fig. 4.15 A schematic diagram of a 193 nm micro-stepper (Courtesy, Exitech, Oxford, UK)

the scanning direction, resulting in better image resolution [35]. Another important observation noticed in scanner performance is its ability to control the synchronization between the reticle stage and wafer stage. Some commercial scanners available on the market show excellent synchronization performance at 250 nm/s scanning speed with standard deviation less than 15 nm.

The 193 nm scanner tools use a KrF (krypton fluoride) laser exposure instead of ArF exposure, though below 130 nm ArF provides greater depth of focus. Thus there is a choice between using a high NA, ArF stepper or a high NA KrF scanner. The greatest disadvantage of using ArF laser exposure is that the optical system requires an expensive CaF_2 material, and the resist material itself is more expensive than KrF resist and very sensitive to contamination.

Figure 4.16 shows the 193 nm step and scan lithography system, which is combined with a 0.75 numerical aperture (NA) lens. The instrument can provide improved leveling performance and leading edge overlay with an exposure capacity >90 wafers/h at a dose of 20 mJ/cm^2.

Fig. 4.16 The 193 nm step and scan lithography system (Photo courtesy ASML, Veldhoven, The Netherlands)

4.1.4 Reticles

Reticles for creating image patterns are produced at a greatly enlarged size, and images are directly created on wafers in step and repeat aligners. It is then reduced to the final stage by a photolithography reduction process to provide accurate pattern dimension. As the design rules dip below 0.18 μm, DUV scanners are used at all critical levels with overlay requirements less than 50 nm.

To optimize the matching of exposure tools, matching reticles with specific overlay patterns are used to calibrate each system, and algorithms have been developed that analyze overlay data. In concentric matching, only the common field area between the stepper and the scanner is available for lithography. To improve overall matching performance, both 4× and 5× features are placed on the same reticle and thus reticle to reticle error is eliminated (Fig. 4.17). Better throughput can be achieved with non-concentric matching; however, it requires a more complex evaluation.

As the exposure wavelengths get shorter the tools become complicated and the lithography becomes time consuming and expensive. One probable cost-effective

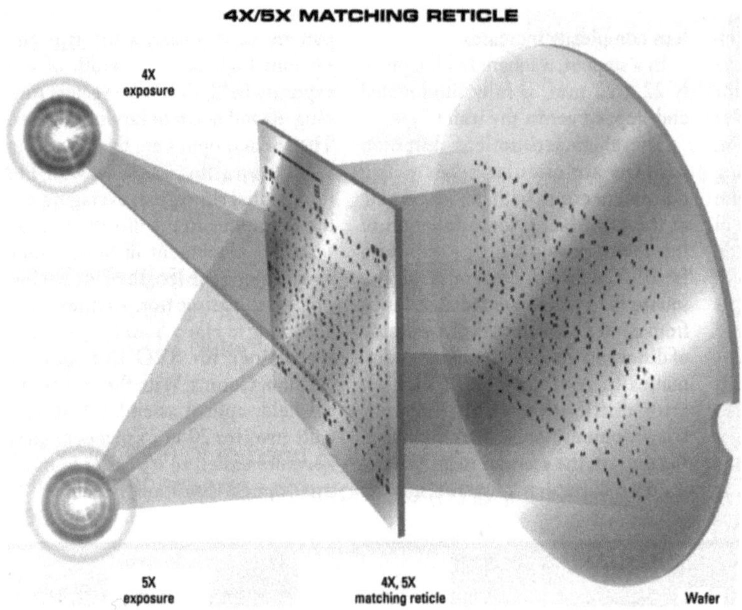

Fig. 4.17 A single matching reticle containing 4× and 5× features is used for non-concentric matching (Reprinted with permission, *Semiconductor International*, Feb. 2000, p. 68)

solution might be *mix and match lithography* though the science behind this technology is complex.

The old i-line (365 nm) steppers are typically 5× reduction tools with 22 × 22 mm field size, while the mid-UV (248 nm) and deep UV (193 nm) scanners are 4× tools with a typical field size of 26 × 33 mm. It seems that matching a scanner field, and the use of i-line tools, will be profitable in the long run, even though the costs of the tools, and of the consumables, are a lot higher than with 193 and 248 nm systems.

The use of lithography tools will depend on the nature of the device, whether it is logic intensive or more memory intensive (Table 4.2). Accordingly, the number of layers will vary from 25 to 30 and the processing will be more expensive.

Table 4.2 Mix-match technology

Logic devices				DRAM, MPU			
100 nm	i	KrF	ArF	rough	middle	critical	
70 nm	i	KrF	ArF EPL	i	KrF	ArF	100 nm
50 nm	i	KrF	ArF EPL	i	KrF	ArF F_2 EPL	70 nm
	Rough	middle	critical	KrF	ArF	F_2 EPL EUV	50 nm

EPL is E-beam projection lithography, EUV denotes extreme UV (Reprinted with permission, *Semconductor International*, Feb. 2003, p. 53)

However, for sub-100 nm node technology *mix and match technology* will be more viable [36].

4.1.5 Enhancement Techniques for Resolution

4.1.5.1 Antireflective Coating (ARC)

To achieve adequate depth of focus (DOF) at 193 nm, the resist thickness must be thinner than it is needed for 248 nm. As the resist layer becomes thin, the need for a reliable method to control reflectivity becomes mandatory. In response to these needs, thin film coatings known as antireflective coatings (ARCs) have been developed. Available ARCs include top-layer coating, (TARC) and bottom layer coating (BARC) materials that are based on organic or inorganic polymers. In general, inorganic BARC materials are CVD based and conformal in nature, but in order to facilitate dual damascene processing, a material that will offer all the advantages of an organic spin-on solution with the advantages of an inorganic BARC is greatly desired [37–39].

Figure 4.18 shows the benefit of optimized resist/ARC for 100 nm lines. The figure also illustrates that the change in ARC with optimized formulation can give higher contrast and finer profile control (as shown in the figure (1 and 2)).

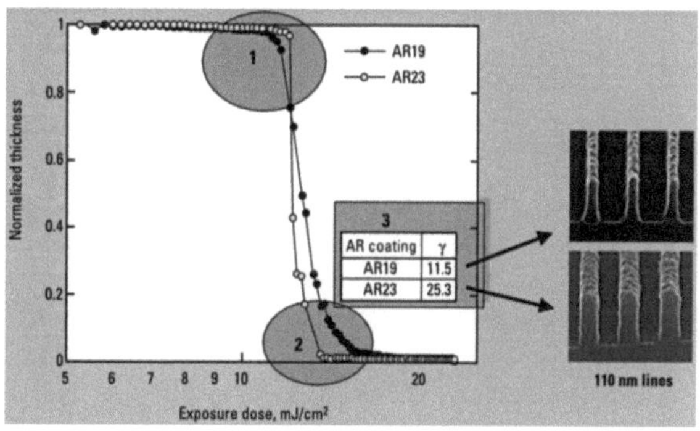

Fig. 4.18 The benefit of optimized resist/ARC for 100 nm lines (Reprinted with permission, *Semconductor International* Feb. 2002, p. 68)

The reflectivity of an ARC material over a substrate can be achieved via two mechanisms: interference and absorption. The interference method requires strict control of the film thickness, which is rather difficult to achieve with non-uniform surface topography. The absorption mechanism, on the other hand, relies on the dissipation of the exposure energy from bottom anti-reflective coating (BARC) to suppress reflection. The primary benefits of BARC are focus/exposure latitude improvement, enhanced critical dimension (CD) control, elimination of reflective notching, and protection of DUV resist from substrate poisoning. The extinction coefficient of BARC depends on the chromophore load, and the pH value that strongly influences the bulk film properties, its density, and the wet-etch rate. The etch process of BARC film is complex, and the implant energy is difficult to adjust. As a result BARC cannot be used for implant layers [40].

Figure 4.19 shows a comparison between dry and wet patterning processes using a back antireflecting coating (BARC). From Fig. 4.19 we can see that the difference between dry and wet patterning processes is that wet DUV BARC is soluble in aqueous based developer, while dry process requires reactive ion etching for opening a window. The BARC cure temperature controls the solubility of the aqueous developer of the wet process, which contains polyamic acid. Thus optimization of the bake latitude is important to attain best resolution. Wet-developable BARCs have been developed specifically for implant layer applications. However, the implementation of a wet-developable BARC process requires optimization of layer thickness, and fine-tuning of the baking temperature.

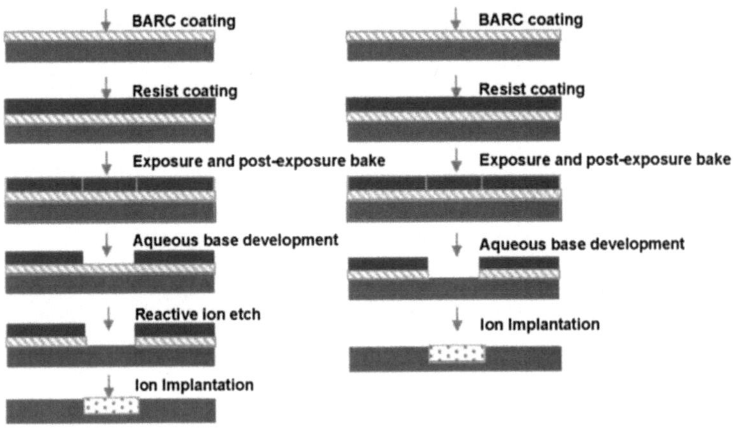

Fig. 4.19 Comparison between dry (**a**) and wet (**b**) patterning process using BARC (Reprinted with permission, Brewer Science, Inc. Rolla, MO)

4.1.5.2 Immersion Technique

Enhancement in 193 nm-ArF systems can be done either by an immersion technique or by phase shift mask (PSM) technology. The immersion technology is based on changing the medium between the lens and the wafer with a material having refractive index (RI) higher than air ($RI_{air} = 1.0$). However, there are number of practical issues involved in implementing immersion lithography that are still needed to be addressed. These are (a) resist immersion liquid (IL) interactions, (b) diffusion of IL, (c) change of resist chemistry, (d) leaching of resist components into the IL, (e) leaching of photo acid generators (PAGs), (f) base and other additives, and (g) photo- acids and solvent.

The immersion technology can be performed by submerging the whole chuck, wafer, and lens in a pool of liquid or limiting the pool size to the top of the chuck, or dispensing the liquid between the lens and the wafer with a nozzle.

Figure 4.20 illustrates the puddle approach where the wafer is recessed into the chuck [41]. It is true that immersion technology is not an easy method to implement, but given the alternatives (e.g. 157 nm with fluorine excimer lasers) with so many odds it might be worth having immersion technology mixed with a *phase shift mask (PSM).*

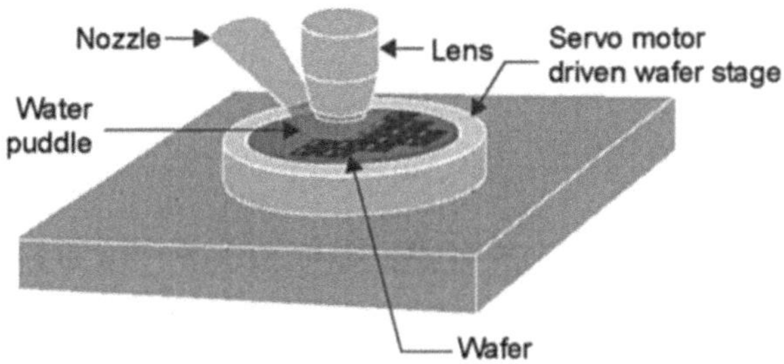

Fig. 4.20 Puddle type immersion lithography (Reprinted with permission, Scotten W. Jones, 2006, IC Technology Report, IC Knowledge, 2006)

4.1.5.3 Phase Shift Mask (PSM)

Today integrated circuits (ICs) are faster and denser, due to continuous reduction of feature sizes. The reduction of feature size has been possible because of the new formulation of resist materials, the exposure system, and the use of advanced optical enhancement strategies such as phase shift masks (PSMs) [42–43], and optical proximity correction [44].

The phase shift mask (PSM) has become a mainstream solution for memory and logic circuits. It enhances the patterned contrast at the edges of small features by destructive optical interference. With its high numerical aperture (NA) in combination with off-axis illumination (OAI) PSM technology can provide very low-resolution factor k_1. There are three types of PSM: (a) complimentary PSM (CPSM); (b) double exposure PSM (DEPSM); and (c) single exposure PSM (SEPSM). Both DEPSM and CPSM use double lenses. In CPSM the primary mask serves as a strong shifter and the second one serves as a trim function.

Phase shift masks can achieve 1:1100 nm lines with proper control of anti-reflecting coating (ARC), soft bake temperature, and airborne chemical contamination [45]. The most serious problem with PSM is evident by image placement error due to coma. However, modified high transmission ($\sim$100%) attenuated PSM (attPSM) has shown promising results with patterning of both poly-gate and contact mask [46], giving rise to k_1 and NA values of 0.2 and >0.7 respectively.

An alternating attenuated mask is fabricated with AlN/CrN multilayers based on optical superlattices [47–48] which are tunable to 248 nm and 193 nm wavelengths. Sputter deposited AlN/CrN mutilayers have π-shift and tunable optical transmission between 5 and 15% at 365, 248 and 193 nm. Figure 4.21 shows the principle of operation of an alternating phase shift mask. Use of a phase shifting mask enhances the fine features of the image by concentrating light diffracted by dense patterns into the most oblique components within the lens numerical aperture (NA).

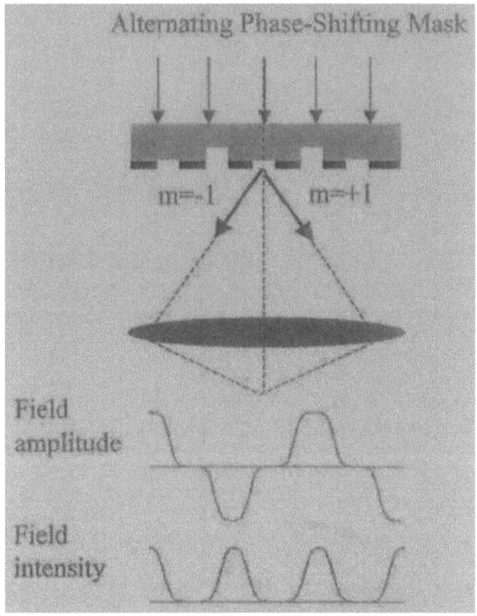

Fig. 4.21 Schematic of the operation principle of an alternating phase shift-mask operation

Figure 4.22 shows a picture of an alternating phase shift mask (ALTPSM), and Fig. 4.23 shows a scanning electron micrograph of a 70 nm gate patterned with 193 nm resist and alternating phase-shift mask.

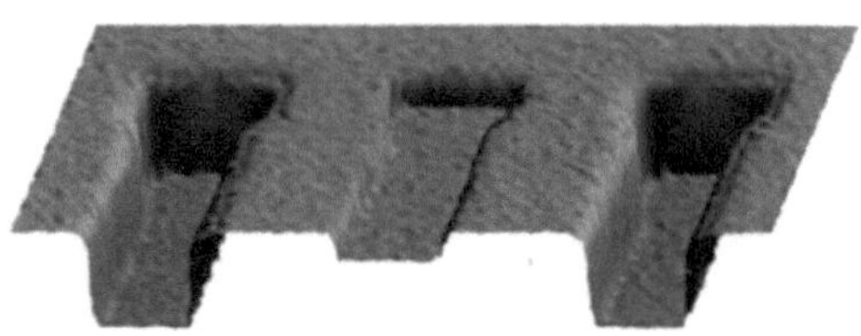

Fig. 4.22 An alternating phase shift mask (Alt-PSM) (Photo courtesy VEECO, Instruments Inc. New York)

There is no doubt that there is a growing need in photolithography to achieve better resolution by using sub-wavelength light sources [13]. But it is also true that they are inherently slow unless they are implemented with massively parallel rays of light sources. Nevertheless, contact mode exposure through rapid chromeless phase shifting masks [42], embedded amplitude masks [43], and elastomeric phase masks [43] provides powerful patterning capabilities that can be applied to large areas in a single step for better resolution.

The elastomeric phase mask uses a casting and curing of a prepolymer to elastomer polydimethylsiloxane (PDMS) against the master. Lithography and

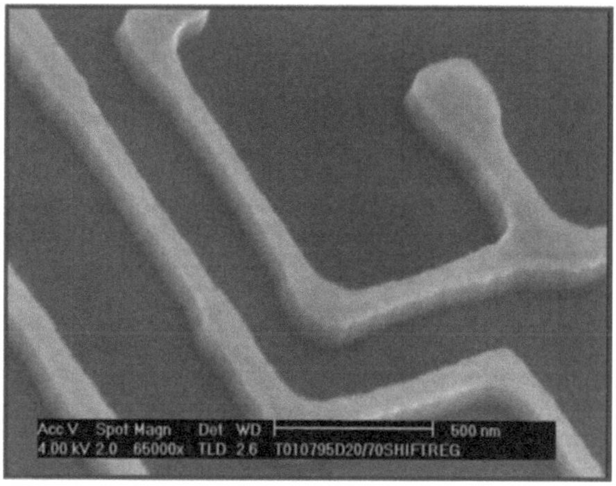

Fig. 4.23 70 nm gate patterned with 193 nm resist and alternating phase-shift mask (Courtesy of IMEC)

anisotropic etching of thermally grown oxide on a silicon-on-insulator (SOI) produce the master, consisting of parallel lines and spaces. Many masks can be made out of PDMS from a single master and can be used effectively as a relatively weak amplitude mask [45].

4.1.6 157 nm Lithography

The drive for small, fast, and low-power semiconductor devices has been pushing optical lithography to go beyond limits. As a result, 365 nm-exposure system has come down to 157 nm exposure. As a matter of fact, the approach to finer resolution has successfully centered on developing new lithography systems with higher NA, shorter wavelength exposure, and modified resist chemistry. At 100 nm node, 157 nm lithography needs a NA of 0.63 to produce k_1 greater than 0.4, whereas longer wavelengths will require a significantly higher NA to achieve a k_1 of 0.4. However, for 65 nm node the k_1 value will be less than 0.4 when 157 nm lithography is used [49]. Thus, in order to fulfill all the requirements of 157 nm lithography, we need to develop proper photoresist, system purging, pellicles, quality and quantity of CaF_2 for lens fabrication [50], attenuated PSM material, and blank mask material [51]. However, the transition to a new wavelength requires an expensive and lengthy development process, which must be considered when looking at the accelerated demands of the *ITRS Roadmap* (Fig. 4.24) [52].

4.1.6.1 Wafer Exposure System

Excimer lasers are considered excellent sources for DUV resist exposure because of the short wavelength and low-coherence of light. However, as the laser source is

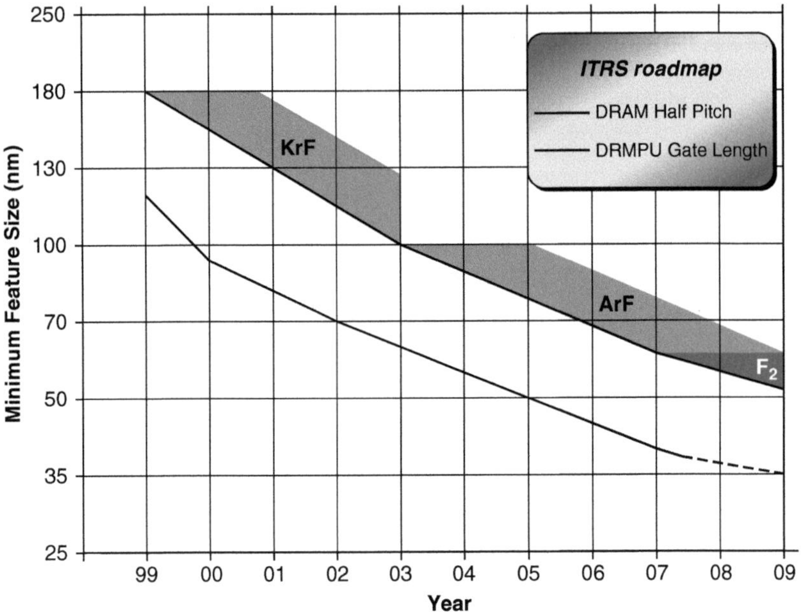

Fig. 4.24 ITRS Roadmap showing the requirements of the lithography system by the end of 2009 (Reprinted with permission ITRS, 1999, SEMATECH, Austin, TX)

moved deep into the UV, use of traditional lenses, pellicles, mask and photoresists is becoming problematic.

The wafer exposure system for 157 nm is still in its development stage though several microsteppers have been built with hyper-NA and a field size of ~2 mm. The optical system is catadioptric, which includes partial reflectors, and refractive elements [53]. Figure 4.25 shows schematic of hyper-NA lens systems based on catadioptric design.

A full wafer exposure system for 157 requires several critical components, such as: lasers with appropriate pulse repetition rate, bandwidth, and power, several supporting materials, e.g. transparent damage resistant pellicle [54], photo-mask, and resist material. The trifluoromethacrylate resist material, which has been developed for 157 nm exposure, has a very low absorbency and it needs a more *transparent resist material* [55–56] and an F_2 laser system (with a standard broadband 600-Hz and 10 W power).

4.1.6.2 Optical System for 157 nm Resist

Figure 4.26 illustrates the experimental set up for a 157 nm-exposure system with four CaF_2 lenses each measuring 2×4 cm^2 in cross-section and 8 cm in length and are mounted on a sample holder inside a vacuum chamber which can be purged

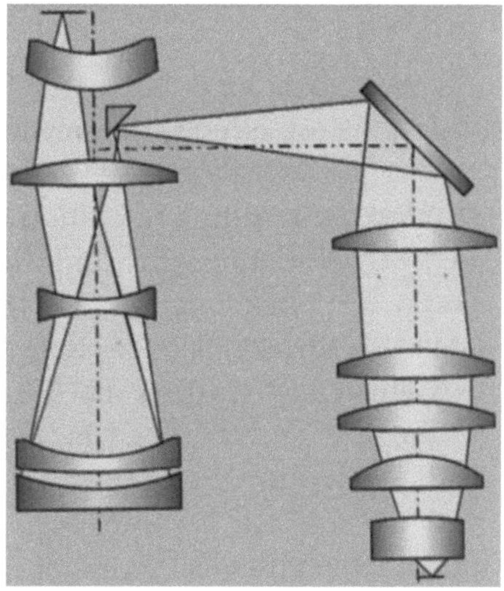

Fig. 4.25 Catadioptric optics with hyper-lens system (Reprinted with permission *Semiconductor International* Feb. 2005, p. 53, courtesy Nikon)

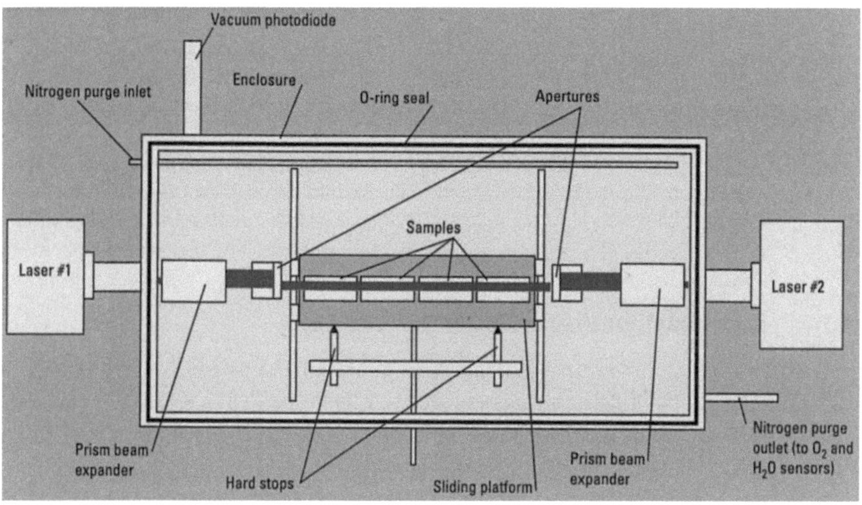

Fig. 4.26 An experimental set up for 157 nm-exposure system (Reprinted with permission, *Semiconductor International*. Feb. 2002, p. 57)

with N_2 before operation. Two 2 kHz F_2 lasers provide pulses that are timed to fire alternately at an interval of 250 μs.

A CaF_2 lens when combined with a second lens with *all-refractive materials* like barium fluoride (BaF_2), strontium fluoride (SrF_2), and lithium fluoride (LiF) can accommodate a larger 157 nm laser line-width. An all-refractive system uses 193 and 248 nm lithography and has the advantage that it utilizes the infrastructure already in place for the exposure tool [13].

Recently, a microstepper for a 157 nm resist system has become available with dual wavelength output (at 157.523 and 175.629 nm) that emits more than 10 W at 1 kHZ repetition rate from a F_2 laser. The beam delivery system of the F_2 laser has a CaF_2 lens combined with mirrors and a discretely variable attenuator. The projection lens is a 0.85 NA catadioptric objective chromatically corrected over the full emission spectrum of the F_2 laser.

Figure 4.27 shows the imaging of 80–100 nm lines/spaces with 1:3 aspect ratio fabricated with 157 nm-resist system. The results at the preliminary stages are encouraging, in spite of the problems with the CaF_2 lens system, like its intrinsic birefringence [57], which causes significant wave-front aberration. However, it is expected that the effects due to birefringence can be eliminated by using pairs of CaF_2 lenses cut along the $\langle 111 \rangle$ direction with transverse axes rotated by 60° [58–59].

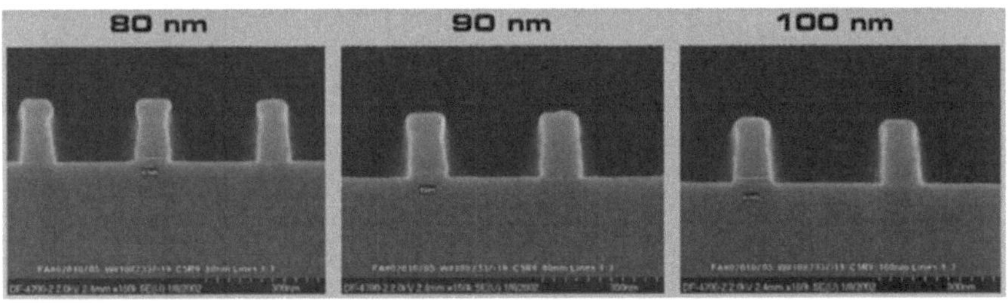

Fig. 4.27 The imaging of 80–100 nm lines/spaces with 1:3 aspect ratio, fabricated with a 157 nm resist system (Reprinted with permission, Semiconductor International, Feb. 2002, p. 70)

4.1.6.3 Reticle and Mask Systems for 157 nm Resist

The pattern tools utilized in VLSI fabrication are known as reticles and masks. In the case of a reticle a pattern image is stepped and repeated in order to expose the entire substrate, whereas in the case of a mask, the pattern is simply transferred from the mask to the substrate. Reticles are to undergo a sequence of dry and wet processing steps during deposition and patterning absorber layers.

The use of low-OH content fused silica incorporated with fluorine has shown better transmission (~82%) with 157 nm wavelength [60]. An ideal photo-mask for 157 nm lithography should show an attenuation coefficient of 0.02/cm, and a loss for each surface of 0.002, with base 10 [61].

The glass plates when coated with chromium oxide (Cr_2O_3) (150–200 μm), its nitride, or combination of both (80 nm thick), do not show promising result at 157 nm. However, experimental observations show that absorption in Cr, CrO, and CrN film stacks can be optimized by adjusting the components of the film [62–63]. But the most challenging task for 157 nm lithography is the phase shifting mask materials [64]. Some commercial photomasks for 157 nm lithography show some encouraging results with chromium film and poly-t-butoxycarbonyloxystyrene

(PBS) photoresist [65]. In recent years efforts have been made to fabricate DUV workable photo-masks without using a micro-ionic laser system. [66–68].

4.1.6.4 Pellicles

The pellicle is defined as a thin transparent (to the wavelength it is exposed) membrane which acts as a tuned etalon [69]. One of the properties of a pellicle is that it should transmit more than 98% of the incident light. Besides transparency, the pellicle film should show radiation tolerance, and good adherence to the substrate. The thickness of the pellicle should not go beyond 0.8 μm, because thicker polymer film becomes too fragile across a 5 in^2 pellicle frame [70]. Figure 4.28 shows the cross-section of a pellicle.

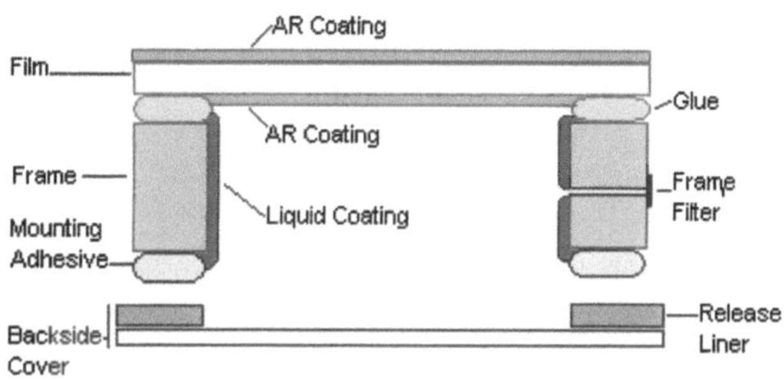

Fig. 4.28 The cross-section of a pellicle (Courtesy, Micro-lithography, Inc.)

Spin coating has been found to be the most suitable process so far for pellicle film deposition. Transmission within the film depends on the film thickness, type of anti-reflecting coating, the wafer absorption and the wavelength of light used. The frame used to mount and support the film should be mechanically rigid, flat, stable and should not introduce any optical interference. There are two different types of mounting adhesive that have been used – a carrier and a non-carrier adhesive.

Amorphous polymers such as $[(CH_2\,CHF)_x\,C(CF_3)_2CH_2]_y$ – or – $(CH_2CF_2)_x$ [2, 2-*bis* (trifluoromethyl)-4,5-difluoro-1, 3-dioxole$]_y$– with chains that alternate fluorocarbon segments with either oxygen or hydrocarbon can be used for 157 nm pellicle material. Besides, a well polished fused silica with proper antireflective coating (ARC) may serve as well as an alternative to an organic pellicle.

4.1.7 Chemically Amplified Resist (CA)

Conventional positive resist can be used in the mid-UV range by changing the photo-active chemical (PAC), but with deep UV, no such easy solution exists. However, there is a negative DUV resist commercially available with the trade name

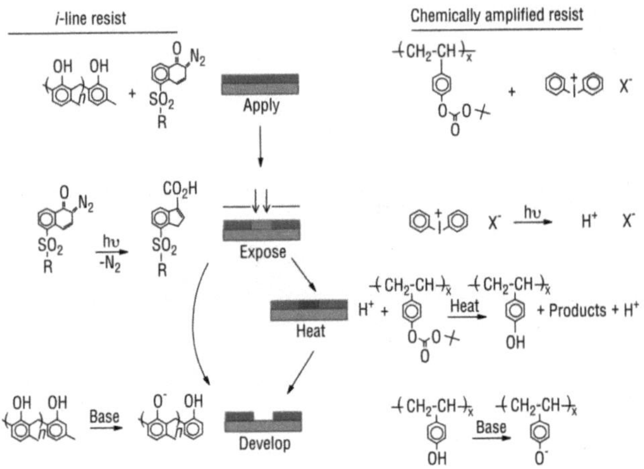

Fig. 4.29 Comparison of the DNQ-novolak resin based i-line resists and chemically amplified resist (Reprinted with permission IBM Research)

of ODUR-110-WR [71–72]. There is another resist material developed for deep UV lithography which is known as *chemically amplified (CA) resist* (Fig. 4.29). The resist is in many ways superior to alternate (UV) materials [73] for their high sensitivity and chemical etch resistance. Modern chemically amplified resists are synthesized with a base resin, a photo-acid generator (PAG), a base, a casting solvent, and dissolution inhibitors. Positive resists have protecting pendent groups that are cleaved off the base resin during a post-exposure. Negative resists, on the other hand, have a linker (or cross-linker) component, which attaches itself to the base resin with one or more of its multiple branches. For CA resist the post-exposure baking (PEB) temperature is very critical (typically 90–140 C for 1–2 min) for the overall imaging quality.

Besides CA resists, two other resists from fluoropolymers and polysiloxanes groups are identified to be potential candidates for 157 nm-exposure systems [55]. These polymers are amorphous, having glass transition temperature (T_g) ~150 °C, and have shown desirable etch resistance and transparency that are better than norborene homopolymer [74]. Recently, a new formulation of 157 nm resist has been reported by Feiring and his co-workers [75], which is a t-BuAc in TFE/norbonene polymer in hexafluoroisopropanol. The resist is highly transparent and shows a glass transition temperature (T_g) in the range of 140–150 °C and having weight average molecular weight of around 10,000 [76].

Figure 4.29 compares the resist chemistries and the process flow of the i-line and CA resist systems. From the figure it is clear that while the i-line resist is rendered soluble in aqueous base, a CA resist requires a post exposure-heating step to be rendered soluble. It has been observed that thermal variation during resist processing is one of the many sources that contributes to the total critical dimension (CD) control of the CA resist (Fig. 4.30).

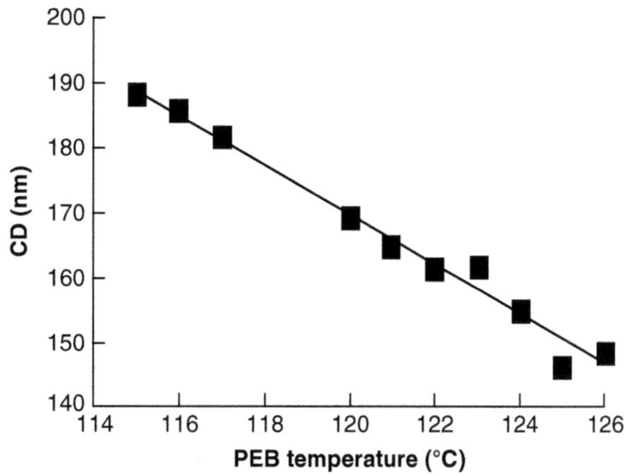

Fig. 4.30 CD sensitivity variations with post exposure bake (PEB) temperatures for CA resist (Reprinted with permission, SST, Sept. 2000, p. 139)

4.1.8 Extreme Ultraviolet (EUV) Lithography

The proposition of extreme ultraviolet lithography (EUVL) as a promising lithographic tool dates back to 1984 [77], 1988 [78], and 1989 [79], while the concept of multilayer reflective mirrors has been known for several decades. However, significant research has been focused on EUVL as the feature size of the integrated circuit has come down to the nano-scale [80].

Extreme ultraviolet lithography (EUVL) is a projection lithography that uses a radiation system in the soft X-ray regime (13.5 nm) and a reduction optical system (4×). As a matter of fact, EUVL is an extension to optical lithography (OL) with the use of shorter wavelengths instead of increasing NA. Significant progress has been made on EUV radiation sources starting from gas discharge-based EUV sources to laser induced hot plasma sources (laser produced or discharge plasma) of ionized xenon (Xe) gas or lithium (Li) vapor (Fig. 4.31).

The projection optics use mirrors with multilayer coatings of molybdenum (Mo)/beryllium (Be), which reflect wavelengths between 11.2–12 nm, or molybdenum/silicon (Si), which reflects between 12.8–14 nm, and have demonstrated ~70% peak reflectivity [81–83].

A schematic of the EUVL multilayer multi-mirror system, shown in Fig. 4.32(a), and (b), shows the concept of band selection in the multilayer mirror with a background cut-off filter and a junction diode detector. The main components of the EUVL system are a soft X-ray source, a condenser optics, a reflective mask, four high precision multilayer surfaces, and a wafer coated with resist. However, illumination at EUV wavelengths can be produced by a variety of sources including a laser produced plasma, synchrotron radiation, gas discharge, discharge pumped X-ray lasers and electron beam driven radiation devices [84–85]. Lately, a gas discharge-based EUV source operated in the back-lighted thyratron (BLT) mode is reported to be simple, compact, and less expensive compared to synchrotron and laser induced plasma sources (Fig. 4.33) [86].

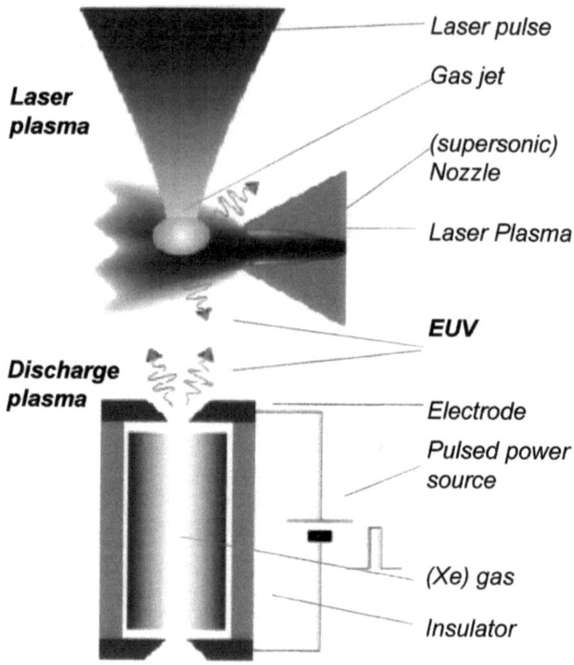

Fig. 4.31 Schematic of plasma sources used to perform EUV lithography (Photo courtesy, Prof. F. Bijkerk, FOM Institute, for Plasma Physics, Rijnhuizen)

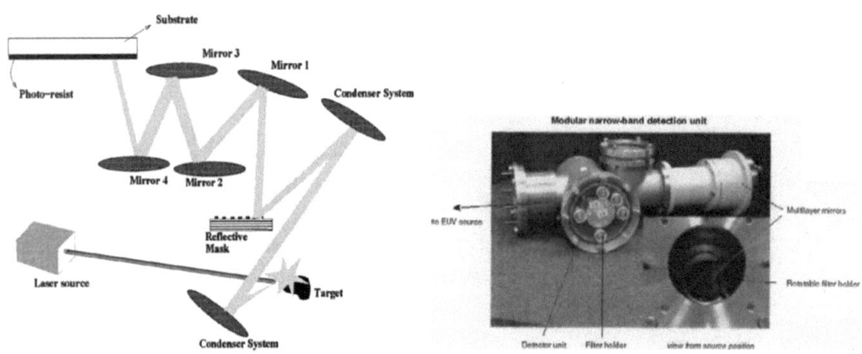

(a) (b)

Fig. 4.32 (**a**) An artist's rendition of the proposed EUVL system (Reprinted with permission, *Solid State Technology*. July, 1997) and (**b**) concept of band selection by a multilayer mirror with a background cut-off filter and a junction diode detector (Photo courtesy, Prof. F. Bijkerk, FOM Institute, for Plasma Physics, Rijnhuizen)

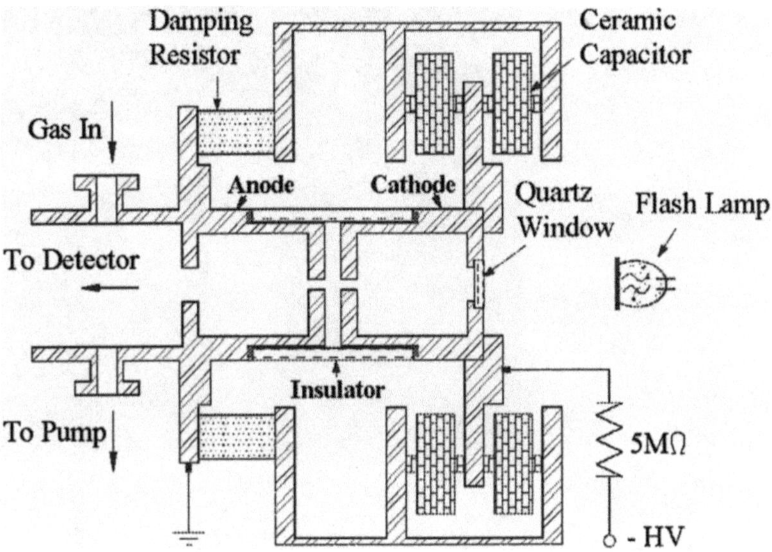

Fig. 4.33 The schematic and associated electronics of a BLT mode EUV light source (Reprinted with permission, *J. Vac. Sci. Technol*, AIP, [86])

Figure 4.33 shows the schematic and associated electronics of a BLT mode EUV system. A quartz window at the end of the hollow cathode allows UV light from a low-voltage xenon flash lamp to enter inside the hollow chambers. Two parallel capacitor rings are arranged symmetrically around the cathode with a series load resistor. The flash lamp optically triggers the pseudo-spark plasma for the EUV generation.

The EUV system with all reflective optical system is capable of printing 50 nm lines when the exposure wavelength is 15 nm with DOF above 400 nm and NA 0.2 (Fig. 4.34).

Figure 4.35 shows the set up of a commercial-EUV lithography system with optics and necessary attachments. A strong emission is generated from highly ionized xenon (Xe) plasma gas under ultra-high-vacuum and low-pressure conditions. The projection optics have mirrors with multilayer coatings of molybdenum/beryllium (Mo/Be) [87]. The key requirement in EUVL is the fabrication of masks with no defect [88].

In the year 2002, ISMT, Austin, TX, and the Extreme Ultraviolet Lithography System Development Association (EUVA) started a four-year project to develop a beta tool. Table 4.3 presents some of the specifications of the beta tool.

In 2004, the basic setup of the discharge plasma source for EUV lithography was upgraded to allow higher repetition rates up to 2 kHz, including modifications in the source and its performance. Figure 4.36 shows an overview of the experimental setup for EUV lithography. The EUV imaging tube is mounted at 40°.

Fig. 4.34 Picture of 50 nm wide printed lines in photoresit by using EUV lithography (Photo courtesy [87])

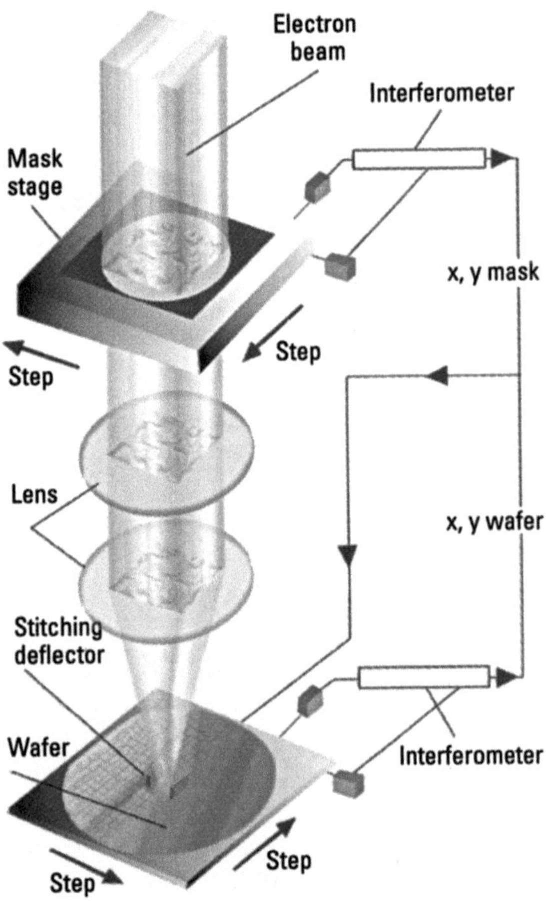

Fig. 4.35 The set up of an EUV exposure system (Reprinted with permission Semiconductor International. Feb. 2000, p. 82)

Table 4.3 EUV specifications for beta tool

Specifications	Beta tool	Production tool
NA	≥ 0.20	≥ 0.25
Sigma	≤ 0.8	0.2–0.8
Magnification	4×	4×
Field size	22 × 26	22 × 26
Light source	LPP or DPP	LPP or DPP
Exposure wavelength (nm)	13.5	13.5
Output power (W)	≥ 4	50–150 ?
Wafer size (mm)	300	300
Throughput (wph @ 50 W)	7	50
CD uniformity	≤ 4	≤ 4
	≤ 11	≤ 11

Reprinted with permission ITRS, SEMATECH, Austin, TX.
Source: EUVA.

Fig. 4.36 Overview of the experimental setup showing the control unit and part of the power supply for EUV lithography (Photo courtesy International SEMATECH, Austin, TX)

4.1.9 e-Beam Lithography (EBL)

Electron beam lithography (EBL) is a process of forming circuit patterns by using a focused electron (e) beam. The special feature of the technology is that the e-beam can be scanned and accurately positioned on the substrate. It has the ability to produce higher resolution (50 nm versus 20 nm) than optical lithography (OPL) because resolution is not diffraction limited as is found in OPL systems. However, in e-beam system resolution is limited by the electrons' forward and back scattering

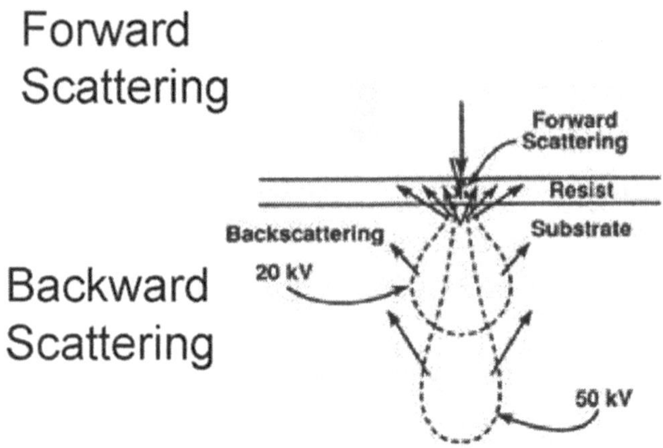

Fig. 4.37 Forward and backward scattering (Courtesy, ECE Department, Carnegie Melon University)

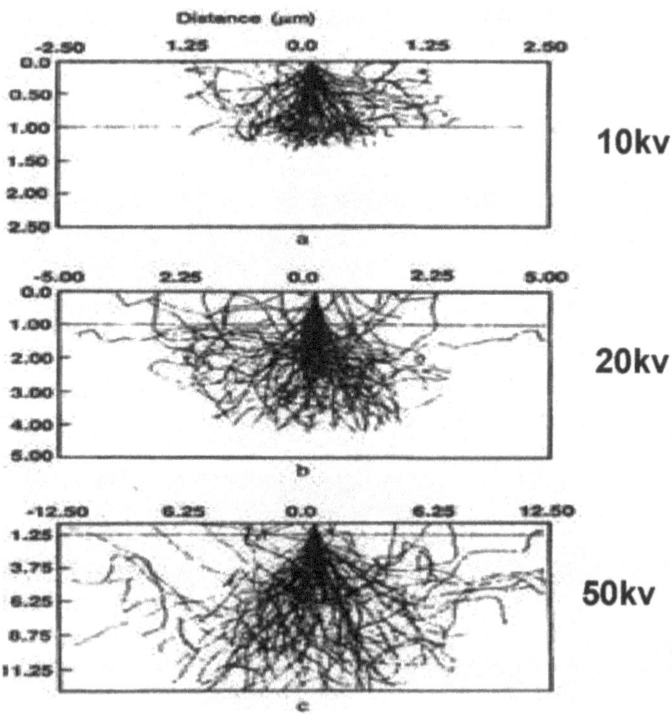

Fig. 4.38 Monte Carlo simulation for scattering in e-beam resist (Courtesy, ECE Department, Carnegie Melon University)

from the substrate (Fig. 4.37). In addition to that, image resolution is affected by resist swelling (negative resist) and the process is extremely slow when compared to optical projection and a step and repeat system.

Figure 4.38 shows the results of a Monte Carlo simulation of 100 electrons from 10, 20, and 50 keV point sources projected on the X-Z plane.

Figure 4.39 shows a schematic of an electron beam system. There are five main sub-systems: (a) electron source and the electrostatic gun lens; (b) gun alignment coils; (c) magnetic lens, blanking cell and aperture; (d) deflectors; and (e) focusing lens on the top of the wafer handling system. There are two types of electron sources

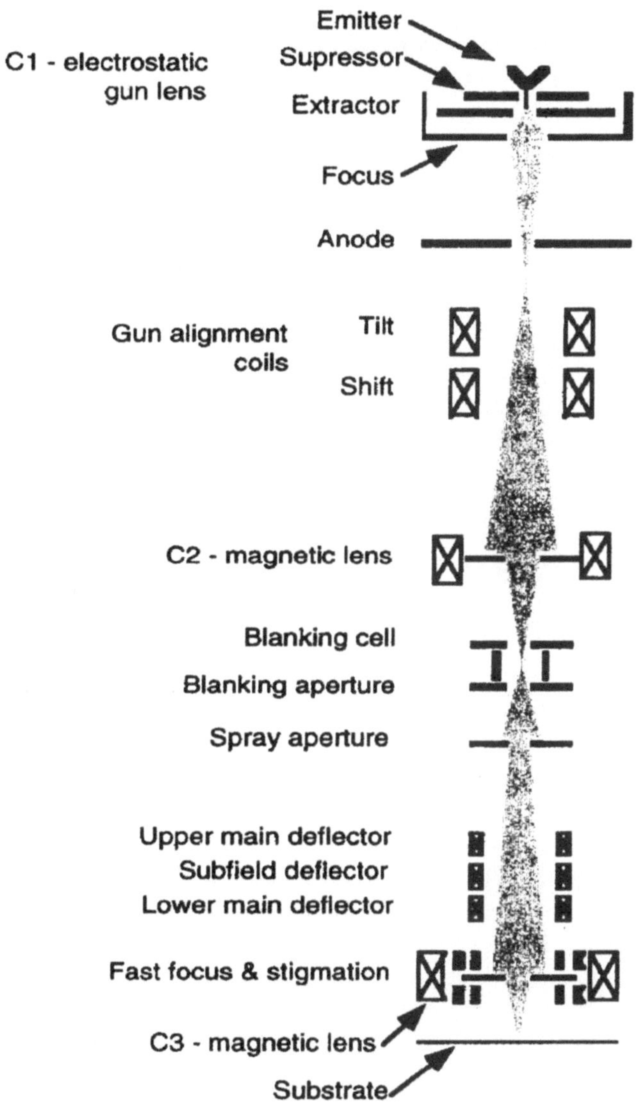

Fig. 4.39 Schematic of an electron beam machine (Courtesy, ECE Department, Carnegie Mellon University)

commonly used in e-beam systems: (a) thermoionic source and (b) field emission source. For thermoionic sources, tungsten (W) and lanthanum hexaboride (LaB_6) are frequently used. The W source provides a stable current and exhibits excellent tolerance to vacuum conditions and has been widely used.

In field emission sources, electrons are collected from the emitter by a high electric field, which forms a fine Gaussian spot. In order to write over the entire wafer, either a laser interferometer controlled stage is used or the pattern is written on the fly. Two alternate types of e-beams are used for EBL: (a) Gaussian shaped round beam; and (b) the variable shaped beam [89–90].

Figure 4.40 shows a three-dimensional picture of an ultra-high-resolution electron beam lithography system and metrology tool. The tool is claimed to write <20 nm wide lines on a 200 mm wafer.

Fig. 4.40 Three-dimensional picture of a EBL system (Photo courtesy, Raith, USA)

4.1.10 Electron-Beam Resist

Resists for optical lithography are designed for specific molecular excitations tuned to the wavelength used. In the case of e-beam resists high-energy electrons randomly ionize all elements in a resist according to atomic cross-sections and concentrations of each element. Generally, polymethylmethacrylate (PMMA) is used for e-beam lithography and Poly (1-butane sulfone), known as PBS, is used to produce photomasks [91]. However, as the feature size has come down to nano-scale PBS type resist can no longer meet the requirements of sub-micron lithography. However, the etch resistance of these materials has been improved by introducing sulfone/novolak resins. Recently there has been significant interest in the use of chemically amplified resist (CAR) for e-beam writing and mask making. Most recently, a low activation energy CAR based on poly (hydroxstyrene) has been developed. It has been

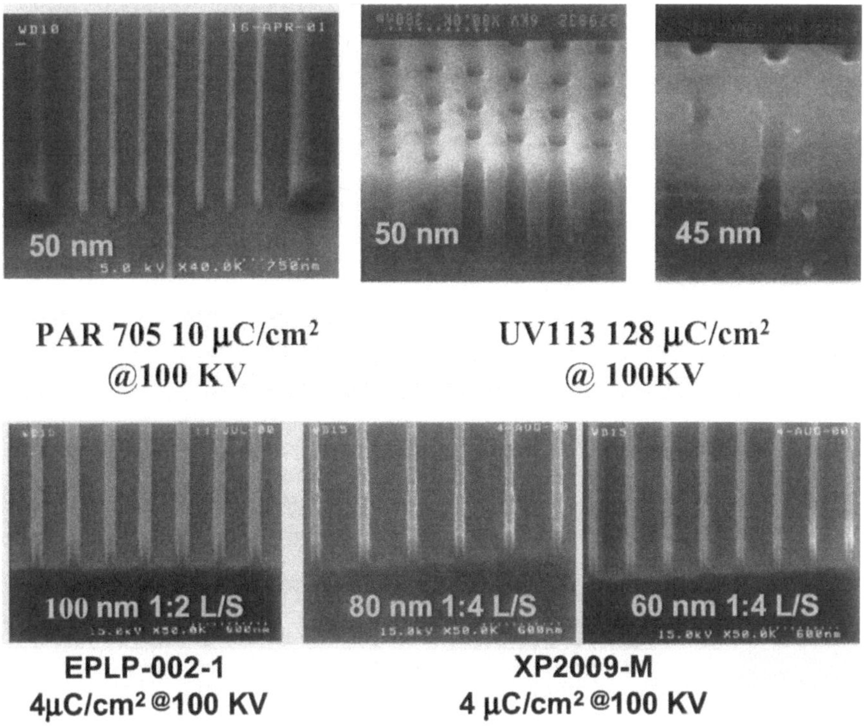

Fig. 4.41 Nano-structures obtained by patterning CAR by using e-beam lithography (Reprinted with permission of MRS, MRS-Symp. Vol. 705, 2002, photo courtesy, Dr. L.E. Ocola, Agere systems)

reported that the modified resist exhibits superior lithography performance and has a high level of processing robustness [91].

Figure 4.41 shows nanostructures created by e-beam (at 100 KV) lithography using chemically amplified resist (CAR).

At the beginning of 1990 some industrial research organizations developed *scattering with angular limitation projection electron beam lithography* (SCALPEL), *projection reduction exposure with variable axis immersion lens* (PREVAIL) (Fig. 4.42), and a different kind of concept for a reduction projection e-beam exposure.

The SCALPEL is the first to implement sequential illumination of the mask in an e-beam reduction projection system by mechanically scanning a reticle and wafer at a 4:1 speed ratio underneath a stationary beam. PREVAIL, on the other hand, combines electronic beam scanning with continuous stage motion [92] to overcome field-limiting off-axis aberration through the use of an innovative *curvilinear variable axis electron lens* (CVAL) (Fig. 4.43). The curve beam path is referred to as the curvilinear variable axis (CVA) and the corresponding lens as CVAL, representing a significant extension of the VAL concept.

Fig. 4.42 A picture of PREVAIL (Reprinted with permission, IBM Research)

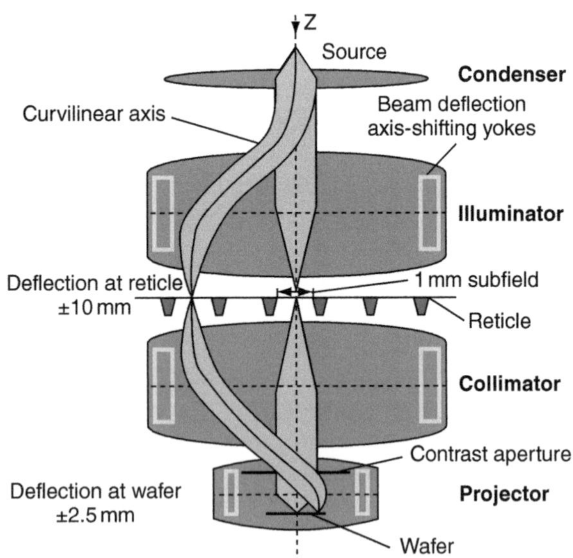

Fig. 4.43 Curvilinear variable axis lenses (CVAL) (Reprinted with permission from *Solid State Technology*, May 200, p. 118)

4.1.11 e-Beam Reticle

The reticle used for e-beam lithography is a 2 μm thick patterned silicon. The reticle may be either a scattering stencil or a scattering membrane (SCALPEL type) type stencil (Fig. 4.44) supported by a minor strut structure.

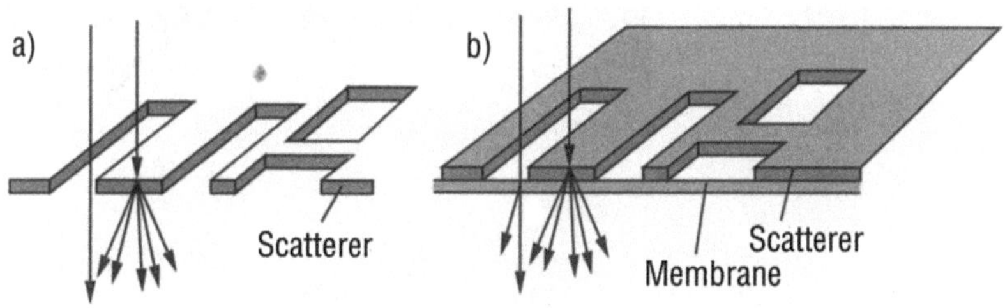

Fig. 4.44 The schematic of a reticle: (**a**) scattering stencil type and (**b**) scattering membrane type for the e-beam stepper (Reprinted with permission from SST, May 2000, p. 121)

There are several important issues regarding the use of the reticle such as image placement, CD control, reticle flatness, *in situ* cleaning and distortion of the membrane due to thermal loading.

Figure 4.45 shows the e-beam writing strategy. Each time the 1 mm^2 e-beam is exposed to a reticle sub-field. The movement of the stages deflected the e-beam to the next sub-field and the projection optics help to place the image properly on the wafer.

4.1.12 Step and Flash Imprint Lithography (SFIL)

The real question for the semiconductor industry is whether any sub-micron lithography can be applied to imprint directly on dielectric material for dual damascene processing and can its benefit be comparable to EUV or e-beam lithography? As a matter of fact, recently, a promising new technology called *Step and Flash Imprint lithography* (SFIL) has shown enough capability of fulfilling the above requirements.

In the standard SFIL process the imprinted mold consists of etched 100 nm structures on a 25 mm square mesa created on a 65 nm square quartz substrate. The mold is prepared from a transparent template processed by photo-polymerization of an organo-silicon material. SFIL-processing is very promising and cost effective. The method simplifies the creation of interconnect structures for the BEOL process although some technical challenges still need to be addressed in terms of defectivity, alignment, and overlay registration [93]. Figure 4.46 shows a schematic of a typical SFIL process.

As a test of the effectiveness of the SFIL process, imprinted wafers were developed at the advanced Technology Development Facility; SEMATECH, Austin, TX

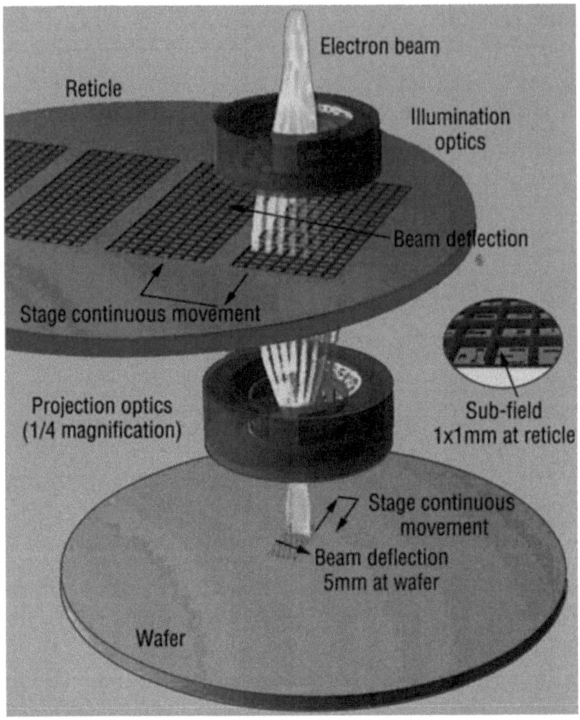

Fig. 4.45 The e-beam stepper writing strategy (Reprinted with permission from *Solid State Technology*. May 2000, p. 120)

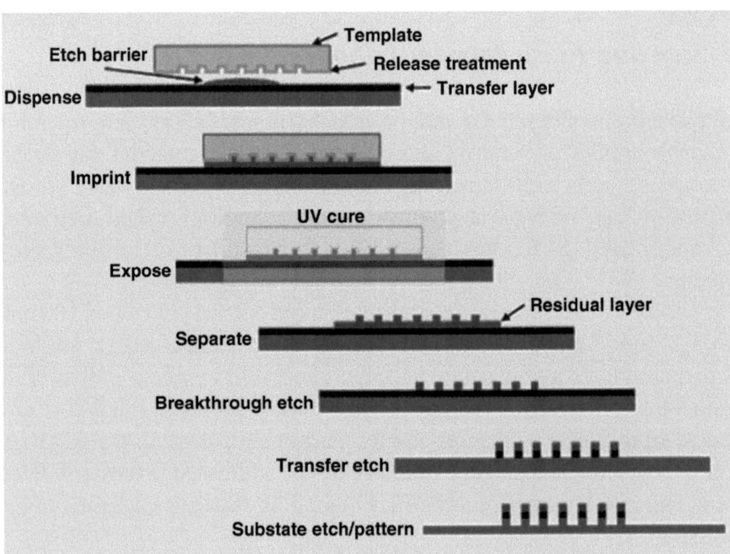

Fig. 4.46 The basic principle of the SFIL process (Reprinted with permission Semiconductor International, photo courtesy Prof. G. Willson, University, Texas, Austin)

Fig. 4.47 Scanning electron micrographs that illustrates the potential of the SIFL technology (Reprinted with permission Semiconductor International, August 2006, photo courtesy, Prof. G. Willson University of Texas, and Dr. J. Wetzel, SEMATECH's ATDF).

and the scanning electron micrographs (SEM) of the imprinted wafers are shown in Fig. 4.47.

4.2 Etching and Cleaning of Damascene Structures

4.2.1 Etching

4.2.1.1 Advanced Process Technology

The etching procedure in dual damascene architecture is a complex process especially for the dielectric components. The dielectric-etch approach is the most widely used process practiced before copper filling inside the etched structures. Interconnecting metal lines are embedded within the dielectric layer by sequential creation of via holes and trenches. Etching of trenches and via holes in the dielectrics and filling them with copper for interconnecting lines avoids difficulties associated with copper etching. However, the process is complex and difficult though progressive research and practice have brought hope and success. Smaller device geometry, exotic materials, and larger wafer sizes are making the etching and wafer handling machines complicated. It has been reported (wafer processing report, *Semiconductor International* March 23, 2007) that some commercial etching units have been designed to meet the rigorous challenges facing etching for 32 nm node devices. Figure 4.48 shows an etching tool, which is reported to offer solutions with a wide operating window for an optimum processing environment.

Fig. 4.48 The inside of an etching process chamber (Reprinted with permission, Semiconductor International, June 2004, p. 43, photo courtesy, Applied Materials)

Dielectric etch in the Cu-damascene (single damascene (SD) or dual damascene (DD)) process is the most complex operation of the copper interconnect manufacturing sequence. A wide variety of applications (e.g. DRAM, ASIC, MPU, and SOC) involve the damascene sequence. Etching the via holes and trenches in the dielectric materials avoids difficulties associated with etching copper, instead filling the etched features with copper through a sequence of processes known as metallization.

The process sequence can be repeated as many times as necessary depending upon the number of metal layers needed to create the specific device. The progressive miniaturization of the devices (130 nm node to 32 nm node and beyond) has brought with it a succession of technological complexities. The variety of film stacks is, in turn, giving rise to different SD or DD etch approaches, e.g. *(i) organic, (ii) inorganic sacrificial layer, (iii) multilayer resists, and (iv) metal hard masks* (adopted from *Applied Materials*). Depending on the approach adopted, several challenges arising at progressively smaller feature scales have been dealt with to a greater or lesser degree. Notable among these are maintaining resist integrity, ensuring damage free processing, especially of low-*K* dielectrics, delivering consistent results over extended runs and productivity.

(i) Organic sacrificial is a dual layer BARC process and is used for 130 nm-lithography process. The method is very straightforward and inexpensive. The incoming wafer is patterned for via etch by removing photoresist and the underlying BARC layer. The layer requires a dual layer back anti-reflecting coating (BARC). The first new layer of BARC is deposited and then removed, leaving the vias partially filled. Following this, a second layer of BARC is deposited and patterned for trench-etch. During trench etching via fill provided by the first BARC protects the sidewalls. The wafer is ashed again after etching of the trench and finally, the barrier layer (which was partially opened during the initial via etch) is completely opened to the copper-filled trench below (Fig. 4.49b). Figure 4.49 (a) shows the etched barrier layer and (b) shows copper filled trench.

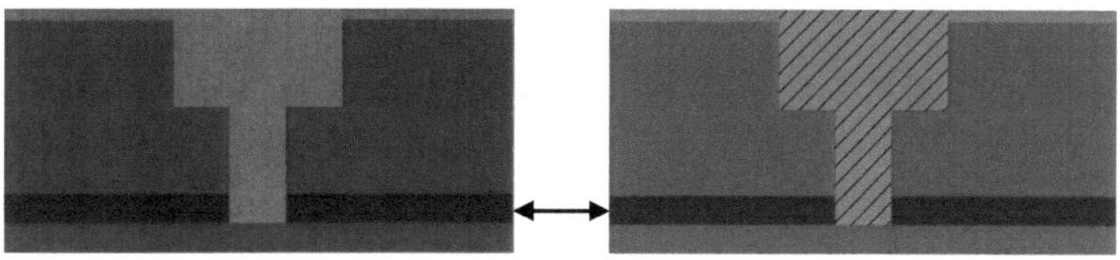

Fig. 4.49 (**a**) *Left*: Open barrier layer (shown by *arrow*) through trench and via and (**b**) *Right*: Cu-metallization

(ii) The *inorganic sacrificial* approach is very similar to the organic sacrificial method except, that it generally involves only one inorganic sacrificial layer. The incoming wafer is patterned for via etch, which is followed by ashing to remove the photoresist and its underlying BARC layer. The ionorganic sacrificial layer is then deposited and patterned for trench etch. Some of the sacrificial layer remains in the bottom of the via following the trench etch and is then removed with a wet clean before the barrier layer (which was partially opened during the initial via etch) is completely opened to the copper filled trench below.

(iii) A *multilayer resist scheme* is used to minimize etch-induced photoresist damage. In this method, the via patterned wafer is etched and ashed to remove the photoresist and its underlying layer of BARC. The layer of BARC is deposited and overlain with successive layers of hard mask and another layer of BARC. A hard mask open etch opens the top of the BARC and hard mask layers, after which the lower BARC is opened and the trench is etched, leaving a partial BARC fill in the via. The remaining BARC and via fill are removed in the subsequent ashing step. The layer is completely opened to the copper filled trench below.

(iv) The *metal hard mask* (shown in Fig. 4.50) method is considered to be the best among all the methods described so far especially when the work involves low-*K* porous dielectric materials. The method also offers protection against resist poisoning but it involves additional metal deposition and complexity of the work involved during deposition. In this film stack, the BARC and the photoresist layers are underlain by a metal layer. The film is hard masked to open the trench through this metal. After this the wafer is exposed to the remaining metal. The next step is to deposit BARC and to pattern for via etch. This protects the trench while the via

width is etched to partially open the barrier layer and the wafer is ashed. BARC is deposited for the second time and patterned for completing the trench etch. During ashing step, the remaining BARC at the bottom of the via hole is finally removed and the trench is filled with copper.

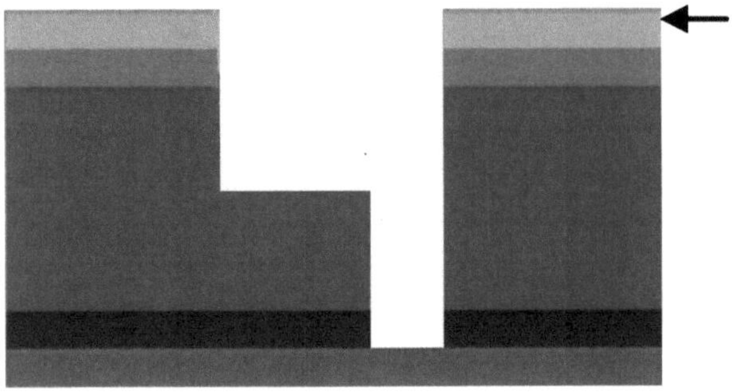

Fig. 4.50 The barrier is open through a metal hard mask (shown by *arrow*)

The development of Cu-damascene technology in the production of complimentary metal oxide semiconductor (CMOS) and dynamic read access memory (DRAM) devices [93–94] has decreased the front end of line (FEOL) (processing that takes place up to and possibly including junction metallization) versus back end of line (BEOL) (processing that creates the metallization and back end of the line) process step ratio [95]. To reduce costs and increase device performance, during implementation of the Cu interconnect with low-*K* in dual damascene (DD) processes, the industry has focused more of its resources on BEOL. As a matter of fact, fundamental innovations must address the various sources of leakage, particularly the gate, the sub-threshold, and the junction. It seems that among all the mechanisms, sub-threshold leakage is perhaps the most difficult to reduce and suppress.

In order to reduce the gate leakage, the industry has addressed the issue by employing high-*K* dielectric/metal gate electrode technology by replacing traditional get oxide/doped poly-silicon (Poly-Si). Indeed, metal gate electrodes have addressed the issue of poly-depletion, boron penetration, and Fermi level pinning effects at the high-*K*/poly Si interface. Initially it was thought that the introduction of high-*K* will require a move from poly-Si to a metal gate, but now it seems that a silicide metal (doped with nickel) gate might be an alternative for the future. However, it is felt that the ultimate solution lies in the structure of a dual-metal gate electrode with different work functions for p and n MOS transistors combined with a high-*K* dielectric.

Figure 4.51 shows the complicated architecture of a 130 nm node CMOS device. It goes without saying that as the device size shrinks, the processing of the devices becomes more and more complicated [96–97].

There is no doubt that the dielectric-etch process needs different procedures for FEOL and BEOL and each process has different requirements. For example, high-*K* and metal gates actually play the main role in FEOL, while BEOL has become a major issue in dealing with low-*K* dielectric materials.

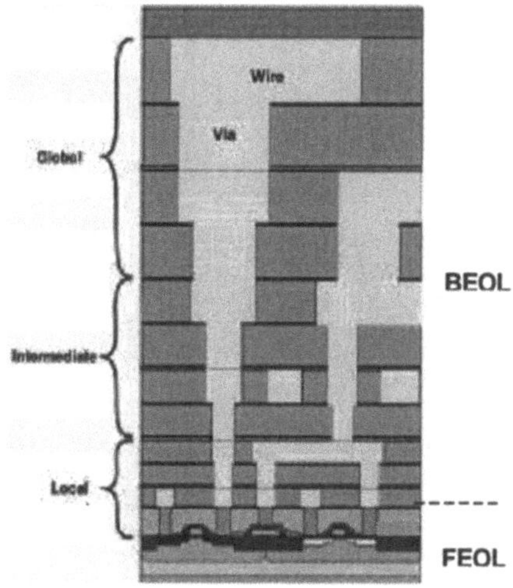

Fig. 4.51 The cross-section of a 130 nm- process structure (Reprinted with permission, ITRS 1999, SEMATECH, Austin, TX)

4.2.1.2 Process Chemistry of Etching

During dual damascene processing, three classes of materials are mainly used. These are metals including transition metals and their compounds for the barrier, and inorganic, organic and hybrid dielectric materials. When the dielectric is an oxide of silicon, the anisotropic profile is controlled first without etching too much of the side wall. As a matter of fact, the silicon of silicon dioxide controls the free fluorine to carbon ratio, which balances the amount of vertical etch, while controlling the amount of side wall passivation. To etch an organic low-*K* dielectric the etch technology uses a thin layer of inorganic material. However, etching process will be different when the material is neither organic nor inorganic but a hybrid, like organosilicate glass (OSG). These materials contain excess amounts of carbon, and significantly impact etch selectivity to photoresist.

Etch rate of porous dielectric is higher than the dense oxide (fluorosilicate glass, FSG) dielectric. As a result, when both of them exist in a stack, patterning time adjustment becomes very difficult because in the difference of etch rate which boosts etch selectivity. Porous oxide alone is etched by using a thin layer of photoresist. Thus when materials to be etched are varied compositionally, and any one of them begins resembling either photoresist or one of the etch-mask materials, the configurations of the etching mask pattern will be different. Figure 4.52 shows different etch mask stack configurations during dielectric etch of the dual damascene process.

Low-K Dielectric Materials

Modern high-speed devices are fabricated with Cu as interconnecting metal over a barrier layer coupled with low-*K* dielectric material, either inorganic, organic (porous/non-porous), or hybrid. Table 4.4 shows some of the dielectric materials that are frequently used in Cu-damascene process architecture.

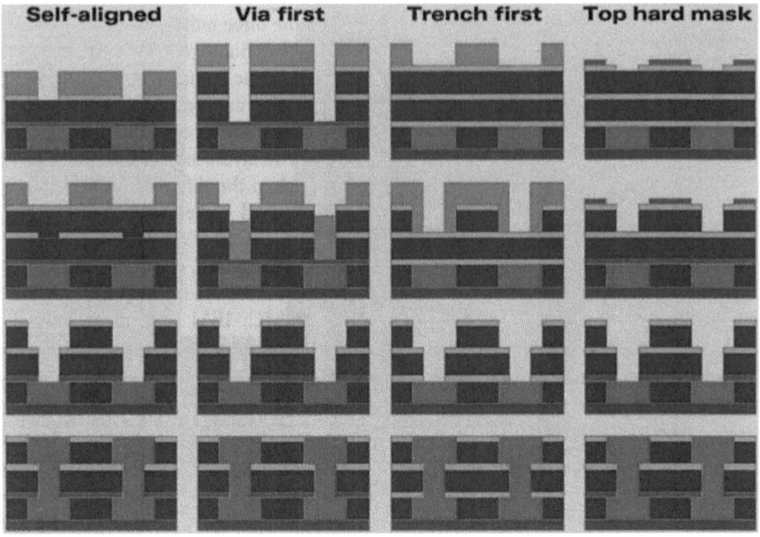

Fig. 4.52 Different etch mask stack configurations during dielectric etch of dual damascene process (Reprinted with permission, *Semiconductor International,* June 2004, p. 44, Courtesy ITRS 2003)

Table 4.4 Low-*K* dielectric materials that are used for the copper damascene process

Inorganic	Organic	Hybrid
SiO_2	Parylene –N	Bezocyclobutane (BCB)
$SiO_{(2-x)} F_y$	Parylene-F	Methyl silsesquioxane (MSQ)
Hydrogen silsesquioxane (HSQ)	Polyarylene ether (PAE-2)	Porous MSQ
Porous HSQ	Polytetrafluoroethylene (PTFE)	Organosilicate glass (OSG)
Xerogels	SiLk/Porous SiLK	
Fluorocarbon	Fluorinated polyimide (FPI)	
Resist mask	FLARE/ Porous FLARE	

Organic polymers require an oxidative reductive chemistry that should not react with copper (Cu). Organosilicate materials, on the other hand, require a more physical component and process technology where the barrier layer serves as the substrate. The oxygen etching chemistry for hybrid dielectric materials that contain a silicon group as an inorganic and a hydrocarbon as an organic group (as in BCB) produces a SiO like passivation layer which leads to low etching rates and surface roughness. Thus the oxidized passivation layer formed during O_2 etching should be removed and/or cleaned by argon (Ar) sputtering. It is also customary to clean the metal lines and the connecting studs following dielectric etch to minimize contact resistance.

Fluorocarbon removal from the metal surface is performed inside a plasma etching reactor employing O_2 following the etch process. During O_2 plasma exposure, the fluorocarbon layer produces a thick oxide layer over Cu, which is subsequently removed by Ar^+ sputtering. However, if the etching layer stack contains MSQ, O_2 exposure may lead to the loss of carbon from the MSQ surface and formation of a SiO like passivation layer [98]. Chou et al. have reported that the formation of the strained oxide layer during exposure to O_2 plasma on the top of polysiloxane fill can reduce the etch rate [99]. Plasma etching with O_2 chemistry provides high reflectivities between Si-containing and non-Si-containing resist but at the cost of considerable lateral etching and poor critical dimension (CD) control [100]. Eon et al. have reported an improvement of selectivity of etching of MSQ non-porous and porous films in C_2F_6 plasma with O_2, Ar+ and H_2. However, they found that best selectivity can be achieved with a C_2F_6/H_2 (10–20%) mixture [101].

Hydrogen doped oxides (e.g. HSQ) are etched using an oxide etch base line process. However, higher level of 'H' doping deaccelerates etching rate that result tapered profile. However, adding a low percentage of oxygen into the system can minimize the effect [102–103].

For etching porous and non-porous SiO_2, inductively coupled plasma system [104–105] with CHF_3 plasma is very effective when the substrate is biased with rf voltage (Fig. 4.53). The etch rate is higher with higher bias and porous SiO_2 shows a higher etching rate than non-porous SiO_2 film due to the lower mass density. Table 4.5 shows the etching rate and selectivity of different dielectric materials.

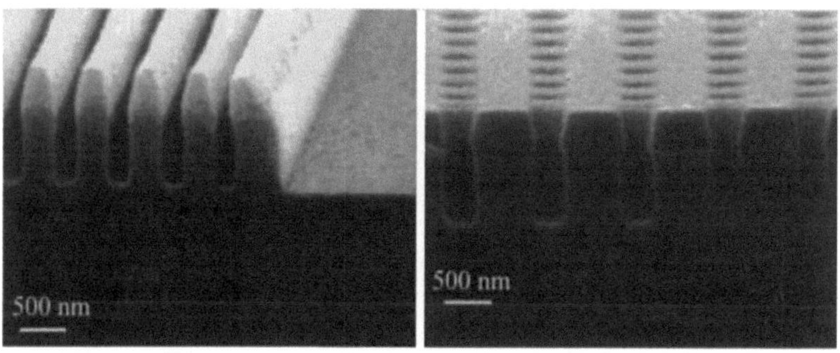

Fig. 4.53 Trenches and via holes etched in 30% porous xerogel films using CHF_3 discharge maintained at 1400 W source power and 10 mtorr pressure (Courtesy SST, May 2000, p. 128)

When the feature size of the device is small and the aspect ratio of the via hole or trenches is high, conventional ion etching can cause significant problems with etch profiles such as *faceting, fencing and trenching* (Fig. 4.54) [106–107]. However, with a special type of *inductively coupled plasma (ICP)* etching system, some of the problems can be minimized [108].

Table 4.6 shows some of the gases that are very frequently used to dry etch some of the materials used in integrated circuits (ICs). The advantage of an ICP system is that it offers a means of controlling independently the plasma density and the accelerating voltage on the lower electrode. As a result, the system can provide improved etch rate, higher selectivity, and lower damage to the etched surface. Figures 4.55 and 4.56 show some commercial plasma etching systems, e.g. *reactive ion*

Table 4.5 Etch rate of different materials achieved using ICP technology

Materials	Etch rate (Å/min)	Selectivity
Polyimide	>1000	>15:1 to SiN_x
Depassivation (SiN_x)	~1000	~1.5:1 to TEOS
SiC	Upto 1370	~1.4:1 TEOS ~0.7:1 to low-*K*
SiOx (ILD) (BPSG, TEOS)	Upto 1000	~0.5:1 to low-*K* ~0.7:1 SiC
SiNx ILD	Upto 4000	~4:1 SiO_x
SiOCH:H low-*K*	Upto 1800	~2:1 to TEOS

Adopted from Oxford Instrum. Plasma Tech. (July, 2004).

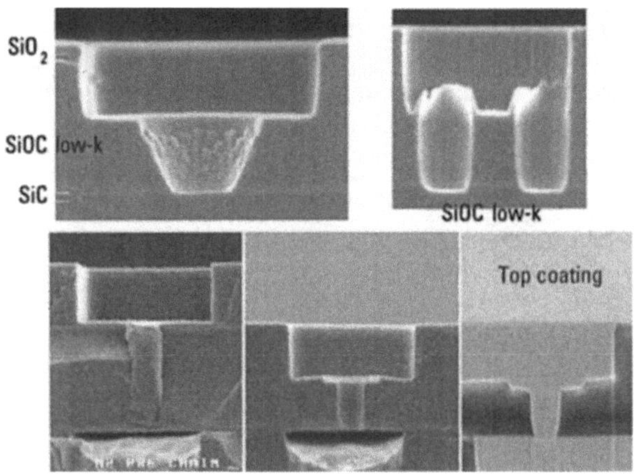

Fig. 4.54 Formation of (**a**) facet (*top left*), (**b**) fence (*top right*), and (**c**) trench (*bottom right*) (Reprinted with permission, Semiconductor International, May 2002, p. 58)

Table 4.6 Etching gases used to dry-etch different materials in semiconductor device processing

Material	Gases
Silicon (Si) including Poly	CF_4, CF_4/O_2, CF_3Cl, SF_6/Cl,Cl_2+H_2
Silicon dioxide (SiO_2)	CF_4/O_2, ClF_3, CCl_3F_5, $SF_{4}//O_2$,
Silicon nitride (Si_3N_4)	C_2F_6, CF_4/O_2, CF_4/H_2, C_3F_8
Tungsten (W), WSi, Mo (moly)	CF_4, CF_4/O_2, C_2F_6, SF_6
Tantalum silicide ($TaSi_2$)	CF_4/Cl, CF_4/O_2, CF_3Cl, SF_6/Cl_2,

etching (RIE), *plasma enhanced* (PE) etching, and *inductively coupled plasma* (ICP) etching.

Atomic oxygen is believed to be a primary neutral active species involved in spontaneous etching of polymers that causes undercut profiles [109]. Experimental data show that anisotropy with O_2 plasma can be attained at –100 ºC [110]. On the other hand, during plasma etching, striation formation on the resist wall can lead to degradation of contact hole resolution. At the same time ion-bombardment will

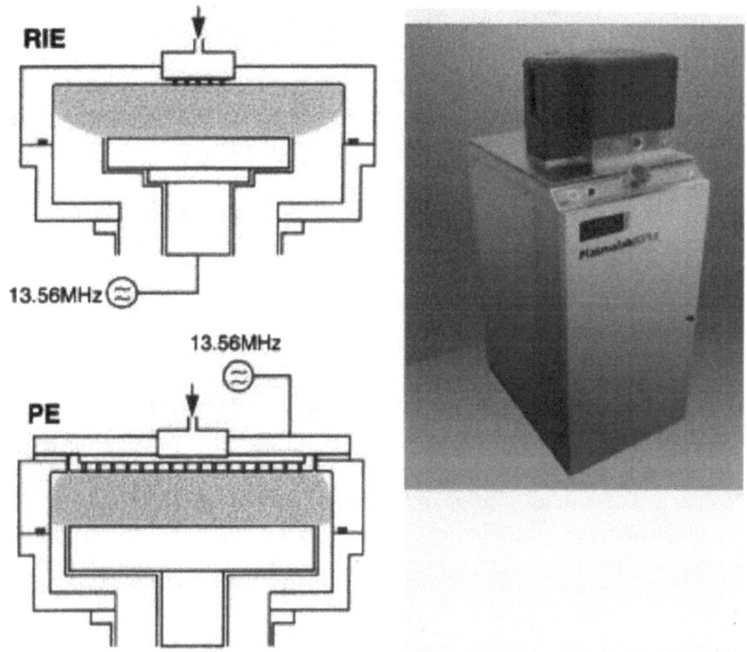

Fig. 4.55 Two types (RIE and PE) of plasma etching systems (Reprinted with permission, Oxford Instruments)

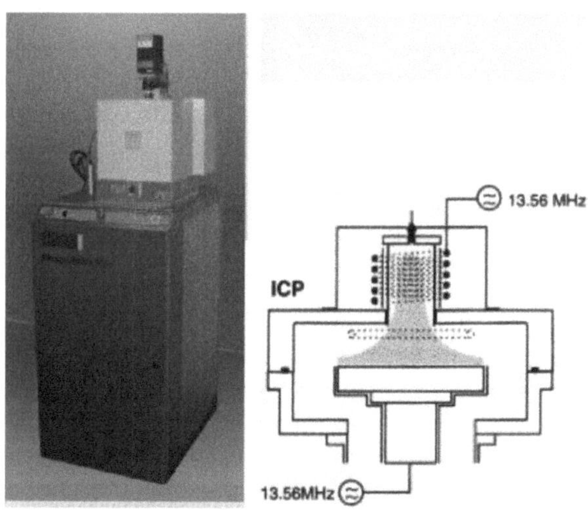

Fig. 4.56 An ICP system (Reprinted with permission, Oxford Instruments)

increase the surface temperature of the polymer and will result graphitization [111–112]. However, the addition of gases like C_2H_4, SO_2 and O_2 has been reported to reduce the micro-roughening and to achieve better anisotropy profiles [113].

For patterning of sub-100 nm features, a clear understanding of the origin and control of line-edge roughness (LER) is extremely desirable, both from a fundamental as well as a manufacturing prospective. Side wall roughness transfer is observed during antireflective coating/hard mask (ARC/HM) open and subsequent oxide etch. This is true for 248 and 193 nm systems with a wide variety of organic ARCs, hard masks, Si-containing ARCs and bi-layers. Oxidation or carbonization of a polymer during micro-masking is believed to be the cause of side wall roughening. This has been noticed during etching of low-*K* and porous low-*K* (e.g. OSG and MSQ) dielectric films containing fluorocarbon. Micro-masking occurs under certain conditions when regions of localized net polymerization form on the surface of the etched film [114]. Experimental observations show that an inductively coupled plasma (ICP) beam with independent control of the ion bombardment energy, incident angle, and proper plasma chemistry can reduce the side wall roughening during etching.

The effects of different etch chemistries on damage of the low-*K* porous dielectric layers (with same hard mask) are shown in Fig. 4.57. Adding an ash to the above process shows further deterioration of the structure, which is shown in Fig. 4.58. The scanning electron micrograph in Fig. 4.58 shows pre-ash and post-ash conditions of the sub-micron structure.

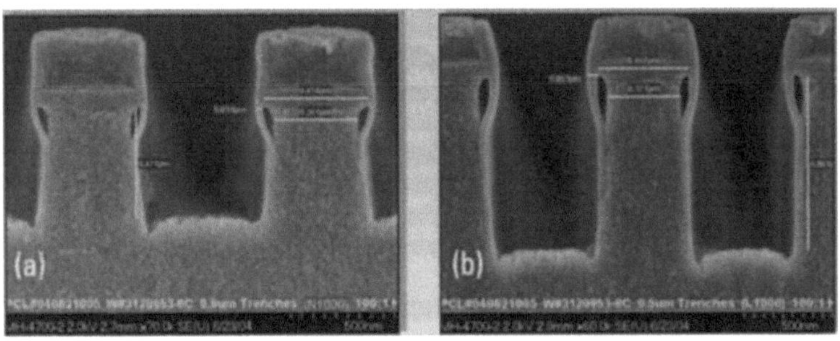

Fig. 4.57 Different under-cut ratios with different etch chemistries and with the same hard mask: (**a**) has an under-cut ratio of 5.8:1 with more damage than (**b**), which has an under-cut ratio of 8.1:1 (Reprinted with permission *Semiconductor International* June 2005, p. 64)

The low-*K* dielectric material is a significant issue in the back-end-of-line (BEOL) process. The biggest challenge is to identify an ideal chemistry that is compatible with low-*K* dielectric materials. However, damage during plasma etching of low-*K* can be minimized by not exposing low-*K* to high-temperature plasma when using hard masks, during conventional stripping and wet cleaning.

There is no doubt that these adverse effects affect the intra-line capacitance, as described by K_{eff} (effective capacitance value). For example, the K_{eff} could be lowered if the etch-stop layers are eliminated [114].

Polysilicon has almost reached its scaling limit as a gate electrode material. Therefore, at 100 nm-node technology metal tungsten (W) or its nitride on poly is used. Tungsten is typically etched with fluorine chemistries, which can also be

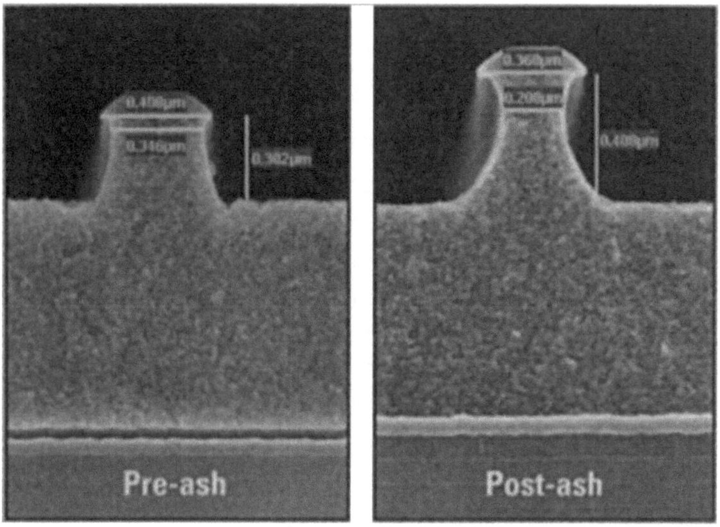

Fig. 4.58 Ashing in $CF_4/O_2/Ar$. *Left hand side* SEM shows pre-ash and *right hand side* SEM shows post-ash (Reprinted with permission *Semiconductor International*, June 2005, p. 64)

used to etch the dielectric and the substrate. A likely solution will be a mixture of fluorine together with chlorine and hydrogen bromide in well controlled strategies [115]. The etching is much more challenging when the etching process involves dissimilar metals like tantalum (Ta), zirconium (Zr), hafnium (Hf), and titanium (Ti) [116].

In the damascene process flow the two most important approaches of etching of trench and via hole are resist masking and the use of an etch stop layer (generally, silicon nitride). Sometimes an intermediate etch stop layer is provided to control trench etch. The etch-stop layer is generally selected from a *high-K* (for oxides ~4, and for nitrides ~7) material and the layer is made thin. During etching if the etch stop layer is eliminated, the etch process becomes robust in terms of etch uniformity, CD control, trench shape, etc. In the self-aligned approach, where the via requires almost perfect trench to via alignment, the challenge will be to maintain a very high selectivity between the dielectric and the intermediate stop layer during etching.

High-K Dielectric Materials

High-K dielectric materials like silicon oxynitride (SiO_xN_y (H)), silicon nitride (Si_3N_4), tantalum pentoxide (Ta_2O_5), hafnium oxide (HfO_2), hafnium silicates ($HfSi_xO_y$), and barium titanate ($BaTiO_3$) are being developed to replace SiO_2 as a gate dielectric material because of the scaling limits of SiO_2 [117–119]. These high-K materials can be etched using CF_4/Ar and C_4F_8/Ar plasma with bias in an ICP etching system. But, as the high-K materials allow thicker gate dielectric, their appearance at 130 and 65 nm nodes creates significant etching problems because of etch-end point detection. Thus techniques for determining the end-point cycle become valuable for reducing the degree of over-etching in order to increase throughput and run to run reproducibility. To detect end-point two methods are generally used: (a) laser interferrometry (LI), and (b) optical emission spectroscopy (OES). LI provides endpoint detection by identifying the change in frequency of the reflectance variations caused by the different refractive indices, whereas OES relies upon the change in the emission intensity of the characteristic optical radiation from either a reactant or product in plasma [120].

In the case of wet etching of high-*K* dielectric materials, the removal must be accomplished without any loss of any silicon, as well as little or no field isolation oxide loss. Experimental observations show that etch selectivity of HfO_2 to Si, as well as HfO_2 to SiO_2, is not greater than unity [120]. However, removal of HfO_2 during wet etching does not damage the underlying silicon (Si). Etch selectivity of thermally grown silicon oxide (TOX) and tetraethylorthosilicate (TEOS) is better with alcohol diluted HF than HF diluted with water. Unfortunately, the viability of alcohol-based chemistries as high-*K* etchants is limited by their flammability [121]. There is no doubt that wet etching is cost effective; however, sometimes the annealed high-*K* films are difficult to wet-etch and low-energy ion-bombardment (sputtering) has been shown to facilitate the etching of these films [122].

Shallow Trench Etching

In the copper damascene process shallow trench isolation (STI) schemes have replaced local oxidation of silicon (LOCOS) isolation. Interrelated factors among the major processing steps like silicon etch, dielectric formation, etching, bottom filling, and chemical mechanical planarization (CMP) determine the actual performance of isolation structures in the STI process module.

The etching of STI is performed in a chlorine plasma environment. The volatile products evolved during etching of silicon in a chlorine-argon (Cl_2-Ar) plasma are some combination of silicon and the reactive halogen atoms, that are deposited on the side walls of the chamber and cause tapering of the trench. The tapering of the trench is positive and may range from 75 to 89°, depending on the amount of re-deposition (Fig. 4.59) which can be avoided by sequential etch and deposition steps using SF_x and C_xF_y plasmas [121].

Control of the STI etch profile is one of the important process capabilities for STI etch module users. In order to control the etch profile in a controllable fashion, some industries are using a low frequency (600 kHz versus 1 MHz) RF power bias for the wafer (Fig. 4.60). The advantage of this process is that the low frequency biasing controls the silicon-etch rate in the Cl_2 plasma.

Challenges Associated with Copper Etching

Copper (Cu) is more difficult to etch and more mobile than aluminum (Al). Copper etch products are non-volatile, remaining on the surface rather than desorbed in the

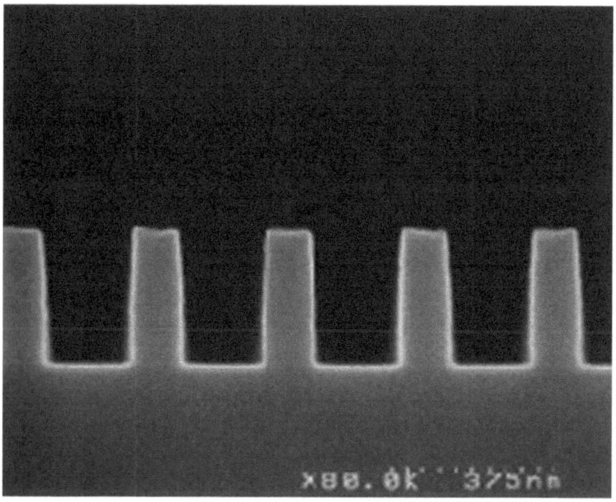

Fig. 4.59 Scanning electron micrograph of an etched shallow trench (Photo courtesy, Hitachi)

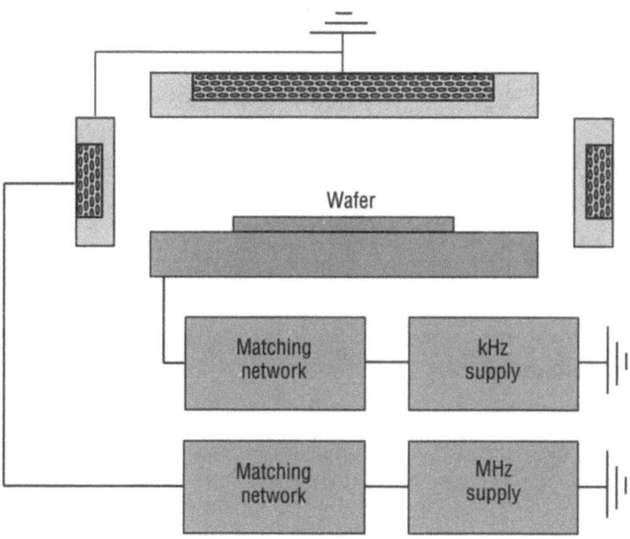

Fig. 4.60 Schematic of a dual frequency plasma reactor (Courtesy, Tegal Corporation, Petaluma, CA)

chamber's gas stream. The wet etching of Cu is generally done in slightly acidic ferric chloride ($FeCl_3$) solutions, typically 36–420 Baume (density at 15 ºC = 1.33 to 1.41 g/mL). The concentration of $FeCl_3$ at this level is very harsh to thin Cu film and the removal rate is very high [122]. The iodine etch solution ($KI+I_2+H_2O$) which is used for copper etching is unsuitable for small feature size devices, because the solution is very dark to detect the end point, and the etch-rate is also very high which frequently results in undercutting.

During photolithography, the presence of copper creates some problems as the sensitizer of a negative resist attacks copper. Thus copper cannot be used with a negative resist formulation. Copper (Cu) is difficult to etch in dry etching processes, because it forms an etch product with chlorine (CuCl), which is relatively non-volatile below 175 ℃ [123]. Thus copper cannot be dry etched with any one of the etchant gases containing chlorine (BCl_3, CCl_4, $SiCl_4$, and Cl_2). Moreover, the reactant gases are toxic and carcinogenic. In order to remove the non-volatile by-products the wafer has to be bombarded with high-energy ions and the temperature of the wafer has to be maintained above 175 ℃ which is detrimental to device yield. [123]. As a matter of fact, the non-volatile materials have become a challenging task for metal-etch in static random access memories (SRAMs) too.

As wet etching is not a viable solution for sub-100 nm feature size and dry etching is difficult with Cu, chemical mechanical polishing (CMP) has been used as the primary method to remove the overburden. The removal rate is controlled in part by mechanical force applied during processing step. CMP may also be employed to remove liner material overlying the dielectric.

In case of copper CMP, it has been observed that the removal of burden depends upon line width, pattern density, slurry chemistry, pad type and degree of conditioning, polishing pressure and rotational speed. Galvanic coupling between copper and the diffusion barrier layer is an additional burden and it is becoming more problematic with new abrasiveless slurry formulations [123]. However, the yield, the simplicity of the methodology, and the cost will determine the future of the etching technology.

4.2.2 Cleaning

4.2.2.1 Introduction

With ever-shrinking dimensions of the devices and increased speed, several new materials such as copper and low-K dielectric materials together with high-K (to meet the EOT challenge) have been introduced. As a result, various wafer cleaning chemistries and technologies have been developed. Many developments are centered on wet and supercritical cleaning, because, that remove the particles from the contaminated substrate surface resulting highest yield.

Contaminant particles adhere to the surface during CMP operation due to van der Waals adhesion forces. Thus removal of particles involves interactions of the forces in macro- and micro-length scales, which are applied externally using machines that can operate either in direct contact or non-contact mechanics. The real contact pressure (P_r) which is a function of the apparent contact pressure (P_a) can be measured from the simple relationship $P_a = f(P_r)$ [124–125]. The apparent contact pressure over the contact area results from the macro-scale force of equilibrium of the two contacting bodies [126].

Experimental

Knowledge of the adhesion force and the onset of a large adhesion force after polishing is necessary for minimization of surface defects. Thus the main objective of the cleaning process after CMP will be to drive away the alumina and the silica slurry particles completely. Oscillating flow rinse has been successful in the removal of the contaminants from the cavity because of its ability to stimulate vortex destruction and regeneration [126]. A solution of benzotriazine (BTA) with tetramethylammonium hydroxide (TMAH) and amonium solution at a particular pH value has also shown very promising results in reducing surface contaminants [126].

Figure 4.61 shows an untreated and post-treated wafer. Various wafer-cleaning chemistries and technologies have been developed for sub-micron damascene technology. The two most common post-CMP cleaning techniques that have been adopted in Cu-damascene process are brush and megasonic cleaning. Brush cleaning is accomplished with de-ionized (DI) water or dilute cleaning agent with some mechanical actions from *rotating brushes,* which can be either *knobby* or *ridged type* (Fig. 4.62). Compared to *knobby* design, the *ridged design* provides 100% coverage and the ridges are more closely spaced. In addition to that, ridged design provides unidirectional airflow facilitating rapid cleaning. Generally, polishing pads and brush scrubbers used in CMP and post CMP cleaning operations are made of highly deformable porous materials.

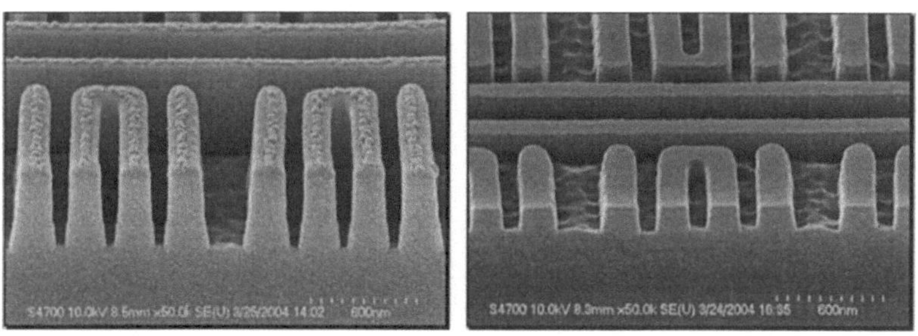

Fig. 4.61 An untreated Cu-low-*K* structure (*left*) and post-clean wafer (Reprinted with permission, Semiconductor International, Dec. 2005, p. 43)

(a) (b)

Fig. 4.62 Roller brushes: (**a**) knobby geometry and (**b**) ridged geometry (Courtesy, ITW Texwipe, NJ)

In brush cleaning the effectiveness of the cleaning depends upon brush pressure, speed, cleaning time, flow rate, and chemistry of the cleaning solution [127].

There is an another *megasonic* non-contact post-CPM cleaning process, which uses high frequency acoustic pressure waves to remove particles from the surface of the wafer (Fig. 4.63) [128]. The process involves use of silica particles with an approximate size of 0.1–1.0 μm with DI water and some commercial cleaning solution. The cleaning efficiency with the megasonic non-contact cleaning procedure with proper processing materials and experience is almost 90–95% (Fig. 4.63b).

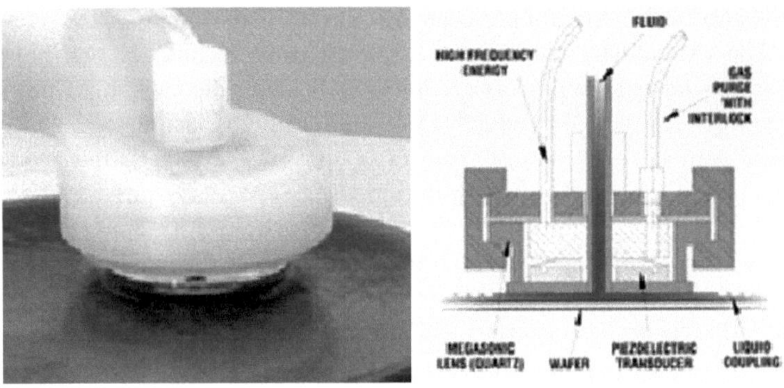

(a) (b)

Fig. 4.63 (**a**) A single wafer megasonic scrub and (**b**) the unique design of a 40 mm transducer with programmable chemical flow rate, spindle speed, and water rinsing system (Reprinted with permission SSE Corp.)

The cleaning process for post-etch polymer residues in sub-100 nm node technology is becoming critical. Conventional etching processes no longer provide the required selectivity, uniformity, corrosion resistance, and repeatability. The traditional wet cleaning approaches for BEOL as used in existing tools currently use hydroxylamine and amine, with process temperatures varying from 60 to 120 °C. One such residue removal process has been performed with an automated tool, which is flexible in terms of the chemical compatibility of the materials of construction, process temperatures and chemical dispense time. Figure 4.64 shows a batch processor which utilizes centrifugal force, center and side chemistry dispense, fresh and re-circulated chemicals and temperature monitoring systems at the heater and process bowl.

The intermediate rinse after residue removal with isopropyl alcohol is incompatible with the Cu- damascene process, and a new generation of cleaning mixture of ammonium fluoride (NH_4OH) with tetramethyl ammonium hydroxide (TMAH), and various organic acids has proven to be a very effective cleaning solution. Most of all, these chemicals are water soluble, work at room temperature, and do not need any rinse in between.

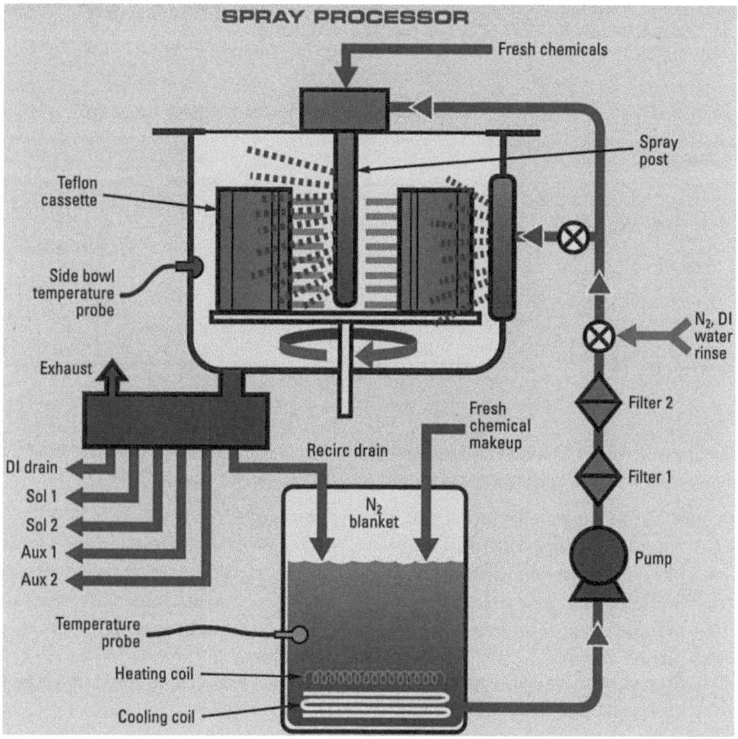

Fig. 4.64 A batch processor to remove residue for BEOL applications (Reprinted with permission Semiconductor International. Aug. 2003, p. 47)

Recently, removal of particulates during post-CMP cleaning by solid carbon dioxide (CO_2) followed by sweeping in CO_2 gas has been successful with low-K dielectric films which are hydrophobic in nature. The method is known as *aqueous cryogenically enhanced* (ACE) technology [129–130]. Most of the research and development work with supercritical fluids (SCFs) for the cleaning procedure has focused on CO_2 because of cost and prior experience. Owing to its high vapor pressure supercritical CO_2 can be readily separated from wafer cleaning by-products and can be recycled. In cryogenic cleaning, liquid CO_2 is used at a high pressure to dislodge particles from the surface.

The major cleaning procedure after photoresist stripping and etch/ash residue removal has attracted much attention. Different cleaning procedures such as dry-cleans, wet cleans, supercritical cleans, and cryogenic aerosol including laser cleans have been tried from time to time to investigate the yields. In fact, far from getting easier, resist stripping is becoming increasingly complex, because of the complexity of so many exotic materials (Fig. 4.65).

From our practical experience, it seems that successful removal of particles after stripping cannot be done with a single wet or dry cleaning process. Meanwhile, the debate between dry and wet cleaning rages on.

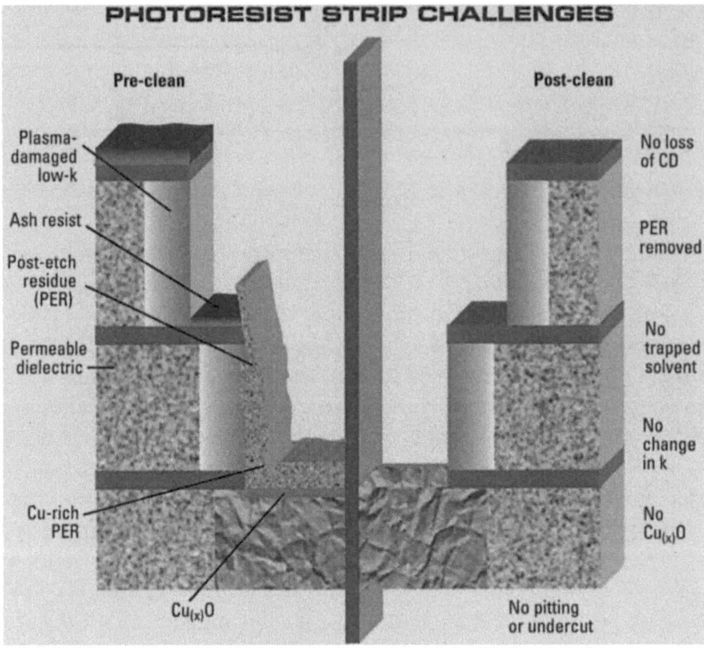

Fig. 4.65 Different kinds of challenges during photoresist striping, and cleaning (Reproduced with permission Semiconductor International, Jan. 2006, p. 30)

4.3 Summary

Due to the interaction of more complex designs with rapidly shrinking process windows, pattern-related yield loss is dominating early production. As a matter of fact, when the device size goes below 130 nm, the critical dimension (CD) limited yield is by far the most challenging problem because it might change the effective focus and exposure dose. Although CD control is absolutely critical for 130, 90 and 65 nm device nodes, the photoresist processing tracks must also provide acceptable process windows (exposure latitude and depth of focus) [131]. It is true that the complexity of track processes has increased five to ten fold as the lithography processes especially for gates, implants, contacts, and dual damascene are customized. A single coater can have 15–20 nozzles to accommodate 15–20 different photoresists (PR). An antireflective coating (ARC), on the other hand, is coated in a separate module. The post-exposure bake is typically between 110 and 130 °C. This process temperature must be ramped within 60–90 s. However, the advanced track system has concentrated on designing better hot plates, which can provide uniform steady heat within a short time. With advanced technology node, it has become increasingly essential to fix the transfer time between the scanner exposure and the start to post-exposure bake [131]. The processing of resist (PR) before post-exposure bake (PEB) is equally important for successful implementation of the resist process. The

PR, which is hydrophobic before PEB becomes hydrophilic after PEB. Thus one should be careful during the developing process, because the polymer becomes soluble in tetramethyl ammonium hydroxide (TMAH). Thus dispensing the TMAH is critical to the performance of the developing step [131].

For production manufacturing, 248 nm lithography with enhancements appears to work well within the range of 120 nm. Beyond this, 193 nm (ArF) lithography will cover 90 nm devices. However, there are too many challenges for 193 nm patterning, including CD control, *overlay accuracy*, and transfer of learning from 248 nm resolution enhancement technologies to the 193 nm system. The following table shows the lithography overlay requirements for the future (Table 4.7).

Table 4.7 Lithography overlay requirements

Year of production	2003	2004	2005	2006	2007	2010	2013	2016
Node (nm)	100	90	80	70	65	45	32	22
Overlay budget (nm)	35	32	28	25	23	18	13	9

Reprinted with permission: ITRS (2003), International SEMATECH, Austin, TX.

A critical component in 193 nm lens systems is CaF_2. Many remain concerned about the supply. Meanwhile a new laser architecture has generated a flurry of interest because it is possible to cover the demand of all the three wavelengths 248, 193 and 157 nm by increasing the power levels of the laser. The idea is that separating the master oscillator from the power amplifier, a much narrower bandwidth can be covered without sacrificing laser power.

In the field of etching, there are several ways to implement the task of etching in the dual damascene process including the via first, line first, and buried etch mask/self aligned via methods. In earlier generations, via and line dielectric layers were separated by a thin intermediate layer, such as SiN or SiC, as a trench etch stop. More advanced processes omit this layer to reduce cost and to improve performance. But the challenge is to prevent fence or veil formations during via first etch. During low-K and FSG (fluoro-silicate)-type processes one has to be careful during etching trench over via [132]. If one is not careful enough it may end up with a veil or a fence. The high-K materials are also problematic not only because of their high K values and etching problems but also because of the associated electrode material compatibility.

In some larger logic devices, some processing engineers are optimistic about dry etching of copper in inter-level metal interconnects. When the feature size is small (less than 0.13 μm) high-density plasma with a fast etching rate may be a better way to achieve a desired etching profile.

It seems that hard masks and etch stop will become the decisive factors in dielectric etch. At the same time, profile control is necessary to meet the critical dimension (CD) and electrical specification [95–133]. However, tapering damascene trenches and via holes by a few degrees enhances metal step coverage and has been shown to impact copper stress migration. Experimental studies show that via etches should be adequately selective to dielectric diffusion barriers covering the lower metal level. Otherwise, copper will be exposed to the sputtering event, which will be detrimental to the interconnect copper layer and to overall yield. At the same time, the etching level of photoresist and the polymer residue should be reduced to a stage where the surface adhesion of the substrate will not impact the step coverage or will not interact with the metal, which has to be deposited later. A well controlled etch with defined edge and desired dimensions is always expected. Thus the dry-etch system has become a very popular and widely used technology although some improvements are necessary with existing dry etching systems.

In the cleaning area, the post-cleaning is very critical because of particle and metallic contaminations from the slurry particles. There are several contact and non-contact procedures available for post-CMP cleaning. The most important goal of the cleaning procedure should be such that no residue is left over after post-CMP cleaning that can affect the surface by contamination and corrosion.

References

1. J.G. Ryan, R.M. Geffken, N.R. Poulin, and J.R. Paraszczak, The Evolution of interconnection Technology at IBM, IBM J. Res. Dev., 39, 371 (1995), and J.D. Meindl, Interconnection limits on XXI century Giga scale integration, Adv. Interconnect and Contacts Mater and process for future ICs MRS Symp., (April 13–16, 1998) (Bell Lab, NJ.) and also A.V. Vairagar et al., J. electrochem. Soc., 153 (9), G-840 (2006)
2. D. Edelstein, J. Heidenreich, R. Goldblatt, W. Cote, C. Uzoh, N. Lustig, P. Roper, T. Mcdevitt, W. Motsiff, A. Simon, J. Dukovic, R. Wachnik, H. Rathor, R. Schutlz, L. Su, S. Luce, and J. Slattery, Full copper wiring in a sub-o.25 μm CMOS ULSI technology, Proc. IEEE, IEDM, pp. 773–776, (1997) and also A.R. Kumar, C.M. Tan, T.K.S. Wong, and C-H Tung, Semicond. Sci. Technol., 21, 1369 (2006)
3. S.J. McNab and R.J. Blaikie, Appl. Opt., 39, 20 (2000) and B. LeGratiet et al., Proc. SPIE, 6922 (33), (Feb., 2008), and Proc. SPIE, 6925, 6925E (April, 2008)
4. J.B. Pendry, Phys. Rev. Lett., 85, 3966 (2000)
5. R.J. Blaike and S.J. McNab, Microelectron. Eng., 61–62, 97 (2002)
6. M.J. Bowden, L.F. Thompson, and C.G. Wilson (Eds.), Introduction to microlithography, Am. Chem. Soc., Washington DC, p. 70, (1994) and also G.M. Wallraff and W.D. Hinsberg, Chem. Rev., 99, 1801 (1999)
7. P.S. Silverman, Intel Technol. J., 6 (2), 55–61 May 16, (2002)
8. H. Liu, L. Karklin, Y. Wang, and Y. Pati, Application of alternating phase shift masks to 140 nm gate patterning (II), Optical lithography XI, Proc. SPIE, 3334 (1998)
9. H. Liu and A. Wong, 150 nm DRAM optical lithography capability, ARCH microlithography interface, Proc., (1999) and also M.D. Levenson, Jpn. J. Appl. Phys., 33, 6765 (1994)
10. R.R. Kunz et al., Outlook for 157 nm resist design, Adv. Resist. Technol. Proc. XVI, SPIE, 3678, 13–23 March 13, (1999)
11. R.H. French et al., Novel hydrofluorocarbon polymers for use as pelicles in 157 nm semiconductor photolithography, J. Fluorine Chem., 122, 63 (2003)
12. International SEMATECH meeting, Los Angeles, (Jan. 30, 2004) and also M. Switkes and M. Rothchild, J. Vac. Sci. Technol., B-19, 2353 (2001) and A.K. Raub et al., Imaging capabilities of resist in deep ultra violet liquid immersion interferometric lithography, J. Vac. Sci. Technol., B-22 (6), 3459 (2004) and R.H. French et al, "Imaging of 32 nm 1:1 lines and spaces using 193 nm immersion lithography...", J. Microlitho. Microfab. Mcrosyst., Topical issue on hyper-NA imaging, 4 (3), 031103, (July-Sep., 2005)

13. A.K. Bates, M. Rothchild, T.M. Bloomstein, T.H. Fedynyshyn, R.R. Kunz, V. Liberman, and M. Switkes, Review technology for 157 nm lithography, IBM J. Res. Dev., 45 (5), 605 (2001) and also S. Kramer, R.R. Fuierer, and C.B. Gorman, Chem. Rev., 103, 4367 (2003) and D. Ginger, H. Zhang, and C.A. Mirkin, Agnew. Int. Ed., 43, 30 (2004)
14. P.M. Montgomery et al., High NA ArF lithography for 70 nm technology, Proc. Opt. Mirolith. XV, SPIE, 4691 (2004) and also R.H. French et al., Novel hydrofluorocarbon polymers for use of pellicles in 157 nm semiconductor photolitholithography, J. Fluorine chem., 122, 63 (2003)
15. S. Wolf and R.N. Tauber, Silicon processing for VLSI era, Vol. 1, Lattice Press, Sunset Beach, CA, p. 486, (1986) and also M. Fritze, 100 nm node lithography with KrF, SPIE Microlith. Conf. Santa Clara CA, (2001)
16. W. Scot Ruska, Microelectronic Processing, McGraw Hill, New York., Chapter 5, p. 163, (1987)
17. M. Bowden, A perspective on resist materials for fine line lithography, In L.F. Thompson, C.G. Wilson and J.M.J. Frechet (eds.), Materials for microlithography, ACS Pub. Washington, DC, (1984) and also C.W. Gwyn, Extreme ultraviolet lithography, J. Vac. Sci. Technol., B 16, 3142 (1998)
18. K. Kotoku, ArF exposure system with 0.85 NA for 90 nm node and beyond, SEMI Tech. Symp. Semicon, Japan, 3–19, (2002) and also S. Thompson et al., Technology Development Status, Semocond. Tech. Manuf. Intel Technol. J., 06 (2), (May 16, 2002)
19. R.R. Kunz, T.M. Bloomstein, D.E. Hardy, R.B. Goodman, D.K. Downs, and J.E. Curtin, J. Vac. Sci. Technol., 17, 3267 (1999)
20. S.C. Fu, H.W. Lin, W.Y. Chou, L.A. Wang, and K.H. Hsieh, Study of the synthesis and characterization of methacrylate photoresists, J. Appl. Polym. sci., 83 (9), 1860 (2002)
21. G.M. Wallraf and W.D. Hinsberg, Chem. Rev., 99, 1801 (1999) and also K. Paterson, M. Somervell, and C.G. Wilson, Solid state Technol., 43 (3), 41 (2003)
22. A.E. Feiring et al., Design of very transparent fluoropolymer resists for semiconductor manufacture at 157 nm, J. Fluorine Chem., 122, 11 (2003) and also A.K. Bates et al., Review of technology for 157 nm lithography, IBM J. Res. Dev., 45 (5), 605 (2001)
23. H. Ito, C.G. Wilson, and J.M. Frechet, US Patent, 4491628 (1985)
24. D.J.T. Hill, J.H. O'Donnell, and P.J. Pomery, Fundamental aspects of polymer degradation by high energy radiation, In L.F. Thompson, C.G. Wilson, and J.M.J. Frechet, Materials for Microlithography, Am. Chem. Soc., Washington DC, (1984) and also T.M. Bloomstein, M.W. Horn, M. Rothchild, R.R. Kunz, S.T. Palmacci, and R.B. Goodman, J. Vac. Sci. Technol., 17, 3267 (1999)
25. E. Reichmanis, O. Nalamasu, and F.N. Houlin, Acc. Chem. Res., 31, 659 (1999) and also D.P. Medeiros et al., Recent progress in e-beam resist for advance mask making, IBM J. Res. Dev., 45 (5), 639 (2001)
26. S.C. Fu, K.H. Hsieh, and K.H. Hsieh, The kinetics of negative tone acrylic photoresist for 193 nm lithography, J. Polym. Sci. A: A Polym. Chem., 38, 954 (2000)
27. R.R. Kunz et al., J. Vac. Sci. Technol., B 17, 3267 (1999) and also T.H. Fedynyshyn et al., Proc. SPIE, 3999, 335 (2000)
28. O. Nalamasu et al., J. Vac. Sci. Technol., B 16, 3716 (1998) and also M. Rithchield et al., J. Vac. Sci. Technol., B-17, 3262 (1999)
29. A.J. Pasquale, Ph.D. Thesis, Virginia polytech. Inst. VA, (April 22, 2002)
30. F.M. Houlihan, T.I. Wallow, O. Nalamasu, and E. Reichmains, Macromolecules, 30, 6517 (1997)
31. U. Okoroanyanwu, J. Byers, T. Shimokawa, and C.G. Wilson, Chem. Mater., 10, 3328 (1998)
32. K. Nozaki and E. Yano, Fujitsu Sci. Tech. J., 38, 1 (2002) and J. van Olmen et al., Integration of 50 nm half pitch single damascene copper trenches, AMC Tech. Dig., 5 (2007)
33. S.C. Fu and K.H. Hsieh, The kinetics of a negative-tone acrylic photoresit for 193 nm lithography, J. Polym. Sci., A: Polym. Chem., 38, 954–961 (2000)

34. Optics and Photonics News, (May, 2003) and also S.D. Berger et al., J. Vac. Sci. Technol., B-9 (6), 2996 (1991)
35. Y. Zhang, P. Ware, K. Kotoku, and Y. Yamada, Potential KrF scanning lithography for 130 nm technology node, Future Fab. Int., ICG Pub., UK, 15 (Nov., 2003) and also R. Richhoeff et al., Patterning curved surfaces, J. Vac. Sci. Technol., B-17 (6), 2965 (1999)
36. M. Noguchi, M. Muraki, Y. Iwasaki, and A. Suzuki, Proc. SPIE, 1674, 92 (1992)
37. M. Hussein, S. Shivakumar, R. Brain, B. Beattie, P. Nguyen, and M. Fradkin, A novel approach to dual damascene patterning, Proc. IEEE-IITC, p. 18 (2002) and also J. Kennedy, T. Baldwin, N. Hacker, and R. Spears, Spin on glass antireflective coatings for photolithography, US patent 6268457 (1999)
38. P. Schiavone, C. Esclope, and A. Halimaoui, SiON-based antireflective coatings for 193 nm lithography, Proc. SPIE-Microlithography, (1999), Advances in resist technology and processing XVI, 3678, 1091–1095 (1999) and also M. Hussein, S. Sivkumar, R. Brain, B. Beattie, P. Nguyen, and M. Fradkin, A novel approach to double damascene patterning, Proc. IEEE-IITC, p. 18, (2002)
39. T.G. Tessier, J.M. Frechet, C.G. Wilson, and H. Ito, The photo-fries rearrangement and its use in polymeric imaging system, L.F. Thompson, C.G. Willson. And J.M.J. Frechet, (eds.), ACS Pub. Washington, DC, p. 287, (1984) and also R.D. Allen, G.M. Wallraff, D.C. Hofer, and R.R. Kunz, Photoresists for 193 nm lithography, IBM J. Res. Dev., 41 (1/2), 10 (1997)
40. X. Shao, A. Guerrero, and Y. Gu, SEMICON, China, (March 17, 2004) and also Q. Lin et al., SPIE, 3678, 241 (1999) and S.A. Rasgoa, Ph.D. thesis, MIT, (Oct. 21, 2004)
41. M. Switkes, M. Rothchild, R.R. Kunz, S.Y Baek, D. Coles, and M. Yeung, Immersion lithography: Beyond the 65 nm node with optics, Microlithography World, p. 4, (May, 2003) and K. Ishimaru, 33rd Euro Solid State Circuit Conf., 11–13, 32 (2007)
42. M.D. Levenson, N. Viswanathan, and R.A. Simpson, IEEE Trans. Electron. Dev., ED-29, 1828 (1982) and also P.F. Cacia, G. Huges, R.H. French, C. Torardi, G. Reynolds, L. Dieu, Thin films for phase shift masks, Vac. & thin films, HIS Pub., 14–21, (Sept., 1999) and R.R. Kunz, R.R. Rothchild, and M.S. Yeung, J. Vac. Sci. Technol., B21, 78 (2003) and H. Dang, J.L.P. Tan, and M.W. Horn, J. Vac. Sci. Technol., B21, 1143 (2003), and P.F. Carcia, R.H. French, M.H. Reilly, M.F. Lemon, and D.J. Jones, Optical superlatices—a strategy for designing phase shift masks for photolithography at 248 and 193 nm, Appl. Phys. Lett., 70 (18), 2471 (1997) and also K.M. Lee et al., Proc. SPIE, 6518, 651514 (2007)
43. M.D. Levenson, Jpn. J. Appl. Phys., 33, 6765 (1994), and also J.G. Goodberlet and H. Kavak, Appl. Phys. Lett., 81, 1315 (2002) and J. Maria, V. Malyachuk, J. White, and J.A. Rogers, Experimental and computational studies of phase shift lithography with binary elastomeric mask, J. Vac. Sci. Technol., B24 (2), 828 (2006) and also F. Sundermann et al., Proc. SPIE, 7028, (April, 2008)
44. M.D. Levenson, N.S. Viswanathan, and R.A. Simpson, Improving resolution in photolithography with phase shifting mask, IEEE Trans. Electron. Dev., ED-29, 1828–1836 (1982) and also P.F. Garcia et al., Thin film for phase shift masks, Vacuum and thin film, HIS publishing group, 14–21, (Sept., 1999) and also T. Schram et al., VLSI Tech Symp., (2008)
45. K.K. Shih and D.B. Dove, Thin film materials for preparation of attenuating phase shift masks, J. Vac. Sci. Technol., B12, 32–36 (1994) and J. Maria, V. Malyachuk, J. White, and J.A. Rogers, Experimental and computational studies of phase shift lithography with binary elastomeric mask, J. Vac. Sci. Technol., B24 (2), 828 (2006)
46. O. Hunderi, Physica, A 157, 309 (1989) and also P.F. Carcia, R.H. French, M.H. Reilly, M.F. Lemon, and D.J. Jones, Optical superlattices—a strategy for designing phase shift masks for photolithography at 248 and 193 nm: Application to AlN/CrN, Appl. Phys. Lett., 70 (18), 2371 (1997)
47. D.J. Jones, Optical superlatices—a strategy for designing phase shift masks for photolithography at 248 and 193 nm, Appl. Phys. Lett., 70 (18), 2471 (1997)
48. D. Schurz, L. Kirklin, X. Li, and T. Weed, Resolution enhancement technology for 157 nm lithography of 100 nm and 70 nm devices, Adv. Reticle Symp., (2001) and also A. Wong,

Resolution enhancement techniques of optical lithography, Opt. Eng. V.TT47 SPIE, (2001)
49. J. Burnett, Z. Levine, and E. Shirley, Intrinsic birefringence in 157 nm materials, 2nd Int. Symp. On 157 nm lithography, Dana Point, CA (2001) and also J. Brunett, R. Gupta, and U. Griesmann, Proc. SPIE, 4000, 1503 (2000)
50. D. Anberg and W. Flack, Optical lithography aims for 157 nm, Photonics spectra, 34 (12), 74 (2000)
51. The International Technology Road map for Semiconductors, Lithography Technology Requirements-Near Term, 2000 update, Int. SEMATECH, Austin TX.
52. A.K. Bates, M. Rothschild, T.M. Bloomstein, T.H. Fedynyshyn, R.R. Kunz, V. Liberman, and M. Switkes, IBM. J. Res. Dev., 45 (5), 605–614 (Sept., 2001) and also J. Webb, Laser Focus World, p. 87, (Sept., 2000)
53. R. Roger et al., J. Fl. Chem., 122, 63 (2000) and P. De Bisschop et al., Proc. SPIE, 6730, 67301G (2007)
54. R.R. Kunz, T.M. Bloomstein, D.E. Hardy, R.B. Goodman, D.K. Downs, and J.E. Curtain, Outlook for 157 nm resist design, J. Vac. Sci. Technol., B17, 3267 (1999)
55. V. Liberman et al., Materials issues for optical components and photomasks in 157 nm lithography, J. Vac. Sci. Technol., B17, 3273 (1999) and also T. Hoffman et al., Revising F_2 laser for DUV lithography, Emerging lithography technologies, SPIE Proc., 3676 (1999)
56. J. Wilson and J.F.B. Hawkes, Optoelectronics, 2nd ed. Prentice Hall, Hemel Hempstead, p. 78, (1989)
57. J.H. Burnett, Z.H. Levine, and E.L. Shirley, Intrinsic birefringence in 157 nm materials, 2nd Int. Symp. On 157 nm lithography, Dana point CA, (May 14–17, 2001)
58. K.A. Pandelisev, Purification and crystal growth of CaF_2 having superior properties, 2nd Int. Symp. On 157 nm lithography, Dana point CA, (May 14–17, 2001) and also C.M. Smith and L.A. Moore, Fused silica for 157 nm transmittance, Emerging lithography, Tech III Proc., SPIE, 3676 (1999)
59. C.M. Smith and L.A. Moore, Proc. SPIE, 3676, 834 (1999) and also A.K. Bates et al., Review of technology for 157 nm lithography, IBM. J. Res. Dev., 45 (5), 605–614 (2001)
60. T.K. Gupta, Effect of temperature on the photosensitivity of a photoresist (AZ 1350 J), Euro. Polym. J., 17, 1127–1130 (1981)
61. L.A. Moore and C. Smith, Properties of fused silica for 157 photomasks, 19th Annual Symp. on Photomask Tech. Proc. SPIE, 3873 (1999) and also W. Flck et al., Simulation of subhalf-micron mask defect printability at XI reticle magnification , Integrated circuit metrology inspection and process control XI Proc. SPIE, 3050 (1997)
62. M. Cangemi, Masking materials for 157 lithography, 19th Annual Symp. Photomask Tech. Proc. SPIE, 3873 (1999) and also T. Bloomstein , V. Liberman and M. Rothchild, Optical materials and coatings at 157 nm emerging lithographic technology III Proc. SPIE, 3676 (1999)
63. T. Bloomstein, M.W. Horn, M. Rothchild, R.R. Kunz, S.T. Palmacci and R.B. Goodman, J. Vac. Sci. Technol., B-15, 2112 (1997)
64. R.E. Miller et al., The development of 157 small field and mid field microsteppers, SPIE Symp., 4000-174, (2000)
65. M. Bohn, H.C. Hamaker, and W. Montgomery, Implementation and characterization of a DUV raster scanned mask pattern generation system, 21st Annual BACUS Symp. On Photomask technology, (2001) and also K. Harda, T. Tamamura, and O. Kogure, J. Electron. Chem. Soc., 129, 2576 (1982) and D.R. Medeiros et al., Recent progress in e-beam resist for advanced mask making, IBM J. Res. Dev., 45 (5), 639 (2001)
66. B.J. Greno (ed.), Proc. SPIE, 4562, 16 (2001) and also H.C. Peifer and W. Stickel, Microelectron. Eng., 27, 143 (1995)
67. M. Born and E. Wolf, Principles of Optics, 6th ed. Pergamon Press, New York, pp. 329–333, (1980) and also K. Nozaki and E. Yano, Fujitsu Sci. Tech. J., 38 (1), 3 (2002)

68. J.L. Mauer and R.A. Carruthers, Selective etching of SiO_2 with CF_4/H_2 plasma, Proc. 21st Electron. Matts. Conf. Boulder CO (1979) and also A.K. Bates, Review of technology for 157 nm lithography, IBM J. Res. Dev., 45 (5), 609 (2001)
69. R.H. French et al., Novel hydrofluorocarbon polymers for use as pellicle in 157 nm semiconductor photolithography, J. Fluorine Chem., 122, 63 (2003)
70. B.J. Lin, J. Vac. Sci. Technol., 12 (6), 1317 (1975)
71. T. Iwayanagi, T. Kohashi, S. Nonogaki, J. Electrochem. Soc., 127, 2759 (1980)
72. H. Ito, Chemical amplification resists: History and development of within IBM, IBM J. Res. Dev., 41, 69–80 (1997) and also M. Kurihara, T. Segawa, D. Okuno, N. Hayshi, and H. Sano, Proc. SPIE, 3412, 279 (1998)
73. M.K. Crawford et al., Proc. SPIE, 3999, 357 (2000) and H. Saitoh, T. Saga, S. Kobu, S. Sanki, and M. Hoga, Proc. SPIE, 3412, 269 (1998)
74. A.E. Feiring et al., Design of very transport fluoropolymer resists for semiconductor manufacture at 157 nm, J. Fluorine Chem., 122, 11 (2003)
75. K.J. Przybilla, H. Roschert, G. Pawlowski, Adv. Mater., 4, 239 (1992) and also I. Ito et al., J. Vac. Sci. Technol., B-19, 2678 (2001)
76. F. Cerrina, H. Guckel, J. Wiley, and J. Taylor, J. Vac. Sci. Technol. B-3, 459 (1984)
77. A.M. Hawryluk and L.G. Seppala, J. Vac. Sci. Technol., B6, 2162 (1988)
78. H. Kinoshita, K. Kurihara, Y. Ishii, and Y. Torii, Soft X-ray reduction lithography using multilayer mirrors, J. Vac. Sci. Technol., B-7 (6), 1648 (1989)
79. M. Singh and J. Braat, Design of multilayer extreme ultraviolet mirrors for enhanced reflectivity, Appl. opt., 39 (13), 2189 (2000)
80. K. Bergmann, O. Rosier, W. Neff, and R. Lebert, Pinch plasma radiation source for extreme ultraviolet lithography with kilohertz repetition frequency, Appl. Opt., 39, 3833 (2000)
81. K. Bergmann, G. Schriever, O. Rosier, W. Neff, and R. Lebert, Highly repetitive extreme ultraviolet radiation source based on a gas-discharge plasma, Appl. opt., 38, 5413 (1999)
82. M.A. Klosner and W.T. Silfvast, Intense xenon capillary discharge extreme ultraviolet source in the 10–16 nm wavelength region, Opt. Lett., 23, 1609 (1998)
83. A.G. Michette, Optical system for soft X-rays, Plenum Press, New York, (1986)
84. W.T. Silfvast, M.C. Richardson, H. Bender, A. Hanzo, V. Yanovsky, F. Jin, and J. Thrope, Laser induced plasmas for soft X-ray projection lithography, J. Vac. Sci. Technol., B-9 (6), 3176 (1991)
85. C. Jiang, Q. Yao, B. Eccles, A. Kuthi, and M.A. Gundersen, Pseudospark discharge-based extreme-ultraviolet radiation source, J. Vac. Sci. Technol., B-21 (6), 2843 (2004)
86. P.A. Kearney, C.E. Moore, S.I. Tan, and S.P. Vernon, Mask blanks for extreme ultraviolet lithography: Ion beam sputter deposition of low defect density Mo/Si multilayers, J. Vac. Sci. Technol., B15, 2452–2454 (1997)
87. D. Sweeney, EUVL Prog. Rept. S8TR Nov. 1999 and C.W. Gwyn, Extreme ultraviolet lithography, J. Vac. Sci. Technol., B-16, 3142 (1998)
88. D.R. Herriot, Electron beam lithography machines, G.R. Brewer (ed.), Electron beam technology in microelectronic fabrication, Academic Press, New York, (1980)
89. H.C. Pfeiffer, G.O. Langer, and M.S. Sturans, Electron beam lithography system, US Patent 5,466, 904; and also H. Yasuda et al., J. Vac. Sci. Technol., B14 (6), 3813 (1996)
90. D.R. Medeiros et al., Recent progress in electron beam resist for advanced mask-making, IBM J. Res. Dev., 45 (5), 639–650 (2001) and also S. Wolf and R.N. Tauber, Silicon Processing in VLSI Era, Vol. 1, p. 501, Lattice Press, Sunset Beach, CA, (1986)
91. R.S. Dhaliwal et al., PREVAIL electron projection technology approach for next generation lithography, IBM J. Res. Dev., 45 (5), 615–638 (2001)
92. C.S. Raffery, H.H. Vuong, S.A. Esharghi, M.D. Giles, M.R. Pinto, and S.J. Hillenius, IEDM Tech. Dig., Issue 14, 311 (1993)
93. D.J. Resnick et al., High-resolution templates for step and flash imprint lithography, J. Microlitho. Microfab. Microsyst., 1 (3), 284 (2002) and also A. Hooper et al., Step and

flash imprint lithography using UV-transparent, electrically conductive templates, Nanotech, Vol. 3, Chapter 1, NISTI Pub., Cambridge, MA (2003)
94. S.W. Crowder, P.M. Rausseau, J.P. Snyder, J.A. Scott, P.B. Griffin, and J.D. Plummer, IEDM Proc. p. 427, (1995)
95. F.H. Baumann, Semiconductor characterization, Present status and future needs, W.M. Bullis, D.G. Seiler, and A.C. Diebold (eds.), AIP Press, Washington DC, (1995) and S. Sankaran et al., IEEE IEDM Tech Dig., Issue 2, 6 (2006)
96. S.K. Krisch, J.D. Bude, and L. Manchanda, IEEE Electron. Dev. Lett., 17, 521 (1995)
97. H.H. Vuong et al., IEEE Trans. Electron. Dev., 43, 1144 (1996) and L. Broussous et al., IEEE IITC,, San Francisco, CA, (June, 2008)
98. F. Watanabe and Y. Ohnishi, J. Vac. Sci. Technol., B-4, 422 (1986)
99. N.J. Chou, C.H. Tang, J. Paraszczak, and E. Babich, Appl. Phys. Lett., 46, 31 (1985)
100. M. Pons, J. Pelletier, O. Joubertand, and P. Paniez, Jpn. J. Appl. Phys. Part-I, 34, 3723 (1995)
101. D. Eon, V. Rablland, G. Cartrt, M.C. Peignon-Fernandez, and Ch. Cardinaud, Eur. Phys. J. Appl. Phys., 28, 331 (2004)
102. N.P. Hacker, Organic and inorganic spin on polymers for low dielectric constant applications, MRS Bull., (Oct., 1997) and also J.C. Witzel, J. Interconnect Technol. Conf. (IITC) Proc., 99 (2000)
103. M.J. Laboda and G.A. Toskey, Understanding hydrogen silesquioxane-based dielectric film processing, Solid State Technol., 41, 99 (1998) and also H.J. Lee et al. Structural comparison of hydrogen silsesquioxane based porous low-K thin films prepared by varying process conditions, Chem. Mater., 14, 1845 (2002)
104. Oehrlein et al., J. Vac. Sci. Technol., A-18, 2742 (2000)
105. T. Standaert et al., J. Vac. Sci. Technol., A 16, 239 (1998)
106. T. Akimoto, H. Nanbu, and E. Ikawa, J. Vac. Sci. Technol., B-13, 2390 (1995)
107. M.F. Doemling, N.R. Rueger, and G.S. Oehrlein, Appl. Phys. Lett., 68, 10 (1996)
108. R. Barnett, L.A. Donahue, A. Newton, A. Baker, and J. Hopkins, Addressing side wall roughness, Solid State Technol., 47, 28 (July, 2004) and S.M. Rossangel and J.P. Hopwood, Appl. Phys. Lett., 63, 3285 (1993) and also M.J. Kushner, IEEE IITC, San Francisco, CA, (June, 2008)
109. M. Pons et al., Jpn. J. Appl. Phys. Part I, 33, 991 (1994)
110. R. Hasio et al., J. Electrochem. Soc., 144, 1008 (1997)
111. O. Joubert et al., J. Vac. Sci. Technol., B-15, 629 (1997)
112. A. Yahata, S. Urano, and T. Inoue, Jpn. J. Appl. Phys. Part I, 36, 6722 (1997)
113. Q. Lin et al., Proc. SPIE, 3678, 241 (1999) and also D.E. Seeger et al., IBM J. Res. Dev., 41, 105 (1997)
114. S.A. Rasgon, Ph.D. thesis, Oct. 21, Dept. of Chem. MIT, MA 02139, (2004) and S. Satyanarayana, R. McGovern, B. White, and S. Hosali, Damage mechanism in porous low-K integration, Semicond. Int., Issue 6, 32, June (2005)
115. S. Morishita et al., Plasma wall interaction in dual frequency narrow gap reactive ion etching system, Jpn. J. Appl. Phys., 37, 6899 (1998)
116. Y. Yonogami et al., Anisotropic etching of RuO_2 and Ru with high aspect ratio, for gigabit dynamic random access memory (DRAM), J. Vac. Sci. Technol., B-18 (4), 1911 (2000)
117. A. Kerber et al., Elect. Dev. Lett., 24, 87 (2003)
118. C. Hobbs et al., Trans. Electron. Dev., 51, 971 (2004)
119. Z. Ren, et al., IEDM Tech. Dig., Issue 4, 793 (2003)
120. M. Sternheim, W. Van Gelder, and A.W. Hartman, A laser interferometer system to monitor dry etching of patterned Si, J. Electrochem. Soc., 130, 655 (1983) and also, P.J. Marcoux and P.D. Foo, Methods of end point detection for plasma etching, Solid State Technol., 19, 115 (April, 1981) and L. Sha and J.P. Chang, Plasma etching of high dielectric constant materials on silicon in halogen chemistry, J. Vac. Sci. Technol. A., 22 (1), 88 (2004)

121. M.Case et al., Selective wet etching of Hf-based layers, Abstract 549, 204th meeting of Electrochem. Soc., (2003) and J.S. Starzynski, Selective removal of Hf-based high-K films, Semicond. Int., 29 (4) 42 (April, 2006)
122. V. Lowaleker and S. Raghavan, Etching of zirconium oxide, hafnium oxide, and hafnium silicates in dilute hydrofluoric acid solutions, J. Mater. Res., 19 (4), 1149 (2004) and K.L. Saenger, H.F. Okorn Schmidt, and C.P. D'Emie, A selective etching process for chemically inert high-K metal oxides, in Novel materials and process for advanced CMOS, Mater. Res. Soc. Proc., 745, 79 (2003)
123. F. Laermer and A. Schlip, Patent No. DE4241045 (US-5501893), (1994)
124. S.K. Ghandhi, VLSI fabrication principles, Wiley, New York, p. 516, (1983) and K. Karuppiah et al., CMP-MIC Tech. Dig., Issue 2, 45–54, Feb. (2006)
125. F. Daniels and R. Alberly, Physical chemistry, Wiley, New York, p. 126, (1981) and also S. Broydo, Important considerations in selecting anisotropic plasma etching equipment, Solid State Technol., p. 159, (April, 1983) Technology News, Solid State Technol., p. 32, (Dec., 2000) and D. Zeidler, Z. Stavreva, M. Plotner, and K. Drescher, The interaction between different barrier metals and copper surface during chemical mechanical polishing, Microelectron. Eng., 37/38, 237 (1997) and L. Wang and F.M. Doyale, Known effects of pattern characteristics on copper CMP and future directions, Proc. 9th Int. Conf. On CMP for ULSI multilevel interactions, Feb. (2004)
126. J.A. Greenwood and J.B.P. Williamson, Contact of normally flat surface, Proc. R. Soc. London Ser. A, 295, 300 (1966)
127. K.L. Jackson, Contact mechanics, Cambridge University, Press, Cambridge, UK, (1985)
128. B. Bhusan (ed.), Hand book of micro/Nano tribology, 2nd ed. CRC Press, Boca Raton, FL, (1999) and J.G. Park and A. Busnaina, Cu-post CMP cleaning, Semicond. Int., 28 (8), 39 (2005)
129. N. Moumen, M. Guarrera, J. Piboontum, and A.A. Busaina, Contact and non-contact Post CMP cleaning of thermal oxide silicon wafers, Proc. 10th Annual IEEE/SEMI. Advanced Semiconductor Manufacturing Conf. Boston, MA, Sept. 8–10, p. 250, (1999)
130. A.A. Busnaina, T.M. Elsawy, Post CMP cleaning using acoustic streaming, J. Electron. Mater., 27 (10), 1095 (1998)
131. V.B. Menon and R.P. Donovan, Handbook of semiconductor wafer cleaning technology, W. Kem (ed.), Noyce Pub., Port Republic, **NJ** (1993)
132. S. Banerjee, H.F. Chung, R.J. Small, and C. Shang, Copper interconnects new contact metallurgies/structures and low-K interlevel dielectrics/2002 G.S. Mathad et al. (eds.), PV 2002-22, p. 293 (2002), The Electrochem. Soc. Pub. Pennington, NJ, (2003)
133. H. Kyodaet al., Improvement of CD control ability in developer process, SPIE microlithography Proc. 5039, (2003) and also G.W. Ray and S. Lassing, Semicond. Int., 24, 96 (2001), and L. Peters, Designing tracks for better CD control, Semicond. Int., 26 (9), 54 (2003), and B. Cohen et al ., Dynamic temperature profiling for post exposure bake, Semicond. Int., 26 (2), 12 (2003) and M. Fritze et al, Nanofabrication with deep ultraviolet lithography and resolution enhancement, J. Vac. Sci. Technol., B17, 3310 (1999)

Chapter 5
Deposition Technologies of Materials for Cu-Interconnects

5.1 Introduction

Scaling of the feature size from 250 nm to sub-100 nm has restricted the use of aluminum (Al) interconnects. At the same time demands for higher speed, better performance of the scaled circuits, and thinner gate material need better dielectric materials other than silicon dioxide (SiO_2). As a result, copper has replaced Al-interconnect and low-K interlayer and high-K gate dielectric materials have replaced SiO_2. Deep submicron copper interconnects cannot be formed by using the conventional cloisonné approach which is ubiquitous in Al metallization. Experimental evidence shows that dry etching of Cu is difficult and photoresist work cannot withstand the temperatures required for Cu-etching (>200 °C). Moreover, wet etching and lift-off techniques of Cu have been attempted without much success. So a new process technology known as the *damascene* process has been introduced to integrate Cu-interconnects in modern integrated circuits (ICs).

Figure 5.1 shows the conventional aluminum (Al) metallization technology versus modern copper (Cu) metallization technology to fabricate interconnecting lines. In Al-technology a thick titanium (Ti)-based shunt (TiN) layer is deposited first after surface treatment with silane (SiH_4) to enhance nucleation of the subsequently deposited tungsten (W) film [1] for via filling. It also uses tungsten (W) as local as the local wiring and contact level metal for the devices in microprocessors, ASIC and DRAMs. On the other hand, in the Cu-damascene technology, a Ta/TaN layer is deposited for the barrier and Cu is used as via filling and interconnecting material.

As the technology shifted from 250 to 100 nm node, the Ti-based shunt layer with tungsten via filled metal became inadequate to support sub-100 nm Cu-interconnects. Moreover, titanium chloride ($TiCl_4$) based titanium nitride (TiN) film was found to develop *micro-cracks* and *stress* due to the presence of higher percentages of chlorine in $TiCl_4$. The *stress* is even higher when Cu-interconnects in the the Cu-damascene process are confined within the barrier layer. Moreover, the presence of chlorine in the $TiCl_4$ can form copper chloride compound, which is not very stable at higher processing temperatures, especially during plasma processing. On the other hand, a Ta-based barrier layer (Ta/TaN) has been very successful for Cu-interconnects, especially when the barrier layer is very thin.

Tapan K. Gupta, *Copper Interconnect Technology*,
DOI 10.1007/978-1-4419-0076-0_5, © Springer Science+Business Media, LLC 2009

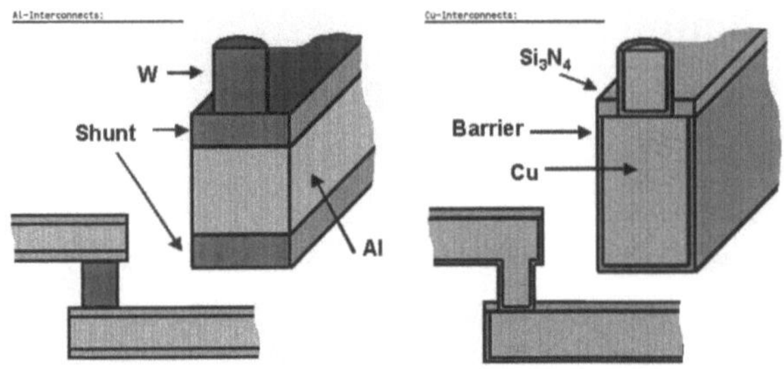

Fig. 5.1 Aluminum versus copper technology (Reprinted with permission Mat. Res. Soc. 2002 Spring meeting)

As the supporting materials for Cu-interconnect have changed, and the feature sizes of via holes and trenches are becoming smaller, the deposition technologies also need to be replaced by either modifying the existing deposition technologies or introducing new technologies. For example, long throw ionized physical vapor deposition (IPVD) has replaced conventional sputtering, the old chemical vapor deposition (CVD) technique has been modified, the old-fashioned recipe and the processing technology for copper plating used in printed circuit boards (PCBs) have been changed, and a new atomic layer deposition (ALD) technology has emerged into the arena as copper became the mainstream interconnecting material. On the other hand, sputter deposited barrier layers of Ta/TaN have been reported from time to time by many investigators. These sputtering systems are mainly magnetron sputtering with special attachments, and are named differently as (a) ion-metal plasma sputtering [2], (b) ionized PVD [3], and (c) reactive sputtering [4]. On the other hand, a CVD barrier layer of Ta/TaN has also shown several filling advantages [5].

5.2 Emerging Technologies

5.2.1 Cu-Damascene Process

In the copper damascene process, copper (Cu) interconnects are formed by depositing a *thin film* of copper on the patterned surface. The Cu-lines forming interconnections must retain their structural integrity [6–7] without deformation, loss of adhesion [8–9], void formation [10], and loss of conformality [11–14]. Scaling of the feature size to sub-100 nm level has introduced additional reliability issues about thinning of interconnects and high aspect ratio via-hole fillings. Damascene Cu-filling is performed by electrodeposition of copper film inside the patterned area. Electrodeposition of copper requires a thin (a few ångstroms thick) Cu-seed layer for nucleation, which is deposited on the top of a *barrier layer*. It is necessary to attain 100% copper coverage with the seed layer to prevent voiding in the acidic plating bath.

5.2.2 Barrier Layer Requirements

The barrier layer is needed to avoid Cu-diffusion inside the dielectric layer (low-K or silicon dioxide (SiO_2)). Copper (Cu) is known to be a fast diffuser in silicon where it can act as a deep level acceptor and can degrade minority carrier lifetime, causing high junction leakage in transistors and short retention time in DRAMs. Cu also diffuses through SiO_2, especially under electrical bias. The barrier layer is generally selected from transition metals and their compounds. The barrier layer thickness must be kept to a minimum in order to preserve the effective conductive advantage of Cu over Al alloy since the bulk resistivity of the barrier layer material is higher than the bulk resistivity of Cu. The barrier layer thickness can vary from 10 to 20 nm and it is hard to deposit continuous, conformal, void and stress free thin dielectric layers of 10–20 nm with existing conventional deposition systems.

The other aspects that should be taken into account are the barrier layer's texture and roughness because, as a substrate, these two parameters will affect the structural growth of Cu-interconnects. The correlation between crystallographic texture and stress induced void formation in tantalum (Ta)-encapsulated Cu-interconnects reveals that <111> texture forms better resistance to void formation. The locally weaker <111> texture of Cu-interconnect near void location has higher diffusivity and twist boundaries [15]. The above requirements have generated much interest in evaluating the properties of the barrier layer before Cu-deposition. As a matter of fact, there exists a strong concurrent effort to develop technologies capable of providing conformal coverage of the barrier layer in very high aspect ratio holes [16].

5.3 Deposition Requirements

Copper interconnect requires depositions of *thin films* mostly for (i) the copper seed layer, (ii) the copper blanket for interconnecting lines, (iii) the barrier layer, (iv) low-K dielectric materials, and (v) a cap layer/passivation layer.

The most important reason for the widespread use of *thin films* in the microelectronics industry is that it can be processed to yield small feature size, low power consumption, high circuit density, and can be applied to a small space. But continuing advances in the fields of very large scale integration (VLSI) of ultra-large scale integration (ULSI) and of the continued development of smaller devices aroused concern about the existing thin film deposition techniques for the copper seed layer, copper interconnects, barrier layers, cap layers, and high and low-K dielectric materials. The most important requirement is the stability of the deposited films during the IC fabrication procedures. As a result, there is ongoing research into the development of a suitable deposition process for a particular class of material (Table 5.1).

During the past decade, advances in the deposition of thin metal film for ultra-large-scale integrated (ULSI) circuits have been limited by the available deposition

Table 5.1 Supporting materials for the Cu-interconnects and deposition methods

Application	Materials	Process technology
Substrate	Silicon (Si), starting wafer	Single crystal, Czochralski
Dielectric	SiO_2, low-*K* (inorganic, organic, hybrid), High-*K* (gate)	CVD, PVD, spinning (SOD), ALD RTCVD
Cap, etch stop	SiN_x, CoWP, SiC	CVD, PVD, ALD, electroless (bath)
Barrier layer	Ta, TaN_x, TaC, WC, TiN, TiC	CVD, PVD, ALD
Cu-seed layer (nucleation)	Cu	ALD, PVD, electroless
Cu-blanket (Cu-interconnect)	Cu	ECD, PVD, CVD

technologies. The device size, the dimensions of via holes and trenches with high aspect ratios and the change in process technology (subtractive Al-interconnect to additive the Cu-damascene process) need more sophisticated deposition technologies, as the existing deposition technologies are not adequate to meet the current needs of modern ICs. We will discuss some of these deposition processes that are very frequently used to deposit thin films during the damascene process such as: (a) physical *vapor deposition (PVD)* [17–18], (b) *chemical vapor deposition (CVD)* [19–21], *including atomic layer deposition (ALD)* [22–25], and (c) *electrochemical deposition (ECD) including electro (-de) less* (bath deposition) deposition [26–31]. As the feature size of the devices has scaled down (from 10^{-7} to 10^{-9} m) communication through these high-speed devices (without compromising yield and cost) has become a challenge [32–34]. Thus an increase in process capability and control has been necessary in the field of thin film deposition and in the area of pattern definition.

5.4 Thin Film Growth and Theory of Nucleation

One of the most important criteria of the Cu-damascene process is deposition of void free, conformal thin films inside a small trench or via hole with high aspect ratio (AR). According to the report of International Technology Road Map for Semiconductors (ITRS), a 12:1 aspect ratio (AR) contact has to be filled for a dynamic random access memory (DRAM) at 90 nm, and 14:1 AR will be needed at 45 nm node technology [1]. However, for logic applications, though the ARs are smaller than DRAM, the requirements for seamless fill and chemical mechanical polishing (CMP) coring resistance to prevent copper contamination of the device are important.

Besides low-K dielectric for the interlayer/intermetal dielectric (ILD/IMD) and high-K dielectric gate material, copper interconnects require a barrier layer to protect the underneath metallization. Each barrier film has a unique solution to a deposition process. *Plasma enhanced chemical vapor deposition* (PECVD) deposits films that have different compositions on the field, sidewalls, and via-hole bottom and creates a difficult situation for film growth initiation. *Physical vapor deposition* (PVD), on the other hand, can deposit uniform films in the field and via holes, but show low step coverage along the sidewalls [35]. Moreover, ordinary sputtered deposition of alloys is difficult, and the presence of nitrogen or any other residual gas during reactive sputtering can produce stress in the film, which can yield increased film resistance. During step coverage, sputtered deposited films sometimes suffer from *thinning* and *cracking* too [36]. However, good trench filling has been reported through reflow after sputtering.

Due to its superior step coverage capacity over PVD, CVD has naturally received much attention, especially the *metal-organic CVD* (MOCVD) process, which can deposit films at much lower temperatures (150–200 °C) by thermal decomposition of the precursor. However, the use of the precursors is critical and the prices of the precursors are high. On the other hand, *rapid thermal CVD* (RTCVD) has been successfully used to deposit dual gates in CMOS processing at low temperatures [37] and is expected to be used for void free thin film deposition inside high aspect ratio trenches and via holes.

More recently, *atomic layer deposition* (ALD), which is based on sequential self-limiting surface reaction, has provided atomic layer control over films on high aspect ratio structures [38]. But the process is very slow, and for thick film deposition, the process can be expensive. However, ALD might be suitable for thin high-*K* dielectric materials and diffusion barrier layer back-end interconnects [39].

Electroplated (EP) copper is known to be self-annealing at room temperature and thermal treatment is necessary to avoid the self-annealing. Self-annealing drastically alters the microstructure of the Cu-film and the reliability of the film [40], such as electromigration, and the change in plasticity.

Electroless (electrode-less) plating, on the other hand, is a relatively cheap and simple process, but lacks process control during deposition. Moreover, the microstructure of electroless Cu-film generally consists of very fine grains implying poor electromigration (EM) resistance.

Experimental evidence shows that solution derived thin films are dependent on the grain growth. Grain growth is again dependent on the average interatomic distance, the average grain radius, the grain radius at the interface with substrate, the film thickness, and the interfacial surface and grain boundary energies, respectively. It is expected also that during Cu-plating additives in the electrolytic solution can influence the surface nucleation and the film structure [41–42]. Thus the study of the *theory of nucleation* is very important for deposition of a void free, conformal thin film over a small area of a sub-100 nm device [43–47].

5.4.1 Nucleation Theory

The process where a minimum number of ions or molecules come together to form a stable second phase is defined as *nucleation* [48]. According to the position of the arrays of the ions or molecules the nucleation can be *heterogeneous* or *homogeneous* [49–52]. Most nucleation theories postulate an equilibrium or steady state to exist between the adsorbed monomers, which diffuse on the substrate surface for a time τ constantly colliding with themselves and other *clusters* and aggregates of various sizes (Fig. 5.3a) [5, 53–57].

The term *cluster* can be best described as aggregates of atoms that are usually too large to be referred to as molecules and too small to resemble small pieces of crystal [5, 58]. Perhaps the most remarkable aspect of clustering, which is also true in the regime of inelastic collapse (Fig. 5.2a), is that it leads to long string-like grain configurations rather than to shapeless blob particles (Fig. 5.2b) [59–61].

Clusters are regarded as crystal defects in silicon and can become electrically harmful by acting as nucleation centers for the *precipitation of copper* as impurity. Studies of device physics have explored the possibility that junction leakage in p-n junction devices and charge storage failure in metal oxide semiconductor (MOS) devices are the results of generation currents originating at precipitates of copper and other metals in silicon [62].

The processes of making thin films always involve phase transformation and, as such, the properties of these films are different from their bulk material [63–64]. Thin film can grow in three well-known growth modes (Fig. 5.3). These modes are: (a) *Frank–van der Merwe (FVD) mode*, where the film grows epitaxially layer by layer and the misfit is zero; (b) *Stranski–Krastanov (S-K)* mode where, films grow in layers and in island fashion; and (c) *Weber–Volmer (W-V)* mode, where the substrate energy is greater than that of the film and island growth prevails [65]. Figure 5.3 illustrates the film growth modes. The vertical axis denotes the difference of surface energy between the film and substrate and the horizontal axis represents the interfacial misfit. The dashed line denotes the separation between the W-V and S-K modes. The circle at the origin denotes the area where a multilevel structure consisting of several layers of two or more materials have met the surface energy and misfit conditions.

In order to achieve epitaxial growth of a thin film having the same lattice structure as that of the underlying substrate, it is crucial to have a close lattice match between the interface and the substrate. Moreover, the surface energy of the film should be less than or equal to the surface energy of the substrate. On the other hand, if the surface energy of the film is greater it tends to agglomerate as it grows and will form a nucleation barrier to the condensation of a permanent deposit [64]. C*rystallization* is regarded as a first-order phase transformation which proceeds by nucleation and growth. Therefore, the associated surface energy information on the kinetics of *crystallization* of an amorphous barrier layer on a dielectric layer can be obtained and manipulated cleverly from the published data [64–66].

The surface energy of a substrate is altered in the presence of impurities or alloying. The study of Melmed shows a strong influence on the nucleation behavior of copper (Cu) when it is deposited on Ta/TaN in presence of residual gas [61, 67]. Deposition of an amorphous barrier layer for Cu-interconnects is currently

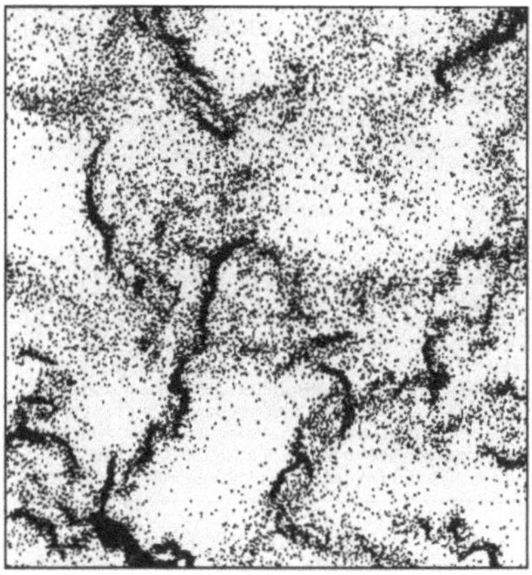

(a)

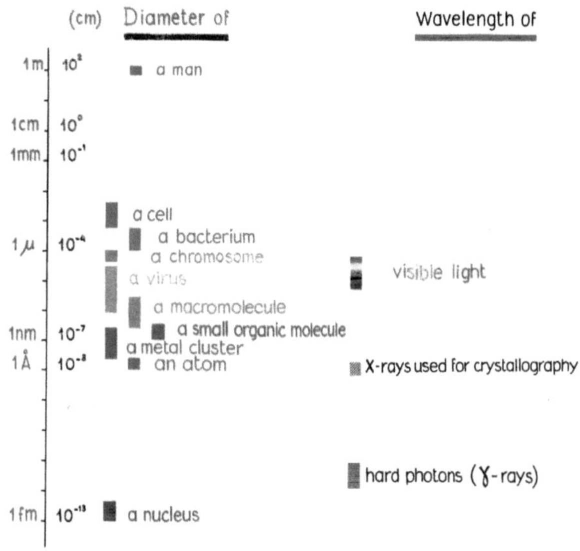

(b)

Fig. 5.2 (**a**) Two-dimensional clustering of particles as a result of inelastic collision (Reprinted with permission from *Physics Today*, APS, April 1996) and (**b**) comparative analysis of the dimensions of the nucleus and the metal cluster (Reprinted with permission, Taylor and Francis Ltd.)

of interest in integrated circuits (ICs), because there will be little or no diffusion of Cu due to the absence of grain boundaries (GBs). Thus the information of the surface energy of the metals, and the mechanisms of nucleation and growth of thin films, are of potential interest for the deposition of a voidless, stress-free, conformal, and stable film.

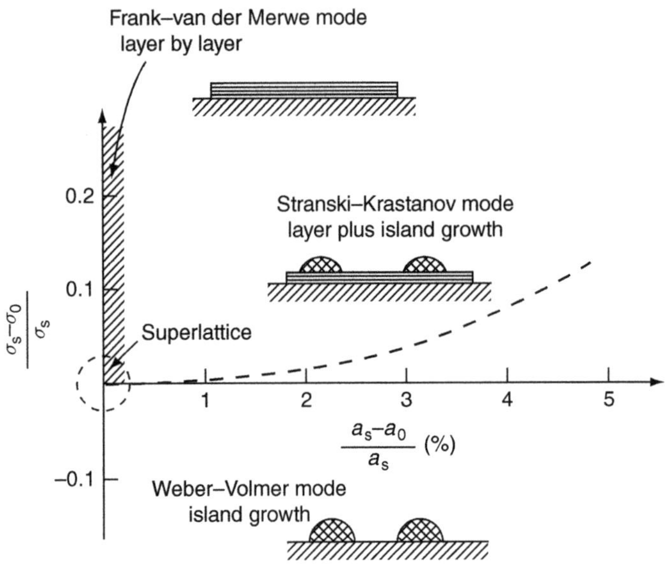

Fig. 5.3 Schematic diagram of thin film growth. a_0, and a_s are lattice parameters of the film and the substrate, and σ_s and σ_0 are the surface energies of the substrate and overlayer, ([65], Reprinted with permission, IBM research)

5.5 Instrumentation

Evaporation theory includes the concept of reaction kinetics, thermodynamics, and solid state theory [67–69]. It also shows the presence of ledge sources in the form of grain boundaries, cracks, crystal edges, and dislocations in polycrystalline materials [70]. During evaporation, as the islands grow, there is a decreasing tendency for them to become completely rounded after they coalesce. In polycrystalline films coalescing pairs of islands are randomly distributed and the grain boundary area is much larger than that in a single crystal film. Impurities present in the grain boundaries and segregation at these grain boundaries contribute significantly to the resistivity of the polycrystalline films. As a result, the stability and performance of an evaporated film will depend upon several factors, e.g. impurity concentration during growth, defects, grain size and its distribution, stoichiometry of the alloys, stress, adhesion, interactions with the substrate, and surface morphology [71–74].

5.5.1 Physical Vapor Deposition

Physical vapor deposition (PVD) is a technique whereby physical processes such as evaporation, sublimation, or ionic impingement on a target, facilitate the transfer of atoms from a solid or molten source on to a substrate. The simplest of all the PVD systems is the resistance heating of the material under vacuum [75–76].

However, the conventional evaporation method is not suitable for deposition of materials inside via holes and trenches having high aspect ratios (ARs).

5.5.2 Sputtering

Sputtering can be described as an energy transfer process from a target bombarded with ions, electrons, etc. [77–78]. The *plasma power,* the main source of heat inside the chamber, is controlled to prevent the degradation of the film properties, especially in a conventional *diode system* [79–80]. To enable the ignition of the plasma usually argon is fed into the chamber. In the Cu-dual damascene process sputter-deposited films of tantalum (Ta) and its nitride (TaN) are used for diffusion barriers. Sputtering is also used to deposit a bi-layer of copper on a Cu-seed layer. In *cathode sputtering,* metal ions are released from a source by bombardment of high-energy ions. These freed ions are then attracted towards the highest positive potential in the system (Fig. 5.4). Sputtering has many advantages over resistive heating PVD systems especially when the source material is a refractory metal (high melting temperatures), or alloys (without any fractionation), or the material has to be deposited without shadowing near steps with better coverage. Figure 5.4 shows a schematic of a low-pressure sputtering system.

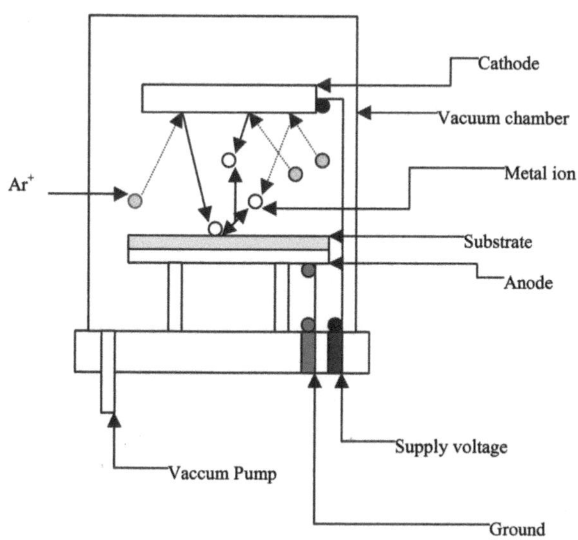

Fig. 5.4 Schematic of a low pressure sputtering system (Courtesy, ICE, AZ)

5.5.2.1 DC and RF Sputtering Systems

Diode or dc sputtering deposits conductive materials (Fig. 5.5) and has two electrodes to feed the working voltage, from 500 to 10 kV. The pressure inside the chamber is generally maintained between 30 and 120 mtorr. The radio frequency (rf∼10 MHz) sputtering system, on the other hand, can sputter both conductive and non-conductive (insulators) materials and the sputtering rate is determined by the flux of the ions and the energetic neutrals. In both rf and dc sputtering systems most of the secondary electrons emitted from the target do not cause ionization events with argon and the methods are not cost effective.

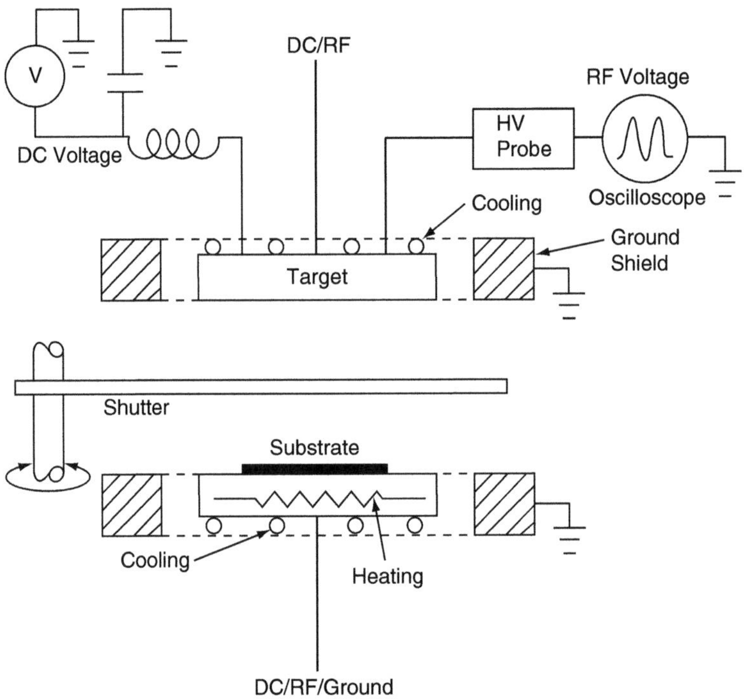

Fig. 5.5 Schematic of a sputtering system with several features (Reprinted with permission from John Wiley, [111])

5.5.2.2 Bias Sputtering

Conformal deposition of thin films (step coverage) inside small via-holes and trenches is very important in the copper damascene process. It is anticipated that bias sputtering will be a suitable process when conformal coverage is an important issue [81–82]. The advantage of this mode of operation is that the rf bias applied to the substrate enhances the acceleration of the impinging ions and transports them to the sidewalls of the steps [83].

5.5.2.3 Magnetron Sputtering

In both dc and rf sputtering, secondary electrons collected near the anode do not cause ionization but they cause unwanted heating. A strong magnetic field is applied to confine most of these secondary electrons near the magnetic field and to facilitate their safe return to the target. The trajectory motion of the secondary electrons is roughly cycloid, until they collide with the Ar atoms of the carrier gas (Fig. 5.6). In magnetron sputtering systems, two principal electrode configurations are generally used: (i) planar magnetron; and (ii) circular magnetron.

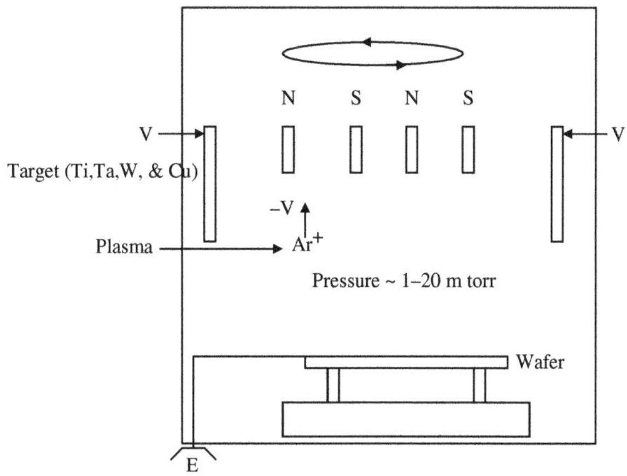

Fig. 5.6 Schematic of a magnetron sputtering system

The working voltage of a magnetron sputtering system is coupled with the magnetic field to concentrate the ions that increase the current density (deposition rate). Step coverage is one of the most important aspects of magnetron sputtering and is achieved by optimizing the shape of the target and by controlling the target regions where maximum sputtering occurs. Recently, a new type of high density aided hollow cathode magnetron (HCM) type deposition system has been developed for deposition of liners and seed layer in the the Cu-damascene process which is expected to deliver excellent sidewall (~12%) and bottom coverage (~33%). The advantages of the magnetron sputtering system are high deposition rate, low process temperature, less plasma contamination, and high yield [84–86].

5.5.2.4 Reactive Sputtering

Reactive sputtering is performed by the introduction of a reactive gas into the reaction chamber, which allows the sputtered atom to combine with gas to form a chemical compound. Nitrides of tantalum (Ta), titanium (Ti), and tungsten are potential candidates for barrier layer in the Cu-damascene process and are deposited by reactive sputtering. It appears that the reactive sputtering process will continue to be the most important PVD process in silicon VLSI fabrication. As the feature size is becoming smaller, the demand for a new and modified sputtering system is very much felt, and a modified version of the sputtering system known as an *ionized PVD* system has been introduced in the market.

5.5.3 Ionized Physical Vapor Deposition (IPVD)

Ionized vapor deposition (IPVD) deposits film from ions by applying an electric field near the surface of the wafer to control the velocity and the directionality of the ions [87]. The system can deposit metal or dielectric films inside a narrow via hole or trench having high aspect ratio (AR) with negligible deposition on the side-walls, which is particularly needed for sub-micron size devices in the Cu-damascene process.

Bottom coverage is determined by the directionality of the incoming material, which is achieved by the ratio between the ions and the neutrals and the angular distribution function of the ions. Thus when the ions are normal to the substrate surface and there is no re-sputtering inside the structure, the bottom coverage can reach almost 100% [88]. In the copper dual (double) damascene process (Cu-DD), overhang and faceting can be controlled by controlling the wafer bias, the ion flux (controlled by controlling the rf power), and ion to neutral ratio (controlled by the RF coil power, pressure, and DC target power) [89].

It has been observed that for conformal wall-coverage (i.e. when base and side-wall coverage are equal) an optimum ratio (1:1.5) between DC: rf values is required. But this has been debated by Rossangel et al., [90] who claim that the conformality of the film is a function of the vertical flux component.

Figure 5.7 shows a schematic of a typical I-PVD system. A commercial hollow cathode magnetron type I-PVD system is available on the market, which is reported to deliver better film composition for the Cu-damascene process. It uses permanent magnets to create a null field region, which forms a cusp mirror that reflects most of the escaping electrons back into the hollow cathode [86–94]. Increased gas pressure

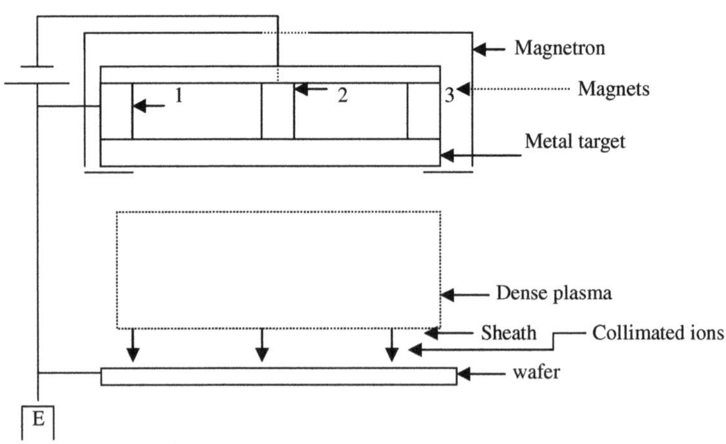

Fig. 5.7 A typical schematic of an ionized physical vapor deposition system

inside the chamber increases the time of flight of the sputtered ions [95–96]. The film coverage inside via holes and trenches depends on re-sputtering of the surface materials, ion enhancement mobility, mobility of surface atoms, and modified sticking coefficients due to the oblique angle of incidence [97]. A high electron density ($n >> 10^{11}$ cm^{-3}) inert gas plasma between the target and wafer is created to ionize the metal vapor. The power source may be either an electron cyclotron resonance plasma or inductively coupled plasma (ICP) [98].

Table 5.2 shows the effects of variable processes and the consequences on the film characteristics deposited by an inductively coupled plasma (ICP) system. The commercial ICP system has four point cusp modules and it employs 60 MHz VHF current to generate plasma. The unit uses either 2.5 kW power with 22 Pa base pressure or 4.0 kW power with 18 Pa base pressure to deposit Ta/TaN or Cu-films, respectively. The processing conditions inside the chamber determine the grain size and the resistivity of the film [99].

Table 5.2 Effect of process variables on film characteristics

Increasing	Bias coverage	Sidewall coverage	Non-uniformity	Deposition rate
Pressure	Increase	Increase	Increase	Decrease
rf coil power	Increase	Decrease	Increase	Decrease
DC target	Decrease	Increase	Decrease	Increase
Substrate bias	Increase	Increase	Variable	Slight decrease
RF power	No change	No change	Decrease	Increase

Reprinted with permission, Aviza Technology, CA, from the article of S.R. Burger, K.E. Buchanan, J. Cresswell, and I. Moncriegg

In Table 5.3, the deposition parameters and the characteristics of Ta/TaN films are listed. It has been observed that when the flow rate of nitrogen is less than 1 mT, β phase Ta changes to α-phase, with a change in resistivity from 175 to 75 μΩ-cm.

Table 5.3 Deposition parameters and characteristics of Ta/TaN films

Parameter	β-Ta	TaN (I)	TaN (II)
N_2 pressure (milli-torr, mT)	0 and 0.2	1–2	> 2
Deposition rate (nm/min)	60	50	22
Tooling factor (Å/kW/min)	600	500	220
Non-uniformity (1 sigma %)	<8	<8	<8
Base coverage %	70	50	56
Sidewall coverage %	50	52	44
Stress (dynes/cm^2)	2×10^{10}	4×10^{10}	2×10^{10}
Resistivity (μΩ-cm)	175 and 75	220–250	2200
Breakdown temperature (C)	550 and <450	650	>750
Texture	<002>		

Reprinted with permission, Aviza Technology, CA, from the article of S.R. Burger, K.E. Buchanan, J. Cresswell, and I. Moncriegg

Not only that, the flow rate of nitrogen also controls the structure of the tantalum nitride (TaN). For example, when the N_2 flow rate inside the chamber exceeds 2 mT, metallic TaN film not only changes to a green/golden color with a mixed crystalline-amorphous phase but a change in resistivity from 200 to 2000 μΩ-cm is also noticed [100]. Further increase of the flow rate of N_2 forms a Ta_2N phase and the film resistivity changes to 200 μΩ-cm [101–103].

5.6 Chemical Vapor Deposition (CVD)

Chemical vapor deposition (CVD) is defined as the deposition process of a non-volatile solid film by vapor phase chemical reactions which require both thermal and chemical reactions for film deposition [104]. The process plays an important role in the semiconductor industry because of its ability to provide high throughput and excellent step coverage, at relatively low temperatures [105]. According to the different variances of the design system, and the reacting gases, the CVD process can be classified into two main categories, namely, *atmospheric pressure CVD (APCVD)* [106], and *low-pressure CVD (LPCVD)* [107–109]. As the complexity of deposition technology has increased, low pressure CVD has been modified to *plasma enhanced CVD (PECVD), metal-organic CVD (MOCVD) and rapid thermal CVD (RTCVD).*

5.6.1 Plasma Enhanced CVD (PECVD) System

The attainment of controlled film properties over an extended surface for large production at relatively low temperatures can be made possible with the plasma enhanced CVD (PECVD) system (Fig. 5.8). The major variables affecting the film deposition rate and uniformity are: (a) rf power density; and (b) distribution of power

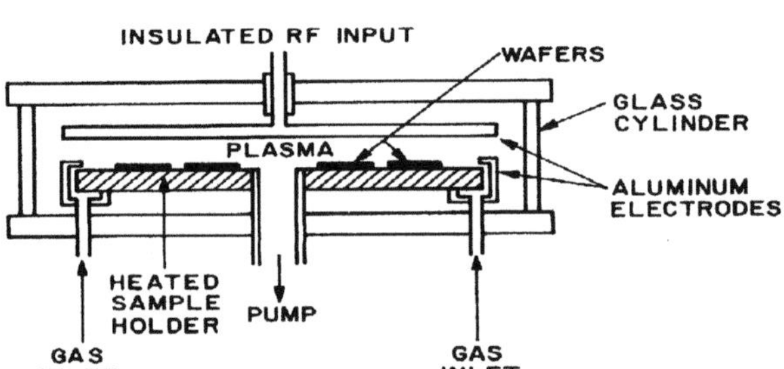

Fig. 5.8 Schematic of a PECVD reactor showing different parts of the reactor and the gas flow system (Reprinted with permission, John Wiley, New York, ref. S.M. Sze, *Physics of Semiconductors*, p. 355)

density [110]. The PECVD system can provide: (i) very good adhesion of the film to substrate; (ii) low pin hole density; (iii) better step coverage; (iv) good electrical characteristics; and (v) compatibility with fine line pattern transfer [111–114].

Figure 5.8, shows a schematic of a PECVD reactor. The power for plasma generation is induced either by an *inductively* coupled rf coil or *capacitively* coupled plates. The film grown by the PECVD method sometimes develops compressive stress, which is a function of the applied rf power, pressure, and the bias. Three types of PECVD systems are available on the market, namely, (i) parallel plate, (ii) horizontal tube, and (iii) single wafer type.

The PECVD method can be used to deposit tantalum nitride (Ta_2N/TaN) or a Ti/TiN liner and Si_xN_y as a hard mask or cap layer for the Cu-interconnect [115–116]. The effectiveness of a film deposited by the PECVD method can be determined by secondary ion mass spectroscopy (SIMS) and C-V techniques [117].

Figure 5.9 shows a commercial PECVD system built to deposit *low-K dielectric* material for the Cu-damascene process. It is expected that to integrate low-*K* dielectric materials into Cu-interconnects, the PECVD method will be simpler than the spin-on deposition (SOD) technique.

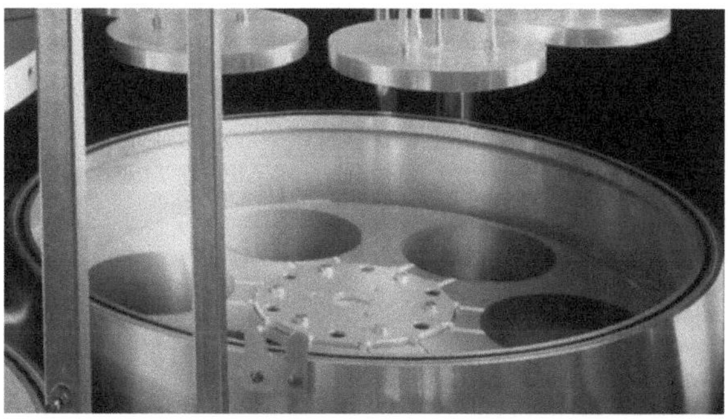

Fig. 5.9 A commercial PECVD system designed to deposit low-*K* dielectric film (Photo courtesy Novellus systems Inc.)

Besides Ta/TaN, the PECVD method has been successfully applied to deposit tungsten (W) and other refractory metals to fill submicron level via and trenches having high aspect ratios [118–120] in the Cu-damascene architecture. In LPCVD and PECVD systems, the reactants have more energy than in the APCVD system, which increases the migratory motion of the adatom and improves the step coverage of the trench or via hole. In the PECVD system, the temperature inside the chamber during deposition is kept below 450 °C, which is very useful for intermetallic dielectric (IMD) film deposition in the Cu-damascene process. It has been reported that integrating APCVD with or without PECVD with tetraethylorthosilicate (TEOS)/O_3 can suppress hillocks and via poisoning. The PECVD underlayer also protects underlying devices from moisture evolving from the IMD at elevated temperatures [121–124]. The RTCVD method, on the other hand, has been used to deposit high-*K* hafnium oxide for dual gate CMOS processing [37]. Table 5.4 shows a comparative analysis of the three different CVD systems.

Table 5.4 Characteristics and applications of CVD reactors

Process	Advantage	Disadvantage	Applications
APCVD	Simple. Fast deposition rate, low temperature operation	Poor step coverage, particle contamination, frequent cleaning necessary	Low temperature oxides
LPCVD	Excellent purity and uniformity, conformal step coverage, large wafer capacity, mass transfer no longer limits the deposition rate	High temperature operation, low deposition rate	High temperature oxides, Si_3N_4, W, WSi_2, Poly-Si
PECVD	Low temperature operation, Fast deposition rate, good step coverage, better film adhesion, low pin hole density, and compatible with fine line pattern transfer process	Chemical (H_2) and particulate contamination	Low temperature insulators, and Passivation layer, inter metallic dielectric (IMD)

5.6.2 Metal-Organic Vapor Deposition (MOCVD)

The metal organic CVD (MOCVD) system is a low temperature process. It can reduce structural defects, and interlayer diffusion, during copper deposition in the Cu-damascene process. The most important criterion of the MOCVD process is that the metal containing precursors are available in a wide range of chemical compounds, from gases through volatile liquids, sublimable solids to relatively non-volatile solids. The precursors can be transported in vapor phase to be condensed on a proper substrate [125–126]. MOCVD has a number of potential advantages over other deposition techniques such as large area growth capability, good composition control, high film density and excellent conformal step coverage [127]. The MOCVD method is capable of depositing a very thin gate with better electrical isolation and reduced leakage current [128–132]. Metal-alkoxide precursors are used more than thermally stable metal-diketonate precursors [133–135] in the MOCVD process. Figure 5.10 shows a picture of a MOCVD system designed to grow high quality dielectric films.

The system has a multizone infrared (IR) lamp as the heating system combined with substrate rotation capability. The unit can handle a 200 mm wafer and can maintain uniform temperature across the substrate (within $\pm$ 0.5 °C).

Figure 5.11 shows a 3-D reactor [136], which has a thermally stable cold wall reaction chamber with load lock, and rf plasma process capability. The base pressure

Fig. 5.10 An MOCVD reactor for thin film deposition (Courtesy AIXTRON, AG, Germany [136])

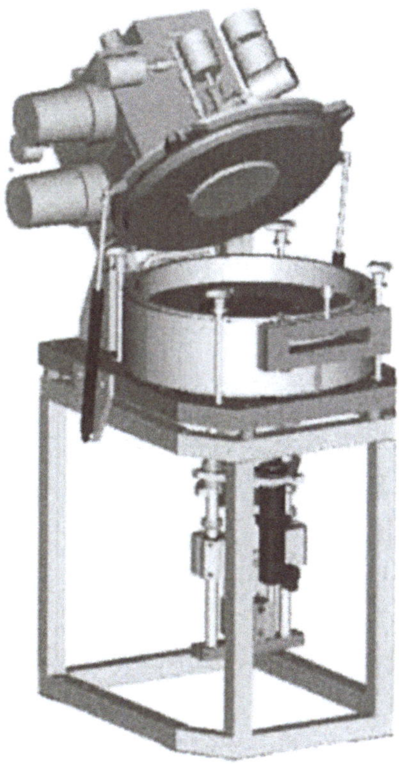

Fig. 5.11 A three-dimensional MOCVD system to grow thin films of dielectric materials of very high quality (Courtesy, AIXTRON, AG, Germany [136])

of the chamber can achieve less than 10^{-2} Pa, and is provided with a water cooling system for the wafers. The unit is capable of depositing a void free blanket of Cu-film using hexafluoroactylactone venyl tetramethylsilane as a precursor [137–139]. The deposition rate and the resistivity of the deposited film are reported to be dependent on the carrier gas, processing temperature, and texture of the substrate [140–142].

5.7 Low Temperature Thermal CVD (LTTCVD) System

Low temperature thermal CVD (LTTCVD) has been successfully used to deposit metallic copper and its compounds like tri-alkylphosphine (cyclopentadienyl) copper (I), or acetyl acetonate, or alkoxide complexes, for multilevel conformal metallization in nano-size devices [143–148].

Figure 5.12 shows an LTTCVD reactor, which has been used to deposit 1.5 μm thick Cu-film inside a 0.7 μm wide and 2.0 μm deep trench. The film resistivity is found to be around 2.0 μΩ-cm (compared to the bulk resistivity of copper ~1.68 μΩ-cm). The growth rate and the substrate temperature are maintained between 200 and 2000 Å/min and below 200 °C, respectively. However, by limiting the rate of mass transport of the precursor and increasing the substrate temperature a smoother film can be grown. Figure 5.13a shows a scanning electron microphotograph (SEM) of a 1.5 μm thick copper film and Fig. 5.13b shows the SEM of the deposited copper film over a tungsten seed inside the trench.

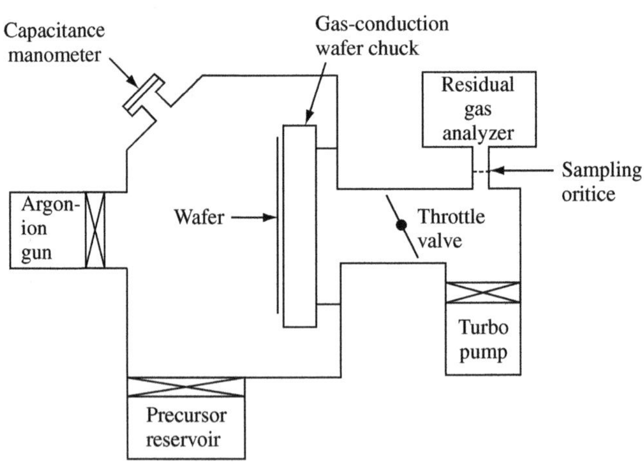

Fig. 5.12 Schematic of a low temperature thermal CVD reactor used for copper deposition (Courtesy of IBM Research, [143])

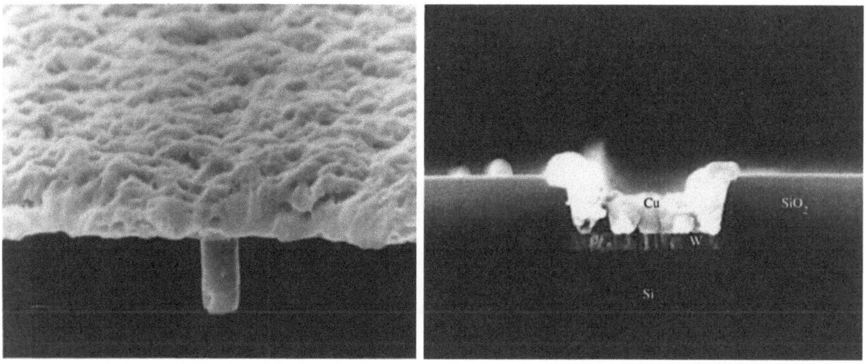

(a) (b)

Fig. 5.13 (**a**) SEM image of a copper film (1.5 μm) inside a trench of 0.7 μm wide and 2.0 μm deep; (**b**) SEM image of a copper film deposited over a seeded tungsten film (Photo courtesy, IBM Research [111])

5.8 Atomic Layer Deposition (ALD)

Atomic layer deposition (ALD) is also called atomic layer epitaxy [149], which utilizes sequential precursor gas pulses (homoleptic *N*, *N*′-dialkylacetamidinato metal compounds) and reactant gas (mostly hydrogen) to deposit stepwise monolayer films. Most of the flow-type reactors operate at pressures from 1 to 20 mbar.

Figure 5.14 illustrates the possible chemisorption mechanisms in an ALD system [150]. The growth rate of the ALD process is very slow (~0.5 Å/cycle, when the metal dose is 4×10^{-9} mol cm^{-2} and H_2 dose is 3×10^{-6} mol cm^{-2}) [151, 152] which is a disadvantage for the system to grow thick films. However, the operational temperature is lower than in the CVD process.

1. Surface of the substrate 2. First precursor molecules 3. Second precursor molecules

4. Un-reacted molecules 5. Deposited monolayer

Fig. 5.14 The possible chemisorption mechanisms in an ALD system: (1) surface of the substrate; (2) first precursor molecules; (3) second precursor molecules; (4) unreacted molecules; (5) deposited monolayer

Figure 5.15 shows a schematic of an ALD system with all accessories (gas delivery systems, speed valves, and vacuum pump). The precursor gases in the vapor state flow past the substrate resting on a susceptor, which in turn is heated by resistive coils. The number of monolayers deposited on the substrate surface after reactions

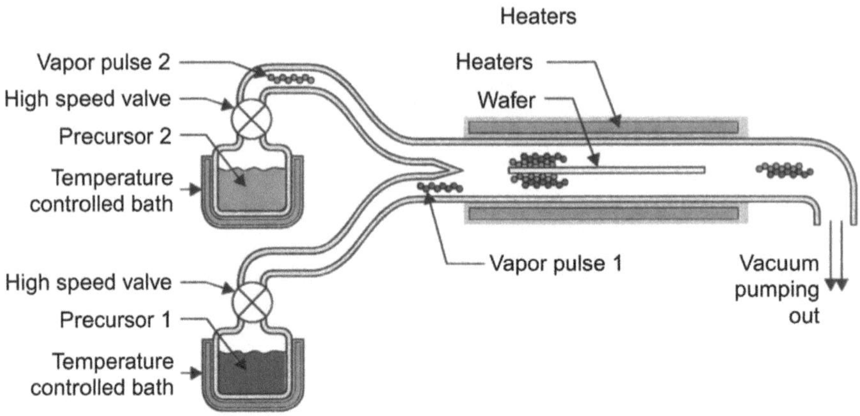

Fig. 5.15 Schematic of an atomic layer deposition system (Reprinted with permission, Wiley-VCH [124])

Fig. 5.16 The inside of a chamber of a commercial ALD system (Reprinted with permission, Semiconductor International, Oct. 2001, p. 52)

will depend upon the number of precursor cycles [153]. Figure 5.16 shows the deposition chamber of a commercial ALD system.

The ALD process has successfully demonstrated atomic layer deposition of many metals, e.g., Cu, Ta, Ti, W, Co (Fig. 5.17) and their compounds within increasingly narrow trenches and via-holes needed for filling the damascene architecture [154–156, 151]. The resistivity of a >60 nm film grown by the ALD process shows near bulk values of the metal [157]. ALD is currently being used to deposit high-*K* gate oxides and diffusion barriers for back-end interconnects [158–159].

Very recently plasma enhanced (PE) ALD has been used to deposit TaN films at 260 °C using *tert-butylimidotris diethylamido tantalum* (TBTDET) and hydrogen radicals (rf plasma 13.56 MHz, power 100 W) [160]. TaN films deposited by the PEALD method have shown lower resistivity, and higher growth rate than the conventional ALD method. Figure 5.17 shows the conformal step coverage with < 15 Å TaN film grown by the ALD method.

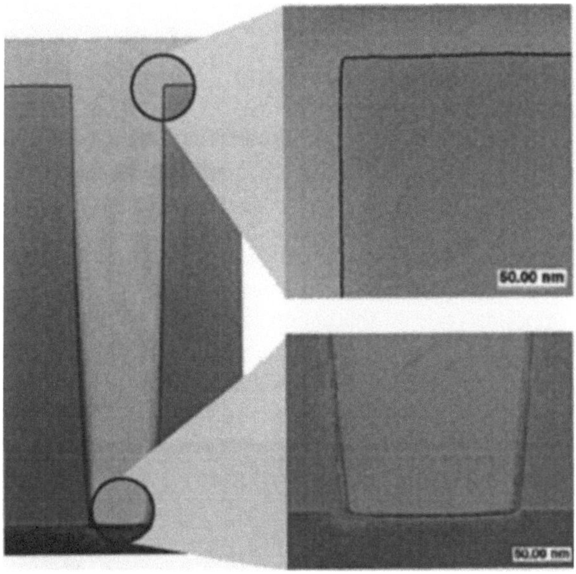

Fig. 5.17 Conformal step coverage with <15 Å ALD TaN film (Reprinted with permission, Applied Materials, Santa Clara, CA)

5.9 Plating

5.9.1 History of Electroplating and Printed Circuit Boards (PCBs)

The first electroplating of metal was reported in early 1800. An Italian chemistry professor, Luigui Brugnatelli, wrote, *I have lately gilt in a complete manner two large silver medals, by bringing them into communication by means of a steel wire, with a negative pole of a voltaic pile, and keeping them one after the other immersed in ammoniuret of gold newly made and saturated*. There is no doubt that the plating industry was thriving long before the printed circuit board (PCB) was introduced on the market (as the metal finishing industry for ornaments and other decorative materials) [161–162]. However, renewed interest in plating technology has increased enormously since the *copper damascene* process was introduced into the IC market by the semiconductor industry [163–164].

Conventional techniques (*DC bath chemistry, air surging, rotating disk electrode (RDE), oscillatory motion of the cathode*) have resulted in non-uniform copper plating inside via holes especially when via holes are in the submicrometer range and the aspect ratio (AR = length/width) reaches six or more. *Plating technologies* are mainly grouped into two categories, namely *electroplating* and *electrodeless* (electro-less) plating. In the *electroplating* (electrode-plating) system, plating is usually done by applying a current through the electrodes dipped into a solution called an *electrolyte*. However, in electro-less (*electorde-less*) plating no external electrical energy is needed to deposit the film.

5.9.2 DC Bath Chemistry

Electroplating is a process of electro-deposition of an adherent material on an electrode by applying direct current (dc) to the electrolyte. The temperature of the bath plays an important role in the conductivity of the bath, current efficiencies, and achieving morphology of the plated material and stability of the electrolyte [165–167]. The electroplating process requires an electrically conductive surface to promote nucleation and growth of the film [168–169]. In an all-cupric bath (containing copper sulfate), the cathode efficiency is practically 100%, while in cyanide copper baths the efficiency can be recovered from 50 to 100% with free potassium salt. The cupric bath typically carries 250 g/L copper sulfate solution ($CuSO_4 : 5H_2O$) and 75 g/L sulfuric acid (H_2SO_4). Copper ions (Cu^{++}) are deposited on the cathode, and sulfate ions ($SO4^{--}$) are deposited on the anode. The amount of copper deposited on the seed plate (metallic cathode) is proportional to the magnitude and duration of the current flow.

Figure 5.18 shows a simple laboratory setup for electroplating copper (Cu) from copper sulfate ($CuSO_4$) solution. In the *copper damascene process* the cathode is a thin layer of copper – called a *seed* [170]. The problems with the conventional plating technique are insufficient transfer of cupric ions and the presence of air bubbles. These effects are mostly related to solution hydrodynamics [171]. Several methods, like PVD, CVD, ALD and electro-bath deposition technologies, have been tried to deposit a void free seed layer [172–174] during the Cu-damascene process.

Plating inside through-holes and crevices is critically important and it is difficult to achieve with conventional plating technology. In order to maintain the concentration of the solution and a better mass transport through the small via-holes or trenches, the solution (electrolyte) is provided with *air sparging*, and occasionally augmented by *forced circulation through a pump* [175–179]. The polarization curves for stationary electrodes indicate that because of limited current density, air sparging alone is not sufficient to plate via holes or trenches uniformly, especially when aspect ratios are high. The limitation can be removed by *rotating the electrodes* during deposition, which partially avoids mass transport current limitation

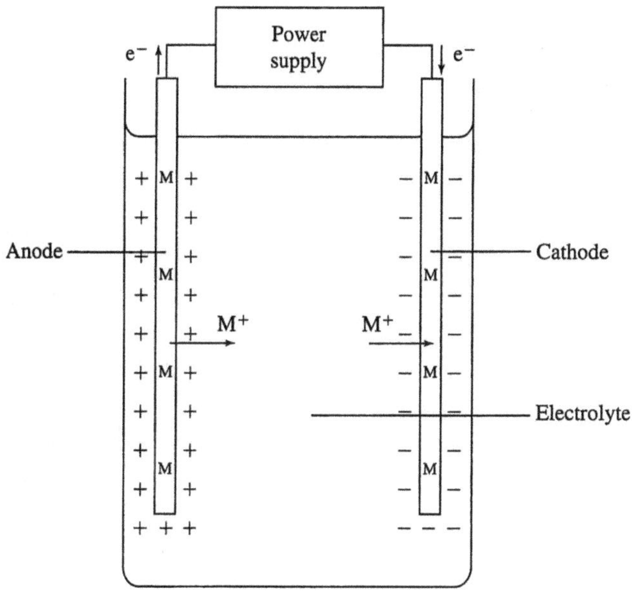

Fig. 5.18 A basic experimental setup for electroplating copper from copper sulfate solution

[180–182]. However, the construction of hydrodynamic electrodes for reproducible mass transfer condition is more difficult than stationary electrodes.

5.9.3 Electroplating of Copper Inside Damascene Architecture

Modified bath chemistry: Conventional electroplating used in printed circuit boards (PCBs) with air surging and a rotating disk cannot be used for conformal copper fillings in the Cu-damascene architecture. Experimental observations supported by bath chemistry theory identified inclusion of additives in the conventional bath, called *suppressor, levelers,* and *accelerator.* Figure 5.19 shows a commercial electroplating unit.

The suppressing polymer (*suppressor)* is a surfactant or wetting agent (sometime also called a *carrier*) and can be a long chain polymer such as polyethylene glycol (PEG) or co-polymers of polyoxyethylene. The second additive, the *leveler* (also called a grain refiner or over-plate inhibitor), is a high molecular weight polymer, with amine (–NH_3) or amide (–NH_2) as the functional group [183]. Mostly, levelers suppress current at locations where the mass transfer rate is most rapid, and adhere preferentially to highly charged areas such as corners and edges to prevent overhang at the mouth of the trench or via [184]. The *accelerator* accelerates the current relative to the *suppressor* and the unsaturated compounds containing a polar sulfur, oxygen or nitrogen as functional group (3-mercapto-propane sulfonate (MPSA))

Fig. 5.19 A commercial electroplating process unit (Photo courtesy Novellus Systems, Inc.)

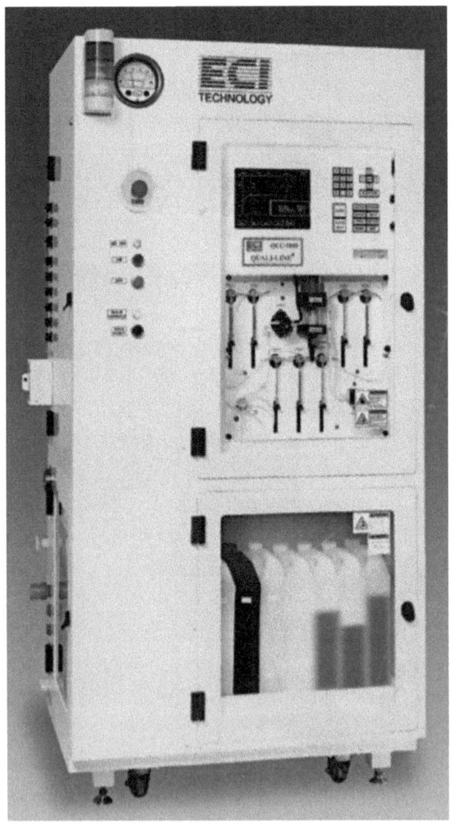

Fig. 5.20 A copper bath control machine (Photo courtesy ECI technology, NJ)

[185–186]. In addition to the additives mentioned earlier, *chlorine* (Cl^-), an inorganic additive is also added to the bath. The theory of additives is based on the adsorption of the additives at the metal-electrolyte interface reducing the active area and thereby inhibiting the metal reduction process. Since the inhibitor concentration is diluted (10^{-3} to 10^{-6} mol/L) the adsorption process is typically a diffusion limited phenomenon [177].

Electroplated copper (EP-Cu) shows higher resistivity than bulk Cu, because of dislocations, vacancies, impurity atoms, and/or second phase particles, and scattering from the grain boundary [187–188]. X-ray diffraction (XRD) studies reveal recrystallization and grain growth of EP-Cu at room temperature that reduce the density of the grain boundary and resistance of the film. Figure 5.20 shows a copper bath control machine, used to control the additives and the electrolyte. Commercial multi-sample programmable systems are being used during electroplating of copper inside a via-hole or trench.

5.10 Process Chemistry for Superconformal Electrodeposition of Copper

During the Cu-damascene process the trench or via geometry results in a lower flux of the inhibitor to the bottom of the features than to the external surface. Consequently, the metal deposition kinetic proceeds more rapidly at the bottom of the via hole and trench which results in *superconformal filling* [189].

The bottom filling (or super-filling) process technology has been the primary driving force for deposition of diffusion gradient suppressing polymers [190]. The filling mechanism has been explained on the basis of the accumulation of the accelerating active species like SPS-*bis* (sodium sulfopropyl) di-sulfide and its product, such as MPSA (marcaptopropane sulfonic acid) [191–192]. The reduction of SPS to MPSA provides a possible catalytic pathway for Cu-deposition through the formation of cuprous thiolate [193].

Figure 5.21 shows the possible profiles of plated copper. In conformal plating, a deposit of equal thickness of metal layer at all points may lead to the creation of a seam or void. On the other hand, subconformal plating will result in substantial depletion of cupric ions in the plating solution inside the feature, which will cause the current to flow preferentially to more accessible locations outside the feature. When the depth of the feature size is large (>50 μm), non-uniformity in the distribution of current is noticed due to the ohmic drop in the electrolyte [194].

From the above figure we can see that the *subconformal* situation leads to the formation of *voids* [195] inside the trench or via, whereas *conformal* deposition of equal thickness in the beginning leads to the creation of a *seam*. *Superconformal* plating, on the other hand, leads to a deposit of equal thickness at all points of a feature [177, 196]. Therefore, in order to attain defect free plating, the plating process should be controlled in such a way to achieve *superconformal* filling.

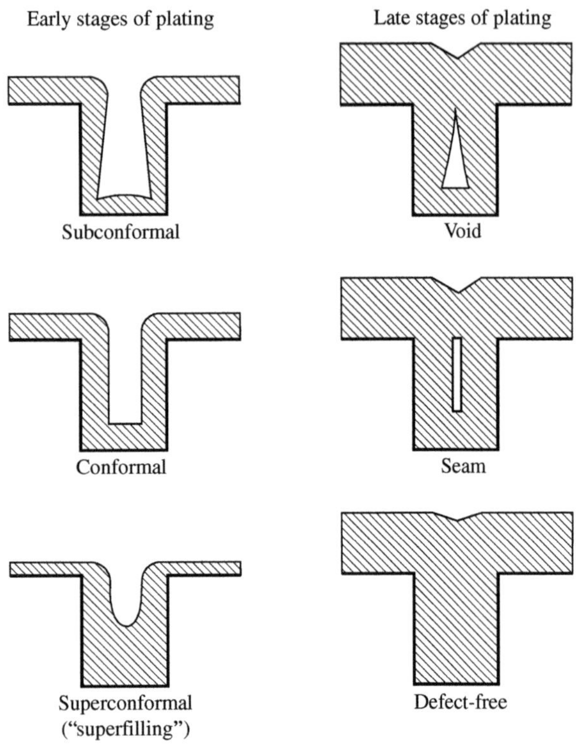

Fig. 5.21 Types of profile in damascene plating ([190] Courtesy of IBM research)

5.11 Electrochemical Mechanical Deposition (ECMD)

In the Cu-damascene process, electrochemically deposited copper (ECD) employs chloride ions, a suppresser, and an accelerator to promote smooth filling of high aspect ratio (AR) via holes and trenches. It has been observed that as copper is being deposited inside the small structures, the additive gradient gives rise to accelerated growth of the film from the bottom of the feature causing voids. During small feature filling, bumps are produced due to overfill, which is interpreted as an *overplating* phenomenon. The overfill is generally avoided by using levelers. However, when the feature has a large step, overfilling cannot be avoided simply by using levelers.

Recently, a novel electrochemical mechanical deposition (ECMD) process has been developed to deposit planar conductor films on a non-planar surface with minimum dishing and erosion [197–198]. The process involves simultaneous plating and polishing/sweeping of the wafer surface. This gives rise to bottom-up filling phenomenon even in the lowest aspect ratio cavities [199–200]. A schematic of the ECMD system is shown in Fig. 5.22.

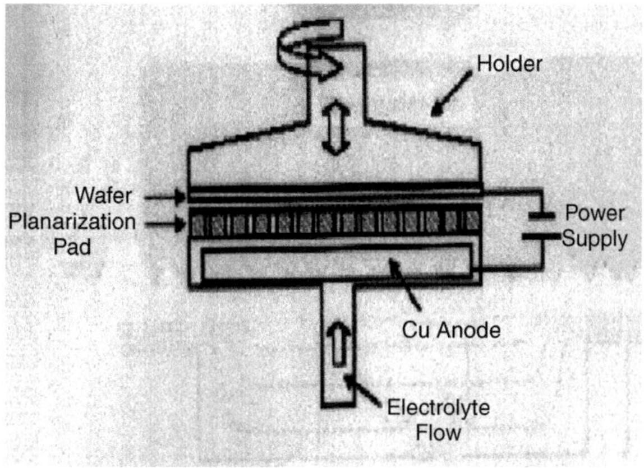

Fig. 5.22 Schematic of an ECMD cell (Courtesy of ASM International, NV)

5.12 Influence of the Seed Layer on Electroplating

Experimental evidence shows that the growth of electroplated Cu is highly dependent on the characteristic of the underlying seed layer. A smooth <111> textured seed layer will help to grow a smooth and <111> textured electroplated Cu-layer. On the other hand, any impurity or surface defect will be detrimental to the growth of <111> texture. Moreover, the presence of any oxide layer on the seed may form

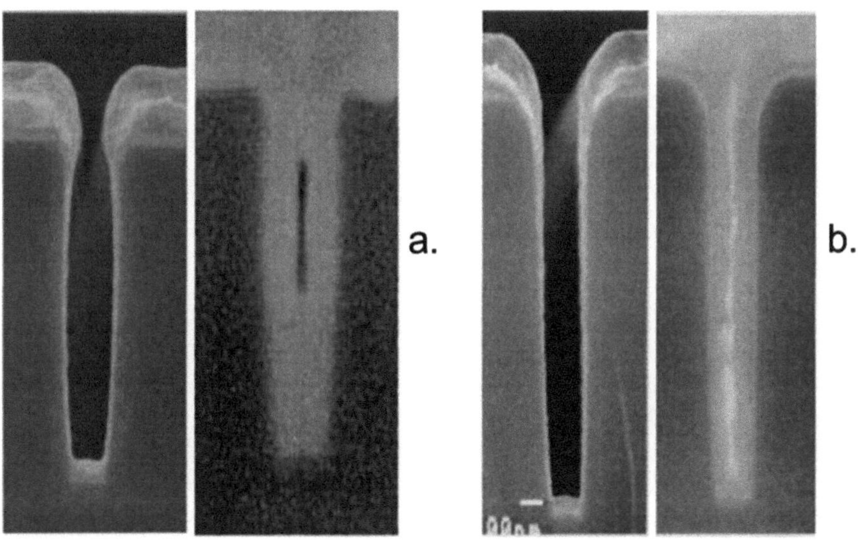

Fig. 5.23 The seed profile inside a submicrometer trench (Reprinted with permission, Novellus Systems Inc.)

highly resistive *black copper,* which will affect the equivalent resistance and adhesion of the EP-copper.

Figure 5.23 shows the influence of the seed profile on copper filling inside a trench. Combined feature and seed profile can produce discontinuous and/or agglomeration of copper filling. When the seed layer grows non-uniformly, it can form a narrow and naked opening in the trench, and can lead to the creation of a seam, which is shown in Fig. 5.23a. On the other hand, Fig. 5.23b shows the generous opening of the trench due to uniform surface deposition of the seed layer. This results a uniform copper deposition over the trench.

5.13 Electroless Deposition of Copper

Controlled auto-catalytic (once there is an initial layer of metal, the reaction continues indefinitely) deposition of thin metal film by the interaction of metal salt in the presence of a catalyst can be categorized as *electroless (or electrode-less) plating*. The metal to be deposited (at present copper) reacts in the presence of the reducing agent and deposition does not require a conductive surface to nucleate the film. The disadvantage of electroless deposition is that the bath is generally smaller and the deposition is not thermodynamically stable. Thus the plated area becomes overactive and decomposes when it stays for a longer time inside the solution. Under proper deposition conditions, the over-potential is large enough to resist any reversible reactions [201–202].

One of the most commonly used bath for electroless deposition of copper consists of 10 g/L $CuSO_4$: $5H_2O$, 50 g/L of potassium sodium tartarate, 10 g/L of sodium hydroxide, and 9.5 g/L of formaldehyde. To deposit a smooth and well-adhered film, a solution of palladium (Pd) in 0.1 wt% of $PdCl_2$ and 0.1 wt% HCl (12 M) is added to the solution as a catalyst [203]. Since sodium or potassium bearing compounds have a detrimental effect on silicon, a well mixed solution of $CuSO_4$ (25 g/L), HCHO (1 mL/L), with a surfactant (GAFRE, 610, 250 mL/L) in tetramethyl ammonium cyanide (1 mg/L), is considered to be a better choice for electroless Cu-deposition over silicon wafer. For better performance of the film, chemisorption of palladium (Pa) is performed in the presence of tin chloride ($SnCl_2$) which acts as a sensitizer [204]. In order to break up the chain of tin chloride shell, an accelerator (hydrochloric acid, HCl) is added in an appropriate proportion. However, according to some experts, $PdSO_4$ will be a better choice for Pd^0 ion as a catalyst instead of $PdCl_2$ [183, 205–211].

In recent years electroless deposition of copper (Cu) for the seed layer over the Ta/TaN barrier layer has emerged as one of the most efficient and easy ways to deposit conformal and void free seeds inside the small trenches and via holes [212–213]. The electroless deposition method has also been successfully applied to deposit CoWP alloys for cap layer in the Cu-damascene architecture by the addition of tungsten (W) ion into Bremer's chloride bath [213].

5.14 Stress in Cu-Interconnects

It is believed that the electrical behavior of electroplated copper (EP-Cu) film is not only dependent on grain size but also on the stress of the film [214]. It is also true that the *stress and orientation* of the seed layer have a great influence on the grain size of the blanket EP-Cu layer [215]. *Stress* in the Cu-seed layer may be due to agglomeration and annealing. However, annealing at high temperature (~400 ºC) shows a relaxed and stress-free EP-Cu layer [216]. Figure 5.24 shows cross-sectional (SEM) via images showing (left to right) agglomerated seed coverage, metal profile following partial fill by electroplating, and the final fill result following electroplating.

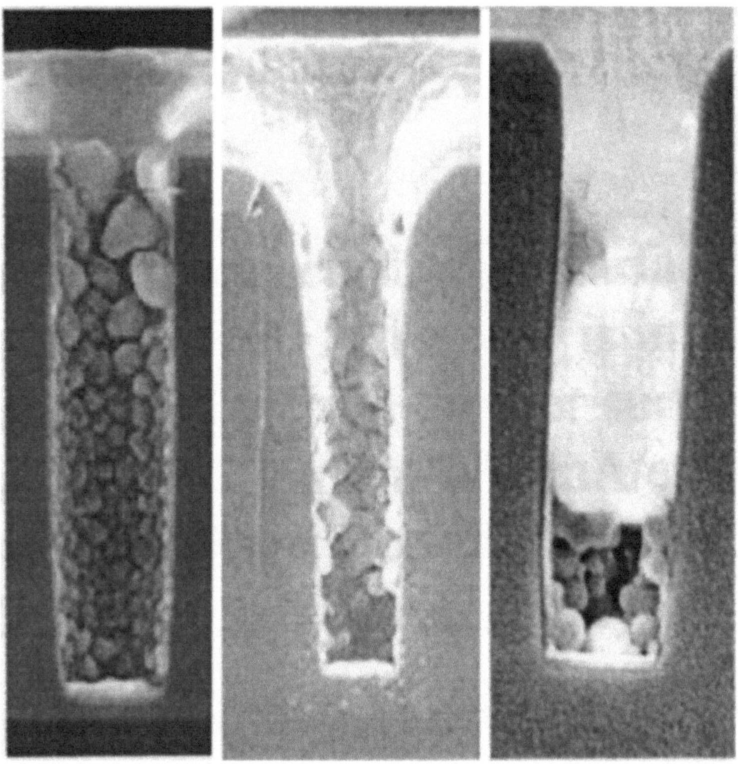

Fig. 5.24 Cross-sectional SEM via images showing (*left* to *right*) agglomerated seed coverage, metal profile following partial fill by electroplating, and final fill result following electroplating (Reproduced with permission Novellus System)

Copper (Cu) interconnects show better mechanical properties, such as Young's modulus, yield strength, and ultimate tensile strength than aluminum (Al) interconnects. However, the higher Young's modulus of copper causes a higher film stress for a given strain, and as a result the adhesion loss is noticed sooner in Cu-interconnects compared to Al-interconnects [217–218].

At room temperature, due to *self-annealing*, EP-Cu develops stress during its growth from the electrolytic solution. *Self-annealing* drastically changes the yield stress, σ_{ys} (σ_{ys} = 180 +0.262 $d^{-1/2}$, where d is the grain size in microns) [219–220] of the electroplated (EP)-Cu layer. It is not unlikely that stress in the Cu-interconnect might have been induced due to the *stress relaxation* of the seed layer.

For instance, when the seed layer is in stress, the stress within the EP-Cu layer increases linearly (from 22 to 32 Mpa). Many methods have been suggested to suppress stress in Cu-interconnects from time to time. Most of these involve either alternate geometries of the line layout and via structure, change of dielectric material underneath to improve passivation, and/or optimizing the thermal cycling process. In addition, it has been theorized that inclusion of a small amount of metal in Cu-film during electroplating and its subsequent segregation at grain boundaries by thermal treatment can suppress stress agglomeration of the Cu-seed layer, which ultimately may lower the stress condition of Cu-interconnects.

The other forms of stresses that are noticed within a thin film are *thermal stress* and *stress due to phase transformation.* The former originates due to the difference of coefficients of thermal expansions (CTEs) of the deposited film and the substrate. It has been observed that the film inside a trench or via hole experiences more thermal stress with significant triaxial component, in contrast to the predominantly biaxial state in blanket film [221]. Encapsulated at high temperature T_0, triaxial tension will be developed upon cooling due to thermal misfit strain ($\alpha_m - \alpha_d$) ($T_0 - T$), where α_m α_d are the thermal expansion coefficients for metal and silicon die, respectively. A similar effect can also be noticed for the metal line embedded within a dielectric.

We know that both stress induced *voiding* and *electromigration* (EM) are diffusion driven phenomena. They share some similarities like interfacial diffusion and are sensitive to the same kinds of process defects. However, there are a couple of key distinctions. Unlike EM a stress induced void is not simply a thermally activated diffusion process. The work of Suo [215] shows that the stress induced void lifetime (t_{life}) for a long metal line can be written as:

$$(t_{\text{life}}) \propto [1/B\{(\alpha_m - \alpha_d)(T_0 - T)\}^2]\exp(Q/kT)$$

where

B = The effective bulk modulus of the interconnect
Q = Diffusion activation energy
K = Boltzmann constant
T= Absolute temperature.

The phase transformation related stress, on the other hand, is observed when the film during deposition changes its phase, e.g. in Ta and TaN films [222–223]. Experimental evidence also shows that a TaN layer forms a very good barrier layer for Cu but during annealing, *stress* in the Cu-seed layer is seen to be higher on the TaN or Ta layer compared to the TaN/Ta bi-layer or on the TaSiN barrier layer [214].

5.15 Summary

The advantage of copper interconnects in integrated circuits is mostly its higher *conductivity and electromigration resistance* compared to aluminum. The small features are mostly beneficial for high performance microprocessors and fast static random access memory (FSRAM). But deposition and processing of thin film inside small vias and trenches are becoming very difficult. However, the Cu-damascene process has potentially reduced the manufacturing cost by eliminating some of the labor-intensive steps required for subtractive aluminum interconnecting process. Indeed, use of the dual-damascene process flow allows construction of a multilevel Cu stack with potentially fewer steps for processing. It is unique among copper deposition methods (electroplating) because it allows filling of high aspect ratio (AR) features from the bottom up, avoiding the introduction of a center seam or void. But viscosity of the electrolyte has an impact on ion migration, which must be *diffusive, migratory,* or *convective*. As a matter of fact, use of organic additives (*suppressors, levelers,* and *accelerator*) in the electrolyte has addressed all of these. There is no doubt that the success of plating to fill high AR features is built upon the achievement of successful *nucleation* followed by accelerated Cu deposition [224–226].

As the choice of metal changes from titanium (Ti) to tantalum (Ta) or from titanium to tungsten (W), deposition technologies also shift from CVD to PVD, or from CVD to ALD. As the dimensions of the films are decreasing, *electron scattering* from the grain boundaries and defects during deposition becomes problematic in that it leads to higher effective resistance of the Cu-interconnects. Solutions could include larger grain size, reduced defects during deposition, and achieving an atomically smooth surface.

Conventional physical vapor deposition (PVD) for thin film deposition inside small dimension via-holes and trenches is not adequate. As a result a modified IPVD system has been introduced. On the other hand, for very thin film deposition, most promising ALD can provide thinner and stoichiometric film depending on the choice of precursors. But some precursors used in ALD can lead to excess carbon deposition and yield high resistive film. A similar situation has been observed with MOCVD, which is operated at a low temperature.

As clock speeds increase and feature sizes decrease, the need for dielectric materials having lower K values than conventional SiO_2 becomes paramount [227]. Most of the low-K materials that are showing promise for the Cu-damascene process generally have lower modulus and hardness, higher coefficient of thermal expansion, lower cohesive strength, and weaker adhesion compared to the conventional oxides. The properties usually get worse when pores are introduced to lower the K further. As a result, many of these low-K materials present the challenge of reliably creating a top metal gate [228]. Besides sputtering, porous films may allow the metal from Cu-interconnects to penetrate and short the capacitor stack. As a result, low-temperature operated CVD, and room temperature operated spin-on coating methods, are gaining momentum for the deposition of dielectric materials, although sealing of the porous low-K materials with poly-xylene and other materials has given better results.

Grain growth of the Cu-layer during deposition is another phenomenon, which is not desirable. It has been observed that in both fiber type PVD-copper (Cu) and pebble type EP-copper films, grain growth takes place at room temperature. Reduction in grain boundary, surface and interface energy, and strain energy are the driving forces for grain growth in electroplated (EP) Cu film beside *self-annealing*. It has been observed that room temperature re-crystallization (*self-annealing*) of EP-Cu film can bring 25–30% reduction in electrical resistance and relaxation of compressive stress [229].

Surface energy reduction favors <111> oriented film whereas <200> textured film is noticed if the strain energy becomes the dominating factor. Thus, one can say that the textural change during grain growth will indicate whether the film is under stress or not. In general, stresses in thin films can be relaxed by multiple mechanisms, such as: dislocation plasticity [230], diffusion creep [231], and voiding. Experimental studies on stress related phenomena show that tensile stress is responsible for cracking whereas compressive stress can delaminate the film from the substrate.

References

1. International technology roadmap for semiconductors (ITRS), (2001) and also S.R. Riedel, S.E. Schulz, and T. Gessner, Microelectron. Eng., 50, 503 (2000), N.F. Wu et al., PECVD Ti-TiNx barrier with multilayered amorphous structure of high thermal stability for Cu-metallization, Electrochem. Solid State Lett., 6 (2), 6–27 (2003), and Y.J. Mei et al., Thin Solid Films, 308, 594 (1997), and J. Hu et al., Thin Solid Films, 308, 589 (1997)
2. R. Roger et al., J. Electrochem. Soc., 146 (9), 3248–3256 (1999) and also B. Chin, Solid State Technol., 41, 141 (1998)
3. J. Hopwood (ed.), Ionized physical vapor deposition, Academic Press, San Diego, CA, (2000) and S.-H. Kim et al., Electrochem. Solid State Lett., 11 (5), H-127 (2008)
4. J. Werner, H.P. Strunk, and H.W. Schock (eds.), Low Temperature deposition of microcrystalline silicon by microwave plasma enhanced sputtering in polycrystalline semiconductors, Schwabisch Gmund, Germany, (1998)
5. L. Chin and T. Ritzdorf, Semicond. Fabr., 12th ed. (July, 2000) and also M.H. Tsai, S.C. Sun, H.T. Chiu, C.E. Tsai, and S.H. Chuang, Appl. Phys. Lett., 67, 1128 (1995)
6. L.B. Freund and S. Suresh, Thin Film materials, Chapter 1, Cambridge university Press, London, (2003) and J.W. Christian, The theory of transformation in metals and alloys, Pergamon Press, London, (1965) and also K. Maex et al., J. Appl. Phys., 98, 8793 (2003)
7. C.V. Thompson, Structural evolution during processing of polycrystalline films, Annu. Rev. Mater. Sci., 30, 159–190 (2000) and also A.R. Grone, Current induced marker motion in copper, J. Phys. Chem. Solids, 28, 347–350 (1962)
8. H.H. Yu, M.Y. He, and J.W. Hutchinson, Edge effects in thin film delamination, Acta Mater., 49, 93–107 (2001) and also D.S. Campbell, Mechanical properties of thin films, In L.I. Maissel and R. Glang (eds.), Chapter 12, McGraw Hill, New York, (1983)
9. R.H. Dauskardt, M. Lane, Q. Ma, and N. Krishna, Adhesion and de-bonding of multilayer thin film structures, Eng. Fract. Mech., 61, 141–162 (1998) and also K.L. Chopra, Thin film phenomena, McGraw Hill, New York, (1969)
10. L.B. Freund and S. Suresh, Thin film materials, p. 571, Cambridge University Press, London, (2003) and also G.A. Bassett, J.W. Menter, and D.W. Pashley, In C.A. Neugebaur,

J.B. Newkirk and D.A. Vermilyea (eds.), Structure and properties of thin films, John Wiley, New York (1959) and also V. Sukarev, E. Zschech, and W.D. Nix, J. Appl. Phys., 102, 053505-1-14 (2007)
11. S. Wolf and R.N. Tauber, Silicon processing for VLSI era Vol-I, p. 185, Lattice Press, Sunset Beach, CA, (1986)
12. M.L. Green and R.A. Levy, Chemical vapor deposition for metals for integrated circuit applications, J. Met., 37, 63 (1985) and also J.M.E. Harper, C. Cabral, P.C. Andricacos, L. Gignac, I.C. Noyan, K.P. Rodbell, and C.K. Hu, J. Appl. Phys., 86, 2516 (2000)
13. T.P. Moffat et al., Superconformal electrodeposition of copper in 500–90 nm features, J. Electrochem. Soc., 147 (12), 4524–4535 (2000) and also T.P. Moffat, J.E. Bonevich, W.H. Huber, A. Stanishevski, D.R. Kelly, G.R. Stafford.
14. J.M. Poate, K.N. Tu, and J.W. Mayer, Thin films interdiffusion and reactions, Wiley, New York, (1978)
15. J.A. Nucci, R.R. Keller, J.E. Sancez, Jr., and Y.S. Diamond, Local crystallographic texture and voiding in passivated interconnects, Appl. Phys. Lett., 69 (26), 4017 (1996)
16. T.K. Gupta, Hand book of thick and thin film microelectronics, Chapter 5, Wiley, Hoboken, NJ, (2003)
17. J.A. Hopwood, Ionized Physical Vapor Deposition, Academic Press, San Diego, (2000) and also D.M. Mattox, Hand book of physical vapor deposition, W. Andrew Pub./Noyes, Park Ridge, NJ, (1998)
18. K.L. Lai, Ionized hollow cathode magnetron sputtering, p. 95, In J.A. Hopwood, Ionized physical vapor deposition, Academic Press, San Diego, (2000) and also R.F. Bunshah, Hand book of deposition technology and applications, 2nd ed. Noyes Pub., Park ridge, NJ, (1994)
19. M.D. Allendorf, F. Maury, and F. Teyssandier, Chemical vapor deposition, Vol. 14, The electrochemical Soc. Pub. Pennigton, NJ, (2003)
20. D.C. Bradley, R.C. Mehrotra, and D.P. Gaur, Metal alkoxides, Academic Press, New York, (1998)
21. J. Huo and R. Solanki, Characteristics of copper films produced via atomic layer deposition, J. Mater. Res., 17 (9), 2394 (2002) and also P.K. Roy and I.C. Kizilyalli, Appl. Phys. Lett., 72, 2835 (1998)
22. S.A. Campbell, D.C. Glimmer, X. Wang, M.T. Hsich, H.S. Kim, W.L. Glandfelter, and J.H. Yan, IEEE Trans. Electron Dev., 44, 104 (1997)
23. M. Copel, M.A. Gribelyuk, and E. Gusev, Appl. Phys. Lett., 76, 436 (2000)
24. B.H. Lee, L.Kang, R. Nich, W.J. Qi, and J.C. Lee, Appl. Phys. Lett., 76, 1926 (2000)
25. J. Sundqvist, H. Hogberg, and A. Harsta, Atomic layer deposition of Ta_2O_5 using the TaI_5 and O_2 precursor combination, Adv. Mater., 15 (20), 245–248 (2003)
26. A.J. Bard and L.R. Faulkner, Electrochemical methods, 2nd ed. Wiley, New York, (2001) and also G.M. Milazzo, Electrochemistry, Elsevier Pub., Amsterdam, (1963)
27. P.M. Hoffman, A. Radisic, and P.C. Searson, Growth Kinetics for copper Deposition on Si (100) from Pyrophosphate Solution, J. Electrochem. Soc., 147 (7), 2576 (2000) and also F.A. Lowenheim, Modern electroplating, Wiley, New York, (1963)
28. Y. Cao, P. Taephaisitphongse, R. Chalupa, and A. West, Three additive model of superfilling, J. Electrochem. Soc. 148(7), C466 (2001) and also A. Brenner, Electrodeposition of alloys, Vol. 1, 2, Academic press, New York, (1963)
29. J.M. West, Electrodeposition and corrosion processes, D. Van Nostrand, Co., Princeton, NJ, (1965) and also T. Moffat et al., Superconformal electrodeposition of copper in 500–90 nm features, J. Electrochem. Soc., 147 (12), 4524 (2000)
30. M. Faraday, Experimental relations of gold to light, Phil. Trans., 147, 145 (1857)
31. J. Duffy, L. Pearson, and M. Paunovic, The effect of pH on electroless copper deposition, J. Electrochem. Soc., 130 (4), 876 (1983)
32. H.S. Nalwa (ed.), Encyclopedia of nanoscience and nanotechnology, Am. Scientific Pub., Stevenson Ranch, CA, (2003)

33. H. Hu, J. Jacobs, L. Su, and D. Antoniadis, A study of deep submicron MOSFET scaling based on experiment and simulation, IEEE Trans. Electron. Dev., ED-43 (4), 669 (1996)
34. C.A. Neugebauer, Condensation, nucleation and growth of thin films, In Hand book of thin film technology, L.I. Maissel and R. Glang (eds.), Chapter 8, McGraw Hill, New York, (1983) and also C. Ji, G. Oskam, and P.C. Searson, Electrochemical nucleation and growth of copper on Si (111), Surf. Sci., 492, 115 (2001)
35. S.M. Rossnagel, Thin solid films, 263, 1–12 (1995) and H. Sakai et al., Adv. Meter. Conf. September 26, The University of Tokyo, Japan (2006)
36. S.N. Wolf and R.N. Tauber, Silicon processing, Vol. I, p. 368, Lattice Press, Sunset Beach, CA, (1886) and also I.A. Bleach, Step coverage by vapor deposited thin aluminum films, Solid State Technol., 26 (12), 123 Dec. (1983)
37. S.J. Lee et al., IEEE Tech. Dig. Int. Div. Meet., 31, (2000) and also A.E. Kaloyeros, A. Feng, J. Garhart, K.C. Brooks, S.K. Ghosh, A.N. Saxena, and F. Luehrs, Low temperature MOCVD of device quality copper films for microelectronic applications, J. Electron. Mater., 19, 271 (1990)
38. S.M. George, A.W. Ott, and J.W. Klaus, Surface chemistry for atomic layer growth, J. Phys. Chem., 100, 1321 (1996) and S.H. Kim et al., J. Electrochem. Soc., 154, D-435 (2007)
39. J.W. Klaus, S.J. Ferro, and S.M. George, Atomic layer deposition of tungsten using sequential surface chemistry with sacrificial stripping reactions, Thin Solid Films, 360, 145 (2000)
40. K. Ueno, T. Ritzorf, and S. Grace, J. Appl. Phys., 86 (9), 4930 (1999) and also A.A. Volinsky et al., Mater. Res. Soc. Symp. Proc., 649, (2000) and H. Lee and S.D. Lopatin, Thin Solid Films, 192 (1–2), 279 (Dec, 2005)
41. R.A. Schwartz, Chem. Mater., 9, 2325 (1997)
42. R. Krumm, J.G. Long, A. Natarajan, and P.C. Pearson, J. Appl. Phys. D: Appl. Phys., 31, 1 (1998), and also C.A. Neugebauer, Condensation, nucleation, and growth of thin film, In L.I. Maissel and L.I. Glang (eds.), Handbook of Thin Film Technology, Chapter 8, p. 5, McGraw Hill, New York, (1983)
43. E.K. Broadbent, Nucleation and growth of chemically vapor deposited tungsten on various substrate materials: A review, J. Vac. Sci. Technol., B5 (6), 1661 (Nov. Dec., 1987) and also H. Dobberstein and R.W. Schwartz, Modeling the nucleation and growth behavior of solution derived thin films, Symp. On Adv. Mater. For next generation, Integrated materials, AIST Chubu, Nagoya, Japan, (May 27, 2002)
44. D.M. Brown D. Gorowitz, P. Piacente, R. Saia, R. Willson, and D. Woodruff, IEEE Trans. Electron. Dev. Lett., 8, 55 (1987)
45. R.S. Blewer (ed.), Tungsten and other refractory metals for VLSI applications, Mat. Res. Soc., MRS Pub., Pittsburgh, PA, (1986) and also T. Smy, K.L. Westra, and M.J. Brett, IEEE Trans. Electron. Dev., 37 (3), 591 (1990)
46. S. Swirhun, K.C. Saraswatand, and R.M. Swanson, IEEE Trans. Electron. Dev. Lett., 5, 209 (1984)
47. S.S. Chen, S. Sivaram, and R.K. Shukla, Properties of $TiSi_2$ as an encroachment barrier for the growth of selective tungsten on Si, J. Vac. Sci. Technol., B5 (6), 1730–1735 (Nov./Dec., 1987)
48. W.K. Burton, N. Cabrera, and F.C. Frank, Phil. Trans. R. Soc., A243, 299–358 (1951) and also M.K. Gobbert, T. Merchant, L.J. Borucki, and T.S. Cale, J. Electrochem. Soc., 1444 (1), 3945 (1997)
49. J.W. Cahn, Acta Metal., 8, 534–561 (1960), and also L.J. Friedick, S.K. Dew, M.J. Brett, and T. Smy, Thin Solid Films, 266, 83 (1995)
50. G.W. Sears, Acta. Metal., 12, 1421–1439 (1964) and also Z. Wong, Y. Li, and J.B. Adams, Kinetic lattice monte Carlo simulation of facet growth rate, Surf. Sci., 450, 51 (2000)
51. S.P. Murarka and M.C. Peckerar, Electronic material science and technology, p. 363, Academic press, San Diego, CA, (1989)
52. M.H. Grabow and G.H. Gilmer, Surf. Sci., 194, 333 (1988)

53. B. Lewis and A.C. Anderson, Nucleation and growth of thin films, Academic press, New York, (1978)
54. W.A. Tiller, Fundamental aspects of film nucleation and growth, J. Vac. Sci. Tech., A7 (3), 1353, (May/June, 1989) and also R.W. Schwartz, J.A. Voigt, B.A. Tuttle, R.S. DaSalla, and D.A. Pyne, J. Mater. Res., 12, 444 (1997)
55. R.A. Broglia, The color of metal clusters and atomic nuclei, Contemporary Phys., 35 (2), 95–104 (1994) and also M. Bernath, C. Yannouleas, and R.A. Brogila, Phys. Rev. Lett., A156, 307 (1991)
56. W.A. de Heer, Rev. Mod. Phys., 65, 611 (1993) and also H. Haberland, Clusters of atoms and molecules, Springer, Berlin, (1994) and E. Barborini, P. Peseri, A. Li, Bassi, A.C. Ferrari, C. Bottani, and P. Milani, Chem. Phys. Lett., 300, 633 (1999)
57. T.P. Martin, In Elemental and molecular clusters, G. Benedek, T.P. Martin and G. Pacchioni (eds.) , p. 2, Springer, Berlin, (1988) and also M. Bruzzi, P. Piseri, E. Barborini, G. Benedek, and P. Milani, Diam. Relat. Mater., 10, 989 (2001)
58. A. Bohr and B.R. Mottleson, Nuclear structure Vol. II, Benjamin, Reading, MA, (1997)
59. I. Goldhirsch and G. Zanette, Phys. Rev. Lett., 70, 1619 (1993)
60. F. Crick, What a mad pursuit, A personal view of science, Basic books Pub., NY (1988)
61. C. A Neugebauer, Condensation, nucleation, and growth of thin films, pp. 8–26, In L.I. Maissel and R. Glang, Handbook of thin films, McGraw Hill , New York, (1983) and also A.J. Melmed, J. Appl. Phys., 37, 275 (1966)
62. S.M. Hu, Defects in silicon substrate, J. Vac. Sci. Technol., 14 (1), 17–31 (Jan/Feb., 1977) and also M. Borner, S. Landau, S. Metz, and B.O. Kolbersen, In Crystalline defects and contamination: Their impact and control in device manufacturing II, B.O. Kolbersen, C. Claeys, P. Stallhofer, and F. Tardiff (eds.), PV 97-22, The Electrochemical Soc. Pub. Pennington, NJ, (1998)
63. W.A. Johnson and R.F. Mehl, Trans. AMIE, 135, 416 (1939) and also R.W. Schwartz, Chem. Mater., 9, 2325 (1997)
64. M. Avrami, J. Chem. Phys., 8, 212 (1940) and also M. Avrami, J. Chem. Phys., 9, 177 (1941)
65. K.N. Tu, IBM J. Res. Dev., 34 (6), 2671–2674 Nov. (1990) and also S.P. Murarka and M.C. Peckrar, Nucleation and growth, pp. 306, and 358 in Electronic materials for science and technology, Academic, San Diego, CA, (1989)
66. A. Bondi, Chem. Rev., 52, 417 (1953) and also B. Lewis and J.C. Anderson, Nucleation and growth of thin films, Academic, London, (1978)
67. C.A. Neugerbauer, Condensation, nucleation and growth of thin films, In L.I. Maissel and R. Glang, Hand book of thin film technology, Chapter 8, pp. 8–29, McGraw Hill, New York, (1983) and also C. Ratsch, A.P. Seitsonen, and M. Scheffler, Phys. Rev. B, 55, 6750 (1997)
68. J.P. Hirth and G.M. Pound, J. Chem. Phys., 26, 1216 (1957), and also H. Brune, Surf. Sci. Rep., 31, 121 (1998)
69. R. Glang, Vacuum evaporation, In Hand book of thin film technology, L.I. Maissel and R. Glang, (eds.), Chapter 1, McGraw Hill, New York, (1983) and also K.A. Fichthorn and M. Scheffler, In Collective Diffusion on surfaces: Collective behavior and role of adatom interactions, M.C. Tringides and Z. Chvoj (eds.), Kluwer Pub., Dordrecht, Netherlands, (2001)
70. J.P. Hirth and G.M. Pound, Condensation, and evaporation, nucleation and growth kinetics, The Macmillan Company, New York, (1963) and also D.M. Saylor, A. Morawiec, and G.S. Rohrer, Acta Mater., 51, 3663 (2003)
71. W. Primak, Phys. Rev., 100, 1677 (1955) and also X.W. Zhou and H.N.D. Wasley, J. Appl. Phys., 84, 2301 (1998) and D.L. Windt et al., Mater. Res. Soc. Symp. Proc., 564, 307 (1999)
72. M.M. Mandurah, K.S. Saraswat, and T. Kamins, Appl. Phys. Letts., 36, 683 (1980) and also F. Nouvertne et al., Phys. Rev. B, 60, 14382 (1999)
73. V. Vand, Proc. R.. Soc. Lond., 55, 222 (1943) and also S. Hamaguchi and S.M. Rossnagel, J. Vac. Sci Technol., B14, 2603 (1994)

74. T.F. Retajczyk and A.K. Sinha, Thin Solid Films, 70, 241 (1980) and also K. Rajan, R. Roy, J. Trogolo, and J.J. Cuomo, Lowenergy ion beam assisted grain size evolution in thin film deposition, J. Electron. Mater., 26 (11), 1270 (1997)
75. J.E. Mahan, Physical vapor deposition of thin films, Wiley, New York, (2000)
76. S. Schiller, U. Heisg, and S. Panzer, Electron beam technology, 2nd ed. Verlag Technik GmbH, Berlin, (1955) and also J. Fu, P. Ding, F. Dorleans, Z. Xu, and F. Chen, J. Vac. Sci. Technol., 17 (5), 2830–2834 (1999)
77. B. Chapman, Glow discharge processes, Wiley, New York, (1980) and also J.P. Hopwood, Phys. Plasmas, 5, 1624 (1998)
78. G. Carter and J.S. Colligon, Ion bombardment of solids, Elsevier Pub., New York., (1968) and also T. Karabacak and T.M. Lu, Enhanced step coverage by oblique angle physical vapor deposition, J. Appl. Phys., 97, 124504 (2005)
79. E.S. Lame and K.T. Compton, Science, 80, 541 (1934) and also G.S. Chen et al., Evaluating substrate bias on phase forming behavior of tungsten thin films deposited by diode and ionized magnetron sputtering, Thin Solid Films, 484 (1–2), 83 (2005)
80. P. Clarke, US Patent No. 3616450, (26 Oct., 1971) and also E.S. Machlin, Materials science in microelectronics, Giro Pub., New York, (1995)
81. B. Chapman, Sputtering Chapter 6, In Glow discharge process, p. 201, Wiley, New York, (1980) and also J.A. Hopwood, Ionized physical vapor deposition, Thin Films, Vol. 27, Academic, Boston, MA (2000)
82. S.N. Wolf and R.N. Tauber, Sputter deposition equipment, In Silicon Processing for VLSI era, Vol. I, p. 359, Lattice Press, CA, (1986) and also M. Malac, R. Egerton, and M. Brett, Vac. Technol. Coat, 2, 48 (2001)
83. I.A. Blech, D.B. Fraser, and S.E. Haszko, Optimization of Al-step coverage through computer simulation and SEM, J. Vac. Sci. Technol., 15, 1856 (1978) and also R.N. Tait, S.K. Dew, T. Smy, and M.J. Brett, J. Appl. Phys., 70, 4295 (1991)
84. W.D. Gill and E. Kay, Efficient low pressure sputtering in large inverted magnetron suitable for film synthesis, Rev. Sci. Instr., 36, 277 (1965) and also E. Klawuhn, G.C. D'Couto, K.A. Asthani, P. Rymer, M.A. Biberger, and K.B. Levy, Ionized physical vapor deposition using hollow-cathode magnetron source for advanced metallization, J. Vac. Sci. Technol., 18 (4), 1546 (2000)
85. G.K. Wehner and G.S. Anderson, The nature of physical sputtering, In Handbook of thin film technology, L.I. Maissel and R. Glang (eds.), Chapter 3, McGraw Hill, New York, (1970) and also J.A. Hopwood, The role of ionized physical vapor deposition in integrated circuit fabrication, Thin Films, 27, 1 (2000)
86. K.N. Tu, Surface and interfacial energies of $CoSi_2$ and Si films, IBM J. Res. Dev., 34 (6), 868 (Nov., 1990)
87. J.A. Thorton and A.S. Penfold, Cylindrical magnetron sputtering in thin film processes, J.L. Vossen and W. Kern (eds.), p. 73, Academic Press, New York, (1978) and also J.C. Helmer, K.F. Lai, and R.L. Anderson, US patent 5482611, (Jan. 9, 1996) and also V. Girault, 9th Int. Workshop on Stress Induced Phenomena in Metallization, Kyoto, Japan, (April, 2007)
88. R.K. Waits, Planar magnetron sputtering in thin film processes, J.L. Vossen and W. Kern (eds.), p. 131, Academic Press, New York, (1978) and also F.B.D. Mongeot et al., Nanocrystalline formation and faceting instability, Phys. Rev. Lett., 91 (1), 016102-1 (2003)
89. C.Y. Ting, V.J. Vivalda, and M.G. Schaefer, J. Vac. Sci. Technol., 15, 1105 (1978) and also S.J. Liu, H.C. Huang, and C.H. Woo, Appl. Phys. Lett., 80, 3295 (2002)
90. S.M. Rossangel, C. Nichols, S. Hamaguchi, D. Ruzic, and R. Turkot, J. Vac. Sci. Technol., 14 (3), 1846–1852 (1996)
91. J.A. Hopwood, Phys. Plasmas, 5 (5), 1624 (1998) and also J.A. Hopwood, The role of ionized physical vapor deposition in integrated circuit fabrication, In J.A. Hopwood (ed.), Ionized Physical Vapor Deposition, Academic Press. San Diego, CA, (2000)

92. J. Forster, Applications and properties of ionized physical vapor deposition films, In J.A. Hopwood (ed.), Ionized physical vapor deposition, Academic Press, San Diego, CA, (2000) and also K. Tao, D. Mao, and J.A. Hopwood, J. Appl. Phys., 91 (7), 4040 (2002)
93. C.A. Nichols, S.M. Rossnagel, and S.Hamaguchi, J. Vac. Sci. Technol., B-14, 3270 (1996) and also C.F. Yeh, T.J. Chen, C.L. Fan, and J.S. Kao, J. Appl. Phys., 83, 1107 (1998)
94. H. Seifarth, R. Grotzschel, A. Markwitz, W. Matz, P. Nitzsche, and L. Rebohle, Thin Solid Films, 330, 202 (1998) and also B. Sun. Proc. MRS, on Adv. Metal. on ULSI applications, 137 (1977)
95. J. Mendonca et al., Proc. MRS on Adv. Metal. on ULSI applications, 741 (1977) and also E. Klawuhn et al., J. Vac. Sci. Technol., 18(4), 1546 (2000)
96. K.F. Lai, Ionized hollow cathode magnetic sputtering, In J.P. Hopwood (ed.), Ionized physical vapor deposition, Academic Press, San Diego, CA, (2000)
97. S.M. Rossnagel et al., J. Vac. Sci. Technol., B-14, 1819 (1996) and also E. Main, T. Karabacak, and T.M. Lu, Continum model for nanocolumn growth during oblique angle deposition, J. Appl. Phys., 95(8), 4346 (2004)
98. S.M. Rossangel and J.P. Hopwood, Appl. Phys. Lett., 63, 3285 (1993)
99. S. Wickramanayaka, Y. Nakagawa, Y. Sago, and Y. Numaswa, J. Vac. Sci. Technol., A18, 823 (2000) and S-H. Kim et al., Electrochem. Solid State Lett., 11 (5), H-127 (2008)
100. K.H. Min, K.C. Chun, and K.B. Kim, J. Vac. Sci. Technol., B14 (5), 3263–3269 (1996)
101. G.S. Chen et al., Thin Solid Films, 484 (1–2), 83 (2005)
102. S.M. Rossangel, C. Nichols, S. Hamaguchi, D. Ruzic, and R. Turkot, J. Vac. Sci. Technol., 14 (3), 1846–1852 (1996)
103. D.R. Cote et al., IBM J. Res. Dev., 43 (1/2), 5 (1999), and also S.R. Burgess, K.E. Buchanan, J. Cresswell, and I. Moncrief, Deposition and characterization of ionised PVD Ta and TaN barrier films for Cu-interconnects, Trikon Tech, Newport South Wales
104. M. Schieber et al., Thick film of X-ray polycrystalline mercuric iodide detectors, J. Cryst. Growth, 225, 118–123 (2001), and IEEE Trans. Nucl. Sci. T-NS 44, 2571 (1997) and also C.H. Kim et al., IEEE IITC, San Francisco, CA, (June, 2008)
105. C.H. Heimer and J.D. Lockard, Life, p. 6, Charles E. Merril Pub. Columbus, OH., (1977) and also M.L. Hitchman and E. Levy (eds.), Chemical vapor deposition, Wiley-VCH, Weinham, Germany, (2002)
106. M.S. Bishop, P.G. Lewis, and B. Sutherland, Earth history, p. 70, Charles E. Merril Pub. Columbus, OH., (1976) and also Y.P. Zhao, D.X. Ye, G.C. Wang, and T.M. Lu, Nano Lett., 2, 351 (2002)
107. W. Kern and V. Ban, Chemical vapor deposition of inorganic thin films, In thin film processes, J.L. Vossen and W. Kern (eds.), pp. 257–331, Academic Press. New York, (1978), and also H. Wolf, J. Rober, S. Riedel, R. Streiter, and T. Gessner, Process and equipment simulation of copper chemical vapor deposition using Cu (hfac) vtms, Microelectron. Eng., 45, 15 (1999)
108. W.A. Johnson, and R.H. Mehl, Trans. AMIE, 135, 416 (1939) and also K. Radhakrishnan, Ng. Geok-Ing, R. Gopalkrishnan, Mater. Sci. Eng., B-57, 224 (1999)
109. M. Avrami, J. Chem. Phys., 8, 212 (1940), 9, 177 (1941) and also J.C. Lin, G. Chen, H.T. Chin, C.E. Tsai, S.H. Chung, Aool., Phys. Lett., 67, 1128 (1995)
110. W. Kern and G.L. Schnable, Low pressure vapor deposition for VLSI processing, A review, IEEE Trans. Electron Dev., ED-26, 647 (1979) and also C. Klein, Chemical vapor deposition processes, In M. Meyyappan (ed.), Computational modeling in semiconductor processing, Artech House, Boston, MA (1995)
111. A. Learn, Modeling the reaction of low pressure chemical vapor deposition of SiO_2, J. Chem. Soc., 132, 390 (Feb., 1985) and D.R. Cote et al., IBM J. Res. Dev., 43 (1/2), 5 (1999) and also S. Sankaran et al., IEEE IEDM Tech. Dig., Issue 6, p. 26 (2006)
112. G. Herbeke et al., Growth and physical properties of LPCVD polycrystalline Si- film, J. Electrochem. Soc., 131, 675 (1984) and also C.R. Klein and C. Werner, Modeling of chemical vapor deposition of tungsten films, Birkhauser, Basel, (1993)

113. N. Matsuki, J. Ohta, H. Fujika, M. Oshima, M. Yoshimoto, and H. Koinuma, Fabrication of oxide gate thin film transistors using PECVD/PLD multichamber system, Sci. Tech. Adv. Mater., 1, 187 (2000) and also M. Tesauro et al., AVS Symp. No. 54, Seattle, WA, (Oct., 2007)
114. F. Ay and A. Aydinli, Comparative investigation of hydrogen bonding in silicon based PECVD grown dielectrics for optical wave guides, Opt. Mater., 26, 33 (2004)
115. E. Eisenbraun et al., Gelest Inc. PA, and J. Sullivan, Integration of CVD-W and Ta based liners for Cu-metallization MKS Instr. Pub., Wilmington, MA 2000, and also Y. Golan, N.A. Alcantar, T.L. Kuhl, and J. Israelachrili, Langmuir, 16, 6955 (2000) and H.J. Wu, US Patent, 10/98007, (June, 2008)
116. W.F. Wu, K.L. Ou, C.P. Chou, and J.L. Hsu, PECVD-Ti/TiN barrier layer with multilayered amorphous structure and high thermal stability for copper metallization, Electrochem. Solid State Lett., 6 (2), G-27 (2003)
117. A.L.S. Lok, C. Ryu, C.P. Yue, J.S. Cho, and S.S. Wong, Kinetics of copper drift in PECVD dielectrics, IEEE Electron. Dev. Lett., 17 (12), 549 (1996)
118. W. Kern and V. Ban, Chemical vapor deposition of inorganic thin films, In thin film processes, J.L. Vossen and W. Kern, (eds.), pp. 257–331, Academic Press, New York, (1978) and also P. O'Brian, N.L. Pickett and D.J. Otway, Development of CVD delivery systems, Chem. Vapor Depos. Adv. Mater., 8 (6), 237 (2002), Wiley-VCH, Weinham, Germany
119. H.B. Nie et al., Structural and electrical properties of tantalum nitride thin films fabricated by using reactive radio-frequency magnetron sputtering, Appl. Phys., A73, 229 (2001) and also S. Wolf and R.N. Tauber, Silicon processing for VLSI era, Chapter 6, Lattice press, Sunset CA, (1986)
120. M. Rand, Plasma promoted deposition of thin inorganic films, J. Vac. Sci. Technol., 16, 420 (1979) and also M. Rossnagel and J.P. Hopwood, Appl. Phys. Lett., 63, 3285 (1993) and J. Lu and M. Kushner, J. Appl. Phys., 89, 878 (2001)
121. E.K. broad bent, tungsten and other refractory metals for VLSI applications, Vol. 1&2, Mater. Res. Soc. Pub. Pittsburgh, PA, (1987), and also M.Y. Kwak, D.H. Sin, T.W. Kang, and K.N. Kim, Characteristics of WN diffusion barrier layer for copper metallization, Phys. Stat. Solids (a), 174, R5 (1999)
122. Y.S. Diamond and A. Dedhia, J. Electrochem. Soc., 140, 2427 (1993) and also R. Nokogaki, S. Nakai, S. Yamada, and T. Wada, J. Vac. Sci. Technol., A 16, 2827 (1998)
123. N. Kobayashi et al., J. Appl. Phys., 73 (9), 4637–4643 (2001)
124. T.B. Gorczyea and B. Gorowitz, PECVD of dielectrics, In VLSI electronics microstructure science, N. Einspruch (ed.), Vol. 8, Chapter 4, Academic Press, New York, (1984)
125. M.H. Tsai, S.C. Sun, H.T. Chim, C.E. Tsai, and S.H. Chung, Appl. Phys. Lett., 67, 1128 (1995) and also C. Blaauw, Preparation and characterization of PECVD silicon nitride, J. Electrochem., 131, 1114 (May, 1984) and C.Y. Li et al., Thin Solid Films, 47 (1–2), 270–279 (2005)
126. C.H. Tseng et al., IEEE Electron Dev. Lett., 23, 333 (2002)
127. J. Yota, M. Janani, L.E. Camilletti, A. Kar-Roy, Q.Z. Liu, C. Nguyen, and M.D. Woo, Proc. IEEE Electron Dev. Conf. San Francisco, CA, (2000) and also C.H. Hoon and Y.T. Kim, The effects of processing conditions and substrate on copper MOCVD using liquid injection of (hfac) Cu (vtmos), J. Electron. Mater., 30 (1), 27 (2001)
128. P. O'Brien, N.L. Pickett, and D.J. Otway, Developments in CVD delivery systems: A chemist prospective on the chemical and physical interactions between precursors, Adv. Mater., 8 (6), 237 (2002)
129. A.C. Jones and P. O'Brien, CVD of compound semiconductors, Wiley-VCH, Weinheim, (1997) and also C. Dubourdiu, M. Rosina, H. Roussel, F. Weiss, J.P. Senateur, and J.L. Hodeau, Appl. Phys. Lett., 79, 1246 (2001)
130. H.C. Aspinall et al., Growth of praseodymium oxide films by liquid injection MOCVD using a novel praseodymium alkoxide precursor, Chem. Vapor Depos. Adv. Mater., 15 (20), 235 (2003), Wiley-VCU Pub., Weinheim, Germany

131. P.A. Pecan, Science, 285, 2079 (1999) and J. Senawiratne et al., MRS Proc. Fall Symp. FF, (2005)
132. G.D. Wilk, R.M. Wallace, and J.M. Anthony, J. Appl. Phys., 89, 5243 (2001)
133. A.C. Jones, J. Mater. Chem., 12, 2576 (2002)
134. R. Kroger, M. Eizenberg, D. Cong, N. Yoshida, L. Chen, R. Ramaswami, and D. Carl, Properties of copper film prepared by Chemical vapor deposition for advanced metallization of microelectronics devices, J. Electrochem. Soc., 146 (9), 3248–3252, 1999
135. K.K. Choi and S.W. Rhee, Effect of carrier gas on CVD of copper with hfac and DMB, J. Electrochem. Soc., 146(7), C-473–478 (2001)
136. H. Wolf, J. Rober, S. Riedel, R. Streiter, and T. Gessner, Process and equipment simulation of copper chemical vapor deposition using Cu (hfac) vtms, Micron. Eng., 45, 15 (1999)
137. G.A. Person et al., J. Electrochem. Soc., 142 (3), 939 (1995) and K.K. Choi et al., Jpn. J. Appl. Phys., 41, 2902 (2002)
138. A.K. Jain, K.M. Chi, T.T. Kodas, and M.J. Hampdensmith, J. Electrochem. Soc., 140 (5), 1434 (1993) and also Y.K. Ko, B.S. Seo, D.S. Park, H.J. Jang, W.H. Lee, P.J. Reucroft, and J.G. Lee, Semicond. Sci. Technol., 17, 978 (2002)
139. H. Wolf, J. Rober, S. Riedd, R. Streiter, and T. Gessher, MOCVD Cu-films using hexafluoroactylactone venyl tetramethylsilane and argon, Microelectron. Eng., 45, 15 (1999)
140. W.H. Lee et al., The effect of carrier gas and H (hfac) on MOCVD Cu-films using (hfac) Cu (1,5-COD) as a precursor, J. Electron. Mater., 30 (8), 3367–3369 (2000)
141. C.H. Jun and Y.T. Kim, The effects of process conditions and substrate on copper MOCVD using liquid injection of (hfac) Cu (vtmos), J. Electron Mater., 30 (1), 27–34 (2001)
142. J.B. Rem, J. Holleman, and J.F. Verweij, J. Electrochem. Soc., 144 (6), 2101 (1997)
143. D.C. Bradley, R.C. Mehrotra, and D.P. Gaur, Metal alkoxides, Academic Press, New York, (1978)
144. D.B. Beach, F.K. LwGoues, and C.K. Hu, Chemical vapor deposition of high purity copper an organometallic source, Chem. Matter., 3, 216 (1990) and also P.M. Jefferies and G.S. Girolami, Chemical vapor deposition of copper and copper oxide thin films from copper (I) teri-butoxide, Chem. Matter., 1, 8 (1989)
145. M. Schumacher, J. Lindner, P. Strzyzewski, M. Dauelsberg, and H. Juergensen, MOCVD processed ceramic thin film layers for future memory applications, Semicond. Fabtech, 11th ed., ICG Pub., UK, p. 227 (2000) and also C. Dubourdiu, M. Rosina, M. Audier, F. Weiss, J.P. Senateur, E. Dooryhee, and J.L. Hodeau, Thin Solid Films, 81, 400 (2001)
146. D.B. Beach, Design of low temperature thermal chemical vapor deposition processes, IBM J. Res. Dev., 34 (6), 800 (Nov., 1990) and also A.E. Kaloyeros et al., Low temperature metal-organic chemical vapor deposition (LTMOCVD) of device quality copper films for microelectronics applications, J. Electron. Mater., 19, 271 (1990)
147. J.M. Janiski, B.S. Meyerson, and B.A. Scott, Mechanistic studies of chemical vapor deposition, Annu. Phys. Chem., 38, 109 (1987) and also F.A. Cotton and T.J. Marks, Systematic preparation and characterization of pentahaptocyclopentadienyl, copper (I) compounds, J. Am. Chem. Soc., 92, 5114 (1970)
148. Preparation of alkylcopper compounds, J. Organomet. Chem., 12, 225 (1968) and also G. Dennler, A. Houdauer, Y. Segui, and M.R. Wertheimer, J. Vac. Sci. Technol., A 19, 2320 (2001)
149. D. Hausmann, J. Becker, S. Wang, and G. Gordon, Rapid vapor deposition of highly conformal silica nanolaminates, Science, 298, 402 (2002) and M.W. Thomson, Philos. Mag., 18, 377 (1968) and also S-H. Kim et al., Electrochem. Solid State Lett., 9, C54 (2006)
150. M. Lapedus, Support grows for atomic layer deposition schemes, EE Times, (Nov. 24, 2003)
151. M. Yamashita, J. Vac. Sci. Technol., A7, 151 (1989) and S.M. Rosengel, A. Sherman, and F.A. Turner, J. Vac. Sci. Technol., B-18, 2016 (2000) and also O.K. Kwon, H.S. Park, and S-W. Kang, J. Electrochem. Soc., 151 (12), C-753 (2004)
152. B.S. Lim, A. Rahtu, and R.G. Gordon, Atomic layer deposition of transition metals, Nature, 2, 749–754 (Nov., 2003)

153. T. Sutola and M. Simpson (eds.), Atomic layer epitaxy, Blackie, Glasgow, (1990), and also C.Y. Li et al., Electron. Lett., 38, 1026 (2002)
154. M. Leskelä and M. Ritla, Angew. Chem. Int. Ed., 42, 5548 (2003), and also J.S. Park, M.J. Lee, C.S. Lee, and S.N. Kang, Electrochem. Soc. Solid State Lett., 4, C-17 (2001)
155. R.G. Gordon, D. Hausnann, E. Kim, and J. Shepard, Chem. Vap. Dep., 9, 73 (2003)
156. J. Huo, R. Solank, and J.McAndrew, J. Mater. Res., 17(9), 2394 (2002) and also J.A. Hopwood, Ionized vapor deposition of integrated circuit interconnects, Phys. Plasmas, 5 (5), 1624 (May, 1998) and K. Ichinose et al., Adv. Metal. Conf. Sept-26, The Univ. of Tokyo, Japan, (2006)
157. T.K. Kodas and M.J. Hampdon-Smith, The Chem. Of Metal CVD, Chapter 4, VCH, New York, (1994), and also S. Lynne et al., Electrical and physical characterization of Atomic layer deposited thin films in copper barrier applications, Proc. Adv. Metal. Conf. Mater. Res. Soc., PA, (2002)
158. J.D. Klaus, A.W. Ott, J.M. Johnson, and S.M. George, Atomic layer controlled growth of SiO_2 films using binary reaction sequence chemistry, Appl. Phys. Lett., 70, 1092 (1997) and H. Kim et al., J. Appl. Phys., 98, 14308 (2005)
159. C.H. Peng et al., A 90 nm generation copper dual damascene technology with ALD TaN barrier, IEDM Proc. (2002) and C.H. Peng, C-H. Hsieh, and S. Lishue, US Patent 20050277, (Dec., 2005)
160. R.L. Puurunen, Growth per cycle in atomic layer deposition: A theoretical model, Adv. Mater., 15 (20), 243 (2003), Wiley-VCH Pub., Weinheim, Germany and D. Jeoung, H. Inoue, and H. Shinriki, IEEE IITC San Francisco, CA, (June, 2008)
161. S.M. George, A.W. Ott, and J.W. Klaus, Surface chemistry for atomic layer growth, J. Phys. Chem., 100, 1321–1331 (1996)
162. J. Huo, R. Solanki, and J. McAndrew, Characteristics of copper films produced via atomic layer deposition, J. Mater. Res., 17 (9), 2398 (2002)
163. R.G. Gordon, D. Hausmann, E. Kim, and J. Shepard, Kinetic model for step coverage by alternating layer deposition (ALD) in narrow hole and trenches, Chem. Vapor Depos., 9, 73–78 (2003)
164. T.K. Gupta, Hand book of thick and thin film hybrid microelectronics, p. 2, Wiley, NJ., (2003)
165. M. Ritala, P. Kalsi, D. Riihela, K. Kukli, M. Leskela, and Jokinen, J. Chem. Mater., 11, 1712 (1999) and S. Maitrejean et al., Adv. Metal. Conf. Sept. 26, The University of Tokyo, Japan, (2006)
166. J.S. Par, M.J. Lee, and S.W. Kang, Electrochem. Solid State Lett., 4, C-17 (2001)
167. V.G. Levich, Physicochemical hydrodynamics, Prentice Hall, Englewood Cliff, NJ, (1962)
168. K.R. Lawless, J. Vac. Sci. Technol., 2, 1 (1965), and also S. Goldbach, B.V. Den Bossche, T. Daenen, J. Deconinck, and F. Lapicque, Copper deposition on micro-patterned electrodes from an industrial acid copper plating bath, J. Appl. Chem., 30 (1), 1 (2000)
169. P.C. Andricacos, Electroplated copper wire on IC chips, The Electrochem. Soc. Interface Spring, p. 2, (1998) and W.S. Shue, IEEE IITC, p. 175, (2006)
170. D. Edelstein et al., International Electron. Dev. Meeting (IEDM), Washington DC, USA, (7–10 Dec.1997)
171. C.J. Coomb, Jr., Printed circuit board, 2nd ed. McGraw Hill, New York, (1988) and also J. Lee and J.B. Talbot, Simulation of particle incorporation during electrodeposition process, J. Electrochem. Soc., 152 (10), C706 (2005)
172. T.W. Dini, In Modern Electroplating, 3rd ed. F.A. Lowenheim (ed.), Wiley, New York, (1974), and also H.S. Ratore, G.S. Mathad, C.Plougonven, and C.C. Schukert (eds.), Interconnect and contact metallization, The Electrochem Soc. Pub. Pennington, NJ, (1997)
173. J. Reid et al., Proc. IEEE Int'l Interconnect Technology Conference (IITC) 1999, 284–286 (May 24–26, 1999)
174. S. Mayer et al., Electrochem. Soc. Proc. 732, (Oct. 17–22, 1999)
175. Y. Shacham-Diamond and V. Dubin, Microelectron. Eng., 33, 47 (1997)

176. T. Nguyen, Y. Ono, D. Evans, Y. Senzaki, M. Kobayshi, L. Charneski, B. Ulrich, and S. Hsu, Electron Chem. Soc. Proc., 97, 120 (1997)
177. D. Denning, G. Braeckelmann, J. Zang, B. Fiordalice, and R. Venkatramen, VLSI Tech. Digest Tech. Papers, Issue 9–10, 22–23 (June 1998)
178. J.M. Huth, H.L. Swinney, and W.D. McCormick, Role of convection in thin layer electrodeposition, Phys. Rev., E-51, 3444 (1995) and also L. Chen and T. Ritzdorf, Semicond. Feb.Tech., 12th ed., ICG Pub., UK, p. 267 (July 2000)
179. M. Goodenough and K.J. Whitelaw, Trans. Inst. Met. Fin., 67, 57 (1989) and also W. Ruythooren et al., Electrodeposition for synthesis of microsystems, J. Micromech. Microeng., 10, 101 (2000)
180. K.S. Oldham, J. Electroanal. Chem., 420, 53 (1997) and also T.P. Moffat et al., Super conformal electrodeposition of copper in 500–90 nm features, J. Electrochem. Soc., 147 (12), 4524–4535 (2000)
181. J. Newman, Electrochemical system, pp. 196–197, Prentice Hall, Englewood, NJ, (1991)
182. P.E. Hoffmann, A. Radisic, and P.C. Searson, Growth kinetics for copper deposition on Si (100) from pyrophosphate solution, J. Electrochem. Soc., 147 (7), 2576 (2000) and also S.C. Goldbach et al., Copper deposition on micro-patterned electrodes from an industrial acid copper plating bath, J. Appl. Electrchem., 30 (1), 1 (2000)
183. P.C. Andricacos and L.T. Romankiw, Magnetically soft materials: Their properties and electrochemistry, In Advances in electrochemical science and engineering, H. Gerischer and C.W. Tobias (eds.), Vol. 3, pp. 227–321, VCH Pub., New York, (1994), and also P.C. Andricacos, C. Uzoh, J.C. Dukovic, J. Horkans, and H. Deligiani, Damascene copper electroplating for chip interconnections, IBM. J. Res. Dev., 42 (5), 567 (Sept., 1998)
184. J. O'M Bockris and M. Enyo, Trans. Faraday Soc., 58, 1187 (1962) and also M.G. Legally and Z.Y. Zhang, Nature, 417, 907 (2002)
185. R.D. Mikkola, Q.T. Jiang, and B. Carpenter, Copper electroplating for advanced interconnect technology, Plat. Surf. Finish., Issue 6, 81–85 (March, 2000)
186. M. Datta et al., Electrochemical fabrication of mechanically robust PbSn C4 interconnections, J. Electrochem. Soc., 142, 3779 (1995)
187. J.M.E. Harper, C. Carbral, P.C. Andricacos, L. Gignal, I.C. Noyan, K.P. Rodbell, and C.K. Hu, J. Appl. Phys., 86, 2516 (1999)
188. C.K. Lingk and M.E. Gross, J. Appl. Phys., 84, 557 (1998)
189. Y. Cao, P. Taephaisitphongse, R. Chalupu, and A.C. West, Three additive model of superfilling of copper, J. Electrochem. Soc., 148 (7), C466 (2001) and also T.P. Moffat et al., Superconformal electrodeposition of copper in 500-90 nm features, J. Electrochem. Soc., 147 (12), 4524 (2000)
190. J. Tafel and Z. Physik, Chem., 50A, 641 (1905) and also C. Mad, M. Matlosz, and D. Landolt, J. Electrochem. Soc., 143, 3936 (1996)
191. V.M. Dublin et al., Int'l. Interconnect Tech. Conf. IEEE Cat No. O1EX461, p. 271, San Francisco, CA, (2001)
192. R.A. Bistead, J. Wu, R. Mikola, and J.M. Calvert, ECS meeting Philadelphia, PS, (2002)
193. E. Farndon, F.C. Walsh, and S.A. Campbell, J. Appl. Electrochem., 25, 572 (1995)
194. P.C. Andricacos, C. Uzoh, J.O. Dukovic, J. Horkans, and H. Deligianni, Damascene copper electroplating for chip interconnections, IBM J. Res. Dev., 42 (5), 567 (Sept., 1998)
195. K.J. Vetter, Electrochemical kinetics, Chapter 2, Academic Press, San Diego, CA, (1967)
196. K.I. Popov, M.G. Pavlovic, and D.T. Lukic, J. Appl. Electrochem., 10, 299 (1980)
197. M.M. Chow et al., Method for producing coplanar multilevel metal/insulator film on a substrate and for forming patterned conductive lines simultaneously with stud vias, US Patent, 4789648, (Dec. 6, 1988)
198. C.K. Hu and J.M.E. Harper, Copper interconnections and reliability, Mater. Chem. Phys., 52, 5 (1998)
199. H. Deligianni, J.O. Dukovic, P.C. Andricacos, and E.G. Walton, Electrochem. Soc. Proc., 267 May, 2–6, (1999)

200. H. Talieh, Method and apparatus for electrochemical mechanical process, U.S. Patent 6, 176, 992 (2001)
201. B.M. Basol, Plating method and apparatus that creates a differential between additive deposited on a top surface and a cavity surface of a work piece using an indirect external influence, U.S. Patent Pub. 2002/002068 A1, (Feb.21, 2002)
202. B.M. Basal, C. Uzoh, H. Taleih, D. Young, P. Lindquist, T. Wang, and M. Cornejor, Microelectron. Eng., 64, 43 (2002) and also W.H. Yu, E.T. Kang, and K.G. Neoh, J. Electrochem. Soc., 149 (11), C592 (2002)
203. C. Ji, G. Oskam, and P.C. Pearson, Electrodeposition of copper on silicon from sulfate solution, Electochem. Soc., 148 (11), C746–752 (2001)
204. E.J. O'Sullivan et al., Electrolessly deposited different barrier for microelectronics, IBM J. Res. Dev., 42 (5), 607–620 (1998) and also J. Duffy, L. Pearson, and M. Paunovic, The effect of pH on electroless copper deposition, J. Electrochem. Soc., 130 (4), 876 (1983)
205. S.S. Tzeng, and F.Y. Chang, Mater. Sci. Eng., A-302, 258 (2001) and also J. Newman, Electrochemical systems, pp. 314–315, Prentice Hall Inc. Englewood Cliff, NJ, (1973)
206. C.Dew et al., Nano Lett., 3, 143 (2003) and also M.A. Leveque, Annals des mines memoires, Ser. 12, 13, 201–239, 305–362, 381–415 (1928)
207. F.A. Lowenheim (ed.), Modern electroplating, 3rd ed. John Wiley, NJ, (1974), and also A.K. Graham (ed.), Electroplating engineering handbook, 3rd ed. Van Nostrad Reinhold Co., New York, (1971)
208. G. Oskam, J.G. Long, A Natarajan, and P.C. Searson, J. Phys. D, 31, 1927 (1998)
209. Y. Sacham Diamand and M. Angyal, Thin Solid Films, 262, 93–103 (1995)
210. M. Paunovic, Plating, 55, 1161 (1968) and also H.P. Fong, Y. Wu, Y.Y. Wong, and C.C. Wan, Electroless Cu deposition process on TiN for ULSI interconnect fabrication via Pd/Sn colloid activation, J. Electron. Mater., 32 (1), 9 (2003)
211. H. Ebneth, In Metallizing of plastics, Handbook of theory and practice, R. Suchentrunk (ed.), ASM International, Materials Park, OH, (1993)
212. H.P. Fong, Y. Yu, Y.Y. Wang, and C.C. Wan, Electroless deposition process on TiN for ULSI interconnect fabrication via Pd/Sn colloid activation, J. Electron. Mater., 32 (1), 9 (2003)
213. N. Petrov, Y. Severdlov, and Y.S. Diamond, Electrochemical study of electroless deposition of Co (P) and Co (W,P) alloys, J. Electrochem. Soc., 149 (4), C187 (2002)
214. S. Balakumar et al., Effect of stress on the properties of copper lines in Cu interconnects, Electrochem. Solid State Lett., 7(4), G68 (2004)
215. T. Hara, K. Kakata, and Y. Yoshida, Electrochem determine grain size of the EP-Cu. Solid State Lett., 5, C41 (2002) and also Z. Suo, Reliability of interconnect structures, Interfacial and nanoscale failure, W. Gerberich, and W, Wang (eds.), Vol. 8, pp. 263–324, Elsivier, Amsterdam, (2003)
216. T. Hara and K. Sakata, Stress in copper seed layer employing in copper interconnection, Electrochem Solid State Lett., 4(10), G-77 (2001)
217. D.Y. Kim, PhD Dissertation, Stanford University, Stanford (Dec., 2003)
218. J.J. Toomy, S. Hymes, and S. Murarka, Stress effects in thermal cycling of copper (magnesium) thin films, Appl. Phys. Lett., 2074–2076 (April 1995)
219. A.A. Volinsky, Mat. Res. Soc. Symp. Proc., 649, (2000)
220. D.E. Kramer, A.A. Volinsky, N.R. Moody, and W.W. Gerberich, J. Mater. Res., 16 (11), 3150 (2001)
221. P.A. Flinn and G.A. Waychuns, A new X-ray diffractometer design for thin film texture, strain, and phase characterization, J. Vac. Sci. Technol., B-6, 1749 (1988)
222. H.B. Nie et al., Structural and electrical properties of tantalum nitride thin films fabricated by using reactive radiofrequency magnetron sputtering, Appl. Phys. A, 73, 229 (2001)
223. Y. Liu and H. Huang, Phil. Mag., 84 (19), 1919 (2004)
224. R.L. Cohen and R.I. Meek, J. Colloid Interf. Sci., 55, 156 (1976) and also R.L. Cohen and K.W. West, J. Electrochem. Soc., 120, 502 (1973)

225. R.L. Jackson, Initiation of electroless copper plating using Pd^{+2} poly acrylic acid films, J. Electrochem. Soc., 135 (12), 3172 (Dec., 1988)
226. W.J. Dressick, C.S. Dulcey, J.H. Georger, G.S. Calabrese, and J.M. Calvert, J. Electrochem. Soc., 141, 210 (1994)
227. E. Budevski, G.S. Staikov, and W.J. Lorenz, Electrochemical phase transfomation and growth, VCH, Weinheim, Germany, (1996)
228. D.M. Kolb, R. Ullmann, T. Will, Science, 275, 1097 (1997)
229. R.M. Stiger, S. Gorer, B. Craft, and R. Renner, Langmuir, 75780 (1999)
230. K. Maex et al., Low dielectric constant materials for microelectronics, J. Appl. Phys., 93, 8793 (2003)
231. F. Lacopi et al., Impact of LKD5109 low-K interfaces in single damascene process and performance, Micrelectron. Eng., 65, 293 (Nov., 2003)

Chapter 6
The Copper Damascene Process and Chemical Mechanical Polishing

6.1 The Copper Damascene Process

6.1.1 Introduction

The *American Heritage Illustrated Encyclopedic Dictionary* defines *damascene* as a native inhabitant of Damascus (*Damaskus)*, which is a major city in the north-west of Syria. The word *Damask* means patterned silk fabrics woven in Damascus, a city notable for manufacturing and shipment of damascened steel sword blades, which were exceptionally hard and resilient. In French, damascene is *damas quiner* means to decorate in the manner of Damascus blades or steel from Damasquine of Damascus. Thus the word *damascene* can be taken literally as the process of decorating a metal with wavy patterns of gold or silver. However, in integrated circuits (ICs) the *damascene process* means an elegant technique of inlaying metal (copper) for interconnect which avoids the complicated process of metal etching.

Figure 6.1 shows the cross-section of an interconnect structure created by: (a) the subtractive aluminum process and (b) the additive dual Cu-damascene process. The damascene process can be either *single damascene* (SD) or *dual damascene* (DD) which are illustrated in Fig. 6.2.

Figure 6.3 shows a tungsten plug that connects silicon to metal using a single-damascene process. The difference between the simple single damascene (SD) process and the simple dual damascene (DD) process is that in the DD process one metal deposition step and one CMP step are eliminated (as well as the dielectric deposition step). The reduction in the number of processing steps has made DD more attractive than its twin single damascene process. In the SD process the plugs are typically tungsten (W), whereas in the DD process via holes are filled mostly with Cu though some industrial houses have used Al in the DD process. Figure 6.4 shows a commercial ASIC product fabricated by the Cu-DD process following 130 nm-node technology. The introduction of low-*K* and Cu-interconnects in the damascene process and with an effective gate length of 80 nm for a MOS device, up to 30% faster computer speed and performance and up to 50% reduction in power have been achieved.

The damascene process consists of etching a trench in the dielectric layer and filling up the trench [1–2] with a barrier layer to stop Cu-diffusion and subsequent

Tapan K. Gupta, *Copper Interconnect Technology*,
DOI 10.1007/978-1-4419-0076-0_6,

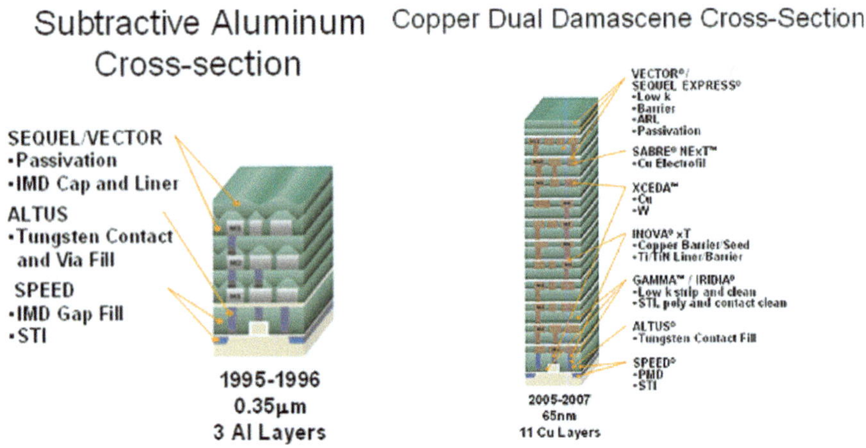

(a) (b)

Fig. 6.1 The cross-section of (**a**) subtractive Al-metallization and (**b**) additive Cu-damascene process (Reprinted with permission Novellus Systems Inc.)

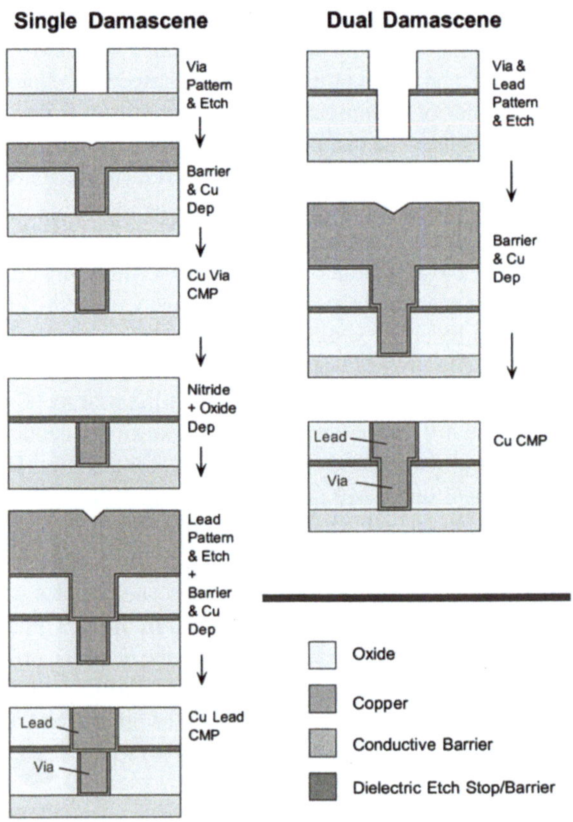

Fig. 6.2 Cu-damascene process flow (Reprinted with permission Dr. K. Saraswat, Stanford University, CA)

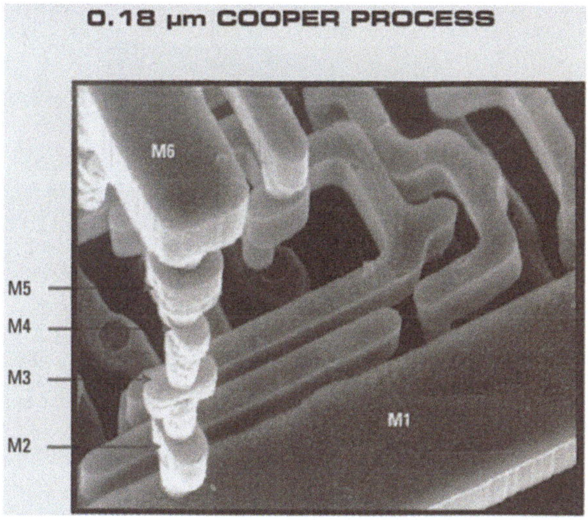

Fig. 6.3 A tungsten plug that connects silicon to metal using a single damascene process (Reprinted with permission, Semiconductor International , June 2000, p. 96)

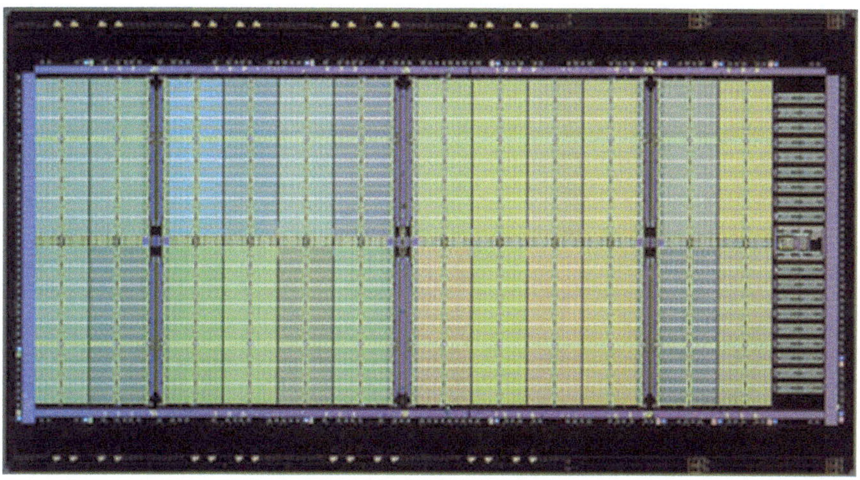

Fig. 6.4 A MOS transistor structure fabricated by following the dual damascene process (Photo courtesy IBM Research)

deposition of the Cu-seed layer as a nucleation center for the electroplated Cu [3]. Finally, the overburden metals are removed by chemical mechanical polishing (CMP). There are five main dual damascene techniques under study though only two principal modes, via first and trench first, are currently in mainstream production. With each of the two approaches, a SiC or Si_3N_4 hard mask is used to provide an etch stop (ES) – appropriately called *embedded via first* or *embedded trench first* [4]. In general, the addition of an embedded hard mask can simplify the etch process by providing an ES during trench formation. Figure 6.5 shows the double damascene

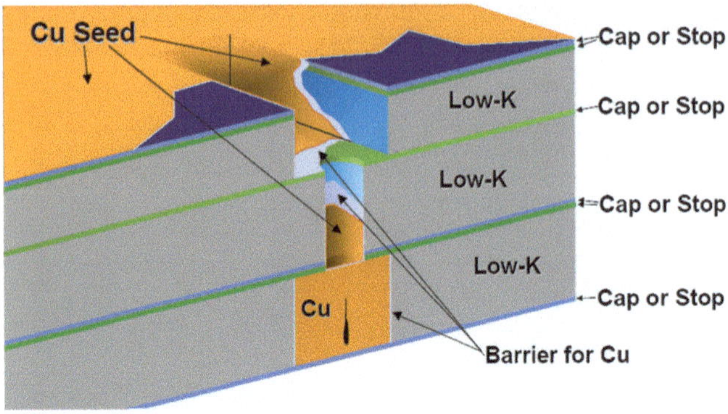

Fig. 6.5 A dual damascene architecture showing the different materials in the unit structure (Photo courtesy, Peter J. Wolf)

unit with different materials embedded in the structure. The clever engineering of the Cu-damascene unit with different layers has fortified the successful operation of the IC with the common goal of achieving the speed of the device.

Damascene processes were developed when efforts to develop an effective Cu-etch process failed. The main theme of the damascene process is to integrate Cu in the integrated circuit (IC) to reduce the *RC* effects of the interconnects. In addition to the use of copper as interconnecting metal, the damascene process requires incorporation of new and often exotic low-*K* interlevel and intralevel dielectric materials, because conventional dielectric silicon dioxide (SiO_2) cannot meet the requirements of the sub-100 nm device [5–7]. As a whole, the damascene architecture has introduced borderless contact via, a decrease in process induced damage regarding the front end part of the line, which addresses soft plasma processes for either dry etching, cleaning or material deposition, improvement in performance (speed) of the devices and *fine line structure*. As a matter of fact, copper damascene processes can fabricate Cu-interconnecting lines less than 1/1000th of the width of a human hair.

Figure 6.6 shows etched holes (trenches) covered by metal during damascene processing [8–9]. The overburden metal layer is polished following chemical mechanical polishing (CMP). In September 1977 the copper damascene process was introduced and in 1999 production and shipment of devices were reported.

6.1.2 Conventional Metallization Technology

The back end process for interconnecting the transistors are done first by contacting the transistor terminals. Later on, different vertically stacking alternate layers of metals, wires, and via holes etc. are connected. The whole operation is performed at a temperature not exceeding 400 °C. The state of the art of the 250 nm technology for conventional aluminum (Al) metallization follows the back-end process where, generally, five levels of metal wires and tungsten via holes are embedded in

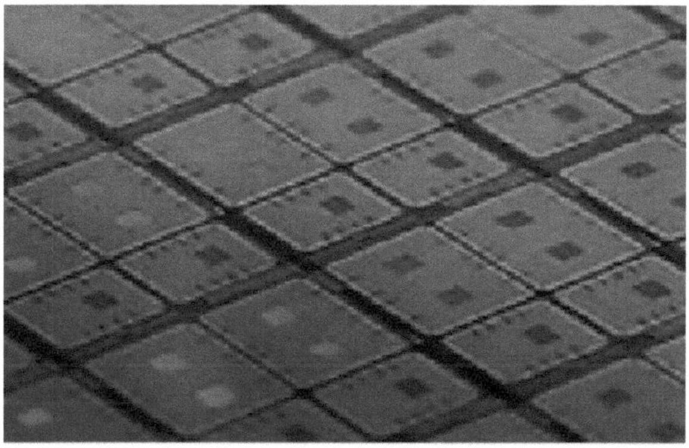

Fig. 6.6 Damascene process application (Photo courtesy of IBM research)

the silicon dioxide (SiO_2) dielectric layer, and the whole process is known as the *Cloisonné* process.

The SiO_2 (dielectric layer) is deposited by the PECVD method with tetraethyl orthosilicate (TEOS) precursor at 350–400 °C [10]. The oxide is patterned following lithography and titanium (Ti), and titanium nitride (TiN) are deposited sequentially by the PVD method (sputtering). Tungsten (W) metal is then deposited by the CVD method at 425–450 °C by reduction of tungsten hexafluoride (WF_6). The metal lines overhanging at the surface of the via hole are polished by a chemical mechanical process (CMP). After via contact, Ti, Al (alloy), and TiN are deposited without breaking the vacuum. The Ti over- and under-wiring is used as a base layer for good adhesion [11–13], and low contact resistance to underlying via holes. TiN film, on the other hand, forms the cap layer to the metal stack to minimize reflectivity, and facilitate control of fine structures during lithography (Fig. 6.7) [14].

Following the 250 nm back-end process, a production level multilevel complimentary metal oxide semiconductor (CMOS) device for high performance logic circuits has been fabricated [15]. The circuit consists of five levels of Al (alloy) wires with tungsten in via holes (as plugs/studs) embedded in SiO_2 (Fig. 6.8).

As the state of the art of the 250 nm technology changed over to 100 nm and less, aluminum (Al) interconnections could no longer support the sub-100 nm node technology, because of its physical limits. Indeed, the smaller line dimensions required in denser microcircuits are not feasible using Al (alloy), because the interconnect resistivity is simply too high (due to its interconnect cross-section). Moreover, the number of metal multi-layers, the pitch, the thickness and width for 250 nm design is longer applicable to sub-100 nm node technology [16] (Table 6.1).

6.1.3 Cu-Damascene Metallization Technology

According to the *International Technology Roadmap for Semiconductors* (ITRS), the number of multilayers is expected to reach 10 by the year 2009. The other

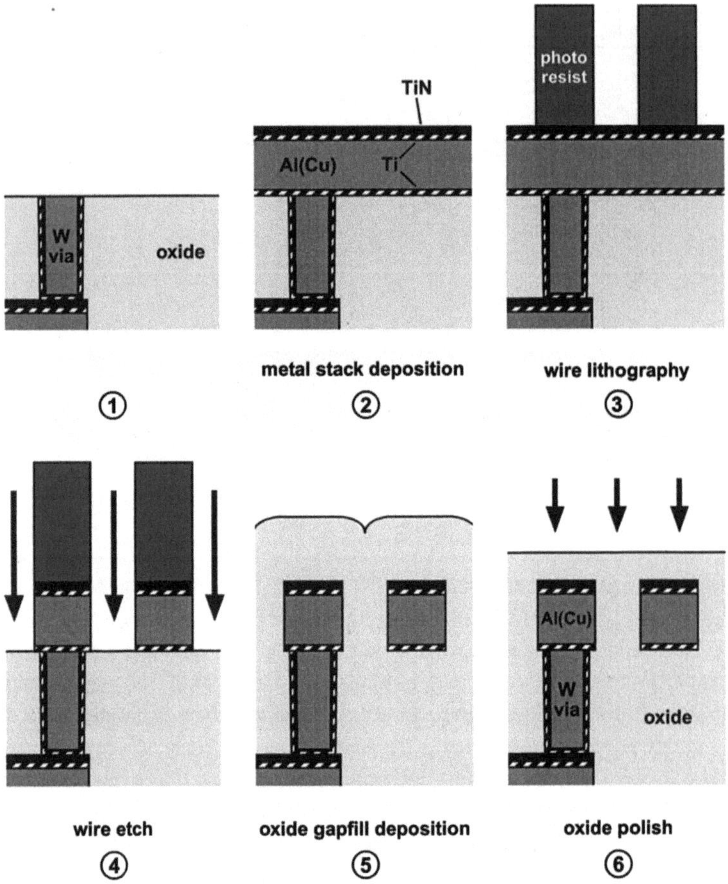

Fig. 6.7 Schematic of conventional aluminum alloy metallization process (Reprinted with permission, [14])

expectations of ITRS includes: (i) pitch ~50 nm, (ii) thickness <50 nm, and (iii) aspect ratio ~2.5/2.3. As a result, Al metal interconnects and the W-via hole fill-up are changed to Cu [17], and these changed the whole process technology of ICs (introduction of barrier layer, low-*K*, and chemical mechanical planarization). The present sub-100 nm fabrication process is based on damascene technology and is known as *the Cu-damascene (single or dual)* process.

Figure 6.9 shows a schematic of a dual damascene (DD) process with copper as interconnecting lines and via connection metal. In the dual damascene process trenches and via holes can be opened in the dielectric layer to the underlying conductor layer by various methods. They are mainly classified as (a) trench first, (b) via first, and (c) buried-via or self-aligned. Each method has its advantages and disadvantages [4].

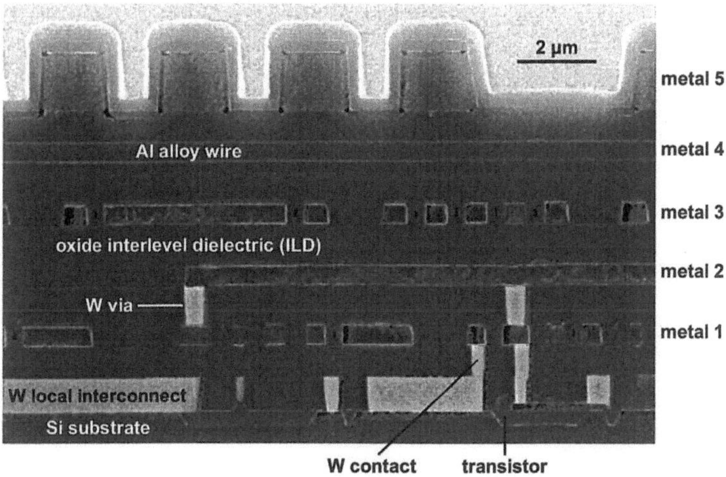

Fig. 6.8 Cross-sectional view of a scanning electron micrograph of a CMOS device (Reprinted with permission, Free-scale Semiconductor, Inc.)

Table 6.1 Interconnect design rules for 250 nm node technology

Layer	Minimum pitch (nm)	Thickness	Wire aspect ratio
Metal-1	480	480	1.5:1
Metal-2	930	900	1.9:1
Metal-3	930	900	1.9:1
Metal-4	1600	1330	1.7:1
Metal-5	2560	1900	1.5:1

Reproduced with permission [14].

6.1.3.1 Different Approaches to Cu-Damascene Metallization

Figure 6.10 shows a schematic of the trench first filling procedure. The top surface of the cap-layer (Si_xN_y) is coated with photoresist (PR) and lihographically patterned (2). Anisotropic dry-etch cuts through the surface hard mask (typically plasma Si_xN_y) and down through the low-K dielectric, stopping on the embeded etch stop layer. The photoresist is then stripped, leaving behind a trench in the inter-level dielectric (ILD) (Fig. 6.3). The photoresist is again applied to the wafers and lithographically patterned (Fig. 6.4). The via etch then cuts through the embedded etch stop layer down through ILD, to the final silicon nitride barrier located at the bottom of the via (Fig. 6.5). The bottom barrier is then opened with a special etch and photoresist is stripped (Fig. 6.6). A thin barrier layer of Ta/TaN is deposited by the PVD/ALD method, which forms the barrier lining for copper. A copper seed layer is then deposited (ALD/PVD/elecroless) prior to the bulk deposition of copper by the electrochemical plating (EP) method [18]. The copper is polished back using CMP and Si_xN_y is deposited on the top to encapsulate the Cu-layer.

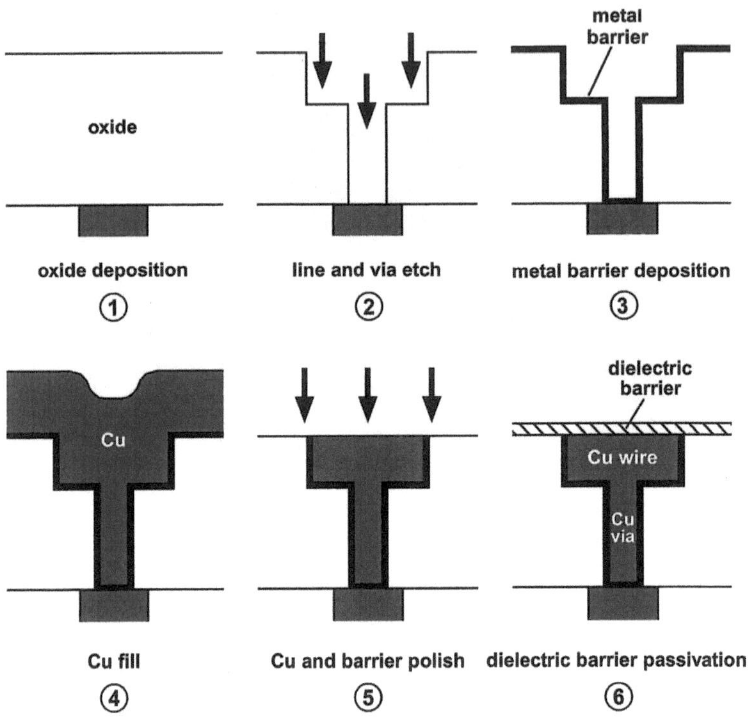

Fig. 6.9 The schematic of a dual-damascene (DD) process for copper interconnect (Reproduced with permission [14])

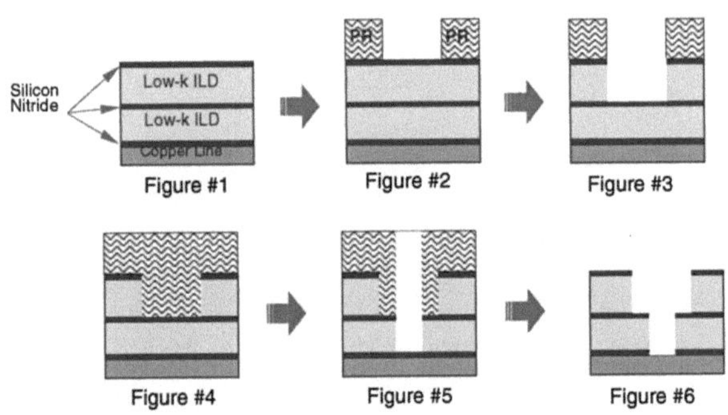

Fig. 6.10 Trench first approach (Reprinted with permission Jerry Healey, 2001)

The major drawback of the trench first approach is that after the trench is etched, the PR that is applied for the via step will completely fill the trench (4). As a result, the PR can be said to be *pooled* in the trench, leaving an extra thick resist layer right in the area where via holes are to be patterned. This creates a difficult situation to fabricate fine structures for via holes. Thus, for feature size of less than 250 nm, the

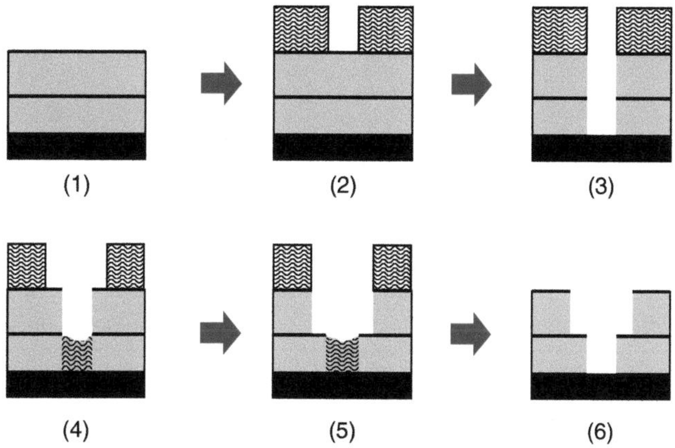

Fig. 6.11 Via first approach (Reprinted with permission Jerry Healey, 2001)

trench first process is not a reliable, and the via first approach has been accepted widely in the industry.

Figure 6.11 presents the via first procedure. In this methodology the wafers are first coated with photoresist and then lithographically patterned (Fig. 6.2). Next, an anisotropic etch is performed to cut through the surface of the hard mask extending down through the ILD, the embedded etch stop, and finally stops on the bottom silicon nitride barrier. It is of primary importance that the via etch does not break through this bottom layer. Otherwise, the via etch will sputter the copper located beneath the barrier up into the unprotected via hole. This will lead to the diffusion of Cu-atoms into the ILD and ultimate device failure. Next, the via photoresist layer is stripped (Fig. 6.3) and the trench photoresist is applied and lithographically patterned. Some of the photoresist will remain in the bottom of the via (Fig. 6.4), and prevent the lower portion via from being over-etched during the trench etch process.

An anisotropic etch is then performed which cuts through the surface hard mask and down through the ILD, stopping at the embedded hard mask. This etch will form the trench. The photoresist is then stripped and the silicon nitride barrier at the bottom of the via is opened (Fig. 6.6) with a very soft, low-energy etch that will not cause copper to sputter into the via. Finally, the tantalum, copper seed, and bulk copper are deposited and planarized using chemical mechanical polishing (CMP).

Figure 6.12 shows the self-aligned or buried via approach, where lithographically the via pattern is created in the embedded Si_xN_y etch stop (1). After the via pattern is etched and cleaned, the top dielectric layer is deposited. Finally, the trench mask is aligned to the via openings in the embedded Si_xN_y layer, and both the trench and via are opened with a single etch step (3). The process requires simultaneous alignment of the via and trench and perfect alignment is very difficult and challenging. In addition, the etch process requires high nitride to oxide selectivity. For these reasons the self-aligned process has not been adopted for production.

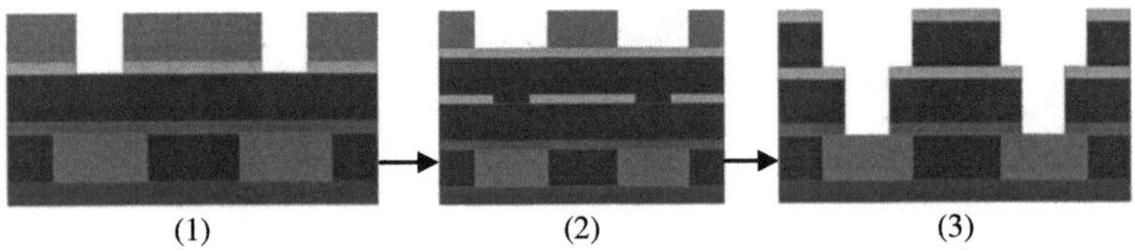

Fig. 6.12 Self-aligned approach (Reprinted with permission Peter J. Wolf)

Table 6.2 Comparison of different etch process of dielectric layer during DD process

Etch process	Advantages	Disadvantages
Buried etch stop	(a) Separate via fill materials are not required (b) Bulk ILD etch is accomplished in one step	(a) Large thickness variation during patterning resist (b) Selectivity of the etch process is critical (c) Multiple hard mask layers effect low-*K* value
Via first	(a) Plasma deposition, CVD , ILD and hard mask can be applied within single cluster tool (b) First hard mask is not required (c) ILD requires only single spin coat (d) Best resolution and CD control can be achieved easily	(a) Etch strip residue can stick at the base of the trench pattern (b) To reduce shadow masking etch rate is high (c) Etch breakthrough step may require via fill material or BARC
Trench first	(a) CVD and plasma depositions of barrier layer, ILD and hard masks can be applied within single cluster tool	(a) Spin-on ILD layer requires multiple passes (b) Multiple hard mask layers effect low-*K* value (c) Large thickness variation is noticed during resist patterning

Table 6.2 shows three main dual damascene techniques that are frequently used to deposit metal layer inside the Cu-damascene architecture and their advantages and disadvantages.

6.1.4 General Objectives and Challenges

Following dielectric etch, via holes and trenches are lined with a conductive barrier layer to prevent copper (Cu) diffusion. Barrier layer encapsulation is also required to ensure that Cu does not diffuse through the surrounding dielectric material. Barrier metals are generally more resistive to current when compared with Cu, and thus the thickness of the barrier layer is kept to a minimum to preserve the effective conductive advantage of Cu over Al- metallization. As a whole the effective resistance of the barrier layer with Cu should be comparable to the Cu-line alone and the Cu-layer should adhere well to the substrate (barrier). In addition, it should not deteriorate during an effective cleaning procedure of via holes after dielectric layer etching. During via etch, the underlying Cu-layer is exposed. Therefore, one must be very careful to see that during cleaning no Cu-metal is re-deposited on via-hole's sidewalls.

The barrier layer is deposited inside the trenches and via holes before a thin seed-layer of Cu is deposited by PVD/ALD as a nucleation layer for electroplated (EP) Cu. The thin seed layer (10–15 Å) should be free from voids and stress besides being *conformal*.

Experimental evidence shows that PVD techniques, even with improved flux directionality, are incapable of achieving void-free copper deposition. Most of the time, *cusping* (Fig. 6.13b) is a common phenomenon observed after metal deposition inside the trench. However, good trench filling has been demonstrated through reflow after sputtering. A similar phenomenon has been observed during *superconformal-filling* of Cu by electroplating method [18].

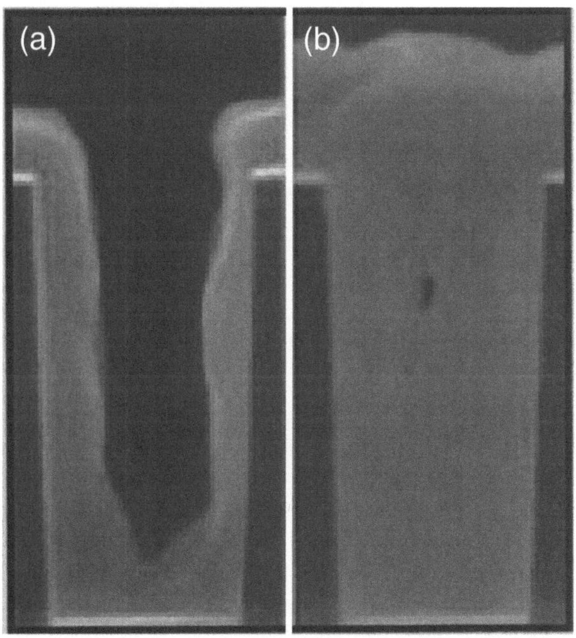

Fig. 6.13 (**a**) Formation of trench for the deposition of copper and (**b**) the formation of a deep cusp above the trench (Reprinted with permission IBM Res. [18])

All three approaches described previously (*trench first, via first*, and *buried via*) for creating damascene structures present difficulties. For example, with the trench first approach, one has to get the resist down in the trench and then pattern the via within the thick resist. It goes without saying that to open a pattern through a thick resist layer is tricky and difficult. Another difficult question to answer is whether an intermediate etch stop (ES) layer is needed to define the trench. Without the ES layer the trench edges can be etched rapidly leading to rounded edges. At the same time, without the ES layer it is difficult to detect the real end-point which will make the critical dimension (CD) and depth control of the trench extremely difficult. Moreover, most of the ES layer materials have high dielectric constant (K), which will significantly impact the overall effective capacitance of the stack.

In the via-self aligned process, good alignment of the trench with the via hole in the hard mask can be extremely difficult. With misalignment the final etched via will be a half-mooned shape. Above all, in practice, it is likely that different materials will be used for the dielectric at the via level and the trench level. The via dielectric is less critical from a capacitance perspective than the trench dielectric. Moreover, some of the low-K materials contain carbon and hydrogen, which requires retuning the process in order to meet the selectivity and etch requirements. The move to the Cu-damascene process with low-K dictates a fundamental change in back-end processing. The move particularly becomes challenging because the IC industry is trying to move to a larger wafer size and smaller device geometry every year.

Advanced microprocessors require several (eight or more) levels of wiring to carry signal and power from transistor to transistor and to the outside world. The dual damascene approach to fabricating these interconnected structures creates a wiring level and via level simultaneously, thereby reducing the total number of processing steps. However, the dual damascene strategy still requires at least 20 process steps to complete wiring. Recently, some IC industries are thinking of replacing the photolithography work involved in the damascene process by step and flash imprint lithography (SFIL) [18].

6.2 Chemical Mechanical Polishing (CMP) and Planarization

6.2.1 Introduction

Planarization is the smoothening of topography over the product die to allow each successive layer to be formed over a flat surface. Device planarization, on the other hand, is the reduction of vertical height during subsequent processing steps, and has come into integrated circuit (IC) technology's forefront as the geometry of the devices began to shrink. Planarization is most critical during the back-end processes where metallization and dielectric layers are used. If the resist material does not cover severe topology, it would have been much easier to image a fine line on a planar surface and to employ anisotropic pattern etch [19]. The main techniques available for planarization of a device surface can be categorized into two basic types: (a) dielectric planarization and (b) metal planarization.

The planarization may be *local* or *global*. *Local planarization* is generally done by repetitive bulk deposition of dielectric materials followed by isotropic etch, whereas *global planarization* is achieved by bulk deposition followed by chemical mechanical polishing (CMP).

In conventional *subtractive aluminum (Al) metallization*, different modes of planarization are adopted according to the type of device and the material on which planarization has to be performed. During local oxidation growth, considerable planarization is achieved because of the thickness of the oxide layer (~10,000–20,000 Å). To etch a window through the oxide layer, either dry etching or wet chemical etching method is adopted. The wet etching system is mostly *isotropic* (which etches all sides equally), and results in a circular contour. The other methods that are applied to achieve planarization are *photolithography* and *taper control*. Photoresist work will be dealt in this book in a separate chapter. *Taper control*, on the other hand, is performed either by means of inactive ion implantation (with argon gas), or using a low melting point glass (phosphor silica glass, PSG) [20–22].

In the Cu-damascene process surface planarization of the inter-level dielectric material is the key technology to realize the multi-level interconnection structure. It involves several approaches, such as *etch back*, *spin coating* of glass or organic films, and bias deposition [23]. The *etch back* process is claimed to provide a very flat dielectric surface all over the device area independent of the shapes and sizes of the lower level metal. The planarization technology is capable of providing three or more levels of interconnections with sub-micrometer metal lines.

To achieve a higher degree of planarization, several methods are being used, such as polishing the front surface of the wafer with an abrasive slurry mixed with chemicals, zapping it with a laser [24–25], and re-flow of materials like boro-phosphor silicate glass (BPSG). Perhaps the most notable challenge in planarization is to completely fill small, high aspect ratio contact holes that connect the first level of metal with the transistor's gate, source and drain, and via holes.

In the copper damascene process, following patterning and deposition (barrier, seed, and Cu-plating) of materials, *chemical mechanical polishing (CMP)* is performed to remove the overburden materials from the surface. The palnarization capability is an important advantage of CMP process. It prevents problems associated with the increase of the number of interconnect layers in multilevel metallization of integrated circuits (ICs), where the topography of device structures increases significantly as additional layers are stacked on top of one another [26].

6.2.2 Chemical Mechanical Polishing (CMP)Technology

There is nothing especially new or exotic about CMP used in the semiconductor industry than the polishing machine used in glass polishing or silicon wafer polishing. The dramatic success of CMP to polish or etch oxides and metals, and its planarization capability, has led the semiconductor industry to push this technology for mass production [27]. In the early stages, CMP technology was used for oxide planarization. The conventional multi-level wiring layers on top of each other, with oxide dielectric in between lines, provide insulation between layers. The oxide topography varies as the oxide layer follows the underlying features and variations accumulate from layer to layer. Meanwhile more stringent resolution specifications have forced micro-lithography systems to use larger numerical apertures and reduced depth of focus (DOF).

Oxide CMP is very successful in the removal of the oxide layer and reducing the topographic variation. As a result, users are naturally curious to use this method for etching other materials, like the tungsten (W)-plug structure. It has been found that CMP could polish the extra tungsten metal exposed to the oxide inter-metallic dielectric (IMD) layer [28].

Figures 6.14 a and b show a schematic of chemical mechanical polishing machines. The polishing pad attached to the platform can have translatory and/or rotary movement during polishing. The polishing pad can be attached to an endless belt guided by two rotating wheels (Fig. 6.14b). There is a provision to replace the pad with a new one or with a pad made of different material. An additional attachment, which conditions the pad from time to time, is also found in some CMP settings (Fig. 6.14b). The wafer surface to be planarized is mounted on a carrier, which can be rotated during polishing. Slurry is fed to the pad upstream of the polishing region, and a diamond grit pad conditioner sweeps across the polishing pad downstream of the polishing region.

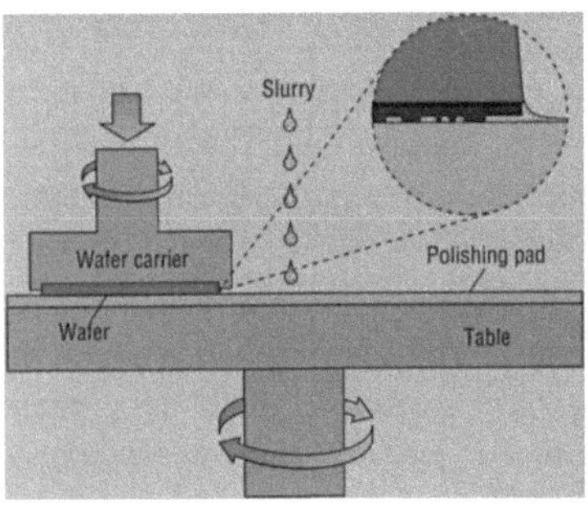

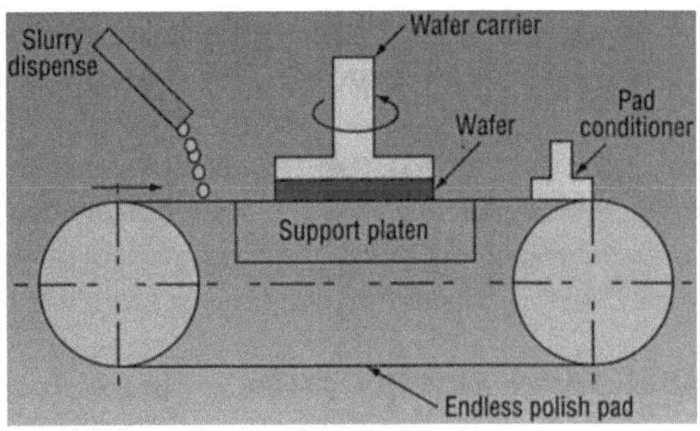

Fig. 6.14 Sschematic of CMP machines (Reprinted with permission from *Solid State Technology*, (a) Oct. 1994, p. 63, and (b) Oct. 1996, p. 108)

Historically, CMP has exploited single slurry processes in ILD and tungsten applications, but none of the single slurry processes developed so far for copper CMP have shown effective results. So a multi-step multi-slurry method has been developed. The process is becoming more attractive, because of the process flexibility, which facilitates process optimization regardless of the equipment design rationale [29–30].

Polishing techniques typically use *rotary, orbital,* or *linear mechanical motion* coupled with an inert abrasive in chemical solution. All the techniques in one way or other try to balance the removal rate, planarization, and uniformity through combinations of pressure, speed, and hardness of the pad. A softer pad promotes better global planariztion, while a harder pad improves local planarization [31]. Higher pressure and speed increases removal rates, and when pressure is reduced, planarization improves but uniformity declines. Most of the CMP technology uses *rotary motion,* though uses of a combination of *orbital and rotational* motions are not rare. The wafer head rotates in a circular path, while the polish head is orbital. The net effect is that each point on the pad traces a spirographic trajectory. The configuration facilitates direct slurry delivery to the wafer surface while maintaining high relative velocities and a small overall tool footprint. In some CMP systems, a variable pressurized ring provides control at the wafer edge and improves removal rate during polishing [32]. The design allows independent control from the main head pressure (Fig. 6.14). Figure 6.15 shows a commercial CMP machine with polishing head and attachments.

Fig. 6.15 A commercial CMP machine with polishing head and the attachments (Reprinted with permission, Semiconductor International, Nov. 1998, p. 57)

In order to achieve ideal planarized structures, the CMP process should be able to remove the metal and the liner material evenly without eroding the underlying dielectric. Therefore a high metal and liner to dielectric selectivity is necessary. Low selectivity can lead to the erosion of patterns during overpolishing, which is necessary to remove larger amounts of metal situated on the top of the spacers. It has been observed that when the metal pattern densities are high, they show higher metal and dielectric removal rates [33–34].

Figure 6.16 a shows a 3-D view of an ideal case of a chemical mechanical polishing surface. Ideally the copper surface should be perfectly flat. Unfortunately in reality the copper surface suffers from *dishing* and *erosion* (Fig. 6.17). *Dishing* is generally defined as a recessed height of a copper line compared to the neighboring oxide line, whereas *erosion* is the difference between the original oxide height and the post-polish oxide height [35–36].

Fig. 6.16 3-D view of (**a**) an ideal case of CMP and (**b**) a realistic case of CMP

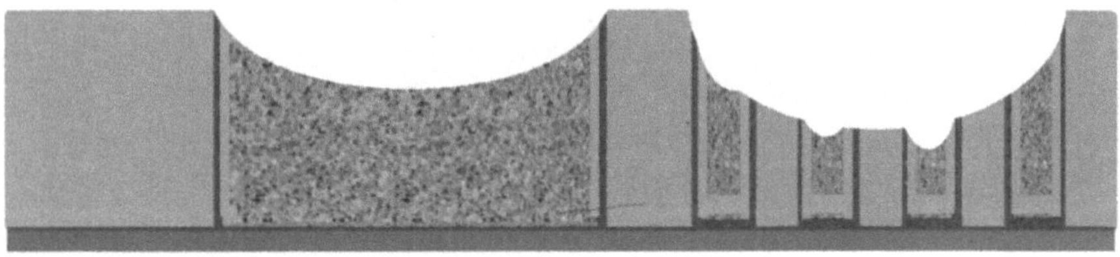

Fig. 6.17 Copper dishing due to over polishing and erosion of dielectric surface (Printed with permission, Semiconductor International. May 2000, p. 74)

6.2.2.1 An Analytical Model

With the advent of shallow trench isolation (STI) and copper interconnects with different exotic materials such as low-K, high-K, chemical mechanical polishing (CMP) has emerged as the most important operation in the fabrication of integrated circuits (ICs). The model proposed by Warnock allows us to predict quantitative wafer topography [36]. However, most of the recent models are mechanics based and can be classified in several groups. One of these model about CMP technology is presented here, which identifies the effects of polishing pressure, platen speed, particle size of the slurry, pad elasticity, and pad asperities on the basis of partial contact node.

During the CMP process, the pad is in partial contact with slurry particles carried by asperities. The total force (F_{T}) present during CMP operation under partial contact mode can be written as:

$$F_{\mathrm{T}} = F_{\mathrm{P}} + F_{\mathrm{A}} = \pi a^2 H \tag{6.1}$$

where,

$$F_{\mathrm{P}} = (2\sqrt{3P_{\mathrm{P}}R^2})/K = \text{Applied polishing force} \tag{6.2}$$

and

$$F_{\mathrm{A}} = (AR/6d^2)(1 + a^2/Rd) = \text{the adhesion force} \tag{6.3}$$

Again,

$$P_p = \eta \int \propto_0 (4/3)\beta^{1/2}(Z-h)^{3/2}\varphi_B\varphi_Z d\beta\, dZ \tag{6.4}$$

$$= \{\sqrt{}(\pi\sigma^3)\}/\{(2\sqrt{}C)\}(E/(1-v^3))\exp.\{-(W-T)/\sigma\} \tag{6.5}$$

for $W–T \geq 0$.

The particle adhesion model for plastic deformations between a particle and a substrate was generalized by Maugis and Pollock and it can be represented in terms of the contact radius between the deformed particle and the wafer as [36]:

$$F = \pi a^2 H. \tag{6.6}$$

From Equation (6.6), we can determine the value of H, when a and F are known. At low platen speed (3 psi), the hydrodynamic pressure P_F is zero and $PA = F = P_p A_p + P_F A_F = P_p A_p$. Thus P_p can be determined when F and A_p are known. Likewise, from analytical Equation (6.2), K can be determined when R, P_p and F_p are known.

Again, Equation (6.5) has to be solved numerically for the pressure distribution and the corresponding pad deflection. From Equations (6.4) and (6.5) we can predict that when P_p increases the dishing ($W–T$) of metal lines will also increase. Thus one can say that recess or dishing of the metal line will increase with increasing polish time. At the same time, the corners of the dielectric will get more rounded. From Equation (6.5) one can predict further that dishing of the Cu-line will increase with increasing values of the roughness parameter (σ) of the Cu-line.

The thickness (x) of the layer removed with time (t) during polishing can be written mathematically as:

$$(\mathrm{d}x/\mathrm{d}t) = K_p P v_R \tag{6.7}$$

If the removable rate across the surface in contact is assumed to be uniform, i.e. the thickness of the layer removed x is constant across the surface, we can write:

$$(\mathrm{d}x/\mathrm{d}t) = (K_w/H)\mathrm{P}v_R \tag{6.8}$$

If the steady state regime is assumed, the amount of Cu-dishing remains constant with over-polishing time. As a result, the material removal on both the Cu and dielectric surface should be at the same rate and we can express the model mathematically as:

$$(\mathrm{d}x/\mathrm{d}t)_{Cu} = (\mathrm{d}x/\mathrm{d}t)_{dielectric} \tag{6.9}$$

Now if we assume that v_R and K_w are the same for copper and the dielectric then combining Equations (6.8) and (6.9) we can write:

$$(P_{cu}/P_{dielectric}) = (H_{cu}/H_{dielectric}) \tag{6.10}$$

To solve P_{cu} and $P_{dielectric}$ with the pattern geometry, the force of equilibrium condition on the interested area across an interconnect and the surrounding dielectric spacing can be employed. However, based on force equilibrium, the pressure on the dielectric will increase with the increase of dishing until it reaches a steady state value. Similarly, the Cu-polishing rate is bounded by the blanket Cu rate and the steady state rate of the surroundings [35].

Again assuming the slurry to be a Newtonian incompressible fluid, lubrication theory can be applied to solve the mean drag force on a spherical particle of radius R in contact with a plane wall as:

$$F_{\mathrm{D}} = \{(32\mu^2)/\rho\}\{\mathrm{Re}^*\} \tag{6.11}$$

Again,

$$\{\mathrm{Re}^*\} = \{(\rho\, v_{\mathrm{R}} D)/\mu\}\{h/D\}^2 \tag{6.12}$$

The value of Re* will be very small because the film thickness is much smaller than the diameter of the wafer. Not only this, the drag force calculated on the basis of laminar or turbulent flow will also be very small compared to the adhesion force (F_{a})

where,

F_{P} = Polishing force
P_{p} = Pad contact pressure
A_{p} = Wafer/pad contact area
P_{F} = Hydrodynamic pressure provided by the fluid layer to support the load
R = Radius of the slurry particle
H = Material hardness (pad/worn material)
K = Particle fill fraction
β = Asperity radius
Z = Asperity height
η = Asperity density
h = fluid film thickness between the pad and the wafer
φ_{B} and φ_{Z} = Asperity radius and height normal density functions, respectively
E = Young's modulus of the pad (for soft pad ~540 × 10^6 Pa, hard pad ~112 × 10^9 Pa)
ν = Poisson's ration of the pad (for soft pad ~0.45, hard pad ~0.28)
W = Deformation of the pad
T = Surface profile
σ = A characteristic roughness parameter
C = Curvature of the top of the asperity
d = Separation distance between particle and substrate
K_{p} = Paterson constant
P = Applied pressure
v_{R} = Relative velocity between pad and wafer

K_w = The wear coefficient
F_D = Mean drag force
D = diameter of the wafer
ρ = Density of the fluid.

Table 6.3 presents some mechanical properties of the materials that are involved in the CMP process.

Table 6.3 Elastic properties of the materials used in the Cu-damascene process

Materials	Young's modulus (Gpa)	Poisson's ratio
Copper	128[a]	0.30
Silicon Oxide	186[a]	0.30
Tantalum	74[b]	0.20
Rodel pad	0.5	0.30

a: *ASM Metal Handbook* and b: *Handbook of Materials Science*, CRC Press.

6.2.3 Copper Dishing Model

Both the damascene and shallow trench isolation (STI) processes introduce stringent limits to erosion and dishing [37–38]. This undermines the CMP process, because of the variations in the topography. When a particular pad with a particular slurry material is used in the CMP process, to polish different materials having different mechanical properties, like hardness, density, etc., the rate of removal of materials will be different on different materials. Copper, which is the mainstream metal for local and global metallization, is softer than tantalum (Ta), titanium (Ti), tungsten (W), and molybdenum (Mo) and their nitrides, borides, and oxides. Thus, during polishing the etching rate of copper is much faster than these refractory metals. As a result, some areas of copper are over-polished compared to these refractory metals, and this results in non-planarity and *dishing*. On the other hand, the pads and slurries want to push right through the barrier and start eroding the dielectric. Experimental observations show that on a 100 μm feature one can expect 200 Å dishing and erosion, less than 100 Å on a high-density feature. The most important issue in achieving super-tight dishing and erosion specs is that these bumps require excessive polishing which can make film thickness variation of 40% across the wafer. One common approach is to use a high-selectivity slurry to minimize dishing and erosion.

Figure 6.17 shows a schematic of copper dishing and erosion of dielectric surface. It is said that success in copper CMP is judged not only by the degree of planarity of the surface, but also by the polishing rate (which impacts throughput), the number of defects, and the left-over residue. Due to corrosion and over-etching, dishing can reach as deep as 100 Å.

The key to success in planarization in the sub-micron level feature size is the understanding of the various parameters of the CMP process, such as (a) slurry chemistry, (b) particle size inside the slurry [39–40], (c) relative velocity of the pad and the wafer, (d) the force applied to the wafer and the pad [41–42], (e) pad elasticity [43–44], and (f) pad conditioning method besides pattern density [45].

6.2.4 Slurry Chemistry

A copper (Cu) CMP slurry is typically made of suspended abrasive particles, an oxidizer, (hydrogen peroxide, hydroxylamine, and potassium iodate), a corrosion inhibitor, plus one or more of a long list of additives including dispersants, chelators, accelerators, colorants, lubricants, and biocids [46].

Copper is comparatively easy to oxidize, and an acidic aqueous solution with pH ~3 can convert copper to cuprous oxide and then to cupric oxide with increasing pH [47]. In neutral basic pH solutions the Cu-oxide film becomes a protective layer [48]. The Cu-oxide layer thus formed is polished with a solution containing abrasive powder, mostly silica or alumina. Experimental observation shows that optimizing the time and using a rotary type polisher with a relative velocity of 0.7 m/s, and normal pressure of 48 kPa, can effectively polish Cu-surface without dishing when α-alumina with pH-7 is used in the abrasive solution.

Due to radically different mechanical and chemical properties of copper and the barrier layer, it is now generally accepted that Cu-CMP can be accomplished more effectively by a *two-step* process [49] (Fig. 6.18). In the first step the abrasive will remove the bulk Cu-surface without disturbing the barrier layer. In the second step, the residual copper and the barrier layer will be removed typically using silica abrasive. Finally, profilometry and atomic force microscopy can be used to confirm the dishing of the Cu-surface.

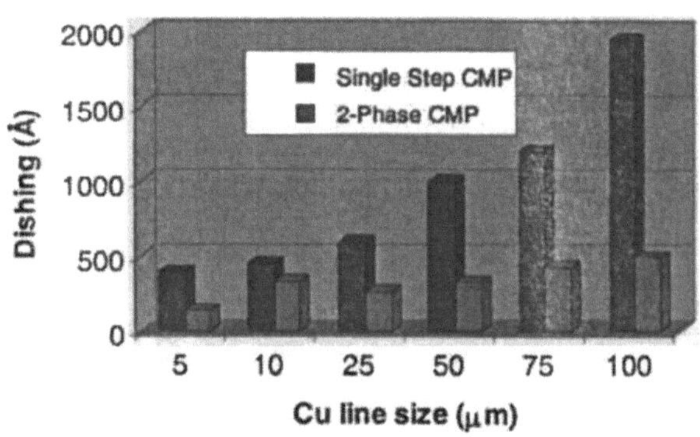

Fig. 6.18 Comparative copper line polishing performance for a typical one-step CMP process and two-step process (Courtesy, EKC Technology)

Removal of the *barrier layer* is more mechanical in nature and typically requires slurry with some abrasive. The requirement must be balanced with the need to tune selectivity both to copper and the barrier. In contrast, certain integration schemes may require the removal or thinning of an oxide cap [50].

Figure 6.19 shows the copper surface after CMP operation. When the slurry chemistry is optimized the copper surface becomes very smooth (Fig. 6.19a).

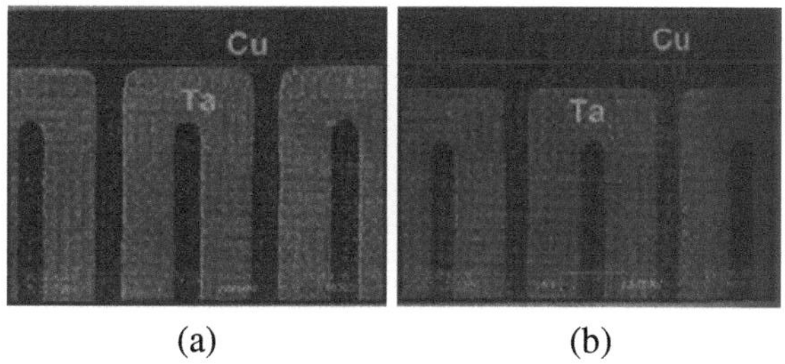

Fig. 6.19 Two SEMs showing surface images of copper (**a**) with and (**b**) without chemistry optimization (Reprinted with permission *Solid State Technol.*, Sept. 2002, p. S10)

Generally, silica or colloidal silica is used for polishing dielectric materials and alumina (Al_2O_3) is used for polishing metals. When slurry uses a silica particle, it remains in the suspension longer providing the slurry with a better concentration consistency. Another important factor of the slurry is the pH value, which should be around 10–12 for dielectric materials and 2–3 for metals. However, in the neutral pH regime the selectivity of copper polishing is favorable with respect to silicon dioxide [51–54].

The differential velocity, pressure and the chemistry of the slurries will ultimately determine the localized removal rate, especially for copper [55]. It has been observed that alkyl contented dielectric materials greatly reduces the removal rate, and it is very difficult to attain a uniform polished surface with silica based slurry, though the silica based CMP process for low-*K* methyl silesesquioxane (MSQ) with tetramethyl ammonium hydroxide (TMAH) works well. Unfortunately, porous low-*K* materials show film fracture and peeling at the substrate's immediate edge (Fig. 6.20).

6.2.5 Particle Size Inside the Slurry

Particle size and the hardness of the particles inside the slurry are responsible for surface damage like scratches and pits. It has been observed that the concentration of the coarser particles increases the surface damage [56], because coarser particles tend to hold the wafer away from the pad reducing the pressure on the smaller size abrasive particles resulting in decreased interaction between particles and substrate. Thus when the feature size is small (0.13–0.10 μm), the particle size within a slurry will be an important issue in the CMP process in achieving better yield [57].

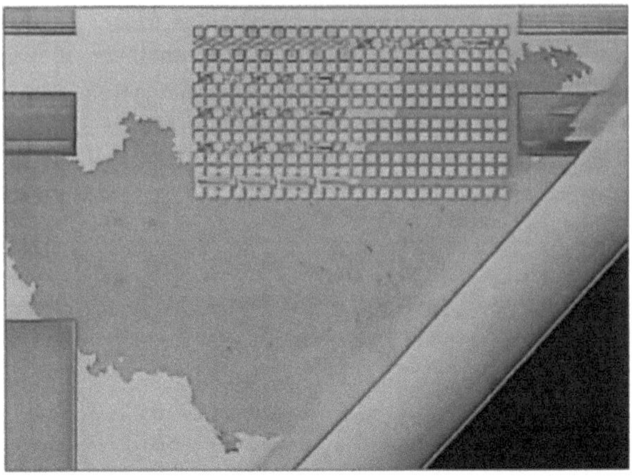

Fig. 6.20 CMP-induced peeling in low-*K* porous dielectric material (Reprinted with permission, Semiconductor International, Nov. 2001, p. 55)

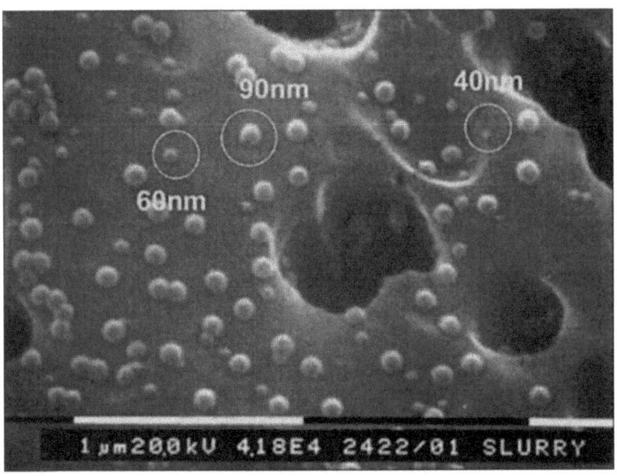

Fig. 6.21 Scanning electron micrograph showing different size particles inside the slurry (Reprinted with permission, Semiconductor International, Aug. 2002, p. 75)

Figure 6.21 shows a scanning electron microphotograph (SEM) of a particular class of slurry with different particle sizes. The smaller (micron size) particles agglomerate, and affect the bonding between particles. Besides size, several other factors, like settling, aging and partial sintering in the container can change the bonding effect too. Strongly bonded particles exhibit non-regular shapes with large angles that may cause damage to the surface during polishing, as is seen with the slurry containing spherical shaped large diameter ($\sim$10 μm) particles.

6.2.6 Relative Velocity of the Pad and Wafer

During linear planarization, the unidirectional motion of the polishing pad provides a uniform velocity vector field of the pad surface across the wafer. The time averaged relative speeds between wafer and pad are constant across the wafer when there is no wafer rotation. When the wafer is rotated during polishing, the time averaged relative speeds become greater at the wafer edge compared to the wafer center. However, linear CMP planarization has a typical pad speed of 400 ft/min with wafer (200 mm dia) rotation of 25 rpm. The relative speed between the edge and the center of the 200 mm wafer differs only by 0.01% [58]. In contrast, conventional orbital polishers using a rotating platen exhibit unequal pad speeds across the platen due to the radial effect. In these polishers the time averaged relative speeds of the wafer and the platen are equal [59]. The single most important criterion of linear planarization is its higher removal rate along the radius; besides, it has higher planarization capacity and overall cost benefits.

6.2.7 Pad Pressure

The removal rate during CMP varies linearly with the pad pressure (or polish pressure) [60] and the pressure distribution on the high feature size particles is found to be non-uniform and elliptical, whereas on the small feature particles it is uniform [61]. Non-uniform pressure distribution results in non-uniform erosion of different surfaces having different physical and mechanical properties [62].

6.2.8 Pad-Elasticity

Young's modulus is a measure of the compressibility of a pad material while the conformity is a measure of the flexibility of a pad. So, a hard pad will show less conformity. For soft material like copper, high compressibility and conformity can increase copper *dishing*. Better planarity can be achieved with a dual layer pad having a hard top layer with a soft bottom layer. However, neither flexibility nor compressibility can determine the polishing rate during the CMP process. The pressure, the particle size, and shape of the slurry, the flexibility and compressibility of the pad, the device feature size and pitch, all of these factors contribute towards the effective planarization of the surface.

6.2.9 Pad Conditioning

The efficiency of a pad in removing the material is seen to degrade with time either (a) due to plugging up of the pores with slurry particles, or (b) due to the development of stress within the pad. Of the two reasons the second one is believed to be more likely responsible for the sluggish efficiency with time. The sluggishness of the pad is called *glazing*. In order to minimize glazing and to recover the polishing rate, the pad is reconditioned every time before and after polishing a new wafer. It has been observed that the stress behavior is concentrated near the pad surface. The normal stress developed within the loaded region will determine the pad's tangential and normal displacement [63].

6.2.10 Shallow Trench Isolation (STI)

Shallow trench isolation (STI) has become the enabling technology for sub-micron level devices because of its higher packing density [64]. It offers improved isolation between devices compared to the traditional approach to local oxidation [65]. STI regions consume less silicon while offering superior latch-up immunity and smaller channel width encroachment than LOCOS isolation (Fig. 6.22). The main theme of the STI process is to remove all of the oxides over the dense active areas while leaving the residual nitride stop layer thickness across all feature sizes.

Fig. 6.22 A stylus nanoprofilometry image of an STI region (Reproduced with permission, Semiconductor International, March 2000, p. 71)

Recently, slurry free fixed abrasive technology has been developed for the STI CMP process [66]. However, the use of abrasive slurries has become routine work for wafer planarization. But *dishing* and *nitride clipping* problems are very common to the STI polish. The planarization processes for STI structures require substantially better process control than is needed for tungsten plug and inter-level dielectric (ILD). The quality of an STI process is measured by the control of trench oxide and nitride thickness. This requires tight control of the belt parameters to control the step height between the trench oxide surface and silicon surface after the nitride strip. It seems that STI polishing with fixed abrasive has the potential to simplify the process integration, which will ultimately lower cycle time and cost.

6.2.11 Abrasive Free Polishing

In order to avoid the problems arising out of the slurry, the abrasive free process seems to be less complicated [67]. The abrasive free (ABF) process consists of a soft pad with a solution of copper oxidizer, a copper-oxide etchant, and a corrosion inhibitor. The oxidizer forms copper oxide, which is protected by the inhibitor and it is removed by the friction of the soft pad. The soft pad does not remove the barrier layer; in addition, it provides a scratch free copper layer. Later the barrier layer is removed by reactive ion etching (RIE) in the presence of SF_6 gas, which does not react with copper. Because chemical action dominates abrasive free polishing (AFP), the slurry used in AFP is highly selective, enabling removal of copper while protecting the underlying barrier layer, regardless of copper thickness. It is expected that low-pressure AFP will provide a wide process window in the near future for the 100 and 65 nm node technologies. However, experimental observation shows that AFP does not work well with traditional rotational tools (Fig. 6.23). Figure 6.23 shows copper residues that are heavy when CMP uses rotational polishers (bottom).

Fig. 6.23 Copper residues are heavy when CMP uses rotational polishers (*bottom*) (Photo courtesy Novelus Inc.)

6.2.12 End-Point Detection

During the CMP process, the end point can be detected either by optical or electrical methods. As a matter of fact, end point detection is problematic especially when there are multiple metal layers (more than four). The optically based system generally adopts the principle of multi-level reflectance spectroscopy. It seems that the detection of the end point by optical methods can be better handled by observing the changes in the spectral shape in the 400–800 nm rather than changes in intensity of the spectrum. The electrical method, on the other hand, can detect the end point

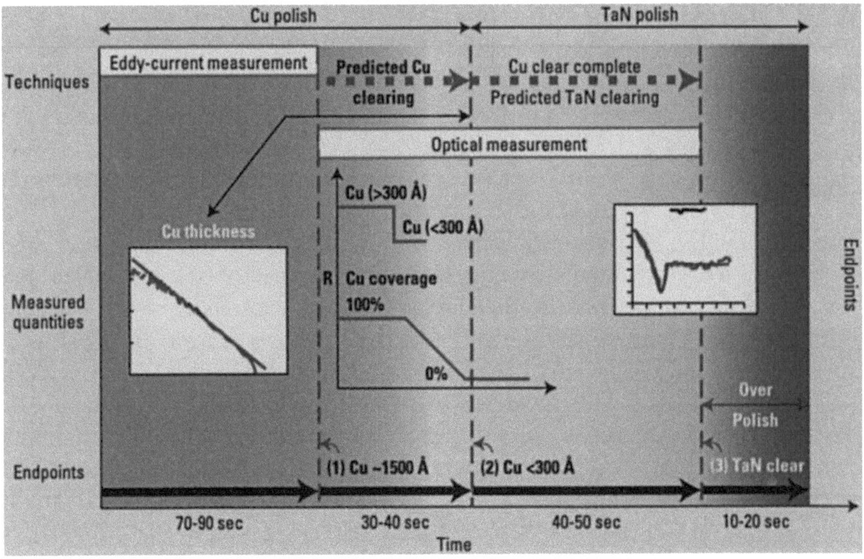

Fig. 6.24 *In situ* measurements during etching to control the end point (Reproduced with permission, Semiconductor International, May 2001, p. 98)

from the change in the electric current observed between the initial time of polishing and at the end of the polishing (Fig. 6.24). The changes in the electric current depend upon several things including the type of metal used for the barrier and the thickness of the copper metal.

Each method has its advantages and disadvantages. There is no doubt that the optical method coupled with electrical measurements will be better than a single electrical or optical method.

Figure 6.25 shows a stylus profilometer based on atomic force microscopy (AFM) for in-line monitoring of the surface characterization and to provide highly accurate non-destructive measurements.

6.2.13 Dry In Dry Out

Dry in and dry out of the CMP process is important especially, when copper CMP is involved. Immediately after the surface of copper is polished it should be cleaned and dried (the film surface should be free from any moisture). The delay in cleaning and drying the surface results in oxidation and corrosion of the copper film. So special care should be taken for the copper CMP process, especially the post-CMP cleaning and drying processes. Post-CMP cleaning needs to be optimized for low-K dielectric materials. Different surfactants are recommended for different material surfaces during post-CMP cleaning as the bonding nature of the particles is different for different materials, which is defined as the *zeta* potential.

Fig. 6.25 An Atomic Force Profiler to monitor the Cu-dual damascene etch process (photo courtesy, Veeco Instruments Inc.)

6.2.14 Multi-Step Processing

In the multi-step process, polishing optimization is achieved simultaneously through the delivery of multiple slurries to each of four autonomous polishing modules. Three or more polishing steps are performed with no additional wafer handling, which result in better throughput and yield [68]. As a matter of fact, the quality of the CMP process depends on both the polishing equipment and the nature of the slurry used. Most of the slurries belong to unstable colloidal systems, which have relatively short shelf lives. Also transport of the slurry from a container to the CMP system is a possible cause of degradation of the slurry due to shear, settling, or agglomeration. So it will be a better idea to monitor the particle distribution inside the slurry from time to time. It is suggested that micro-filtering the slurry before delivery to the CMP system, to obtain even-size shape of the particles, will minimize the defects and micro-scratches on the surface [69].

6.2.15 Post-CMP Cleaning

Once the polishing is over, a post-CMP cleaning is essential to remove residual slurry and metal particles that are detrimental to device yield (Fig. 6.27). Post-copper (Cu) CMP cleaning presents challenges in addition to slurry removal. It is necessary to remove Cu contamination from some areas of the wafer (front side dielectric, edge, and backside) while avoiding or limiting Cu removal from the other areas.

It has been reported that for 16 Mb DRAM there should be only 10^{13} atoms/cm^2 of metal ions [70]. During cleaning the abrasive particles develop charged surfaces that cause particles to be repelled from one another. For this reason, when alumina is used as an abrasive it is designed to keep pH less than neutral. In contrast, silica-bearing slurries are stable at high pH, and highly charged particles do not tend to agglomerate. Based on these facts, CMP cleaning solutions are prepared with many additives that include surfactants, chelators, and other compounds that either interact with wafer or particle surfaces, or react with metals that are dissolved in solution or adsorbed on the wafer surface. Figure 6.26 shows a single wafer non-contact post-cleaning CMP system.

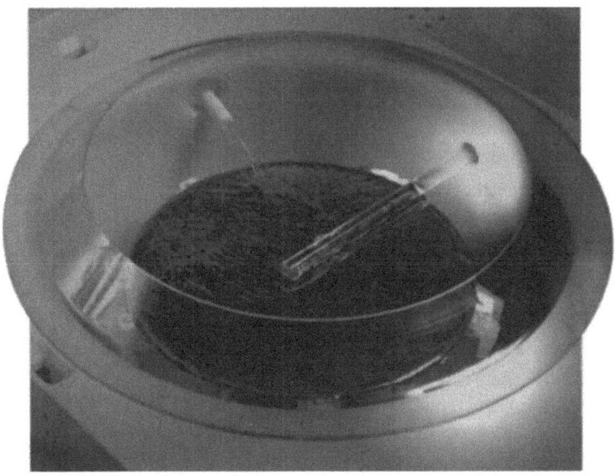

Fig. 6.26 CMP system for non-contact single wafer post-cleaning (Photo courtesy VERTEQ/AKRION Inc., PA)

Megasonic cleaners are widely used because they offer an effective solution while eliminating problems such as particle loading and brush maintenance associated with brush scrubbers. They also offer a better and much wider chemical pH operating range that is less expensive than traditional brush scrubbing. A buffered chelating treatment with some commercially available cleaning solutions is performed followed by RCA-2 (1:1:5, HCl: H_2O_2: DI) chemistry cleaning. Final cleaning is done in de-ionized (DI) water and the surface is dried immediately after cleaning. It has been observed that in-line detection of metal residue on the surface of the polished layer prior to post-CMP-cleaning is important. Figure 6.27 shows metal residues after CMP.

Atomic force microscopy (ATM) has been used to characterize post-CMP cleaning and time of flight secondary ion mass spectroscopy (SIMS) technique has been used to measure Cu-contamination. Recently, employing oblique illumination,

Fig. 6.27 Metal residues after the CMP process (Reprinted with permission Semiconductor International, October 2000, p. 72)

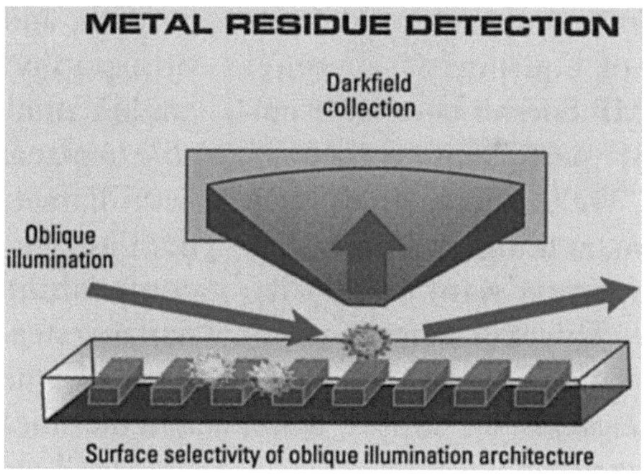

Fig. 6.28 Surface selectivity of oblique illumination architecture (Reprinted with permission, Semiconductor International, Oct. 2000, p. 72)

in situ detection of the metal residue has been performed after post-cleaning. Figure 6.28 shows an optical set-up for the measurement of the metal residue.

6.2.16 CMP Pattern Density Issues

CMP is seen to be very critical as the device size becomes smaller and the polishing wafer has step features. The pressure on the polishing pad is seen to concentrate only on the top features, which increases the polishing rate above that of a blanket wafer. In addition, it causes non-uniformity in the removal rate across the pattern of different densities due to variations of pressure distribution across the pattern. The way to prevent this is to create dummy shapes around the sparse pattern to try and match the higher pattern density.

6.3 Summary

The Cu-damascene (single or its twin dual damascene) is an elegant technique of inlaying copper for wiring throughout the back-end-of-line. In a dual damascene structure, only a single metal deposition step is used to simultaneously form the main metal and the metal in the via holes. That means both trenches and via holes are formed in a single dielectric layer. The damascene process provides a solution to the problems arising from the lack of a directional metal etch capability [71–74], but introduces new problems that arise from the need to fill the high aspect ratio structures with metal without any void and stress. It has been observed that vapor deposition of copper is not adequate and the deposition of metal on the corners of the trenches leads to pinch off and void formation. However, control of key experimental parameters such as deposition/dissolution times deposition/dissolution current densities and off time of the ECP-Cu can result in high quality electrodeposition from the plating bath. Trenches are typically etched to a depth of 4000–5000 Å whereas the depth of the via holes generally vary between 5000 and 7000 Å. After the via and trench recesses are etched, the via and the trench are filled with copper. The excess metal outside the trench is removed by chemical mechanical polishing (CMP).

Chemical mechanical planarization/polishing (CMP) is a *critical technology* for the integrated circuit (IC) industry for the next generation of device wafers. Indeed, it is a critical technology in the sense that there are too many critical parameters to be identified and optimized. The critical parameters are the size and shape of the abrasive particles, their distribution, the chemistry and physics of the slurry (pH, density, viscosity, and particulate impurities), polishing pad (roughness/topography, including hardness and Young's modulus), and finally the pressure applied to the pad and the wafer. During polishing, a wafer is mechanically pressed into the polishing pad that elastically deforms in response. As a result, structural damage such as *erosion*, *dishing*, *scratches*, *pits*, and *delamination* of film surface are frequently noticed.

The presence of *scratches* and *pits* on the surface during planarization makes post-CMP cleaning very difficult. Experimental observation shows that hydrodynamic force is not sufficient to remove smaller (<0.2 μm) slurry particles when the penetration depth inside the wafer is deep (~0.5 μm). Though CMP has become routine work for the copper damascene process, yet, people who are involved in the processing do not feel very comfortable and confident about the process. However, some experts believe that a *new generation of CMP is risky but not without opportunity* [75].

References

1. A. West, Theory of filling of high aspect ratio trenches and vias in presence of additives, J. Electrochem. Soc., 147, 227 (2000) and J. Van Olmen et al., AMC Tech Dig., 5 (2007)
2. P.C. Andricacos, C. Uzoh, J.O. Dukovic, J. Horkans, and H. Deligiani, IBM, J. Res. Dev., 42, 567 (1998)
3. E.K. Broadbent et al., Experimental and analytical study of seed layer resistance for copper damascene electroplating, J. Vac. Sci. Technol., B-17, 2584 (1999)

4. S. Wolf, Silicon processing for VLSI Era, Lattice Press, Sunset Beach, CA, Vol. 4, Chapt-15 (1986) and also R.L. Opila and D.W. Hess, A century of dielectric science and technology, J. Electrochem. Soc., 150 (1), S1–S10 (2003)
5. T.K. Gupta, Hand book of thick and thin film hybrid microelectronics, Wiley, NJ, Chapter 4 (2003)
6. S. Wolf, Deep Sub-micron process technology, in Silicon processing for VLSI era, Lattice Press, Sunset beach, CA, Vol. 4 (2002) and also R.J. Guttman et al., Integration of copper multilevel interconnects with oxide and polymer interlevel dielectrics, Thin Solid Films, 270, 472 (1995)
7. J.C. Maisonobe, G. Passemard, C. Lacour, T. Motte, T. Noel, and J. Torres, Silk compatibility with IMD process using copper metallization, Microelectron. Eng., 50, 25 (2000) and J. Van Olmen et al., IEEE IITC, Tech Dig., 49 (2007)
8. Y. Gotkis and S. Guha, Cu-CMP for dual damascene technology, J. Electronic Mater., 30 (4) (2001)
9. T.N. Theis, The Future of interconnection technology, IBM J. Res. Dev., 44 (3) (2000)
10. A.C. Adams, Dielectric and polysilicon film deposition, in VLSI Technology, 2nd ed. In S.M. Sze (ed.), McGraw Hill, New York, p. 251 (1988) and also N. Matsuki, J. Ohta, H. Fujika, M. Oshima, M. Yoshimoto, and H. Koinuma, Fabrication of oxide gate thin film transistors using PECVD/PLD multichamber system, Sci. Tech. Adv. Maters., 1, 187 (2000)
11. T. Hara and K. Sakata, Electrochem. Solid State Lett., 4, G-77 (2001)
12. Y.S. Kim, and Y. Shimogaki, J. Vac. Sci. Technol., A-192642 (2001)
13. S. Gandikota et al., Microelectron. Eng., 50, 547 (2000)
14. O.K. Kwon, J.H. Kim, H.S. Park, and S.W. Kang, Atomic layer deposition of ruthenium thin film for copper glue layer, J. Electrochem. Soc., 151 (2), G-109 (2004)
15. A.L.S. Loke, Process integration issues of low-permittivity dielectrics with copper for high performance interconnects, Ph.D. thesis, Stanford University, March, (1999)
16. Motorola XCM63R836RS3.3 M SRAM copper metallization Report: The chipworks Inc. Ottawa, ON, CA (1998)
17. S.P. Murarka and R.J. Gutmann, Advanced multilayer metallization schemes with Cu as interconnection metal, Thin Solid Films, 236, 257 (1993)
18. R. Solanki and B. Pathangey, Electrochem. Solid State Lett., 3, 479 (2000) and T.P. Mofat, D. Wheeler, M.D. Edlestein, and D. Jossel, Superconformal film growth: Mechanism and quantification, IBM J. Res. Dev., 49 (1) (2005) and also M. Stewart et al., Interconnect patterning in a single step with multilevel nanoimprint lithography, VLSI multilevel interconnect Conf. (VMIC) October, (2005)
19. J.M. Steigerwald, S.P. Murarka, and R.J. Guttman, Chemical Mechanical Planarization, Wiley, New York, (1977)
20. J. Gotzlich and H. Rysell, Tapered windows in SiO_2 and Si_3N_4 and polysilicon layers by ion implantation, J. Electrochem. Soc., 128, 617 (1981)
21. W. Kern and R.S. Roster, Advances in the deposition processes for passivation films, J. Vac. Sci. Technol., 14, 1082 (1977)
22. L.K. White, Bi-layer etching of fixed oxide and passivation layers, J. Electrochem. Soc., 127, 2687 (1980) and also S.N. Wolf and R.N. Tauber, Silicon Processing, Lattice Press, Sunset Beach, CA, Vol. 1, pp. 185–189 (1986)
23. G. Nanz and L.E. Camilletti, Modeling of chemical mechanical polishing: A review, IEEE Trans. Semiconductor manufacturing, 8 (4) Nov. (1995) and also P.Y. Wu and F.C. Chou, Complete analytical solutions of film planarization during semicoating, J. Electrochem. Soc., 146 (10), 3819 (1999)
24. D.B. Tuckerman and A.H. Weisberg, IEEE Electron Dev. Lett., EDL-7, 1 (1986)
25. T.S. Magee, J.S. Osborne, P. Gildea, and C.H. Leung, U.S. Patent No. 4758533, July 19, (1988) and also E. Ong, H. Chu, and S. Chen, Metal planarization with an excimer laser, Solid State Technol., 63, Aug (1991)

26. P. Wrschka, J. Hernandez, G.S. Oehrlein, and J. King, Chemical mechanical planarization of copper damascene structures, J. Electrochem. Soc., 147 (2), 706 (2000)
27. M. Stell et al., Characterization of chemical mechanical planarization processes, In S.P. Murarka et al., (ed.), MRS Pub. Washington DC., (1994) and L. Lang, Modeling CMP for copper dual damascene interconnects, Solid State Technol., 111 June (2000)
28. F.B. Kaufman, D.B. Thompson, R.E. Broadier, M.A. Jaso, and W.L. Guthrie, J. Electrochem. Soc., 138, 3460 (1991) and L. Karuppiah et al., CMP-MIC Tech Dig. Feb., 45 (2006)
29. J. Tichy, J.A. Livert, L. Shan, and S. Danyluk, Contact mechanics and lubrication hydrodynamics of CMP, J. Electrochem. Soc., 146 (4), 1533 (1999) and also J.M. Steigerwald, S.P. Murarka, and R.J. Guttman, Chemical mechanical planarization of microelectronic materials, John Wiley, New York, (1997)
30. Y. Homma et al., Control of photocorrosion in copper damascene process, J. Electrochem. Soc., 147 (3), 1193 (2000) and A.B. Kahng, Adv. Met. Conf. UC Berkeley, Sept 22 (2008)
31. S. Sundararajan et al., Two dimensional wafer-scale chemical mechanical planarization models based on lubrication theory and mass transport, J. Electrochem. Soc., 146 (2), 761 (1999) and also T. Nakamura, K. Akamatsu, and N. Arakawa, Bull. Jpn. Soc. Precis. Eng., 19, 120 (1985)
32. W.T. Tseng and Y.L. Wang, Re-examination of pressure and speed dependencies of removal rate during chemical mechanical polishing processes, J. Electrochem. Soc., 144 (2) (1997)
33. W.J. Patrick et al., Application of chemical mechanical polishing to the fabrication of VLSI interconnects, J. Electrochem. Soc., 138, 1778–1784 (1991)
34. P.A. Burke, Porch. IEEE 8th Int. VLSI interconnect. Conf. (VMIC), Santa Clara, CA, p. 379, June 11–12 (1991)
35. C. Lou et al., Dishing effects in CMP planarization process for advanced trench isolation, Appl. Phys. Letts., 61, 1344 (1992) and also J. Yu Lai, N. Saka, and J.H Chun, Evolution of copper-oxide structure in chemical mechanical polishing, J. Electrochem. Soc., 149 (1) G41 (2002)
36. C. You, A. Lulus, M. Grief, and T.T. Dan, Porch. IEEE, VLSI Multilevel Interconnect. Con., Santa Clara, CA, p. 156, (1992) and also J. Warnock, J. Chem. Soc., 138, 2398 (1991) and F. Zhang, A. Busnaina, and G. Ahmadi, Particle adhesion and removal in chemical mechanical polishing and post CMP cleaning, J. Electrochem. Soc., 146 (7), 2665 (1999) and also M. Sugiyamal, K. Ishikawa, M. Nakaishi, K. Yamashita and T. Ohba, Adv. Met Conf. The University of Tokyo, Japan, Sept 26 (2006)
37. J. You Al, N. Aka, and J.H. Chop, Evolution of copper-oxide damascene structures in CMP, Contact mechanics modeling, J. Chem. Soc., 149 (1) G-031 (2002)
38. T. Vo, T. Buley, and J.J. Gagliardi, Improved planarization for STI, Solid State Technol. (SST), 123 (June 2000) and also J.Y. Lai, N. Saka, and J.H. Chun, J. Electrochem. Soc., 149, G-31 (2002)
39. G.B. Basim, J.J. Adler, U. Mahajan, and B.M. Moudgil, Effect of particle size of CMP slurries for enhanced polishing with minimal defects, J. Electrochem. Soc., 147(9), 3523–3528 (2000)
40. A.S. Dublin and P.J. Goetz, J. Colloids Surf., 158, 343 (1999)
41. D.J. Chen and B.S. Lee, J. Pattern planarization model of chemical mechanical polishing, Electrochem. Soc., 146, 3420 (1999)
42. L. Shan, S. Danyluk, J.A. Levert, Interfacial pressure measurements at chemical mechanical polishing interfaces, In S.V. Babu et al. (eds.) Chemical mechanical polishing, fundamental challenges, Proc. MRS., 187 (1999)
43. S.R. Runnels and P. Renteln, Dielectric Sci. Technol., 110, (1993) and also S.R. Runnels, F. Miceli, and I. Kim, Validation of a large area three dimensional erosion simulation for chemical mechanical polishing, J. Electrochem. Soc., 146 (12), 4619 (1999)
44. D. Wang, J. Lee, K. Holland, T. Bibby, and T. Cale, J. Electrochem. Soc., 141, 2843 (1997)
45. J. Warnock, J. Electrochem. Soc., 138, 2398 (1991) and T. Park, et al., Proc. 4th Int. Chem. Mech. For ULSI multilevel Interconnection Conf., CMP-MIC, Santa Clara, CA, p. 184 (1999)

46. P.T. Liu et al., Improvement of post CMP characteristics on organic low K MSQ as IMD, J. Electrochem. Soc., 147 (11) 4313–4317 (2000)
47. Y. Feng et al., Corrosion mechanisms and products of copper in aqueous solutions at various pH values, Corrosion, 53 (5), 389 (1997)
48. K. Kobayashi, K. Shimizu, G.E. Thompson, and G.E. Wood, Direct observation of the mosaic structure thermal oxide films on copper, Revue de metallurgie, 90 (12), 1627 (1993)
49. S.V. Babu, Y. Li, and A. Jindal, Chemical mechanical planarization of Cu and Ta, J. Materials (JOM), 50, March (2001)
50. S. Lassig, S. McClatchie, and A. Kiermasz, Selective removal of strategies for low-K dual damascene, Sem. Fab. Tech., 185 (2003)
51. S. Aksu and F.M. Doyle, Electrochemistry of copper in chemical mechanical planarization (CMP) slurries containing glycene and hydrogen peroxide, in Chemical mechanical planarization V, PV-2002-1, In S. Seal (ed.), The Electrochem. Soc. Pub. Pennington, NJ, (2002)
52. P. Suphantharida and K.O. Asare, Cerium oxide slurries in chemical mechanical polishing electrophoretic mobility and adsorption investigations of ceria/silicate interactions, J. Electrochem. Soc., 151 (10), G-658 (2004)
53. K.O-Asare and K.K. Mishra, J. Electronic Mater., 25, 1599 (1996), and also H. Hirabayashi, M. Huguchi, M. Kinoshita, H. Haysaka, K. Mase, and J. Oshim, US Patent number 5575885 (1996)
54. Q. Luo, D.R. Campbell, and S. Babu, Thin Solid Films, 311, 177 (1997)
55. Q. Luond and S.V. Babu, Dishing effects during CMP of copper in acid media, J. Electrochem. Soc., 147(12) 4639–4644 (2000)
56. A.S. Dukhin and P.J. Goetz, Colloid Surf. A, 158, 343 (1999) and E.E. Ramsen et al., J. Electrochem. Soc., 153, G453 (2006)
57. S. Kondo et al., Abrasive free polishing for copper damascene interconnections, J. Electrochem. Soc., 147(10) 3907–3913 (2000)
58. S.P. Murarka, Advanced metallization for devices and circuits science and technology and manufacturability, MRS Pub. Philadelphia, PA, (1994) and also R. Jairath et al., Solid State Technol., 107 Oct. (1996) and Z. Stavreva, D. Zeidler, M. Plotner, and K. Drescher, Characterization in chemical mechanical polishing of copper: Comparison of polishing pads, Appl. Surf. Sci., 108, 39 (1997)
59. J.J. Colacine et al., Analysis of velocity as a cause of thickness variations in a CMP process, Solid State Technol., 30 (Aug 1973) and also C. Rogers, J. Coppeta, L. Racz, and D. Bramono, Analysis of flow between a wafer and pad during CMP processes, J. Electron. Mater., 27 (10), 1082 (1998) and also Y. Li et al., 23rd Int Conf. on VLSI/ULSI Multilevel Interconnect, Sept-26, (2006)
60. O.G. Checknia, L.M. Keer, and H. Liang, J. Electrochem. Soc., 145, 2100 (1998)
61. J.U. Lai, N. Saka, and J.H. Chun, Evolution of copper damascene in chemical mechanical polishing, Electrochem. Soc., 199 (1), G31–G40 (2002)
62. T. Yu et al., A statistical polishing pad model for CMP, Proc. IEEE Int. Electron Dev., (1993) and also S.H. Ng, R. Height, C. Zhou, I. Yoon, and S. Danyluk, Pad soaking effect on interfacial fluid pressure measurements during CMP, J. Tribology, 125 (3), 582 (2003)
63. L.A. Galin, In contact problems in the theory of elasticity translated by H. Moss and I. Snedon, (eds.) N.C. State College Raleigh, NC, (1961) and also D. Wang, J. Lee, K. Holland, T. Bibby, and T. Cale, J. Electrochem. Soc., 141, 2843 (1997) and Y. Li et al., 23rd Int. Symp. on VLSI/ULSI multilevel interconnect, Sept. 26 (2006)
64. Y. Homma et al., Control of photocorrosion in the copper damascene process, J. Electrochem. Soc., 147 (3) 1193 (2000)
65. J.M. Steigerwald et al., Chemical mechanical planarization of microelectronics materials, Wiley, New York, (1997)
66. J.J. Gagliardi, STI polishing with 3 M's fixed abrasive, 16th Intl. VLSI multilevel interconnection Conf. pp. 223–228 (1999) and B. Lee, D.S. Boeing, and L. Economikos, Proc. CMP-MIC Conf., Santa Clara, CA, p. 395 (2001)

67. M. Fayolle, J.F. Lugandi, F. Weimar and W. Bruxvoort, Proceedings of the CMP-MIC Conf., San Diego CA, p. 128 (1998) and also T. Park, T. Tugabaw, D. Boeing, S, Hymes, T. Brown, K. Samekalin, and G. Schwartz, In Chemical Mechanical Polishing in IC Device Manufacturing, III, In R.L. Olipa et al. (eds.), Electrochem. Soc. Pub. Pennington, NJ (1999)
68. U. Mahajan, M. Bielman, and R. Singh, Mater. Res. Soc Pub., Pittsburgh, PA, Vol. 566 (1999)
69. L. Doyen, D. Vacher, K. Tarutani, R. Bouard, F. Picore, and D. Girad, Analysing large particles in CMP slurries, Semcond. Intl., p. 75, (August, 2002) and also R.S. Subramanian and R.M. Appat, Electrochem. Solid State Letts., 4 (12), G-115 (2001)
70. P. Wrschka, J. Hernandez, G.S. Oehrlein and J. King, Chemical mechanical planarization of copper damascene structures, J. Electrochem. Soc., 147(2), 706–712 (2000) and Y. Uozumi et al., IEEE IITC, p. 25 June (2007)
71. N. Toyama, Copper impurity levels in silicon, Solid State Electron., 26 (1), 37 (1997)
72. J.M. Steigerwald, R. Zirpoli, S.P. Murarka, D. Price, and R.J. Gutmann, J. Electrochem. Soc., 141, 2842 (1994)
73. J.M. Fayolle and F. Romagna, Microelectron. Eng., 37/38, 135 (1997)
74. Z. Stavreva, D. Zeidler, M. Ploetner, and K. Drescher, Appl. Surf. Sci., 108, 39 (1997)
75. J.U. Lai, N. Saka, and J.H. Chun, Evolution of copper damascene in chemical mechanical polishing, Electrochem. Soc., 199 (1), G31–G40 (2002) and also K. Witt and L. Cook, Getting an edge with CMP, Semicond. Int. p. 74 Oct. (2000)

Chapter 7
Conduction and Electromigration

7.1 Conduction

7.1.1 Introduction

The requirements of high-speed sub-100 nm devices are higher conductive interconnecting lines for local and global interconnections and lower associated capacitive reactance of the interconnecting lines. The conductivity of copper ($\sigma_{Cu} = 0.598$ ohm-cm and $\sigma_{Al} = 0.374$ ohm-cm) is higher and it offers higher resistance to electromigration (EM) compared to aluminum (or Al-alloy). At the same time, copper has the advantage of making finer wires with lower resistive loss which will subsequently reduce *RC* delay and increase the speed of the device (Fig. 7.1). Therefore,

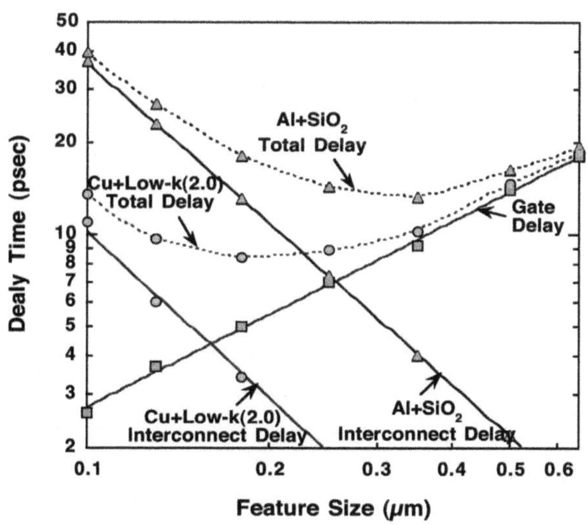

Fig. 7.1 A comparative analysis of feature size versus delay time of a conventional Al-SiO_2 coupling with a modern Cu-low-*K* coupling (Reproduced with permission, Dr. K. Saraswat, Stanford University)

Tapan K. Gupta, *Copper Interconnect Technology*,
DOI 10.1007/978-1-4419-0076-0_7, © Springer Science+Business Media, LLC 2009

recently copper has become the primary choice for local and global interconnecting lines for modern integrated circuits (ICs) [1].

Now, in order to integrate the Cu-lines in modern ICs, the conventional metallization processing associated with Al-alloy has been changed from a subtractive process to an additive bottom fill-up process [2]. Indeed, as the device size is becoming smaller, the interconnecting line dimensions are also becoming smaller and the electrical properties of aluminum (Al) are becoming inadequate to meet the new requirements of sub-100 nm metal lines. Thus the conventional Al-SiO_2 system in ICs has been replaced by Cu-low-*K* material to increase the speed of the devices (Fig. 7.1).

The problem of higher *resistivity* of a thin metallic film compared to the bulk material is of great interest and the interest is increasing steadily as the feature size of the device is steadily decreasing. For sub-micron processes (100 nm and below) minimization of the size of the interconnect cross-section and of the oxide thickness has become critical to meeting the increasingly stringent electrical requirements of a modern IC. The pitch-matched local via holes and contacts to the metal layer generally have much smaller cross-sectional area compared to the interconnects themselves. Let us assume that the contact area of the Cu-line is very small and circular, having radius r_1 and resistivity ρ, and there are *N* number of such contacts. Now we can model the resistance (*R*) of the Cu-line, when it is bonded to a Cu-pad as [3]:

$$R = \rho\,\{(1/2Nr_1) + (1/2r_1)\}. \tag{7.1}$$

From the above theoretical model, we can assume that the effective resistance of the Cu-line will show different resistivity than the bulk resistivity of copper (Cu) [3]. Figure 7.2 shows the effect of shrinking cross-sections on the resistivity of Cu-interconnects.

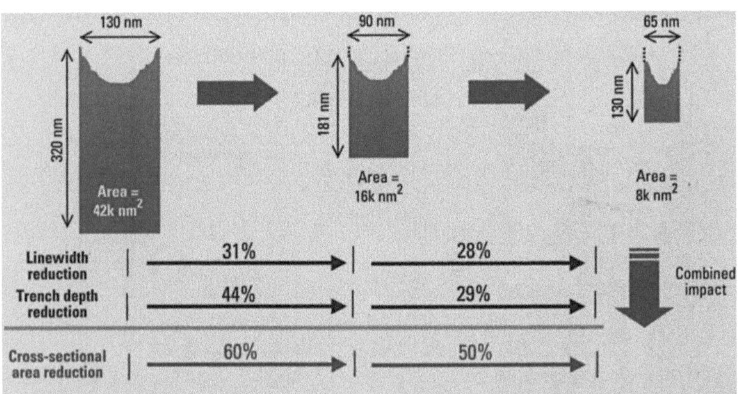

$$R = \rho\frac{l}{w\cdot t} = (\rho\cdot l)\cdot\frac{1}{(0.72\cdot w)\cdot(0.71\cdot t)} = (\rho\cdot l)\cdot\left(\frac{1}{(0.50\cdot w\cdot t)}\right) = 2\cdot\left(\rho\frac{l}{w\cdot t}\right)$$

Fig. 7.2 From the 90 to 65 nm node, the interconnect cross-sectional area is reduced by 50%, resulting in a 100% increase in resistance for a line of the same length (Reprinted with permission, *Semiconductor International*, July 2005, p. 63, Courtesy ITRS)

Studies on scaling effects are not only important from the standpoint of the electrical resistance of the Cu-lines, but with the decreased feature size, interface area fraction increases, and the ultra-fine lines suffer from electromigration (EM) damage. However, measurements demonstrate that a bamboo-like microstructure developed, in a narrow Cu-line, relatively slower surface diffusion, resulting in increased EM lifetimes.

It has been theorized that in a metallic constriction, in nanoscopic cross-section, the transverse motion is quantized, leading to a finite number of sub-bands below the Fermi-energy ε_F. A striking consequence of these discrete sub-bands is the phenomenon of conductance quantization, which exhibits a universal force of oscillations of the order of ε_F/λ_F, where λ_F is the Fermi wavelength (which is 0.46 nm for Cu and 0.36 nm for Al). On the other hand, experimental observations based on the theoretical model confirm that the tensile force (*F*) existing in such a system will influence the *conductance*, *scattering phenomenon* and the *mechanical properties* of the *nanoscopic* metallic contacts [4].

7.1.2 Conduction Mechanism and Restrictions

Electrical conduction (σ) in a thin metal film can be modeled by representing the film in terms of a finite number *M* of monolayers or perfect atomic planes arranged periodically in the direction perpendicular to the boundaries of the film. Applying this model the resistivity of copper (lines) embedded in copper (seed layer) has been found to be (116/M) μΩ-cm [5].

Conduction through a medium is a process of transmitting *energy* through the medium without perceptible motion of the medium itself. The *transmitted energy* may be electrical where the carriers are *electrons* and/or *holes* or may be heat where energy is being transported due to the oscillations of the thermally *excited atoms* (*phonons*). Thermally activated phonons introduce a temperature gradient along the way of their oscillations. The medium through which the electric or the heat energy is transmitted easily is called a *conductor*. A good conductor of electricity will transmit maximum energy with less resistance, and there will be a minimum loss of electrons due to *scattering, diffusion*, and *stress (current and/or heat)*. When the dimension of the conducting line is very small (of nano-order) electron–phonon transport is characterized by *interface effects* [6–10], and the conduction through the line will be different from the conduction through the bulk material.

The conduction mechanism can be modeled after the Boltzmann transport equation as:

$$-\{(e/m)\}\{E+(v/c)H\}\,\text{grad}\,v f+v\,\text{grad}\,r f \tag{7.2}$$

$$=-\{(f-f_0)/\tau\} \tag{7.3}$$

where

$e =$ Electronic charge
$m =$ Effective mass

E = Electric field
v = Electron velocity
c = Velocity of light
H = Magnetic field
f_0 = Electronic distribution at equilibrium
f = Non-equilibrium electronic distribution functions
τ = Relaxation time for return to equilibrium.

The z-axis is taken to be perpendicular to the plane of a thin film of thickness d, and current flows through the film in the x-direction. The second term of equation is zero for bulk material but not in the z-direction of a thin film. Thus the non-equilibrium electronic distribution function f can be written as:

$$f = f_0 + f_1(v,z) \tag{7.4}$$

and substitution of (7.4) in (7.2) gives rise to

$$\{(eE/mv_z)(\partial f_0/\partial v_z)\} = (\partial f_1/\partial z) + (f_1/\tau v_z). \tag{7.5}$$

The solution to (7.5) can be used to calculate the current density (J) across the film in the form:

$$J = 2e(m/h)^3 \int vf\,\mathrm{d}v \tag{7.6}$$

where h is Planck's constant.

By averaging the current density over all the values of z from 0 to d, we can derive an equation for resistivity as:

$$(\rho/\rho_0) = \{1/[1 - 3/2k \int_1^{\infty}\{(1/x^3) - (1/x^5)\}\{1 - \exp.(-kx)\}\,\mathrm{d}t] \tag{7.7}$$

When the width of the Cu-line is comparable to or less than the mean free path (λ) of the conducting electrons, Equation (7.7) will transform to Equation (7.8) due to *scattering* effects (in Equation (7.7) we have assumed $k = d/\lambda$, d is the film width, x is the integration variable, ρ is the resistivity of the Cu-line, and $\rho_0$is the resistivity of bulk Cu at a given temperature).

7.1.2.1 Scattering and Its Influence on Coductivity

The scattering of electrons may be elastic or non-elastic. Elastic scattering does not affect the *resistivity* of a polycrystalline film. However, inelastic scattering, which depends on the mean free path of the charge carriers and the *scattering coefficient* p, increases the resistivity of the film [11–13]. According to the Fuchs–Sondheimer (FS) model the value of p is constant, but according to the quantum mechanical model (Mayadas–Shatzkes model or MS-model) the value for p is p cosθ, which indicates that the value of p is dependent on the angle of scattering.

The *resistivity* of a conductor increases substantially when the size of the interconnecting lines reaches 50 nm and below, because of scattering mechanisms that are controlled by quantum mechanical phenomena and structural/morphological effects [14].Surface and interface scattering become predominant in thin films when the thickness is comparable to the mean free path (λ) of the conduction electron [15]. Based on the semiclassical scattering model, several comprehensive expressions have been put forward to show the relation between resistivity and line-width of the interconnects when scattering due to surface, interface, and roughness of the metal line is taken into account [16]. The *surface scattering* dependent resistivity ρ_S can be represented mathematically as:

$$(\rho_S)/(\rho_0) = [(1)]/[1 - \{3(1-p)\lambda\}/\{2d\}] \int_1^{\infty} [(1/x^3 - 1/x^5)] \\ [(1 - \exp. - kx)]/[(1 - p(\exp. - kx)]dx \tag{7.8}$$

where ρ_0 is the bulk resistivity at a given temperature, x is the integration variable, λ the mean free path of electrons, d the smallest dimension of the film (in our case it is the width), and k is equal to (d/λ). As a matter of fact, since the thickness of the diffusion barrier for copper interconnect does not scale with technology scaling, and the effective area through which the current conduction takes place is reducing, the effective resistive element of the interconnect is increasing.It is clear from the above equation that since copper (Cu) has a smaller intrinsic resistivity than aluminum (Al), it will be more affected by surface scattering than Al [17].

The impact of the metal line-width on the resistivity of a Cu-line due to the influence of surface, electron, and grain boundary scattering is shown in Fig. 7.7 below [18–19]. Figure 7.7 also shows the standard bulk resistivity of the metal. Apart from induced scattering due to surface roughness, the increased resistivity due to thickness reduction is affected also by surface and interface induced scattering phenomena, which become predominant when the thickness of the film becomes comparable to the mean free path (λ) [20].

In the copper damascene process interconnect lines are laid on the top of a copper seed layer where the barrier layer forms the bottom surface (substrate) of the seed layer. According to the model set up by Hebard et al., [26] the scattering of conduction electrons at grain boundaries or at planar interfaces will be defined by the top and bottom surfaces of the Cu-seed layer. The seed layer is very thin (a few ångstroms thick), and we can take it for granted that it is in the coalescence regime (50–200 Å thick) with island morphology.The deposition of such a thin film, which will uniformly cover the barrier walls, is challenging besides its reliability from the standpoint of *stress migration* and *EM failure*. In order to mitigate *stress migration* and *EM related phenomena,* the Cu-seed layer is doped with a metal [25]. However, excessive dosing or the wrong type of alloying elements can produce adverse effects by decreasing electrical conductivity, weakening adhesion, modifying diffusion paths, sensitizing corrosion, and consequently degrading interconnect performance and reliability.

Figure 7.3 shows the relation between line-width and the average grain size of doped and non-doped Cu-film. The top line represents the pure Cu-seed line and the

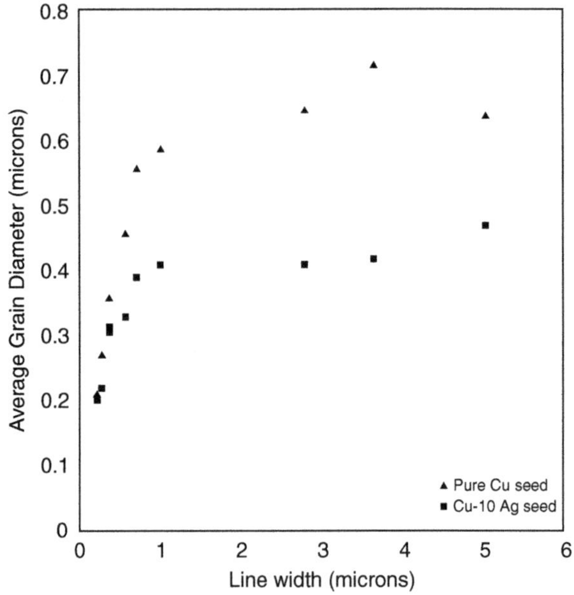

Fig. 7.3 Comparison of the average grain diameter between Cu and Cu-Ag lines (Reproduced with permission from Elsevier, *Thin Solid Films*, 2005)

bottom line is that of the Cu-seed line doped with Ag. The alloy line has a smaller average grain diameter than pure Cu. From the graph we can also notice that when the line width is smaller than 1 μm the grain sizes of the two lines are almost the same. However, when the line-width is above 1 μm the grain diameter of the Cu-seed layer is very large. In polycrystalline metal film, when the grain size is larger than 1 μm, it contains a relatively low volume fraction of *grain boundaries*, and their effects on the electrical conduction are only significant at low temperatures [21–23].

We know from our past experience that smaller grains will produce longer grain boundaries that will effectively cause more *grain boundary* diffusion and increase the electrical resistivity (ρ) of the film even at room temperature [24]. As a matter of fact, the specific grain boundary resistivity gives a relative value for the electron *scattering power* of a grain boundary, averaged over all the grain boundaries in the metal [25].

Figure 7.4 shows the resistivity of doped (with 0.5 at% of Ag, Al, Sn, and Ti) and un-doped sputtered Cu-seed films. We can see from Fig. 7.4 that the doped films have higher resistivity than the 6 N copper film. X-ray diffraction (XRD) pattern analysis further shows that the orientation of these doped and un-doped copper films are different. As a matter of fact, secondary ion mass spectroscopy (SIMS) analysis reveals that the alloying elements in the Cu-seed layer readily diffused through the electroplated (EP) copper lines and increase Cu-line resistance.

A similar diffusion process can be envisioned between the underneath mono-layer of the diffusion barrier (e.g. Ta/TaN) and the Cu-seed layer. The process can dope the interface layer due to the charge transfer from the seed-layer, and ultimately can create a shunting path, which will ultimately decrease the resistance of

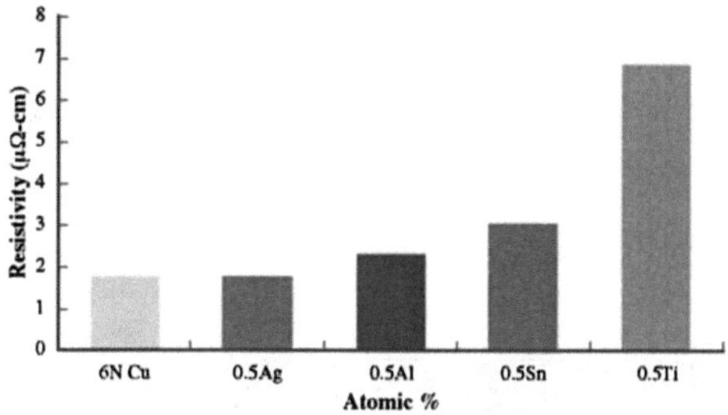

Fig. 7.4 The electrical resistivity of thin films deposited using pure copper and four Cu-alloy targets during sputtering (Reprinted with permission, Honeywell, 2001 Vertilog, V-EMT 1:16, 12th July, 2004)

the copper line [26–27]. As a matter of fact, Auger Electron Spectroscopy (AES) depth profile reveals that the composition of the interface layer has a predominant intermetallic phase [27]. However, in the sub-100 nm regime, the size effects are very pronounced and they account for the increase in electrical resistance of the interconnects. Indeed, in very fine Cu-line the charge carrier scattering from the sidewall is a more dominating factor (Fig. 7.5).

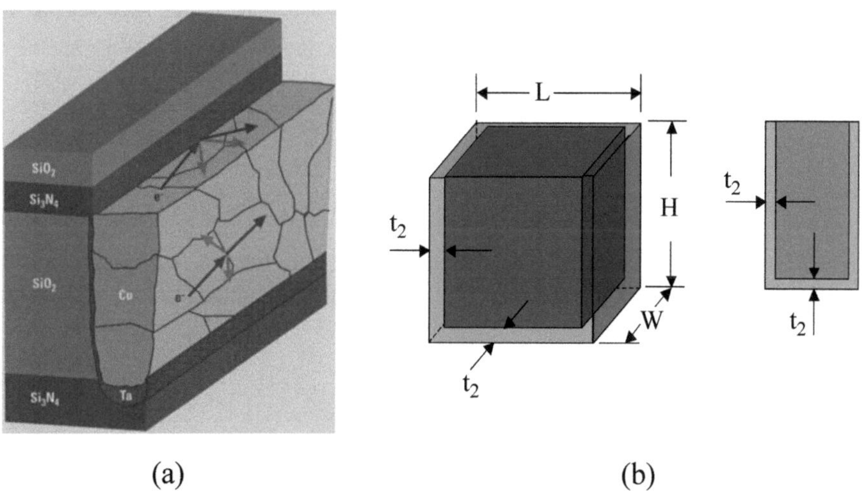

Fig. 7.5 (**a**) Sidewall scattering (*top arrows*) from Si_3N_4 and grain boundary scattering center (*arrows* from copper) (Reprinted with permission, Infineon Tech. Munich, Germany) and (**b**) confinement of Cu-line inside barrier (*t* is the thickness of the barrier)

Figure 7.5 shows a Cu-damascene structure with small (nanometer size) copper lines embedded inside the dielectric trench.Due to the confinement of the Cu-line inside the barrier layer (*t* is the thickness of the barrier layer, *H* and *W* are the height and width of the trench, Fig. 7.5b) the resistivity of the Cu-line will be affected by the sidewall scattering. As a result, the effective resistivity of the Cu-line will be different from the bulk resistivity. The sidewall scattering effect can be modeled as [28]:

$$\rho = \rho_0[+\{3/4(1-p)\lambda/W\}] \tag{7.9}$$

where ρ is the resistivity of the interconnecting line, ρ_0 is the bulk resistivity, p is the specular reflectivity, λ is the mean free path (for Cu it is ≈ 40 nm), and W is the width of the interconnect. Equation (7.9) is only valid when the structure of the lines has a square or circular cross-sectional area and the line-widths are significantly larger than the mean free path of the electrons. The value of p can vary from 0 to 1. When the value of p is 0 it is called diffuse scattering and when it is 1 it is called elastic scattering (Fig. 7.6) (p=1 means elastic scattering, and no resistivity increase caused by sidewall scattering occurs).

Fig. 7.6 Diffuse and elastic scattering

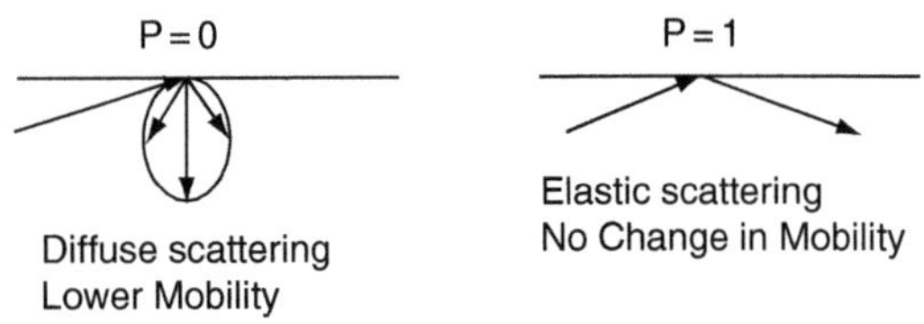

Figure 7.6 shows the schematics of diffuse scattering ($p = 0$) and elastic scattering ($p = 1$).The typical value of p for inelastic scattering as reported in some literature is 0.47 [11].

7.1.2.2 Grain Boundary (GB) Scattering

As the width of the interconnecting Cu-line decreases, it shows a long distribution of grain size, and a short distance between scattering barriers. As a result, the effective resistivity of the Cu- line increases due to the scattering from the sidewall and from the grain boundary. The scattering of electrons at grain boundaries has been modeled by Mayadas and Shatzkes as [1, 29]:

$$\rho = \rho_0(1/3\ /[1/3-\alpha/2+\alpha^2-\alpha^3\text{In}\,(1+1/\alpha)]-0.45\,\{(1-p)\,(1+\text{AR}\,\lambda)/(\text{ARW}) \tag{7.10}$$

where $\alpha = (\lambda/s)(\Re/1-\Re)$

where ρ_0 is the bulk resistivity, p the scattering coefficient, AR, the aspect ratio, λ, the mean free path of the electrons, W the line-width, s the average distance between grain boundaries and the reflectivity coefficient. The first term inside the bracket represents grain boundary scattering.The combined effects of these phenomena (different types of scattering) along with the presence of a finite diffusion barrier layer reduce the effective copper conducting area, and cause a sharp fall in the conductance of the Cu-lines. We will calculate the effective resistance of the Cu-line due to the presence of the diffusion barrier and the grain size of the Cu-lines in a later part of our discussions.

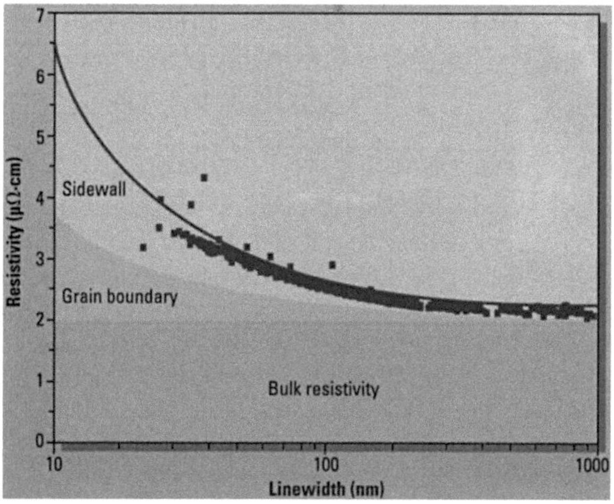

Fig. 7.7 The dependency of resistivity of damascene copper lines (150 nm height) on line-width (Reprint with permission Infineon Tech., Munich, Germany)

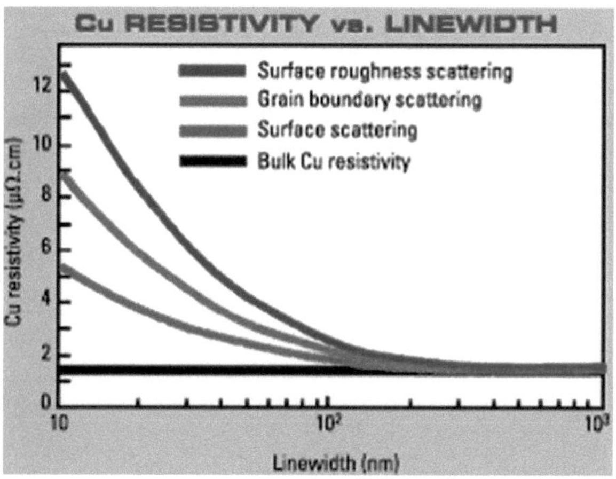

Fig. 7.8 The variations in electrical resistivity with line-width of Cu-interconnect with and without the effects of surface scattering, grain boundary scattering, and surface roughness scattering (Reprinted with permission, *Semiconductor International*, Jan., 2003).

Figure 7.7 shows the dependency of resistivity of Cu-interconnecting lines on line-width [1–2]. The fitted curve is obtained from the compact model of Equation (7.10).

Figure 7.8 shows the variations in electrical resistivity with line-width of Cu-interconnects with and without the effects of surface scattering, grain boundary scattering, and surface roughness scattering. In the case of global interconnecting lines, the line resistance will decrease further due to inductive effects. However, accurate inductance modeling still remains a challenging problem [3].

7.1.2.3 Effective Resistance of Cu-Interconnect Due to Different Size of the Grains and the Presence of Barrier Layer

The resistivity of a Cu-line can be significantly different depending on the line-width because the grain size of the line changes with the line-width. In addition, the thickness of the sidewall (barrier metal) can decrease the narrow Cu-lines. If the total Cu-line-width becomes less than about 4× the thickness of the barrier layer on a flat surface, it is likely that the sidewall thickness of the barrier layer will be less than the thickness of the Cu- film on the flat surface due to self-shadowing effects. Figure 7.9 shows a schematic of a damascene trench with Cu-line over the barrier layer.

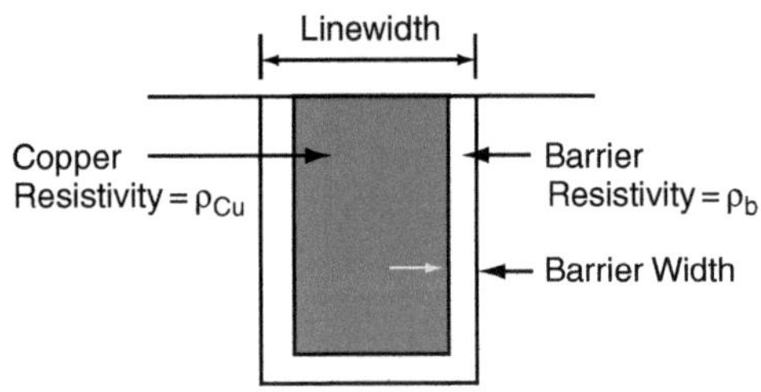

Fig. 7.9 Schematic of the cross-section of a Cu-interconnect showing the metal and the barrier

Now we can determine the effect of the Cu-grain size on the Cu-line resistance using two metal lines having two different line-widths. The first line is of width >5× the nominal Cu-grain size, and the second line has a width <2× the nominal Cu-grain size (there will be grain size shrinkage due to smaller line-width). However, both the lines are >6× the barrier metal thickness over a flat surface (no sidewall thinning). Now we can express the effective resistance of the Cu-interconnect (Cu-line-width >5×, the nominal Cu-grain size) as [33]:

$$1/R_{eff} = 1/R_{cu} + 1/R_B \tag{7.11}$$

$$= \left[(W - 2t_B)/\rho_{cu}L\right] + 1/R_B$$

Now when the Cu-grain size is reduced (due to reduced line width) to <2× the nominal Cu-grain size, the value of $(1/R_B)$ will be modified to:

$$(1/R_B) = \{[(\Delta W/\rho_{cu}) + (hc/\rho_B)/L]\}$$

where

R_{eff} = Effective resistance of the Cu-line having width <4× the barrier layer thickness
R_{cu} = Resistance of the Cu-line
ρ_B = Resistivity of the barrier metal
ρ_{cu} = Resistivity of Cu

W = Width of the Cu line
L = Length of the line
t_B = Thickness of the barrier
h = Height of the metal line
ΔW = Difference between the designed and actual metal line-width
c = Difference factor between barrier layer thickness over a flat surface and sidewall barrier thickness.

7.1.3 Effect of Grain Boundary (GB) Resistance on the Conductivity of Cu-Interconnects

Grain boundaries (GBs) have long been known to affect the electrical properties of metals [30]. As a matter of fact, a GB shows two orders of magnitude higher resistance than that of the grain interior [12]. Experimental observations show that for both Cu and Al films electrical resistivities are proportional to the grain boundary area per unit volume [21–23, 31].We know that smaller grain size gives rise to larger numbers of grain boundaries per unit volume. The grain boundaries impose additional resistance on the total grain resistance. Following the *brick-layer model* [32–33] (applicable to relatively narrow grains that are homogeneously distributed) the total resistance offered by a thin copper line can be written as: $R_{total} = R_g + R_{gb} = \left[R_g + \left\{t/(<d> R_1)\right\}\right]$. Here we have considered R_{total} as the total resistance, R_g the resistance of a Cu-grain, R_{gb} the resistance of the grain boundary, t the thickness of the line, $<d>$ the average grain size, and $R_1$the resistance of the individual grain boundary layer. As a matter of fact, with certain limitations the *brick layer model* can be thought of as an appropriate model to estimate the grain boundary properties of polycrystalline material. Following the above equation, the increased electrical resistivity $\left(R_{total} - -R_g\right)$ caused by the grain boundary has been measured for Cu and Al as 3.12×10^{-12} and 3.26×10^{-12} Ω-cm, respectively [21,34].

7.1.4 Effect of Grain Size and Morphology of the Substrate

In dual damascene process Cu-interconnecting lines are laid on the barrier layer. The fabrication procedure of ICs with conventional aluminum (Al) metallization does not need a four-sided barrier, but the Cu-damascene process does. As a result, the sidewall effect of the barrier will be pronounced with Cu-interconnects [35]. Figure 7.9 shows orientation maps of electroplated Cu film on (a) Ta, (b) TaN, (c) TaN/Ta, and (d) Ta/TaN/Ta.

From Fig. 7.10, we can see that the grain sizes of the Cu-lines having the same width are different over different diffusion barrier layers.Experimental data show that Cu-film deposited over a TaN/Ta layer has the largest grain size with the highest

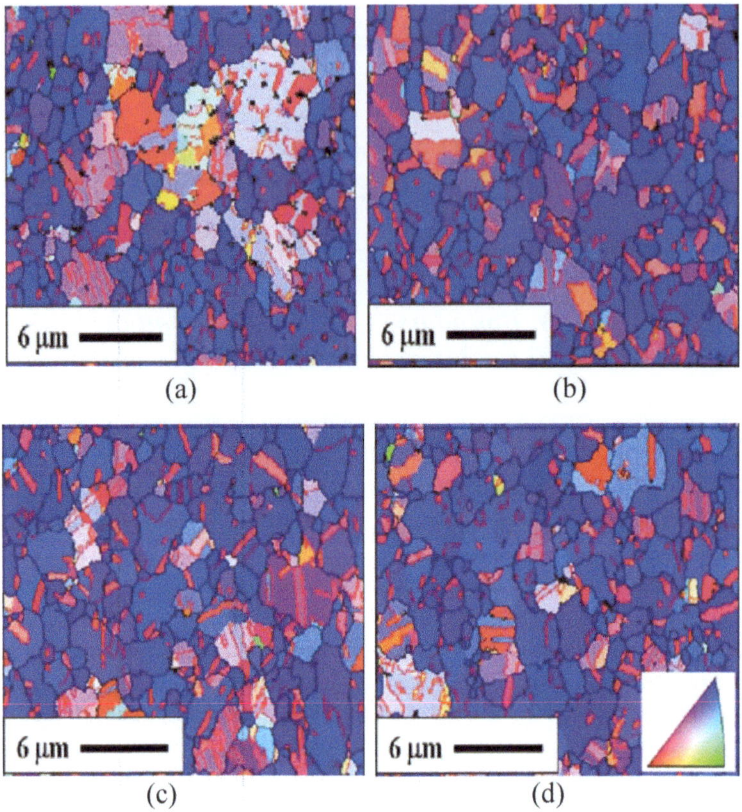

Fig. 7.10 Electroplated Cu-film on (**a**) Ta, (**b**)TaN, (**c**) TaN/Ta, and (**d**) Ta/TaN/Ta (Reproduced with permission from Elsevier, *Thin Solid Films*, 2005)

twin boundary (TB) fraction [35]. As the grain size influences the grain boundaries (GBs) the films will have different numbers of diffusion paths and line resistivity. An electroplated copper film over-blanket generally shows <111> grains but over the sidewalls it shows [011] in-plane orientation of the grains due to the constraint imposed by the sidewall. The effect of the barrier layer on the Cu-line resistivity will be more pronounced on the sidewall.

7.1.5 Morphology of the Cu-Film and Its Influence on the Conduction (Electrical) Mechanism of Cu-Interconnects

The Cu-interconnecting lines formed by Cu-damascene processes are very small in dimensions and they are mostly *polycrystalline*. The transport properties of a *thin polycrystalline film* are significantly influenced by the topology of the microstructured film. As a matter of fact, polycrystalline films can have different *grain sizes* and *grain boundaries*. As a result, it can affect the film in two ways: (i) the large grains can serve to shunt the flux pass adjacent to smaller grains, and (ii) it can increase the length of the equipotential contours due to the fact that they follow the pre-existing grain boundary. Both of these effects can lead to a decrease in the total resistance of the Cu-line as the microstructure departs from a regular *brick–layer* structure [9, 11, 18].

In the copper (Cu) damascene process interconnecting lines are fabricated from electrochemically deposited (ECD) copper. The ECD method has been experimentally verified to be the most practical and promising technology to deposit interconnecting lines for a Cu-damascene structure. It has been observed that if the lateral dimension of Cu-lines is scaled below 50 nm, the conductivity of Cu-lines becomes roughly a factor of 2 lower than the bulk value of the metal. In addition to the size effect, electroplated copper undergoes grain growth even at room temperature (Fig. 7.11). Thus understanding of the relationship between the microstructure of the metal interconnects and its electrical properties will help in optimizing the process flow and increasing device yield and reliability [36, 1].

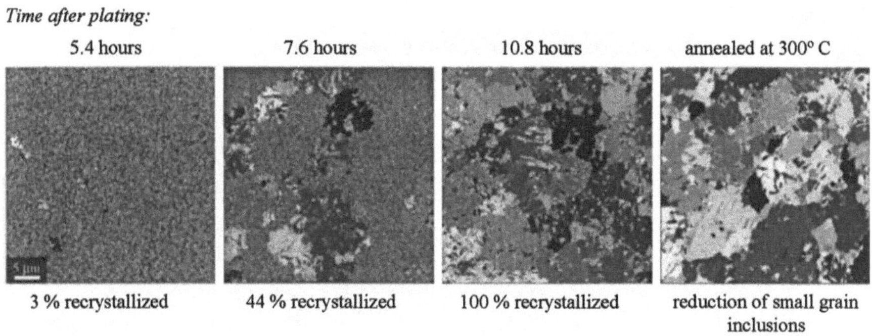

Fig. 7.11 Plan-view focused beam secondary electron images showing re-crystallization of electroplated copper at room temperature with time (Reprinted with permission, *Mat. Res. Soc. Bull.* 612, 2000)

Annealing or self-annealing of electro-chemically plated copper (ECP-Cu) at room temperature (Fig. 7.11) introduces more grain boundaries (GBs). The GBs are barriers to intragrain dislocation motion and they increase the electrical resistivity (ρ) of the metal line [37–38]. Indeed, the GB has shown two orders of magnitude higher resistance than that of the grain interior [12]. Grain boundary effects are important, especially when the width of the interconnecting line shrinks.

Figure 7.11 shows plan-view focused beam secondary electron images showing re-crystallization of electroplated copper at room temperature with time. From experimental data one can theorize that key to preserving the bulk conductivity in thin film is to optimize the microstructure so that dislocation motions are arrested while scattering of the conduction electrons is minimized [39].

Figure 7.12 shows the temperature coefficient of resistance of three different Cu-films. The nano-twined Cu-line shows a similar temperature coefficient of

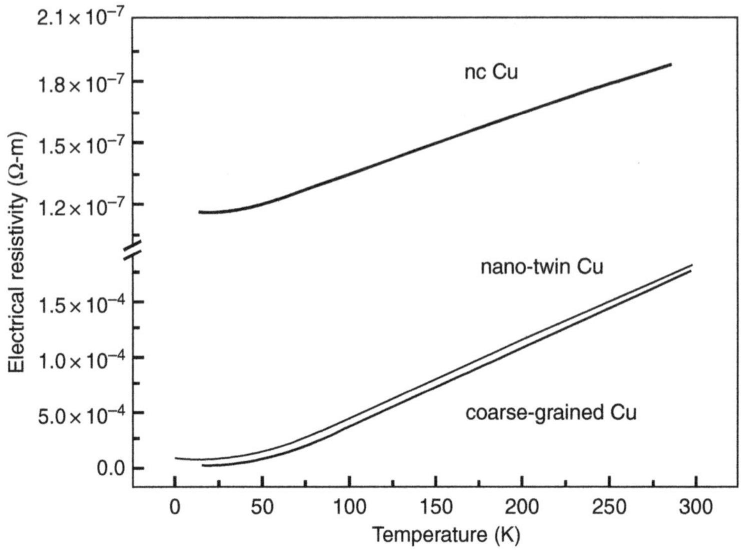

Fig. 7.12 Temperature coefficient of resistance of nano-grained with and without twins and coarse-grained Cu-line (Reprinted with permission from AAAS, *Science*, April, 2004)

resistance (α) to the coarse-grained Cu-line, while the nano-grained without twin shows a higher temperature coefficient of resistivity (α). From the graph, we can infer that the introduction of more twin boundaries (TBs) in a metal film during deposition will preserve the intrinsic conductivity of the metal. A literature survey suggests that such a film can be deposited by the pulse electrodeposition method, which ultimately increases the nucleation site for the formation of twin boundaries (TBs) [40–41]. Indeed, the twin boundary is a special kind of coherent boundary, and is able to block dislocation motion, like conventional GBs [42–43]. But electrical resistivity of coherent TBs is about one order of magnitude lower than conventional GBs [44].

The higher temperature coefficient of resistance (α) of the nano-grained Cu-line without TB compared to the coarse grained and nano-grained (with TB) Cu-lines can be explained in the spirit of the scattering phenomenon, as has been applied to the previous resistivity equation (7.8) and can be expressed mathematically as:

$$\left\{\left(\alpha/\alpha_0\right)\right\} = \left\{1/\left(\ln\left(1/k\right) + 0.423\right)\right\} \tag{7.14}$$

where $k \ll 1$, α and α_0 are the TCR of the nano-grained Cu-line without TB and that of the Cu-line (coarse grained) respectively.

Experimental evidence further shows that the presence of a TB can change the surface roughness, electrical conductivity [45], and mechanical behavior [46], and can prevent crack nucleation [47] in Cu-interconnect.

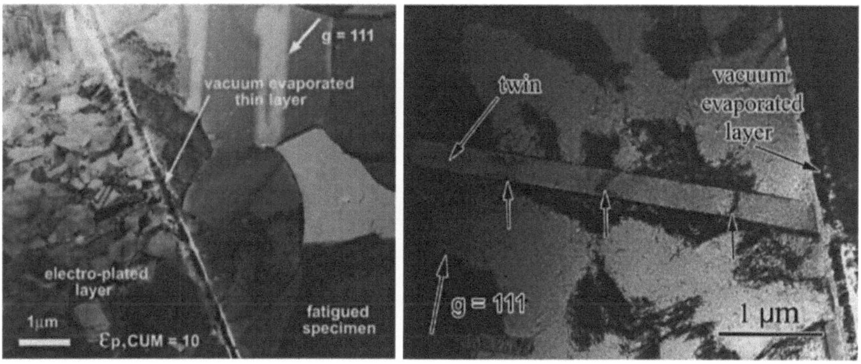

(a) (b)

Fig. 7.13 TEM pictures of (**a**) the interface structure in vacuum evaporated and electroplated copper deposited on a vacuum deposited seed layer and (**b**) PSB-like structure developed in the narrow twin and matrix-like sub-structure in the neighboring twin-related grain. The *arrows* indicate the dislocation rungs formed inside the twin grain (Reprinted with permission Taylor and Francis, *Phil. Mag*. Jan–Feb., 2004)

Figure 7.13 shows transmission electron microscope (TEM) pictures of the interface structures when electroplated Cu is deposited over a vacuum evaporated thin layer of Cu. In Fig. 7.13(a) the electroplated layer shows finer nano-sized grains and the absence of any epitaxial layer. The interface structure of a vacuum deposited seed layer when electroplated with copper develops more strain compared to a substrate electroplated with copper. The fatigued samples show the development of *surface roughness* under constant *plastic strain*. In Fig. 7.13(b) the volume fraction of persistent slip band (PSB)-like structures are formed by the dominant slip system and three ladder rungs are clearly visible inside the twin grain. The results are thought to be due to the size effect that allows one dominant slip system to operate in the plane parallel to the coherent twin boundary (TB).It has been observed that the stress that appears next to the TB produces a secondary slip that causes local strengthening of the zone. It has been observed further that the formation of a PSB-like structure is suppressed in fine-grained electro-plated (EP) copper.

Figure 7.14 is a plot of the fraction of the twin boundaries in a pure copper seed layer as a function of line-widths.The highest twin fraction is observed when the line-widths are of medium sizes. Figure 7.15 shows (111) and (011) pole figures obtained by a discrete binning method for three different line sections of a pure Cu-seed layer which acts as a nucleation center for the EP (electroplated)-Cu layer.

The pole figures (Fig. 7.15) obtained by the discrete binning method show that the deposited Cu-film (seed layer) blanket on a barrier layer has grains which are predominantly (111) textured whereas on the sidewall inside the trench the Cu-film has (011) texture. The probable reason for the (011) texture on the sidewalls is due to the decrease in the grain size as a result of an additional energy minimization other than the surface and interface energies. Therefore, we can expect different line resistivities when the texture of the line is different [48].

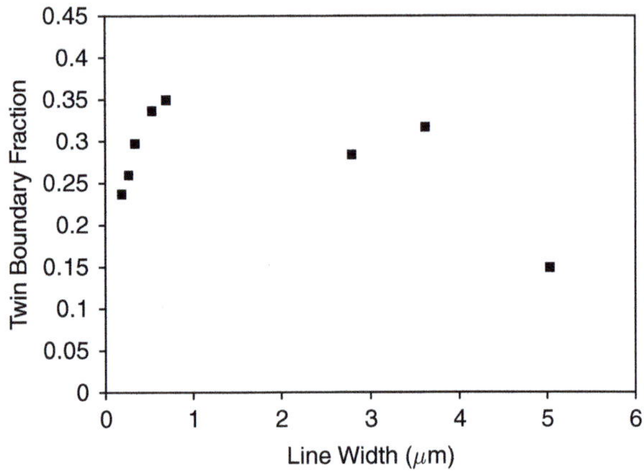

Fig. 7.14 The fraction of twin boundaries in pure Cu seed wafer grains as a function of line-width (Reproduced with permission from Elsevier, *Thin Solid Films*, 2005)

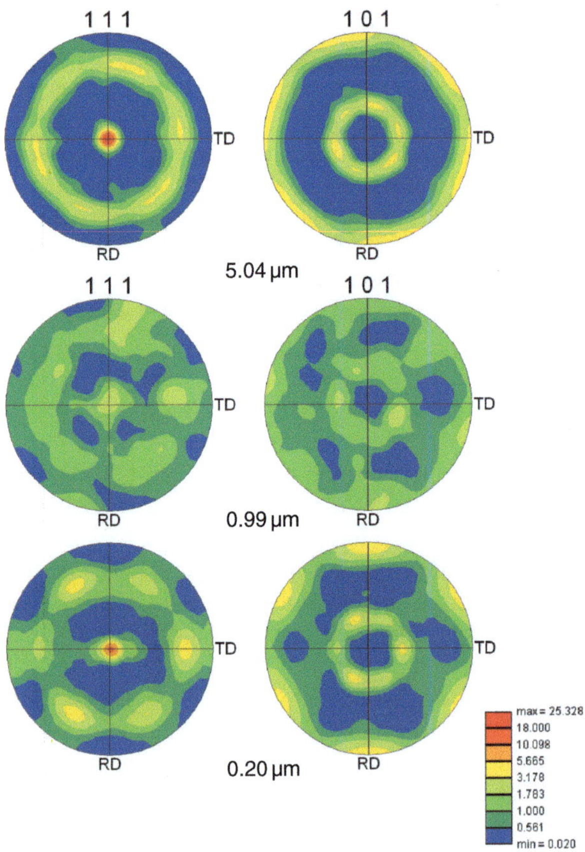

Fig. 7.15 (111) and (011) pole figures obtained by a discrete binning method for three different line sections of pure Cu-seed (Reproduced with permission from Elsevier, *Thin Solid Films*, 2005)

7.1.6 Effect of Film Thickness on the Conductivity of Cu-Interconnects

Recent advances in the manufacturing of complex integrated circuits (ICs) have led to different techniques for laying metal interconnection on the chip. Since the major factor determining the chip size is the metal interconnect size and spacing, one way to conserve the *real estate* while achieving highly complex circuits is to employ more than a single layer of interconnect metal. In multilayer metallization, the interconnecting line thickness is kept small in order to keep the die size small, and the line above is connected to the bottom line through via to reduce the stray resistance and capacitance.

Figure 7.16 shows the resistivity of a typical copper line as a function of thickness. The metal film is deposited by atomic layer deposition (ALD) and selective dry etching (reactive ion etching) is performed to create fine lines. It can be seen from Fig. 7.16 that the resistivity of the film is comparable to the bulk value (1.68 $\mu\Omega$-cm for Cu) when the film has thickness greater than 75 nm, [49–51]. On the other hand, when the film thickness reaches 75 nm or less, the resistivity of the film becomes significantly different from the bulk value, due to *interface effects*. As the film thickness approaches the mean free path (for Cu, $\lambda \approx 40$ nm) the variation of resistivity with thickness becomes very high (as discussed earlier) [52–53, 14–15].

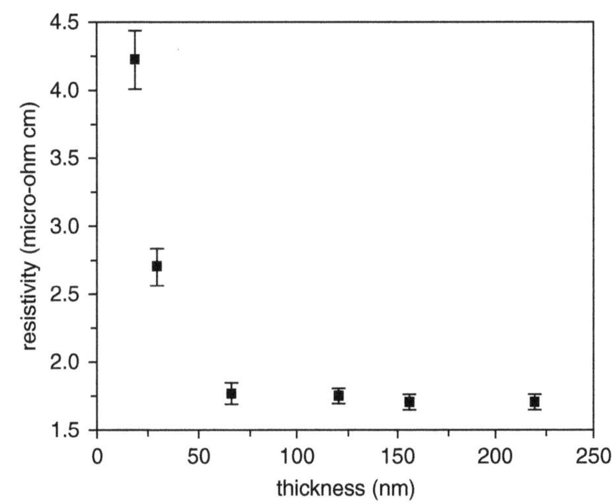

Fig. 7.16 The resistivity of a copper film as a function of thickness (Reprinted with permission, from Material Research Society, Warrendale, PA, ref. *J. Mat. Res.* **17**, 9, Sept., 2002)

It has been observed that the average grain size within a metal film decreases with line thinning . It is also true that the current carrying capacity of the metal line decreases as the width of the metal line decreases. But in order to work with the same power capacity with the thinner conducting lines, the lines are put under *stress (thermal and current)*. The *stress* forms voids due to metal migration (*electromigration*).

However, the phenomenon of *electromigration* (EM) depends upon a complex set of factors including *physical, morphological,* and *structural* properties of the conducting line.

7.1.7 Diffusion Related Impacts on the Conductivity of a Cu-Line

Diffusion induced inelastic scattering changes the resistivity of a ploycrystalline film [49–52]. In the absence of *diffusive scattering* (Fig. 7.6) at the boundaries (100% specular reflection) $p = 1$. This means that the boundaries are ideal and have no appreciable roughness. But in reality, thin films have sufficient roughness and they simulate current flow only in the finite layer subject to the $p = 0$ boundary condition. As a matter of fact, charge transfer and diffuse scattering cannot be simply separated. The charge transferred from a metal to an adsorbate implies charge separation and simultaneous creation of a localized static impurity potential from which conduction electrons can scatter without conserving momentum. Alternatively, the presence of an adsorbate layer may induce a rearrangement of surface and near-surface bonds and thereby creating a non-conducting dead layer and a corresponding increase in resistance.

As the width of the integrated circuit elements decreases, width dependent surface diffusion becomes dominant over bulk diffusion. On the other hand, anisotropy in the surface diffusivity plays a key role in the formation of slits and voids that collapse into a slit perpendicular to the direction of current flow and cause open circuit interconnect [54, 55]. It has been observed that at moderate temperatures stresses in polycrystalline thin films can be relaxed by diffusional flow of matter between the free surface and the grain boundaries (GBs) [56]. On the other hand, electromigartion studies in Cu-lines show that mass transport is dominated by diffusion at the interface between the copper interconnect and cap layer [57].

Experimental evidence shows that in sub-100 nm copper lines with bamboo-like microstructure, Cu/cap layer interface diffusion is the dominant path for electromigartion (EM) and the activation energy determined from the EM lifetime is found between 0.8 and 1.2 eV [57]. However, in bamboo-like microstructure, mass transport along grain boundary is neglected because of the lack of a continuous GB path. Table 7.1 shows the comparative analysis of diffusion of copper (Cu) and aluminum (Al). From the table we notice that the activation energy required for lattice diffusion is higher in Cu than in Al.

Table 7.1 Comparative analysis of diffusion in copper and aluminum

Activation energy required	Copper (Cu)	Aluminum (Al)
Lattice diffusion	1.7 eV.	1.4 eV
Grain boundary diffusion	0.7–1.2 eV	0.4–0.8 eV.

It has been observed that when Cu-interconnects are at high stress level, the diffusion mechanism dominates the deformation process. However, at high stress conditions, dislocation glide and power law creep may intervene in the diffusion process [58–61]. It is well known that microelectronic circuits fail often because of voids and cracks that cause open circuits in the interconnects, and the magnitude of the voids and cracks depends upon *surface diffusion.*

7.1.8 Cu-Line Stress and Its Consequences

Stress induced voiding in passivated copper interconnects has been examined as a function of line-width, current induced *Joule heating*, temperature cycling [62], and crystallographic orientation [63]. As the width of the metal line decreases, the current stress increases, which causes *Joule heating* and weakening of the metal interconnects. As a matter of fact, the contacts and local via holes have smallest cross-sectional dimensions and are more vulnerable than the interconnecting lines [64].

When the reliability of a metal interconnect is taken into account, it is found that Cu is less vulnerable to electromigration than Al, and less likely to fracture under stress. Figure 7.18 shows the excellent reliability of copper metal compared to metal aluminum (Al) when stress induced effects (due to current and temperature) are taken into account. The stress-induced effects (voids) (Fig. 7.17) are the prime cause of electromigration in metal wire [13].

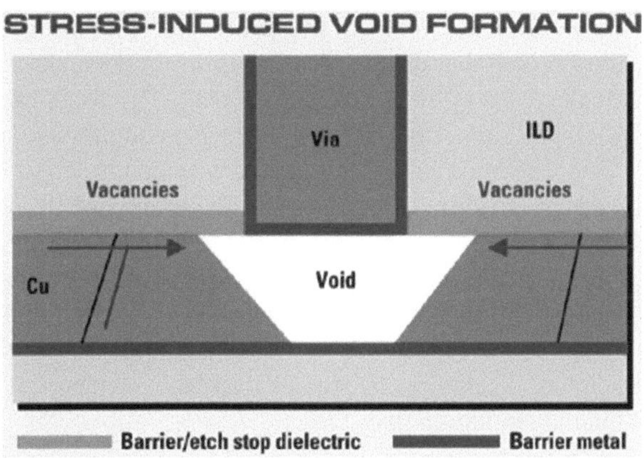

Fig. 7.17 The effect of stress induced void formation in Cu-line (reprinted with permission, *Semiconductor International* July, 2004)

In the copper damascene process it has been observed that a PVD deposited Cu-seed layer is always in stress unless some agglomeration occurs at the interface [65]. Further studies reveal that the resistivity of the electroplated Cu layer deposited over the seed layer is affected by the properties of the seed layer [66].

The damascene structure, confinement of submicron Cu-lines within the dielectric, and inclusion of low-*K* and other exotic materials, have introduced misfit strain due to the difference in thermal expansion between these materials. Moreover, any change in interconnect geometry such as Cu-line-width, or via connection and local

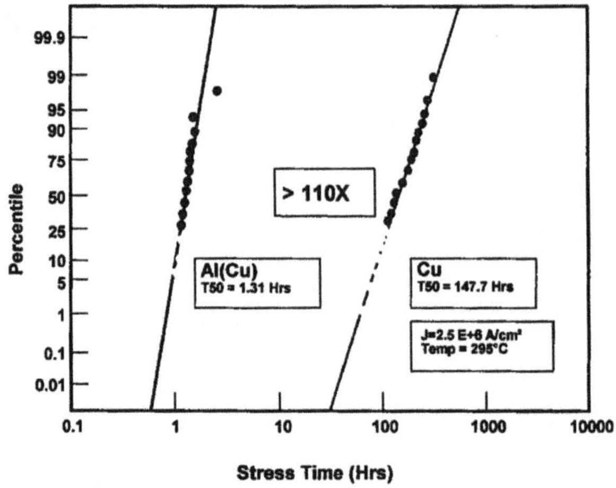

Fig. 7.18 The stress (current and temperature) induced effects on metal lines (Reproduced with permission, IBM research)

flaws, could create a stress gradient and act as a void nucleation site. Further, the void volume that is required to cause an observable resistance increase is expected to be minimum when the nucleation occurs under and within via similar to those observed in electromigration (EM) tests.

Figure 7.18 shows the stress (current and temperature) induced effects on metal lines. It has been observed that deposited thin films introduce stress and the stress increases with the decrease of film thickness [67–68]. The stress may be *tensile* or *compressive*. The tensile stress can produce cracks and compressive stress can result in buckling [69–70]. The grain size of the electroplated Cu is determined mainly by the stress and orientation of the Cu-seed layer. The grain size of the electroplated copper also affects [71] the resistivity of the Cu-interconnects [66].

There is an another type of stress, known as *mechanical stress,* which arises from a difference in thermal expansion coefficient between copper lines and the substrate [72]. The mechanical stress at room temperature is hydrostatic and tensile, and is linear with temperature. Experimental studies show that mechanical stress is lower in copper lines than in aluminum lines [73].

7.1.8.1 Stress Evolution Related Resistivity

In the model based on Clement and Thomson, the resistivity (ρ) of the line is related to the stress evolution according to the piezoresistive effect [74]. For vacancy fluxes we can present a mathematical model as:

$$\left(\partial C/\partial t\right) = \nabla J_V + G \tag{7.15}$$

where C is the relative vacancy concentration ($C = C_V/C_L = C_0$ exp. $\left(\sigma\Omega/kT\right)$, where C_V and C_L are the vacancy and lattice site concentrations, respectively, C_0 is the initial relative vacancy, σ the mechanical stress, Ω the atomic volume, k the Boltzmann constant, and T is the absolute temperature), J_V is the flux vacancy, and G is a source/sink term describing the annihilation and creation of vacancies $(= (\partial C_L/\partial t) = -(C_L/B)(\partial\sigma/\partial t)$, where B is the bulk modulus, and can be defined as $\left(1/B = 1/B_m + 1/\mu_{eff}\right)$, where B_m and μ_{eff} are the bulk modulus of metal and the effective shear modulus of the surrounding linear and dielectric materials, respectively).Now, the piezoresistive effect relates the change in the resistivity (ρ) of the Cu-line due to the change in stress (σ) as:

$$\left(1/\rho\right)\left(\partial\rho/\partial\sigma\right) = \alpha_{PR} \tag{7.16}$$

where α_{PR} is the piezoresistive coefficient of copper. By replacing the value of σ of the above equation we can write:

$$\left(\rho/\rho_0\right) = \left(C/C_0\right)^y \tag{7.17}$$

where ρ_0 is the initial resistivity of the Cu-line. Now the change in resistivity at saturation for the Cu-line with blocking boundaries at both ends can be related to the threshold product $(jL)_C = (\Omega\partial\sigma_c)/\,Z{*}\,e\,\rho$, where j_c is the critical current density, L_c is the critical length of the line (Blech length), Ω is the atomic volume, σ_c is the critical stress, Z^* is the effective charge number, e is the electronic charge, and ρ is the resistivity of the Cu-line [75].

7.1.9 Conduction of Heat Through Cu-Interconnects

The metallic characteristics of a conductor are predetermined jointly by the number of electrons free to participate in the transport processes, and their mobility. The other important criterion of a conductor is its ability *to conduct heat* (heat dissipation capacity) efficiently, which is measured by the thermal resistance R_{th} of the metal. The values of R_{th} for aluminum (Al) and copper (Cu) are $(R_{th})_{Al} = 5.88$ K/W and $(R_{th})_{Cu} = 3.57$ K/W, respectively, which indicate that Cu will dissipate more heat per square area than Al. At a given temperature, the thermal and electrical conductivities of metals are proportional, and the rise of the temperature (T) will increase the thermal conductivity (K) while decreasing the electrical conductivity (σ). This behavior is quantified in the Wiedemann–Franz law with a constant of proportionality L, called the Lorenz number $\left(K/\sigma T = L\right) = \left(\pi^2/3\right)\left(k/e\right) = 2.44 \times 10 - 8\,W\Omega/K^2$.

The Wiedemann–Franz (W-F) law holds when the collisions are elastic. The dominant scattering mechanism at low temperatures is due to static defects, and because these collisions are elastic the W-F law holds. At high temperatures, electron–phonon collisions only cause a small change in the electron's energy relative to kT after the collision. Thus the law also holds at higher temperatures.

As the device size decreases, the current generation inside integrated circuits (ICs) has become comparable to the phonon mean free path in the substrate. It is projected that the device size will be much smaller within ten years. As a result of scaling, the heat conduction on nano-scales and in ultra-fast processes will deviate from the prediction of Fourier's law due to boundary and *interface scattering* and the finite relaxation time of heat carriers [76–78]. The effect of lattice scattering on the thermal resistance of the Cu-line can be modeled similarly to the model presented in Equation (7.8).

Table 7.2 shows a list of the thermal conductivities of some metals, which are very frequently used in integrated circuit fabrication.

Table 7.2 Characteristics of the metals of interest for metallization in IC circuits

Metal	Heat conductivity Cal/s-cm^2	Thermal expansion X10^{-6} °K	Temperature Coefficient of resistance (TCR)	Lorentz number 10^{-8} W-ohm/K^2 373 K
Aluminum	0.504	23.1	0.0034	
Copper	0.918	16.5	0.00393	2.33
Gold	0.700	14.2	0.0034	2.40
Silver	0.79	18.9	0.0038	2.37
Molybdenum	0.346	4.8	0.0033	2.79
Platinum	0.1664	8.8	0.003	2.60
Tantalum	0.130	6.3	0.0031	
Tungsten	0.476	4.3	0.0045	3.20

7.1.10 Thermal Cycling (Annealing) Related Phenomena

Annealing is routine work after thin film deposition to remove structural defects. In certain cases, moderate heat treatment may lead to an increase in resistivity because of oxidation and/or agglomeration [79–80]. The change in resistivity during annealing is thought to be due to migration of vacancies and interstitials towards each other, and the change is dependent on the contribution to the residual resistivity [81–82]. However, annealing in the presence of nitrogen decreases the number of distortions. It is believed that nitrogen atoms act separately on the individual atoms that compose the distortion, and thus allow it to be destroyed [83]. Improved stress voiding resistance following heat treatment has been observed when mis-orientation angles at voided and no-voided line segments are analyzed for two different heat treated Cu-samples [84].

The amorphous state of some metals is considered as supercooled liquids having low electrical resistances. When they are annealed they show increased electrical resistance [85]. Similar results have been observed with copper, when it is deposited on an amorphous $Ta_{2.3}N$ barrier layer [86]. As-deposited copper film on amorphous TaN shows small peaks around (111) and (220). After annealing at 500 °C and above, the (220) peak becomes more intense, which is due to the change in phase of TaN (from amorphous to crystalline). As TaN changes its phase and forms crystallites at higher temperatures, the copper atom diffuses through the grain boundaries.

Experimental study shows that when Cu-interconnects undergo thermal cycling between –180 °C and room temperature they exhibit a longitudinal strain and transverse tensile strain of approximately 0.10 and 0.05%, respectively [87–88].

Figure 7.19 shows a [233] zone axis pattern obtained from a single grain in a copper-interconnect when the specimen is held at –180 °C. Thermal cycling can also introduce stress inside Cu-interconnecting lines, which can result in voids and ultimately lead to open circuit failure [89], which can be a reliability problem.

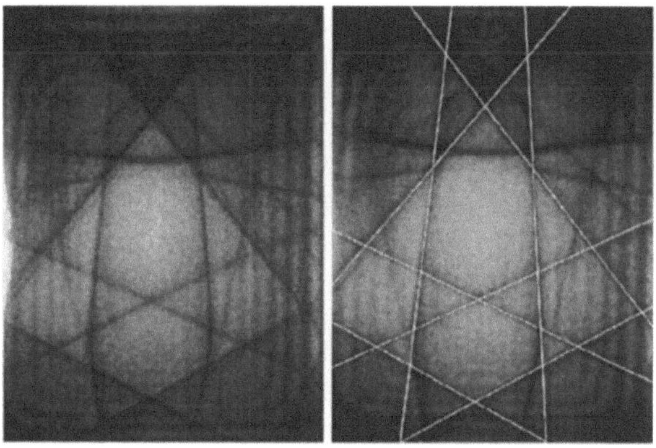

Fig. 7.19 [233] zone axis pattern and the same pattern with kinematical simulation (Reprinted with permission, Materials Research Society, Warendale, PA)

To study the variation of resistivity (ρ) with temperature (*T*), it is necessary to calculate the collision time, the only term in electrical conductivity which varies with temperature. Let us consider that *P* is the probability for an atom having thermal energy higher than that necessary to become an obstacle. Among the N_i atoms there are constantly creation and annihilation of obstacles. Let p_c and p_a be the corresponding probabilities. In the stable condition of current flow, there will be a balance between creation and annihilation of obstacles $\{(N_i\ P\ p_a = N_i\ (1–P)\ p_c\}$. As the temperature increases, the value of *P* will also increase and the resistivity (ρ) will be equal to $\rho_T = \rho_0\{(P)/(1–P)\}$, where ρ_0 is the resistivity at room temperature, and ρ_T is the resistivity at a higher temperature.

During thermal cycling, we can expect that the temperature dependent resistivity (ρ_T) of the Cu-interconnect will follow the Matthiessen–Vogt equation [90–91]. This equation can be written as $\rho_T = \rho_0\ (C)+\ \rho N_{ep}\ (T)$ and will be influenced by (a) electron–electron scattering, (b) electron–phonon scattering, (c) inelastic electron–impurity scattering, (d) localization and electron–electron interactions with weak disorder, (e) surface scattering, (f) scattering from dislocations, and (g) scattering from metallic impurities.

In addition, encapsulated at high temperature T_0, triaxial tension will be developed upon cooling due to thermal misfit strain $(\alpha_m–\alpha_s)\ (T_0–T)$, where $\alpha_m$and α_d are the thermal expansion coefficients for metal and substrate, respectively. The triaxial tension together with the mechanical constraints arising from the liner and the surrounding dielectric materials can change the resistivity of the submicron copper lines.

Failure of the Cu-interconnecting lines under current load can also be related to I^2R losses (Joule heating). The current carrying capacity of a conductor primarily depends upon: (i) the resistance offered by the metal in the flow of electrons; (ii) the temperature coefficient of resistance (TCR); and (iii) thermal transfer where the conductor is surrounded by a thermal insulator.

7.2 Electromigration (EM)

7.2.1 Electromigration (EM)

Electromigration is the result of momentum transfer from the electrons to the ions which make up the lattice of the interconnect metal. In aluminum (Al), electromigartion (EM) is a *grain boundary diffusion* phenomenon whereas in copper (Cu) it is a *surface diffusion* phenomenon, because of the damascene structure. As a matter of fact, in conventional metallization with aluminum, electromigration failure mechanisms were established for wider interconnect lines, typically 5–20 grains across the line-width, which require grain boundary diffusion pathways for the electron wind driven mass flux [92].

As the line-widths become comparable to the grain size, the electrical, mechanical, and physical characteristics of the metal lines can no longer retain the properties of the bulk material. In addition to that, metallic lines experience very high current densities, which lead to reliability issues like, *EM, void formation, hillock formation, track thinning,* and *stress* (Fig. 7.20).

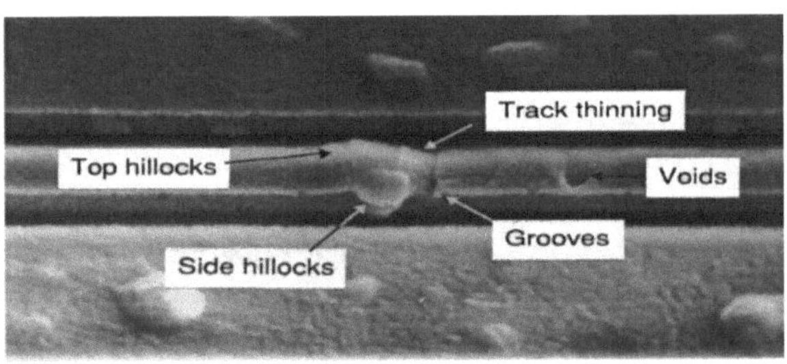

Fig. 7.20 Formation of voids, grooves, hillocks, and metal thinning due to metal migration

EM behavior on on-chip interconnect structures has been significantly affected by recent changes, such as replacing aluminum with copper [93], replacing silicon dioxide (SiO_2) with low-permittivity dielectric materials [94–95], removing shunts [96–97], and adding caps.

The Cu-damascene process is very different from the old generation Al-metallization process. In the old generation, the TiAl layer used to provide a shunt, but in the new generation (damascene process) TaN liners are ineffective as a shunt layer [98]. In the Cu-damascene process electroplated copper is deposited inside the patterned structure and is confined within the barrier layer. It has a strong <111> texture, which is weakened due to *twin* formation (115 in Fig. 7.21) [99] and it undergoes abnormal grain growth or re-crystallization at room temperature [100].

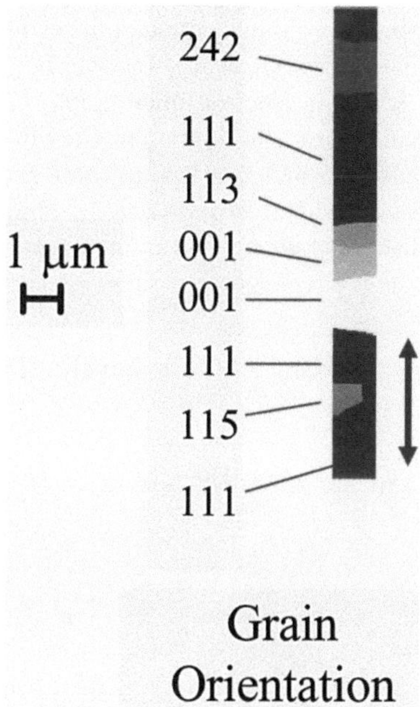

Fig. 7.21 The grain orientation of copper-interconnect with a twin (115) (Reprinted with permission, N. Tamura, LBNL)

Grain growth at room temperature leads to a non-uniform distribution of large grains and clusters of small grains in the interconnect. For copper, EM is a surface related phenomenon, and the formation of twins rather than grain boundaries is important [101].

7.2.2 Mechanism of Electromigration (EM) and Its Effects

As the interconnecting Cu-line shrinks in dimension, the interface EM is likely to be the most active EM failure mode in Cu interconnecting lines [102]. Experimental evidence shows that under a current stress of 10^7 A/cm^2 the grain size changes from 0.03 to 1 μm [103] and the current stress helps in mass migration and the formation of hillocks and voids. The void volume is seen to be a function of time [104]. For dual-damascene (DD) Cu-interconnects, EM failures are modulated by the direction of electron flow relative to the via/line arrangements and can be summarized into two distinctive categories: via above and via below. For *via above* Cu-lines, the electrons flow from via through liner into the line in the lower metal layers. Because of the zero atomic flux boundaries imposed by the liner, EM voids are always located in the proximity of the via bottom and predominantly nucleate at the Cu to etch stop interface due to poor adhesion. On the other hand, in *via below* Cu-interconnects, where electrons flow from via into the interconnect above, the exact location of void nucleation is the result of competition between flux divergence at the bottom of the via and lower critical stress for void nucleation at the etch-stop/Cu interface on top of the line.

Figure 7.22 shows a scanning electron micrograph of a hillock formed due to positive mass flux divergence or metal migration. Growth of hillocks (Fig. 7.22) is a kind of surface perturbation, which can lead to short-circuiting with the adjacent interconnect. Indeed the surface perturbation can be thought to be due to both the applied electric field and stress gradient. The atom flux due to EM can be written as:

$$F_a = \left(D/\Omega KT\right)\left(Z^* e\rho j + \Omega\left(\partial\sigma/\partial x\right)\right) \tag{7.18}$$

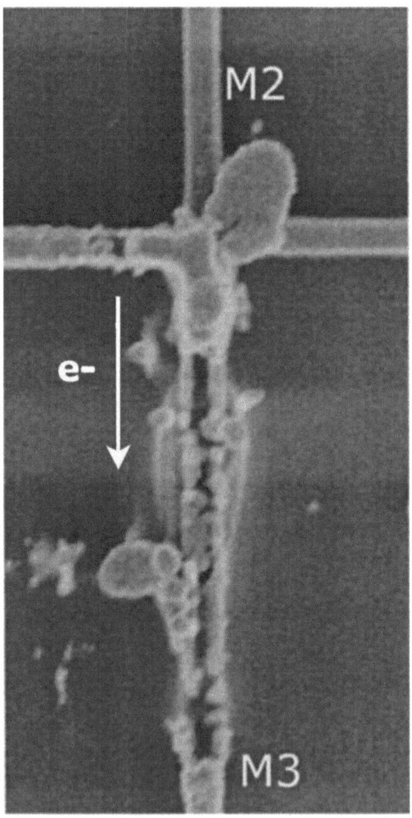

Fig. 7.22 Out-diffusion of Cu and formation of a hillock (Photo courtesy, Agere Systems)

where j is the current density, D the diffusion coefficient of atoms, Ω the volume per atom, k Boltzmann's constant, T the absolute temperature, Z^* the effective valence, e the elementary charge, ρ the resistivity, and σ the stress which is equal to $B\theta$. B is the effective elastic modulus (as described earlier), and θ is the volume fraction of atoms depleted from a segment of the line. The first term of Equation (7.18) is due to the wind force and the second term is due to the capillary force [105].

The migration of the metal is due to the direct force and the wind force. The direct force is the result of the electrostatic interaction with the field, whereas the wind force is due to momentum transfer from the carriers (electrons). When the electrons of the metal gain momentum they interact with the imperfect lattice and scatter. The scattering phenomenon causes more vibration of the adjoining atoms, which is a source of resistance and *Joule heating*. As the interconnecting metal lines shrink in dimensions, the resistance and Joule heating are expected to increase.

A finite element based on compact thermal modeling methodology has been developed to obtain the temperature rise in Cu-interconnects due to *Joule heating*. This approach uses continuum based thermal modeling based on Fourier's law of heat conduction. The compact elements include both metal and dielectric. The temperature drop across the cross-section is ignored and thermal conduction along the length has been given significant importance. The error in compact model predictions for three-dimensional interconnects terminated by via holes has been found within 5–10% of the detailed simulation [79].

Now we can define *Joule heating* in the fine metal as the thermal accelerating process, which causes *current crowding* and local heating that in turn further accelerates the void growth [106]. The heating effect refers to those caused by the atomic flow in the direction opposite to the electron flow. The back flow sometimes acts as a healing process in EM.

The EM failure mechanisms are thought to be due to the metallurgical statistical properties of the interconnects and the thermal acceleration process.The metallurgical statistical properties of the metal depend on the microstructure of the interconnecting wire. The microstructure of the interconnecting line depends on the grain size, the distribution of the grains along the grain boundaries, and the orientation of the grains with respect to current flow.

Electron back-scattered diffraction (EBSD) studies reveal that a bamboo-like microstructure is found in the narrow line whereas a polygranular structure is developed in wider lines. The fraction of $\sum 3$ boundaries is increased as the line-width increases but is decreased in blanket film [107]. The *bamboo structure* (Fig. 7.23) eliminates the high diffusivity grain boundary transport paths [108], which result in slower vacancy diffusion, because in a perfect bamboo structure there are no triple junctions and there is no continuous grain boundary pathway down the interconnect line.

However, the liner/Cu interfaces at the lateral and the bottom sides of the lines have been observed to be coherent and robust and are not believed to be diffusion paths. Some observers [109] noticed that as long as the liner for Cu-interconnects is not very thin and the effective modulus of the material remains intact, the liner does not introduce any threat due to EM. Thus when the Cu-line is coherent and robust,

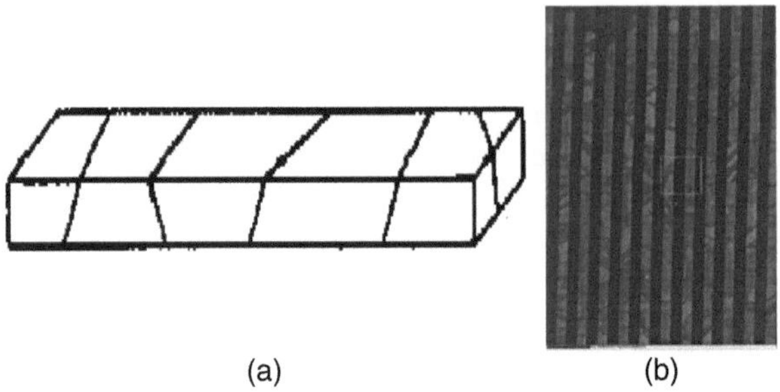

Fig. 7.23 (**a**) Schematic of a bamboo structure and (**b**) EBSD map of a Cu-interconnecting line showing the grains inside a bamboo structured film (Photo courtesy HKL Technology Inc., Germany)

Fig. 7.24 Migration of copper from Cu-interconnects along the cap dielectric-layer (Photo courtesy, IBM Research)

the *cap* dielectric material is believed to be the source of EM in Cu-interconnecting lines (Fig. 7.24).

As a result, the liner has a strong influence on EM reliability, and studies show that via/liner contact should be robust in order to have better EM reliability of the Cu-interconnects [110].

7.2.3 Void Formation

Experimental observations show that when other fast diffusion paths such as grain boundaries (GBs) are not available interface diffusion can lead to global thinning of the wire [111], which can lead to lateral void growth over a large area. This is in contrast to the catastrophic open circuit failure in near bamboo Al-interconnects [112, 113].

As electron flow drives copper atoms to drift in a line, a void forms at the upstream via [113, 96]. Moreover, there might be a discontinuity between the via hole and the line beneath it. The current crowding effect and the associated increase in temperature make the via a more vulnerable site for EM induced voiding [114–115].

Void formation (Fig. 7.25) occurs in the interconnecting metal lines due to a combination of *stress* and *electromigration*. The formation energy of voids in bulk aluminum (Al) and copper (Cu) has been calculated on the basis of the embedded atom method (EAM) and is found to be ~5.2 and 7.7 eV, respectively [116–119]. Thus voiding can lead to an increase in resistance or open circuit failure in copper interconnects.

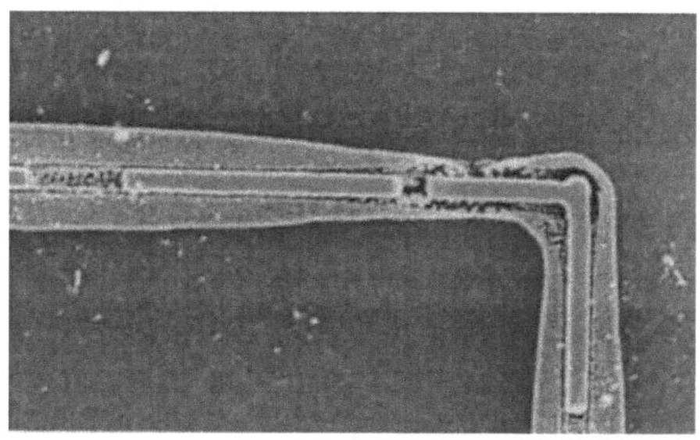

Fig. 7.25 Void formation in a Cu-interconnecting line (Photo courtesy Agere Systems)

As the feature size becomes smaller and smaller and the aspect ratio (AR) of the trench and via holes increases, unsuccessful filling inevitably leaves seam lines and transforms into nano-size voids. The nano-size void grows with time and consumes surrounding nano-size voids. This is known as the *Oswald ripening process* [120]. As the vacancy at a higher potential migrates to a vacancy at a lower potential the larger voids become larger, and smaller voids become smaller.

In a copper damascene interconnect there are two sites where generally voids are formed (a) top interface above via hole (A) and (b) bottom of via hole (B) (shown in Fig. 7.26).

Copper (Cu) atoms at the top of the via hole (top interface, A) are forced to move by the electron drag force through the top interface, and a vacancy is created. In order to compensate the annihilated volume of the vacancy, a tensile stress is built up along the perpendicular direction, and with time the tensile stress at point A or B reaches a certain limit to nucleate the void. It has been estimated that the difference in stress between point A and B is only 8% [121]. However, as the interface energy

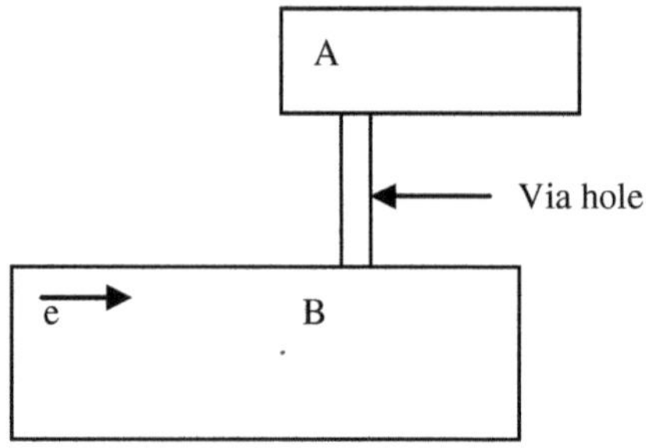

Fig. 7.26 Diagram of the damascene structure (cathode side)

release rate of Cu/SiN is less than Cu/TaN, we can expect that void formation at point A will be earlier than at point B [122].

Recently, Lane et al., [123] have demonstrated that the EM lifetime can be improved by optimizing the interfacial bond and by adding different cleaning processes. However, the lifetime of the interconnecting lines is difficult to predict, because it depends on a complex set of factors including microstructures (grain size, grain boundaries, etc.), the geometry of the lines, the current density, the temperature (Joule heating), and the substrate material. The distribution of voids within an interconnecting line has been calculated as a function of time using a transient model and the calculated median critical volume per unit area and has been found to be ~145 nm [124, 114].

7.2.4 Analytical Model on Stress Related EM

Electromigration (EM) induced stress evolution in interconnecting metal lines has been described by the Korhonen model [74]. Now we can take it for granted that EM is the process of self-diffusion due to momentum exchange between electrons and atoms. The dislocation of atoms causes stress build-up which can be represented by:

$$\left\{\left(\partial\sigma/\partial t\right)\right\} - \partial/\partial x\left[D_a\left(B\Omega\Big/kT^2 \in\right)\right]\left[\partial\sigma/\partial x - elE/\Omega\right]\}$$

where

$\sigma\ (x, t)$ = The stress function
D_{a} = Diffusivity of atoms
B = Elastic modulus
Ω = Atom volume
k = Boltzmann's constant
T = Temperature in absolute
$\in$ = Ratio of the line cross-sectional area to the area of the diffusion path
e = Electronic charge

l = Length of the effective diffusion path of atoms
E = Applied electric field.

If we assume a uniform temperature across the interconnect characteristic length (l), and $\beta T = \left[D_a\left(B\Omega/KT^2 \in\right)\right]$ and $\alpha\,(j) = \left(qlE/\Omega\right)$, we obtain the following simplified version of Equation (7.19) as:

$$\left\{\left(\partial\sigma/\partial t\right)\right\} - \left\{\left(\partial/\partial x\right)(\beta T)\right\}\left\{\left(\partial\sigma/\partial x - \alpha\,(j)\right)\right\} = 0. \tag{7.20}$$

According to the above model (Equation 7.20) the incremental change in the hydrostatic stress (dσ) per unit stress (σ) can be related to the change in the number of available lattice sites per unit volume (dV) as:

$$\left(dV/V\right) = -\left(d\sigma/\sigma\theta\right) \tag{7.21}$$

where θ is the volume fraction of atoms depleted from the segment of a Cu-line, and $\sigma\theta$ is the effective elastic modulus. The hydrostatic stress (σ) for a given set of free strains (ϵ_1, ϵ_2, ϵ_3) has been calculated using finite element analyses (FEM) and has been related to the dilation as:

$$\left(dV/V\right) = (\varepsilon_1 + \varepsilon_2 + \varepsilon_3) \tag{7.22}$$

where $\epsilon_1 = \epsilon_2 = 0$ for damascene Cu-lines with Si_3N_4 as an interlayer diffusion barrier. The study shows that the EM induced stress change has little effect on the effective modulus of the interlayer dielectric (ILD), because in the Cu-damascene architecture, Cu is not in direct contact with the ILD, but is surrounded by a liner material (Fig. 7.27) [108].

From the above discussions, we can assume that as a result of stress (Fig. 7.27) in the Cu-interconnects, voids and hillocks will be found along the Cu-lines. Consequently, these will result in grain growth near the hillocks, and the majority of them overlay triple junctions, which causes the climb of grain boundary (GB) dislocations. The compressive stress due to climb of GB dislocations can be relieved by lattice dislocations, which is emitted from the GB steps and glide to the free surface [104].As a matter of fact, partial dislocations play an important role in the plasticity of nano-structured thin films and the grain boundary structure has a significant influence on dislocation density in neighboring grains.

There are two kinds of stresses developed in a Cu-interconnect: tensile stress at the cathode, which can cause voiding; and compressive stress at the anode end, which can cause extrusion. Extrusion is not the primary failure mode during service conditions or in accelerated testing. Stress in copper interconnects has been measured by using a bending beam system [75].

In addition to the current stress (Equation 7.18), large stresses can also develop in thin metallic Cu-lines attached to the barrier and cap layer due to large differences in thermal expansion coefficients between metals (Cu/Ta) and the dielectric materials

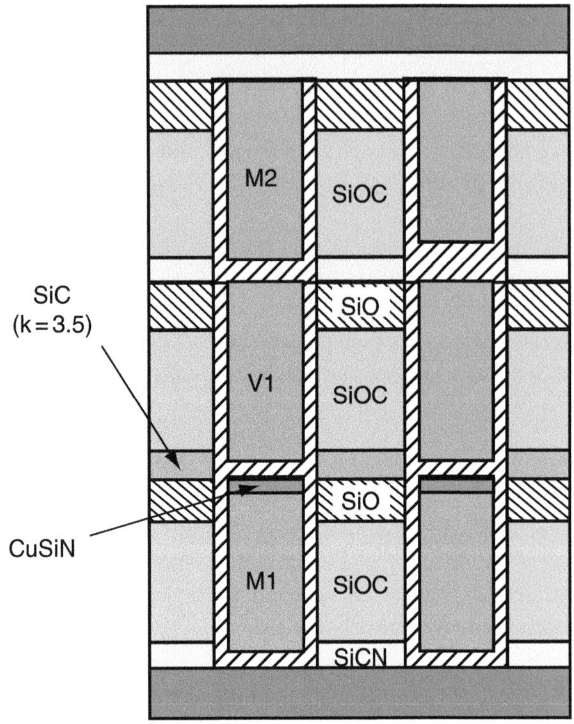

Fig. 7.27 Schematic of a Cu-damascene structure with Cu-lines and ILD. The Cu-line is embedded within the barrier layer and not in direct contact with the ILD (Courtesy NEC)

(TaN and Si_3N_4). As a matter of fact, during thermal cycling the interface layers experience thermal extrusions and induce plastic deformation accompanied by creep and interfacial sliding [125].

For passivated films, stress relaxation is more sensitive to interface diffusivity than in unpassivated films where surface diffusivity is faster than grain boundary diffusion. In electroplated Cu-films, within a certain range of temperature the stress diffusion mechanism dominates the deformation process. However, when the stress is higher, dislocation glide, and power law creep may intervene in the diffusion process. Figure 7.28 shows that failure times on mesh-type structures with Cu-Cu (DD) build are distributed log-normally within a temperature range 225–275 ºC. The failure times obtained at 225 ºC are about 5× larger compared to Cu-W-Al indicating that Cu-Cu (DD) designs are more robust with respect to wear-out by stress migration.

When Equation (7.20) is solved with different time dependent α functions, one finds that a time varying current cannot create a stable counterbalancing stress gradient for EM.On the other hand, when Equation (7.20) is solved with β (*T*), variable with time, EM stress build-up can be derived as [123]:

$$\sigma_0\ (t) = \sigma_1 [\ (1/\beta_1)\ \int_0^t \beta\,(T)\,(t)\ )dt \qquad (7.23)$$

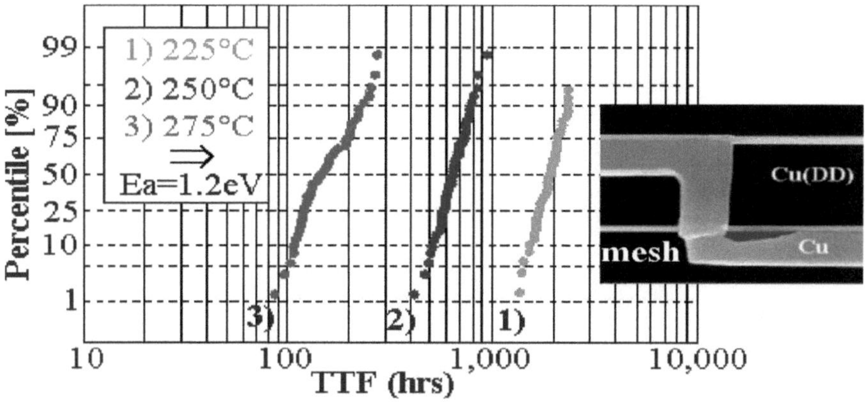

Fig. 7.28 Failure times on mesh-type structures with Cu-Cu (DD) build are distributed log-normally within a temperature range 225–275 °C (Courtesy, Infineon, Tech. AG, Munich)

If we assume that the stress build-up reaches a certain threshold value (σ_{Th}) at which an EM failure occurs, we have

$$\int_{failure\ 0}^{t} \beta\,(T)\,(t)\,dt = \varphi_{Th} \tag{7.24}$$

7.2.5 *Effect of Microstructure of the Film on Mass Migration*

During annealing, an abrupt local change in grain size is observed in the metal lines. The local variations in grain size create steps in between different grains that result in current crowding. As a result, an abrupt change in grain size superimposed on current crowding accounts for degradation of the mean time to failure (MTTF) [126]. At the same time the stress migration failure rate increases as the line-width decreases below 3 μm [127–129].

EM induced voids can be shallow edge notch type, equiaxed type, or narrow slit type, according to the variation of the length to the width of the line. In the first case, the void length increases as the width remains constant; in the second case both the length and the width increase with time, and the last one occurs when a notch is produced. It has been observed that prior to open circuit due to electromigration, numerous events take place, e.g., grain growth, precipitation, erosion, etc. All these contribute towards the change in resistance and 1/*f* noise of the film [130–131].

In the copper-damascene process the seed layer is deposited either by chemical vapor deposition (CVD), physical vapor deposition (PVD), or atomic layer deposition (ALD). Over the seed layer, copper (Cu) is electroplated. CVD/ALD grown Cu-interconnects have fine grain structures (with CVD the median grain size is 0.29 μm and with ALD it is ~47 nm), whereas electroplated copper has a near-bamboo structure [132]. As a result, electroplated copper is much more reliable than CVD/ALD grown Cu [133–136].

Figure 7.29 shows a scanning electron microphotograph of (a) a 180 nm thick Cu-seed layer and (b) a 1500 nm thick electroplated Cu-film. The grain size of the copper seed layer is much smaller than the bamboo (columnar) structured electroplated copper film [137].

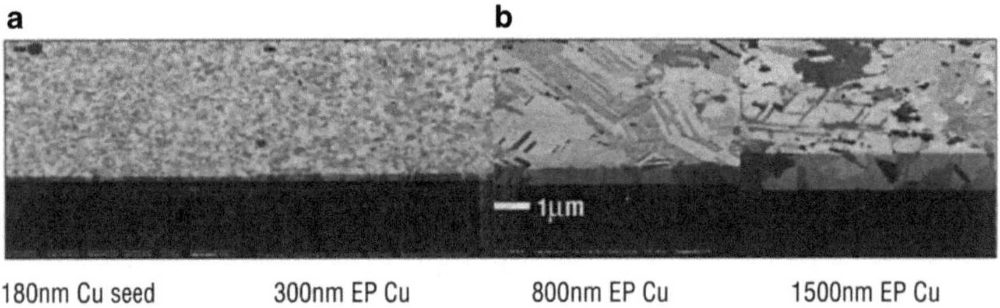

Fig. 7.29 Scanning electron photomicrograph of (**a**) thin seed layer and (**b**) the electroplated thick copper (EC) layer (Reprinted with permission, *Solid State Technology*, August 2000, p. 64)

The electroplated Cu-film is observed to recrystallize at room temperature, a phenomenon known as *self-annealing*. As a result the as-deposited Cu-film, which generally shows a grain size of 100 nm, starts to grow larger in size and within a week it transforms into a lump size grain (Fig. 7.11). This feature is comparable to a highly deformed metal piece that has a high density of dislocations. The *self-annealing* of the grains introduces a change in the texture of the film. The overall texture changes from <111> to <200> and <422> [138]. The EM in sub-100 nm Cu-interconnections shows a dependency on the grain structure [139]. The debatable result of the study shows that the well-textured lines with <111> grains are more EM resistant than those with poorly textured <100> grains [138].As a matter of fact, the dynamic recrystallization leads to a balance between grain size reduction and grain growth processes set up in the neighborhood of the boundary between the dislocation field and the diffusion creep field.

The microstructure of Cu-interconnecting lines is dominated by the abundance of grain boundaries. The higher angle of the grain boundary (>10 μm misorientation) with a $\sum 3$ twin boundary may have significant benefit on the electrical properties and lifetime of Cu-interconnecting lines, while the microstructure of Cu-interconnecting lines with non-twins may be a source of defects and can help in EM. As a matter of fact, the fraction of $\Sigma 3$ twin boundaries is increased as the line-width increases but is decreased in blanket film [140]. It has been observed that <111> growth structure has a minimum surface energy, and <200> texture has minimum strain energy. Thus as-deposited film will prefer a texture that minimizes the energy. But after annealing between 300 and 400 °C the film develops compressive stress and <100> texture changes to <200>.

7.2.6 Effect of Solute on Electromigration

The addition of a small amount of copper (Cu) in aluminum (Al) showed much less electromigration (EM) [141]. As a matter of fact, solute atoms of Cu in bulk Al are known to enhance the lattice diffusion. The possible explanation for the experimentally verified fact is that, in polycrystalline metal film, the addition of Cu to Al film forms Al_2Cu, which precipitates in Al grain boundaries. These precipitates dissolve and serve as sources of Cu to replenish the loss of Cu in grain boundaries when EM depletes them by driving them into the anode [142]. Slowing down of diffusivity of Cu has been found in the Cu-damascene process, when tin (Sn) is added to Cu [143]. A possible explanation for this is due to the strong binding force between Cu and Sn atoms. However, above 350 °C the effective EM of Cu-atoms through grain boundaries is not reduced much.

7.2.7 Melting Temperature of a Metal and Its Effect on Grain Growth

Melting points of metals have a profound effect on the mobility of the atoms. Materials with low melting temperatures, like aluminum (Al), have high atomic mobility, and will show high grain growth during deposition. On the other hand, metals with high melting temperatures will have small grain size during film growth (Fig. 7.30). According to this theory, copper (Cu) should have an intermediate grain structure during initial growth of the film, but unfortunately it has big driving force for grain growth even at room temperature [138, 144].

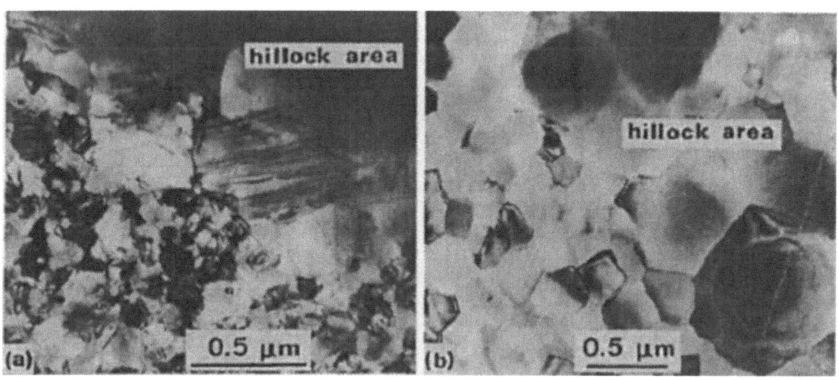

Fig. 7.30 Different grain growths (**a**) in Cu and (**b**) in Al film (Reproduced with permission, AIP)

Figure 7.30 shows different grain growths in Cu-film and Al-film during current stress. The grain growth in Cu-and Al films is observed from 0.2 to 1 μm and from 0.1 to 1 μm, respectively.

7.2.8 Effect of Temperature on EM

The ever increasing speed and functionality of silicon based integrated circuits (ICs) have resulted in a dramatic reduction of interconnect metal pitch and have increased the number of metallization levels to accommodate the ever increasing numbers of wired circuits per chip. As a result, thermal effects (due to self- or Joule heating) have affected the maximum allowable RMS value of the current density (J), the lifetime (reliability) due to electrtomigration (EM) (which has an exponential dependence on inverse metal temperatures) [145], and induction of open circuit metal failure with short duration of high peak currents including *electrostatic discharge* (ESD) [146].

7.2.9 Current Density and Its Effect on EM

It has been observed that when the interconnect Cu-lines have a turn of 90°, a very high current exists at the inner edge of the turn and the current density gradient decreases from the inner edge to the outer edge of the turn. As a result, scattering on the higher current density side becomes higher than that on the side having lower current density, and a net force (F_{net}) pushes the vacancy down the gradient (Fig. 7.31) [142]. On the basis of the scattering, the force is expected to have a square dependence on current density, and the vacancy flux driven by the gradient of the force (F_{net}) can be written as:

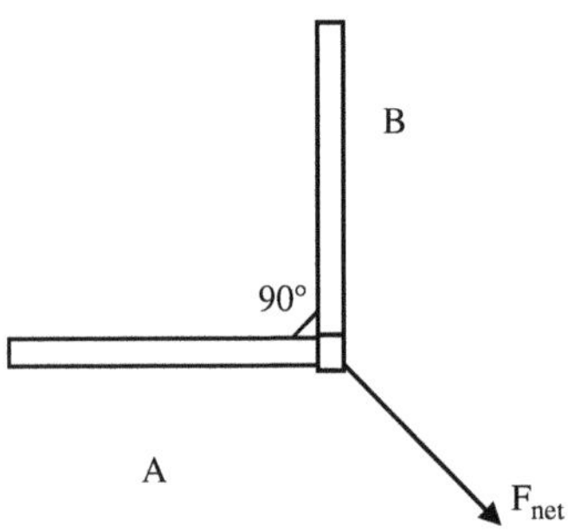

$F_{flux} = C_v\,(D/kT)\,(-dP/dr)\ldots.$

Where, $\Delta C_v$is the excess vacancy concentration, D the diffusion coefficient, k the Boltzmann constant, T the absolute temperature

$P = (e\,j\,A\,\Delta\rho)$, where e is the effective charge,

j the current density, and $\Delta\rho$ the resistance in the vacancy in the lattice and A is the scattering cross section

Fig. 7.31 Two interconnect lines A and B turned at an angle 90°. F_{net} is the net current density force

7.3 Summary

Bulk copper (Cu) has higher conductivity and higher resistance to electromigration (EM) compared to bulk aluminum (Al). But as the feature size decreases to the sub-100 micron level, the thin lines cannot retain the electrical, mechanical, and physical characteristics of the bulk metal. When the lateral dimensions of the Cu-lines are scaled to values ~100 nm and less, the resistivity of the lines increases roughly to a factor of 2 higher than the bulk value. At the same time the self-heating of the interconnect is further aggravated because of Joule heating. In the Cu-damascene process, Cu-lines are confined within the barrier layer and as a result the charge carriers are scattered at the sidewalls and affect the resistivity of the lines. Additional changes to the resistivity of the fine lines come from grain boundary (GB) scattering.

The small dimensions of the lines produce stress on the metal lines. The stress can be a source of mass migration, which can form voids and hillocks and ultimately can be a source of device failure (due to open or short circuits). The stress may be due to excessive current density, or self-heating. Due to thermal or electrical stress, the fine Cu-lines suffer the same EM-reliability problem as Al-lines. However, due to the higher bulk conductivity in Cu, EM related reliability problems might not be as severe in Cu-interconnects as is observed in Al-interconnects at a particular current density.

The additional burden with the Cu-interconnects is the diffusion barrier layer. This has added complexity, like nucleation, morphology (grain size, grain boundaries, etc.), interface layer, and overall the effective resistance of the conducting lines [147–149].

References

1. W. Steinhoegl, G. Schindler, G. Steinlesberger, M. Traving, and M. Englehardt, Comprehensive study of copper wires with lateral dimensions of 100 nm and smaller, J. Appl. Phys., 97, 023706 (2005) and A.F. Mayadas and M. Shatzkes, Electrical resistivity model for polycrystalline films: The case of arbitrary reflection at external surfaces, Phys. Rev., B, 1 (4), 1382 (1970) and also M. Armacost, Materials 7 Processing. IEEE IITC, San Francisco, CA, (June 2008).
2. G. Steinlesberger, M. Englehardt, G. Schindler, and W. Steinhoegl, Processing technology of the investigation of sub-50 nm copper damascene interconnects, Solid State Electron., 47, 1237 (2002) and K. Banerjee, S. Souri, P. Kapur, and K. Saraswat, 3-D ICs: A novel chip design for improving deep-submicrometer interconnect performance and system-on-chip integration, Proc. IEEE 89, 602 (May 2001) and also M. Leong et al., Transistor scaling with novel materials, Mater. Today, 9(6), 26 (2006)
3. C. Blass, P. Weinberger, L. Szunyog, P.M. Levy, C.B. Sommers, and I. Mertig, Phil. Mag., B-78, 549 (1998) and N. Srivastava and K. Banerjee, A comparative scaling analysis of metallic and carbon nano tube interconnections for nanometer scale VLSI technologies, Proc. 21st Int. VLSI multilevel interconnect Conf. Waikoloa, HI, 393, Sept. 29–2 Oct. (2004) and J.A. Green wood, Br. J. Appl. Phys., 17, 1621 (1966) and R. Holm, Electrical contacts. Theory and applications, Springer-Verlag, Berlin, (1967) and Y.H. Jang, and J.R. Barber, J. Appl. Phys., 94 (11), 7215 (2003) and also J. Van Olmen et al., IEEE IITC Tech Dig.,, 241 (2007)
4. H.E. Camblong and P.M. Levy, Electrical resistivity of a thin film, Cond. Matter., 9911093 V1, Nov. (1999) and also J.Y. Cho et al., J. Electron. Mater., 34 (1), 53 (2005) and N. Michael et al., J. Electron. Mater., 32 (10), 988 (2003) and C.A. Stafford, D. Baeriswyl, and Burki, Jellium model of metallic nanocohesion, Phys. Rev. Letts., 79 (15) 2863 Oct. (1997) and also J. Gouldstone et al., Acta Mater., 55, 4015 (2007)
5. C. Blass, P. Weinberger, L. Szunyog, P.M. Levy, and C.B. Sommers, Phys. Rev., 60, 492 (1999) and H. Kudo, IEEE IITC Burlingame, CA, April (2008)

6. Y. Imry and R. Landauer, Conductance viewed as transmission, Rev. Mod. Phys., 71, S306–S312 (1999) and K.F. Var, F.A.P. Osorio, A.N. Borges, and P.C.M. Machado, Atomistic simulation of ultra submicron MOSFETs, Microelectron. J., 34 (7–8), 599 (2003)
7. G. Chen, Thermal conductivity and ballistic transport in cross plane direction of super lattice, Phys. Rev. B, 57, 14958–14973 (1998) and A. Roy et al., Semicond. Sci. Technol., 21, 1369 (2006)
8. Y.S. Ju and K.E Goodson, Phonon scattering in silicon film with thickness of 100 nm, Appl. Phys. Lett., 74, 3005–3007 (1999) and T. Sun, B. Yao, A. Warren, V. Kumar, K. Bormak, and K.R. Coffee, IEEE IITC, San Francisco, CA, June (2008)
9. C.L. Tien, A. Majumdar, and F.M. Garner, Microscale energy transport, Taylor and Francis Pub., New York, (1998) and O. Muscato, Physica, 365 (2), 409 (2005)
10. D.A. Young, and H.J. Maris, Lattice-Dynamical calculations of the Kapitza resistance between fcc lattices, Phys. Rev. B, 40, 3685–3693 (1989) and G. Ta, C. Ouvrad, H. Chauveau, and S. Nath, Microelectron Relib., 47 (5), 610 (2007)
11. E.H. Sondheimer, Adv. Phys., 1, 1 (1952), and K. Banerjee and A. Mehrotra, Analog IC and signal processing, 35, 97 (2003) and F. Chen and D. Gardner, Influence of line dimensions on the resistance of Cu-interconnects, IEEE Electron Dev. Letts., 19, 508 (1998) and also M. Hauschidt et al., J. Appl. Phys., (2007)
12. K.L. Chopra, Phys. Rev., 155, 660 (1967) and also B. Dieny, J. Phys. Cond. Matter. 4, 8008 (1996)
13. M.P. Oxley and L.J. Allen, ICSC: a program for calculating inelastic scattering cross sections for fast electrons incident on crystal, J. Appl. Cryst., 36, 940–943 (2003)
14. J.A. Davis et al., Interconnect limits on giga scale integration (GSI) in the 21st century, Proc. IEEE, 89 (3), 305 (2001) and J.Y. Cho, M. Mirkuri, D.N. Lee, J.K. An, and J.A. Szpunar, J. Electron. Mats., 34 (1), 53 (2005) and V. Sukharev, E. Zschech, and W.D. Nix, J. Appl. Phys., 102, 3505 (2007)
15. K. Fuchs, The conductivity of thin metallic films according to the electron theory of metals, Proc. Cambridge Philosophical Soc., 34 (1938), and also S. Datta, Electronic transport in mesoscopic systems, Cambridge University Press., Cambridge, London, (1995)
16. K.C. Elsom and J.R. Sambles, Macroscopic surface roughness and the resistivity of thin films, J. Phys., F11 (647) (1981) and also S.N. Chandler-Wilde and R Potthast, The domain derivative in rough-surface scattering and rigorous estimates for first order perturbation theory, Proc. Royal Soc. Lond., A, 1 (2002)
17. P. Kapur, J. Mcvitte, and K.C. Saraswat, Technology and reliability constrained future copper interconnects, IEEE Trans. Electron Dev., 49, (4) 590 (2002)
18. K.A. Dun, Reliability implications of ultra-thin copper lines and low-k dielectrics, SEMATECH, TRC, Oct. 25–27 (2004) and also S.N. Chandler-Wilde and R. Potthast, The domain derivative in rough surface scattering and rigorous estimates for first order perturbation theory, Proc. R. Soc. Lond. A, (2002)
19. S.M. Rossnagel and T.S. Kuan, Alteration of Cu-resistivity in size effect regime, J. Vac. Sci. Technol., B-22, (1), 240 (2004)
20. A. Gladkikh, M. Karpovski, A. Palevski, and Y.S. Kaganovski, J. Appl. Phys., D31, 1626 (1998) and also J.A. Davis et al., Interconnect limits on giga scale integration in the 21st century, Proc. IEEE, 89 (3) (2001) and also T. Tonegawn, M. Hiroi, K. Motoyama, K. Fuji, and H. Miyamoto, Proc. IEEE, IITC, 216 (2003) and Y. Igarashi, and T. Ito, J. Vac. Sci. Technol., B-16 (5), 2745 (1998)
21. A.M. Song, A. Lorke, A. Kriele, J.P. Katthaus, W. Wegscheider, and M. Bichler, Non-linear electron transport in an asymetric microjunction: A ballistic rectifier, Phys. Rev. Letts., 80 (17), 3831–3834 (1998) and S. Yokogawa and H. Tsu Vhiya, J. Appl. Phys., 101(1), 013513 Jan. (2007)
22. K. Takayanagi, Y. Kondo, and H. Ohnishi, Jpn. J. Appl. Phys., Int. (3), Jan. (2000)
23. N. Agraiet, J.G. Rodrigo, and S. Vieira, Phys. Rev., B-47, 12345 (1993) and also S.B. Cronin, Y.M. Lin, O. Rabin, M.R. Black, G. Dresselhaus, M. S. Dresselhaus, and P.L. Gai, Microsc.

Microanal., 8, 58 (2002) and also F. Chen, J.P. Gambino, J.P. Gill, B. Li, and T.D. Sulivan, US Patent 7231617 (2007)

24. J.L. Pascual et al., Science, 267, 1793 (1995) and also D.M. Saylor, A. Morawiece, and G.S. Rohrer, Acta Mater., 51, 3663 (2003)
25. G.L. Costa-Kramer et al., Phys. Rev., 855, 5416 (1997) and also U.F. Kocks, C.N. Tome and H-R. Wenk, Texture anisotropy: Preferred orientations in polycrystals and their effect on materials properties, Cambridge University Press. New York, (1998)
26. A.F. Hebard, R.R. Ruel, and C.B. Eom, Charge transfer and surface scattering at Cu-C_{60} planar interfaces, Phys. Rev. B, 54 (19) (1996) and S.N. Chandler-Wilde and C.R. Ross, Scattering by rough surfaces, Math. Methods Appl. Sci., 19, 959 (1996) and also K. Barmark, C. Cabral Jr., K.P. Rodbelt, and J.M.E. Harper, J. Vac. Sci. Technol. B., 24 (6), 2485 Nov. (2006)
27. P. Wissmann, Surface physics, In G. Hohler (ed.) Tracts in modern physics, Springer, New York, Vol. 77 (1975) and G.P. Beyer et al., Al speed Fill, Maters. Sci. Semicond. Process, 2, 75 (1999) and FJJ.V. Loo and G.D. Reck, Acta Metal., 21, 61 (1973) and also S. Zhang et al., IEEE Int. Symp. Adv. Packaging Mater., March (2006)
28. E.H. Sondheimer, The mean free path of electrons in metals, Adv. Phys., 1, 1–42 (1952) and also S. Datta, In R.H. Friend and M.A. Reed (eds.) Physics of electronic transport in single atoms, molecules and related nano-structures, IOP publishing, England, (2004) and also W. Steinhogl, G. Schindler, G. Steinlesberger and M. Engelhardt, Size dependent resistivity of metallic wires in the mesoscopic range, Phys. Rev. B, 66, 075414 (2002)
29. W. Steinhoegl, G. Schindler, G. Steinlesberger, M. Traving, and M. Englehardt, Scaling laws for the resistivity increase of sub-100 nm interconnects, Proc. Int. Conf. simulation of semiconductor processed devices, p. 27 (2003)
30. K. Hansen, E. Laegsgaad, I. Stensgaad and F. Besenbacher, Phys. Rev., 856, 2208 (1997) and W. Zhang et al., Surface and grain boundary scattering studied in beveled polycrystalline copper films, J. Vac. Sci. Technol., 22 (4), 1830 (2004) and also K-C Chen, W-W Wu, C-N Liao, L-J chen and K.N. Tu, Science, 321 (5892), 1066 Aug. (2008)
31. K. Schwab, E.A. Henriksen, J. Woriock, and M.L. Roukes, Nature, 404, 974 (2000)
32. G. Chen, Ballistic diffusive heat conduction equations, Phys. Rev. Letts., 86, 2297–2300 (2001)
33. K.E. Goodson and Y.S. Ju, Heat conduction in novel electronic films, Annu. Rev. Mat. Sci., 29, 261–293 (1999) and also T. Turner, Requirements for dual-damascene Cu-line width resistivity measurements, Solid State Technology (SST), April (2000)
34. N.D. Lang, Phys. Rev., B52, 5335 (1995) and F. Chen, J.P. Gambino, J.P. Gill, B. Li, and T.D. Sullivan, U.S. Patent 7231617 June (2007)
35. T. Ishibashi and Y. Yamauchi, A possible near ballistic collection in AlGaAs/GaAs HBT with a modified collector structure, IEEE Trans. Electron. Dev. Ed-35, 401 (1988) and also R.N. Tait, S.K. Dew, T. Smy, and M.J. Brett, J. Appl. Phys., 70, 4295 (1991) and T. Muppidi, D. Field, and J.E. Sanchez, Barrier layer, geometric and alloying effects on the microstructure and texture of copper films and damascene lines for interconnect applications, Thin Solid Films, 471, 63–70 (2005)
36. Q.T. Jiang, Analysis of copper grains in damascene trenches after rapid thermal processing or furnace anneals, J. Electron. Mater., 31 (1), 10 (2002)
37. J.M.E. Harper, C. Cabral, Jr., P.C. Andricacos, L. Gignac, I.C. Noyan, K.P. Rodbell, and C.K. Hu, J. Appl. Phys., 86, 2616 (1999) and K-C Chen, W-W Wu, C-N Liao, L-J. Chen, and K.N. Tu, Science, 321 (5892) 1066 Aug. (2008)
38. D. Walther, M.E. Gross, K. Evans-Lutterodt, W.L. Brown, M. Oh, S. Merchant, and P Naresh, Room temperature recrystallization of electroplated copper thin films, Proc. MRS., Vol. 612 (2000)
39. L. Lu, Y. Shen, X. Chen, L. Quin, and K. Lu, Ultrastrength and high electrical conductivity in copper, Science, 304, 422 (2004)

40. J. Stohr et al., Science 259, 658 (1993) and also Kh. M. Manan, Kh. R. Karin, J. Phys. F Metal Phys., 5, 1687 (1975) and P.L. Gai, R. Mitra, and J.R. Weertman, Structural variations in nanocrystalline nickel films, Pure Appl. Chem., 74 (9), 1519 (2002)
41. S.S. Brenner, J. Appl. Phys., 27, 1484 (1956) and also H. Yoshinga, Phys. Stat. Sol., 18, 625 (1966) and also V. Dureuil, C. Ricolleau, M. Gandias, and C. Grigis, Phase transitions in Co nanoclusters grown by pulsed laser deposition, Eur. Phys., J. D., 14, 83 (2001)
42. J.W. Christian and S. Mahajan, Prog. Mater. Sci., 39 (1) (1995) and also R.A. Masumura and I.A. Ovid'ko, Enhanced diffusion near amorphous grain boundaries in nanocrystalline and poly crystalline solids, Mater. Phys. Mech. 1, 31 (2000)
43. S.D. Dahlgren et al., Thin Solid Films, 40, 345 (1977) and also D. Farkas, Fracture mechanisms of symmetrical tilt grain boundaries, Phil. Mag. Letts., 80 (4), 229 (2000) and K-C Chen, W-W Wu, C-N Liao, L-J. Chen, and K.N. Tu, Science, 321 (5892) 1066 Aug. (2008)
44. A.P. Sutton and R.W. Balluffi, Interfaces in crystalline materials, Clarendon Press, Oxford, London, (1995)
45. I. Nakamichi, Mat. Sci. Forum, 207, 47 (1996) and also L. Lu, Y. Shen, X. Chen, L. Qian, and K. Lu, Science, 304, 422 (2004)
46. M.A. Meyers and K.K. Chawla, Mechanical behavior materials, Prentice Hall, Upper Saddle River, NJ, (1999) and A. Gouldstone et al., Acta Mater., 55, 4015 (2007)
47. H. Kawazoe and M. Niewczas, Dislocation microstructures and surface roughness in fatigued fine grained copper polycrystals, Phil. Mag., 84, (3–5), 381 (2004) and T. Zhu, J. Li, and S. Yip, Proc. R. Soc. Lond. A., 462, 1741 (2006)
48. T. Masaru, K. Nobuhiko, and M. Brandbyge, Theory of ballistic electron transport through atom bridge between tip and surface, Proc. APS., Los Angeles, March 18 (2003)
49. M.P. Anderson, G.S. Crest, and D.J. Srolovitz, Phils. Mag., B59, 293 (1989), and also R. Potthast, Point sources and multipoles in inverse scattering, Chapman and Hall, New York, (2001) and L. Lu et al., Acta Mater, 53 (70), 2169 (2005)
50. S.N. Chandler-Wilde, and B. Zhang, A uniqueness result for scattering by infinite rough surfaces, SIAM, J. Appl. Math. 58, 1774 (1998a) and also N. Bonanos, B.C.H. Steele, and E.P. Butler, Applications of impedance spectroscopy, In J.R. McDonald (ed.) Impedance spectroscopy, John Wiley, Chapt. 4, p. 191 (1987)
51. M.P. Anderson, D.J. Srolovitz, G.S. Grest, and P.S. Sahni, Acta Metal., 32(5) 783 (1984) and also H.E. Camblong, and P.M. Levy, Electrical resistivity of a thin metallic film, Cond. Matter. 1, 1 Nov. (1999)
52. K.L. Chopra and M.R. Randlett, J. Appl. Phys., 38, 3144 (1967) and also D.J. Keavney, S.K. Park, C.M. Falco, and J.M. Slaughter, J. Appl. Phys., 86 (1), 476 (1999) and C.A. Schuh, Mater. Today, 9 (5), 32 (2006)
53. J.C. Maxwell, A treatise on electricity and magnetism, 2nd (ed.) Clarendon Press, Oxford, London, (1954) and M.P. Anderson and S. Ling, Computer simulation of transport of thin films, J. Electron. Mater., 19 (11), 1161 Nov. (1990) and A.M. Coratolo, J. Mater. Sci., 41, 3105 (2006)
54. D.R. Fridlind and A.F. Bower, Influence of anisotropic surface diffusivity on electromigration induced void migration and evolution, J. Appl. Phys., 85 (6), 3168 (1999)
55. O. Kraft and E. Artz, Acta Mater., 45, 1599 (1997)
56. R.M. Keller, S.P. Baker, and E. Artz, Stress temperature behavior of unpassivated thin copper films. Acta Mater., 47, 415 (1999)
57. C.K. Hu, R. Rosenberg, and K.Y. Lee, Electromigration path in Cu thin films, Appl. Phys. Lett., 74, 2945 (1999)
58. L. Zang and H. Gao, Coupled grain boundary and surface diffusion in a polycrystalline thin film constrained by substrate, Z. Metallkd., 93, 417 (2002)
59. J.R. Rice and T.J. Chuang, Energy variation in diffusive cavity growth, J. Am. Ceram. Soc., 64, 46 (1981) and R.P. Vinci, E.M. Zielinski, and J.C. Bravman, Thin Solid Films, 262, 142 (1995)

60. Y.A. Antipov and H. Gao, Atomic diffusion from a material interface into grain boundary, Proc. Royal. Soc. Lond., A458, 1673 (2002)
61. M.D. Thouless, Effects of surface diffusion on the creep of thin films and sintered arrays of particles, Acta Metal Mater., 41 1057 (1993), and also M.J. Buchler, A. Hartmair, and H. Gao, Atomistic and continum studies of crack like diffusion wedges and dislocation submission thin Films, J. Mech. Phys. Solids, 51, 2105 (2003) and K-C Chen, W-W Wu, C-N Liao, L-J. Chen, and K.N. Tu, Science, 321 (5892) 1066 Aug (2008)
62. J.A. Nucci, Y.S. Diamond, and J.E. Sanchez Jr., Appl. Phys. Lett., 66, 3585 (1995) and also V.R. Hunter, IEEE Trans. Electron. Dev., 44, (2), 304 (1997)
63. J.A. Nucci, R.R. Keller, J.E. Sanchez, and Y.S. Diamond, Appl. Phys. Lett., 68, 4017 (1996)
64. T.M. Shaw, L. Gignac, X.H. Liu, and R.R. Rosenberg, Stress voiding in wide copper lines. In S.P. Baker, M.A. Korhonen, E. Artz, and P.S. Ho, (eds.) Stress induced phenomena in metallization, 6th Int. Workshop, 177–183 (2001)
65. T. Hara and K. Sakata, Stress in copper seed layer employing copper interconnections, Electrochem. Solid State Lett., 4 (10) G-77 (2001)
66. S. Balakumar et al., Effect of stress on the properties of copper lines in Cu interconnects, Electrochem. Solid State Lett., 7 (4), G-68 (2004)
67. O. Kraft, M. Hommel, and E. Artz, Mat. Sci. Eng., A 288, 209 (2000) and H-J Lee, H-H Han, S-H Kong, J-Y Sun, and K-H Oh, MRC Proc. Symp. F, Spring (2006)
68. D. Chocyk et al., Evolution of stress and structure in Cu thin films, Cryst. Res. Technol., 40 (4/5), 509 (2005)
69. M.F. Doerner and W.D. Nix, Crit. Rev. Solid State Mater. Sci., 14, 225 (1988) and also A.A. Volinsky, Mechanical reliability and characterization of modern microelectronic interconnect structures, Int'l Eng. Consortium (IEC), Royal Plaza Hotel & Trade Center, Marlborough, MA, June 23–25, (2003)
70. H. Windischmann, Criti. Rev. Solid State Mater. Sci., 17, 547 (1992)
71. T. Hara, K. Sakata, and Y. Yoshida, Electrochem. Solid State Lett., 5, C41 (2002) and D. Xu, W-L Kwan, K. Chen, X. Zhang, V. Ozoli and K.N. Tu, Appl. Phys. Lett., 91, 254105 Dec. (2007)
72. P. Besser et al., MRS. Symp. Proc., 563 (1999) and P. Besser, Stress induced phenomena in metallization, AIP Conf. Proc., 491, 229 (1999) and also D. Rodney et al., Mat. Sci. Eng. 13, 427 (2005)
73. P. Besser, SRC /TRC on Reliability, SEMATECH, Austin, TX, Nov. (2000) and A. Gouldstone et al., Materialia, 55, 4015 (2007)
74. M.A. Korhonen, P. Boergesen, K.N. Tu, and C.Y. Li, J. Appl. Phys., 73, 3790 (1993) and also Z. Lu, W. Huang, M.R. Stan, K. Skadron, and J. Lach, Interconnect life time prediction for reliability aware systems, IEEE Trans. On VLSI systems, 15 (2), 159 (2007) and J.J. Clement and C.V. Thompson, J. Appl. Phys., 78 (2), 900 (1995) and N.L. Beverly, G.B. Alers, and J.A. Prybyla, Appl. Phys. Lett., 68 (17), 2372 (1996)
75. D. Gan, P.S. Ho, R. Huang, J. Leu, J. Maiz, and T. Scherban, Isothermal stress relaxation in electroplated Cu-films. I Mass transport measurements, J. Appl. Phys., 97, 103531 (2005) and also I.A. Bloch, J. Appl. Phys. 47, 1203 (1976)
76. J.L. McCrea, K.T. Aust, G. Palumbo, and U. Erb, Mat. Res. Soc. Proc., 581 (2000) and also D. Li, Y. Wu, P. Kim, L. Shi, P. Yang, and A. Majumdar, Appl. Phys. Lett., 83, 2934 (2003)
77. A. Seeger and G. Schottky, Acta Met. 7, 495 (1959) and also K.E. Goodson and Y. Ju, Ann. Rev. Mater. Sci., 29, 261 (1999) and W. Liu and M. Ashegi, IEEE 21st Semicond. Therm. Meas. Management Symp., p. 243, March (2005)
78. M.B. West and C.R. Robenson, Phil. Mag., 19, 887 (1969) and also D.G. Cahil et al., J. Appl. Phys. 93, 793 (2003)
79. N.F. Mott, Adv. Phys. 16, 49 (1967) and also B. Yang and G. Chen, Thermal conductivity, T.M. Tritt (ed.) Pergamon Press, Oxford, (2003) and S. P. Gurrum, PhD thesis, Thermal modeling and characterization of nano-scale metallic interconnects, Georgia Institute of Technology, May (2006)

80. A.K. Jonscher, Nature, 256, 566 (1975), and also L.I. Maissel, Electrical properties of metallic thin films, In L.I. Maissel and R. Glang (eds.) Hand book of thin film technology, McGraw Hill, New York, pp. 11–18 (1983) and Z. Lu, W. Huang, M.Q. Stan, K. Skadron, and J. Lach, IEEE Trans. On VLSI syst., 15 (2), 159 (2007)
81. E. Emin, Adv. Phys., 22, 57 (1973) and also T. Chang and G. Chen, Phonon heat conduction in thin films, J. Heat Transf., 123(1), 340 (2001)
82. Z.A. Weinberg and R.A. Pollak, Appl. Phys. Lett., 27, 254 (1975) and also Annealing disorder, In J.W. Mayer and S.S. Lau (eds.) Electronic materials science, Macmillan Pub. New York., p. 239, (1990)
83. A. Rose, Performance of photoconductors, in Photoconductivity conference, R.G. Breckenridge, B.R. Russel, and E.E. Hahn, Wiley, New York, (1956) and also J.B.D. Soole and H. Schumacher, InGaAs metal semiconductor photodetectors for long wavelength communications, IEEE Quant. Electron., QE-27, 737 (1991) and also M.C.M. Lee and M.C. Wu, J. Microelectron Mech. Syst. 5(2), 338 (2006)
84. J.A. Nucci, R.R. Keller, D.P. Field, and Y.S. Diamond, Grain boundary mis orientation angles and stress induced voiding in oxide passivated copper interconnects, Appl. Phys. Lett., 70 (10), 1242 (1997) and H.J. Lee, H.-N. Han, J.-Y. Sun, and K.-H. Oh, MRS Proc. Symp F, Spring (2006)
85. J.M. Ziman, Principles of the theory of solids, Cambridge University press, Cambridge, London, p. 236 (1964)
86. H. Ehrenreich, Optical properties and electronic structure of metals and alloys, F. Abbles (ed.) North Holland, Amsterdam, (1965) and also H.B. Nie et al., Structural and electrical properties of tantalum nitride thin films fabricated by using radio-frequency magnetron sputtering, Appl. Phys. A, 73, 229 (2001)
87. J.A. Nucci, R. Keller, S. Kramer, C.A. Volkert, and M.E. Gross, Localized measurement of strains in damascene copper interconnects by convergent beam electron diffraction, MRS Symp., 612 D 8.5.1, (2000) and also S-H Rhee, C.E. Murray, and P.B. Besser, MRS Proc. Symp. F, spring (2006)
88. S. Kramer and J. Mayer, In O. Kraft, et al., (eds.) Proc. 5th Intl. Workshop on stress induced phenomena in metallization, 491, 289 (1999) and also M. Huang, Z. Suo, and Q. Ma, J. Mech. Phys. Solids 50, 1079 (2002)
89. Y. Shen and U. Ramamurty, Temperature dependent inelastic response of passivated copper films, J. Vac. Sci. Technol., B-21 (4), 1258 (2003)
90. A. Gangulee, J. Appl. Phys., 45, 867 (1972) and also R. Monig, R.R. Keller, and C.A. Volkert, Thermal fatigue of thin films, Rev. Sci. Instrum. 75 (11), 4997 (2004)
91. M.J. Aus, B. Szpunar, U. Erb, A.M. El-Sherik, G. Palumbo, and K.T. Aust, J. Appl. Phys., 75, 3632, (1994) and also C.U. Kim, J. Park, N. Michael, P. Gillespie, and R.N. Augru, Study of electron scattering mechanism in nano-scale Cu-interconnection, 32 (10), 982 (2003) and J. Bonevich, D. van Heerden, and D. Rossel, J. Mater. Sci. 14, 1977 (1999)
92. A. Zehe, A selection rule for void-resistant crystalline metallic alloys exposed to electromigration, Cryst. Res. Technol., 37(8), 817 (2002)
93. R. Rosenberg, D.C. Edelstein, C.K. Hu, and K.P. Rodbell, Annu. Rev. Mater. Sci., 30, 229 (2000)
94. S.P. Hau-Riege and C.V. Thompson, J. Mater. Res. 15, 1797 (2000)
95. Z. Suo, In W. Gerberich and W. Yang, (eds.) Comprehensive structural integrity, Elsevier, Amsterdam, Vol. 8, p. 265 (2003)
96. S.P. Hua-Riege, J. Appl. Phys., 91, 2014 (2002)
97. C.K. Hu et al., Appl. Phys. Lett., 63, 869 (2003)
98. S.P. Hau-Reige, J. Appl. Phys., 91, 2104 (2002) and also C.S. Hau-Reige, A.P. Marathe, and V. Pham, Proc. 41st Annual IEEE Int. Reliab. Symp., Dalas, p. 173, (2003)
99. J.M. Paik, K.C. Park, and Y.C. Joo, Relationship between grain structures and texture of Cu-damascene liner, J. Electron. Mater., 33 (1) (2004) and also B.C. Valek et al., Combined high voltage and X-ray microdiffraction studies of damascene copper interconnects, LBNL, NIH and DOE funded research, (2000)

100. C. Lingk, M.E. Gross, and W.L. Brown, J. Appl. Phys., 87, 2232 (2000) and also E. Linger, L. Gignac, C.K. Hu, and S. Kaldor, In situ study of void growth kinetics in electroplated Cu-lines, J. Appl. Phys., 92 (4), 1803 (2002)
101. K.N. Tu, Recent advances on electromigration in very large scale integration of interconnects, J. Appl. Phys., 94 (9), 5451 (2003)
102. N. Michael, C. Kim, P. Gillespie, and R. Augur, Electromigration failure in ultra-fine copper interconnects, J. Electron. Mater. 32 (10), 983 (2003)
103. A. Gladkikh, Y. Lereah, E. Glickman, A. Karpovski, A. Plevski, and J. Schubert, Hillock formation during electromigration in Cu and Al three dimensional grain growth, Appl. Phys. Lett., 66 (10), 1214 (1995) and also X. Pang, A.M. Kroman, and G.H. Bernstein, Electromigartion in nanometer Al-Cu interconnect lines, J. Electrochem. Soc., 148(2), G-103 (2002)
104. J. He, Z. Suo, T.N. Marieb, and J.A. Maiz, Electromigration lifetime and critical void volume, Appl. Phys. Lett., 85 (20), 4639 (2004)
105. J.H. Choy, and K.L. Kavanagh, Effects of capillary force on copper/dielectric interfacial void evolution, Appl. Phys. Lett., 84 (25), 5201 (2004)
106. S.K. Ghandhi, VLSI fabrication principle, Si & GaAs, Wiley, New York, p. 455 (1983) and also F. Braud, G. Tartavel, J. Palleau, J. Torres, A. Perisco, and G. Reimbold, Quality and Reliab. Eng. Intl., 12, 305 (1996) and also C. Chen, 9th Int. Workshop on Stress induced Phenomena in metallization, University of Japan, April 2007
107. J. Cho, K. Mirpuri, D.N. Lee, J.K. An, and J. Szpunar, Texture investigation of copper interconnects with different line width, J. Electron Mater. 34 (2), 53 (2004)
108. A.W. Hunt, S.P. Riege, and J.A. Prybyla, Appl. Phys. Lett., 70, 2541 (1997)
109. S.P. Hau-Riege and C.V. Thompson, The effects of the mechanical properties of confinement material on electromigration in metallic interconnects, Mat. Res. Soc. Symp. 612, D10.2.1 (2000)
110. C.K. Hu, R. Rosenberg, H. Rathore, D. Nguyen, and B. Agarwala, Scaling effect on electromigration in on-chip Cu-wiring, Proc. Int. Interconnect Technol., IEEE Conf., p. 267 (1999)
111. A. Gladkikh, M. Karpovski, A. Palevski, and Y. Kaganovski, J. Phys. D, 31. 1626 (1998)
112. O. Kraft and E. Artz, Acta Mater., 45, 1599 (1997) and also, J.M. Paik, K.C. Park, and Y.C. Joo, Relationship between grain structures and texture of damascene Cu-line, J. Electron. Mater., Jan (2004)
113. J. He, Z. Suo, T. N. Marieb, and J. Maiz, Electromigration lifetime and critical void volume, Appl. Phys. Lett., 85 (20), 4539 (2004)
114. C.K. Hu, L. Gignac, S.G. Malhotra, R. Rosenberg, and S. Boettcher, Mechanisms for very long electromigration lifetime in dual damascene Cu-interconnections, Appl. Phys. Lett., 78 (7), 904 (2001)
115. E.T. Ogawa, K.D. Lee, V.A. Blasche, and P.S. Ho, Electromigration reliability issues in dual damascene Cu-interconnections, IEEE Trans. On Relib., 51 (4), 403 (2002)
116. A.C. Damask and G.J. Dienes, Point defects in metals, Gordon Breach, Sci. Pub. New York, (1963) and also V.M. Dwyer, An analysis of the weakest link model for early E-M failure, J. Phys. D, 37, 2935 (2004)
117. V. Vand, Proc. Phys. Soc. Lond. 55, 222 (1943), and also N. Michael, C. Kim, P. Gillespie and R. Odauger, Mechanisms of E-M failure in submicron Cu-interconnect, J. Electron. Mater., 31 (10), 1004 (2002)
118. P.G. Wilkinson, J. App. Phys., 22, 419 (1951) and also P. Kim, L. Shi, P. Pang, and A. Majumdar, Appl. Phys. Lett., 83, 2934 (2003)
119. W. Buckel, R. Hilsch, and Z. Physik., 138, 109 (1954) and also C. Dames and G. Chen, J. Appl. Phys., 95, 682 (2004)
120. W.C. Johnson, P.W. Voorhees, and D.E. Zupon, The effects of elastic stress on the kinetic of Oswald ripening: Two particle problem, Met. Trans. A., (Phys. Met. Mater. Sci.) 20A (7) 1175 (1989) and also A. Raab and G. Springholtz, Oswald ripening and shape transformations of self assembled PbSe quantum dots on PbTe (111) during annealing, Appl. Phys. Lett., 77 (19), 2991 (2000)
121. D.Y. Kim, Ph.D. Thesis, Mat. Sci. Eng., MIT, Dec. (2003)

122. A. Sekiguchi, J. Koike, S. Kamiya, M. Saka, and K. Maruyama, Void formation by thermal stress concentration at twin interfaces in Cu-thin film, Appl. Phys. Letts., 79 (9), 1264 (2001)
123. M.W. Lane, E.G. Liniger, and J.R. Lloyd, J. Appl. Phys., 93, 1417 (2003)
124. L.W. Chu, W.K. Chim, K.L. Pey, J.Y.K. Yeo, and L. Chen, J. Electron. Mater., 30 (12) 1513 (2001)
125. A. Sekiguchi, J. Koike, S. Kamiya, M. Saka, K. Maruyama, Void formation by thermal stress concentration at twin interfaces in Cu-thin films, Appl. Phys. Lett., 79 (9), 1264 (2001) and A. Gouldstone et al., Acta Mater., 55, 4015 (2007)
126. F. Erclessi and J.B. Adams, Europhys. Lett., 26, 583 (1994) and also C.N. Liao, and K.C. Chen, Effect of interfacial resistance and contact size on current crowding at Ni/poly Si junction, Semi. Sci. Technol., 20, 659 (2005)
127. T.O. Odurtani and E.E. Oren, A comparative simulation of the voids dynamics under the action of electromigration and capillary forces in narrow thin interconnects, Adv. Metallization Conf., San Diego, CA, Oct 3–5, (2000)
128. A.P. Sutton and V. Vitek, Phil. Trans. R. Soc. A., 309, 37 (1983) and also T. Wang et al., Stress migration and electromigration improvement for Cu-double damascene interconnections, J. Electrochem. Soc. 152 (1), G-45, (2005)
129. D. Farkas, Fracture mechanism of symmetrical tilt grain boundaries, Phil. Mag. Lett. 80 (4), 229–237 (2000)
130. F. d ' Heurle and R. Rosemberg, Electromigration in thin films, In G. Haas, M. Framcombe, and R. Hoffman (eds.) Physics of thin films, Academic Press, New York, Vol. 7, (1973) and also A. Gladkikh, M. Karpovski, A. Palevski, and Y.S. Kagnovski, Effect of microstructure on electromigration kinetics in Cu-lines, J. Phys. D. Appl. Phys., 31, 1626 (1998)
131. A, Christou, Electromigration and electronic device degradation, Wiley, New York, (1994) and also R. Frankovic, and G.H. Bernstein, IEEE Trans. Electron. Dev., 43, 2233 (1996)
132. E. Jiran and C.V Thompson, Capillary instability in thin films, J. Electron. Mats., 19 (11), 1153 (1990) and also S. Hoffmann et al., Effect of catalyst film thickness on plasma enhanced nanotube growth, J. Appl. Phys., 98, 034308 (2005) and Y.C. Choo and C.V. Thompson, Analytic model for grain structures of near bamboo interconnects, J. Appl. Phys., 76 (11), 7339 (1994)
133. J.R. Lloyd and S. Nakahara, Thin Sold Films, 72, 451 (1988) and also C. Ji, G. Oskam, and P.C. Searson, Electrodeposition of copper on silicon from copper sulfate solution, J. Electrochem. Soc., 148 (1), C746 (2001)
134. R.A. Sigsbee, J. Appl. Phys., 44(6), 2533 (1973) and also A.A. Pasaand and W. Schwarzacher, Phys. Status. Solidi. A, 173, 73 (1999)
135. I.A. Blech and E.S. Meieran, Appl. Phys. Lett., 11, 263 (1967) and also R. Krumin, B. Guel, C. Schmidt, and G. Staikov, Electrochim. Acta, 45, 3255 (2000)
136. I. Blech, H. Sello, and L. V. Gregor, Thin films in integrated circuits, In L.I. Maissel and R. Glang, (eds.) Hand book of thin film technology, McGraw Hill, New York, (1983) and also J.C. Ziegler et al., Surf. Sci., 452, 150 (2000)
137. Y.E. Strausser, B.L. Euzent, R.C. Smith, B.M. Tracy, and K. Wu, 25th annual Proc. Int. Rliab. Symp. IEEE, p. 140, (1987) and also A.A. Volinski et al., Microstructure and mechanical properties of electroplated Cu thin films, Mat. Res. Soc. Symp., 649, Q5.3.1 (2001)
138. J.E. Sanchez, L.T. McKnally, and J.W. Morris, Morphology of electromigration induced damage and failure in alloy thin film conductors, J. Electron. Mats., 19 (11), 1213 (1990)
139. S. Vaidya, T.T. Sheng, and A.K. Sinha, Appl. Phys. Lett., 36, 464 (1980) and also J.M. Paik, K.C. Park, and Y.C. Joo, Relationship between grain structures and texture of damascene Cu-line, J. Electron. Mater., Jan. (2004)
140. J.Y. Cho, K. Mirpuri, D.N. Lee, J.K. An, and J.A. Szpunar, Texture investigation of Cu-interconnects with different line width, J. Electron. Mater., 34 (1), 53 (2005)
141. J.W. Mcpherson and C.F. Dunn, J. Vac. Sci. Technol., B5, 1321 (1987) and also C.K. Hu, K.P. Rodbell, T.D. Sulivan, K.Y. Lee, and D.P. Bouldin, Electromigration and stress induced voiding in fine Al and Al-alloy thin films, IBM J. Res. Dev., 39(4) (1995)

142. R.H. Koch, J.R. Lloyd and J. Cronin, Phys. Rev. Lett., 55, 2487 (1985), and also E. Glickman and M. Nathan, J. Appl. Phys., 80, 3782 (1996)
143. S.R. Shatynski, IEEE J. Solid State Circuits, SC-19, 98 (1984) and also H.P. Kong. Y. Wu, Y.Y. Wang, and C.C. Wan, J. Electron. Mater., 32 (1) (2003)
144. D.S. Campbell, The deposition of thin films by chemical methods, In M.I. Maissel and R. Glang (eds.) Hand book of thin film technology, McGraw Hill, New York., pp. 5-5, 5-6 (1983) and also R. Ryu et al., Microstructure and reliability of copper interconnects, IEEE Trans. On Electron Dev., 46 (6), 1113 (1999)
145. D.Y. Kim, Study of Reliability of VLSI interconnection structures, Ph.D. thesis, Stanford University, Stanford, CA, Dec. (2003)
146. C.S. Hau-Riege and C.V. Thompson, Electromigration in Cu-interconnects with very different grain structures, Appl. Phys. Lett., 78 (22), 3451 (2001)
147. C. Ryu, A.L.S Loke, T. Nogami, and S. Wong, Effect of texture on the electromigration of CVD copper, IEEE Int. Relib. Phys. Symp., Proc. 201–205 (1997)
148. C. Ryu, K.W. Kwon, A.L.S. Loke, V.M. Dubin, R.A. Kavari, G.W. Ray, and S.S. Wong, Electromigration of submicron damascene copper interconnects, 1998 Symp. On VLSI technology, June 8–11 (1998)
149. N.L. Michael, C. Kim, P. Gillespie, and R. Augur, Electromigration failure in ultrafine copper interconnects, J. Electron. Mat., 32 (10) (2003)

Chapter 8
Routing and Reliability

8.1 Routing

8.1.1 Introduction

Increased complexity and functionality in semiconductor devices have placed a greater emphasis on interconnect routing, where *RC* (*R* is resistance and *C* is capacitance) delay has become one of the most significant factors affecting chip performance. Interconnect routing within a block is called local routing, which is short and where *C* effects dominate. In global routing, interconnect lines are long and both *R* and *C* determine chip performance. Efforts to minimize routing delay have largely concentrated on rapid introduction and integration of *new materials* (copper-low-*K*), *new processes* for achieving dimensional control (atomic layer deposition and advanced lithography) (Fig. 8.1 a), change in *design layout* (Fig. 8.1b), and maintaining physical and electrical *reliability* (noise figure, cross-talk, etc.) of the routing structures. These changes must be met while maintaining manufacturability and defect management targets that will meet overall cost and performance requirements.

Fig. 8.1 (**a**) 70 nm NMOS transistor and (**b**) damascene copper interconnect (Photo courtesy Intel)

Tapan K. Gupta, *Copper Interconnect Technology*,
DOI 10.1007/978-1-4419-0076-0_8, © Springer Science+Business Media, LLC 2009

On-chip routing is somewhat immune from electrical problems when the length is much smaller than the wavelength of the signals within the chip. On the other hand, as the operating frequency of the chip increases beyond 1 GHz, long routing on large chips behaves like *distributed* electrical lines. For example, a microprocessor unit (MPU) (Fig. 8.2) can carry almost 1 billion transistors on a chip [1], and a giga-transistor chip can carry more than 10,000 I/Os off the chip, thus making interconnect routing between the pads of the chip require more than 40,000 microvias.

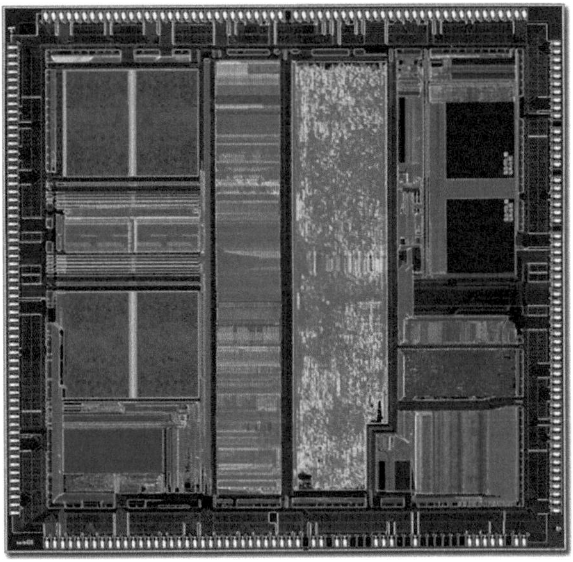

Fig. 8.2 A 486DX4 microprocessor with 1.6 million transistors (Photo courtesy AMD)

In order to accommodate these interconnecting lines and vias within a single chip, the size of the chip has to be increased and the feature size of all the elements within the chip has to be scaled down. The scaling incentive has been fortified by the added bonus of increased *device speeds* resulting from smaller channel lengths. *Device speed* (the latest supercomputer is expected to sustain speeds of 36 teraflops, and will model the folding of human proteins; a teraflop is 1 trillion calculations per second), which was formerly insignificant, is becoming a bottleneck because of its tendency to degrade with scaling [2–4]. Larger chips demand longer interconnecting lines and smaller devices demand closely packed routing (smaller pitch). As a result, resistance and capacitance (*RC*) within a circuit have increased causing more *RC delay* (Fig. 8.3) [5–6].

As we move into the sub-100 nm regime, the *RC* crisis is severe, because the interconnect *RC* delay largely determines the chip performance instead of the intrinsic gate delay (Fig. 8.3). Implications of rising *R* and *C* values in a circuit can go far beyond the time delay, creating problems with noise tolerance, power dissipation, and electromigration [7–8].

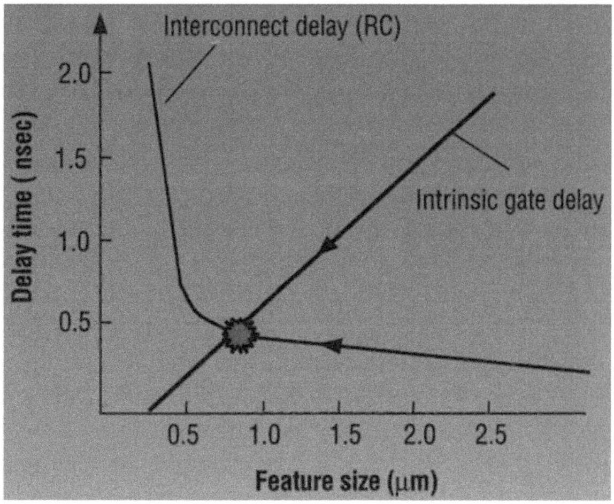

Fig. 8.3 The impact of the feature size on the *RC* delay of the circuit (Reprinted with permission *Solid State Technology*, Oct. 1998, p. 63)

8.1.2 Methods of Improving Interconnect Routings

8.1.2.1 Materials

At the beginning of the 21st century, the technology node has reached 0.1 μm [9–10]. To maintain the same *RC* delay at 0.1 μm design rules with conventional metal aluminum and dielectric silicon dioxide (Al/SiO_2) we may need more than 10 metal levels (*N*). To limit *N* and to keep the metal pitch tight, the *R* and *C* values of the circuit have been reduced by using copper (Cu), which is about 35% lower in resistivity than aluminum (Al), and low-*K* materials having lower *K* than conventional SiO_2. This has benefited in designing low-power Static Random Access Memory (SRAMs) to reduce the bit line voltage swing for reads [11–12] and the data bus voltage swing [13]. It is also expected that copper lines will reduce the power consumption and *RC* delay of long interconnects (Table 8.1).

Table 8.1 Distributed *RC* time constants for three different metals commonly used in IC circuits

Material	Resistance, in ohms	Distributed RC (time constant)	Delay (90%)	Delay (50%)
Aluminum (Al)	245	0.49 ns	0.49	0.19
Copper (Cu)	156	0.31 ns	0.31	0.12
Silver (Ag)	151	0.30 ns	0.30	0.12

In the early 1970s, all interconnections were pure capacitive loads and the size of the driver and length of the interconnections was used to determine the delay time. As the chip grew larger and the minimum feature size shrunk (between 0.18 and 0.1 μm), R_{int} (Fig. 8.6) gained importance, because the parasitic resistance of the interconnecting routes becomes comparable to the channel resistance. In order to minimize the *RC* delays due to R_{int} contributions, two approaches are becoming very much popular. The first one deals with the reduction of interconnecting resistance by using copper, with *multilevel* interconnections, and the second one deals with the use of *repeaters* that divide the interconnections into smaller subsections [14]. It can be seen that delay becomes dominant by R_{int} C_{int} as dimensions are scaled down.

It is also true that the R_{int} C_{int} constraint is not the only factor driving the evolution of the interconnecting systems. However, the introduction of copper and low-K dielectric materials has been observed to improve the RC delay. A six-level copper wiring process in complimentary metal oxide semiconductor (CMOS) logic circuits is a typical example of a hierarchical interconnecting system in which the successive RC constant delay has been reduced by a factor of $(1/\lambda^2)$ [15]. In the year 2000 the CMOS technology was updated with eight levels of Cu-interconnects (with wire bond pitch of 60 μm) together with a low-K (SiLK) interlayer dielectric (ILD) to reduce the resistance and capacitance values of the circuit. The low-K dielectric material (SiLK) is introduced with dual hard mask patterning (Fig. 8.4)

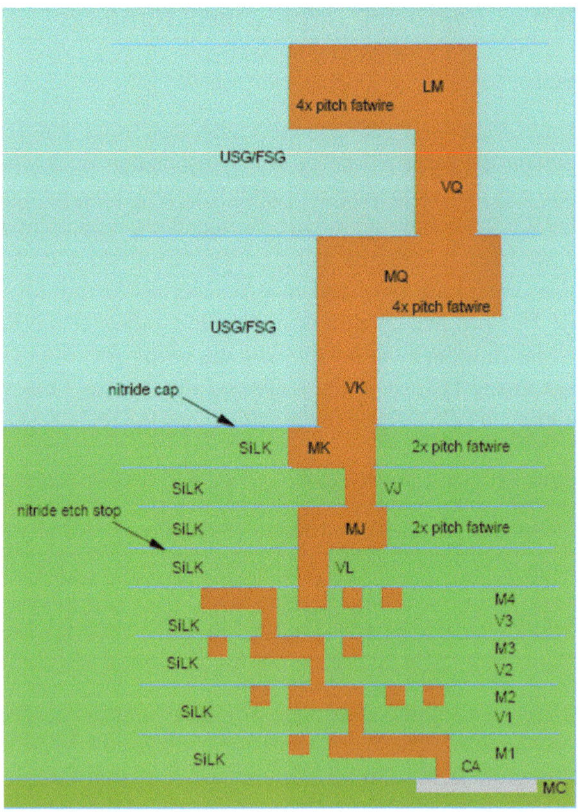

Fig. 8.4 The eight-level Cu-interconnecting lines with low-K dielectric (SiLK) (Courtesy IBM Research, E. Birth, IBM Microelectronics, Hopewell Jn., New York, 9S BEOL stack)

8.1.2.2 Process Integration

Interconnect routing within the chip is achieved by deposition of thin films (metals and dielectrics) by physical vapor deposition (PVD), chemical vapor deposition (CVD), sputtering, atomic layer deposition (ALD) and spinning (in the case of dielectric polymers). According to the requirements, the deposition procedure can vary from time to time. Most of the time these films are polycrystalline and/or amorphous in nature. Resistance or sheet resistivity of a polycrystalline or an amorphous film is very much different from the bulk material.

The effective resistance ($R_{\mathrm{eff}} = (1/R_1)+(1/R_2)$) of the Cu routing ($R_1$) will depend on the resistance of the substrate (R_2). For example, Cu-interconnecting lines are deposited on thin barrier layer (mostly transition metals and their compounds) of higher resistivity. Thus the effective resistivity of the Cu-routing on the barrier layer will be different from the resistivity of the deposited Cu-lines alone (Fig. 8.5) [16–17]. Moreover, if there is an interaction between the substrate and the metal routing (formation of alloy/s) the electrical and mechanical properties of the interface layer will change drastically.

8.1.2.3 Advanced Lithography

There is a growing need for lithography materials and systems that will provide higher resolution. One method of achieving an improvement in resolution is through the use of shorter wavelength radiation with higher energy. As a result the industry has moved from 365 to 257 nm (deep UV) exposure from high-powered excimer lasers (especially KrCl and KrF). For lithography materials, the mid-UV conventional positive resist can be used by changing the photoactive chemical (PAC), but with deep UV, no such easy solution exists. Thus the demand for high resolution depends on proper formulation of the resists, design of the optical system, and the power and wavelength of the radiation source.

8.1.3 Interconnect Routing Design

8.1.3.1 Repeater Design

In order to address the odds that are hindering the progress of sub-100 nm technology, several mitigating solutions have been proposed for the physical design of interconnect routing. In high-speed systems, the distribution of signal and power and containment of noise are the major design issues. On the other hand, in global interconnections long wire length not only degrades the signal propagation delays (*RC*) but also makes it very difficult to design a good power distribution network. A typical global interconnection route generally has a length of ($\sqrt{A_c}/2$), where A_c is the chip area, and the interconnect length in the circuit, l (Fig. 8.6), can be related to the circuit area A as $\Gamma = (\sqrt{\pi}/5)(\sqrt{A}) \cong (\sqrt{A}/3)$ [18–19].

It has been observed that the propagation delay increases when the resistance of the interconnect routing is comparable to or larger than the resistance of the driver

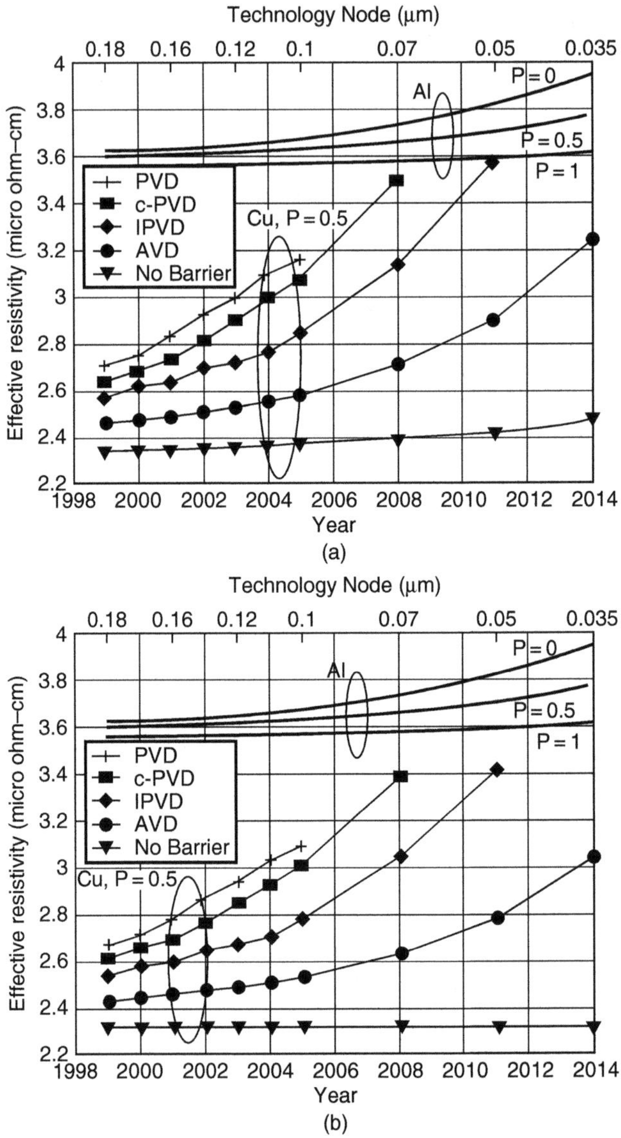

Fig. 8.5 The effect of barrier technology and the scattering factor P on the effective resistivity of copper line. The data of the aluminum lines are also shown for comparison (Reprinted with permission *IEEE Trans. Electron. Dev.* 2002, [16, 17])

and the use of *repeaters* (long deposited interconnections are broken into parts) makes the time delay linear with length [20].

If we want to analyze the circuit of Fig. 8.6, we will find that the repeaters will introduce an input impedance of Z_R, which will give rise to a delay time (τ) as:

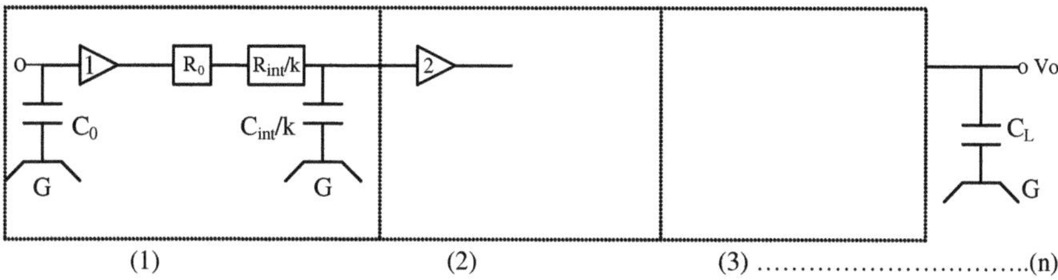

Fig. 8.6 The equivalent circuit of an interconnection route with minimum size (k) and n number of repeaters. G means ground connection

$$n\{(Z_R + ((0.5\,R_{\text{int}})/n)\}\,((C_{\text{int}})/n) + (Z_{R+}(R_{\text{int}}/n))C_L\} \tag{8.1}$$

where n is the number of repeaters and is equal to $\{(0.4\ R_{\text{int}}\ C_{\text{int}})\ /\ 0.7\ R_0C_0)\}^{1/2}$ (to achieve shortest time delay, the delay of the segments connected by the repeaters should be equal to that of a repeater (assuming 50% delay)), C_L is the load capacitance, and R_0 and C_0 are the output resistance and capacitance of the film, respectively. The R_{int} (total interconnect-resistance) of Equation (8.1) can be written as :

$$\rho(l_{\text{int}}/W_{\text{int}}\ H_{\text{int}}) \tag{8.2}$$

and C_{int} (the interconnection capacitance) of Equation (8.1) is

$$\begin{aligned}\{1.5(W_{\text{int}}/t_{\text{ox}}) + 2.8(H_{\text{int}}/t_{\text{ox}})^{0.222} + [0.06(W_{\text{int}}/t_{\text{ox}}) + 1.66(H_{\text{int}}/t_{\text{ox}})\\ - 0.14(H_{\text{int}}/t_{\text{ox}})^{0.222}](t_{\text{ox}}/W_{SP})^{1..34},\ldots\end{aligned} \tag{8.3}$$

where W_{int} is the width , l_{int} the length, and H_{int} the height of interconnects. t_{ox} is the oxide or any insulting layer thickness, and W_{sp} the width of separation between two conducting lines (Fig. 8.7).

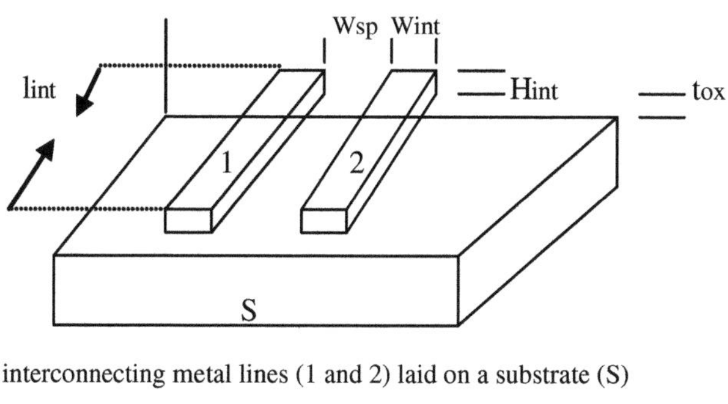

Fig. 8.7 Two interconnecting metal lines (1 and 2) laid on a substrate (S)

8.1.3.2 Bit Line

The use of a repeater as described above is not applicable to bit lines because repeaters are incompatible with memory operation. Therefore, Cu-bit lines are formed for memory circuits. It has been observed that 20% delay in memory circuit and 12% in read and write are reduced when interconnecting lines are changed from Al to Cu. Not only that, the long Cu-interconnect has been broken into shorter pieces to reduce delay. In addition copper lines have been used in low-power operation [20].

8.1.3.3 Scaling and Design

Interconnecting lines, both local and global, are reduced by the same factor as the devices, to reduce the gate delay by $1/s$, where s is the ideal scaling factor [21]. Long interconnect not only increases *IR* drops, but it also increases the noise level of the circuit. Scaling interconnections and insulator thickness by smaller than s, will lower the values of $R_{\text{int}} \times C_{\text{int}}$ by a factor of s^2 ($s \times s$) and the global interconnection length l_{int} by s_c, where s_c is the chip scaling factor. In quasi-ideal scaling of local interconnections, the horizontal dimensions are scaled by $1/s$ to improve the overall packing density by a factor of s. In order to maintain a small *RC* (time constant) value, the vertical dimensions are reduced by $(1/\sqrt{s})$ and thus the delay is decreased by $(1/\sqrt{s})$. This actually degrades the packing density, as the pitch of the wire is not reduced as much as the transistor dimensions. However, in constant dimension scaling, cross-sectional interconnection dimensions are held constant.

In a transistor a straightforward approach called *ideal scaling* is applied, where the dimensions (width, length, height, and the oxide thickness) of the device are reduced in equal amounts. Under ideal scaling, the gate delay decreases by 30% from one generation to the next, the transistor density doubles, and the dynamic power per transistor decreases by 50% (assuming constant electric field scaling where voltage scales down by 30%) [21]. In quasi-scaling of local interconnections, the horizontal dimensions are scaled by $1/s$ to improve the overall power density by a factor of s. Scaling of local interconnections is shown in Table 8.2 . The dimensions of the lines used in Table 8.2 are shown in Fig. 8.7.

Table 8.3 presents the scaling for global interconnections. Ideal global interconnections are always difficult to achieve, because their *RC* constitutes a negligible portion of the total delay. As the chip size becomes bigger, the problem becomes more severe. In constant-dimension scaling of global interconnections, all cross-sectional dimensions are held constant, and the propagation delay rises by ${s_c}^2$ and in constant delay scaling the total delay remains constant as a result of the increased interconnection dimensions.

As *RC* delay is an important parameter in the recent sub-100 nm level devices, innovative methodologies and architectures eventually account for signal delay at all levels of the wiring hierarchy and at all stages of the design process. Thus the important and inescapable challenge for future interconnection technology is to continue

Table 8.2 Scaling of local interconnections

Parameters	Ideal scaling	Quasi-ideal scaling	Constant-*R* scaling	Generalized scaling
Thickness H_{int}	$1/s$	$1/\sqrt{s}$	$1/\sqrt{s}$	$1/S_H$
Width (W_{int})	$1/s$	$1/s$	$1/\sqrt{s}$	$1/S_w$
Separation (W_{sp})	$1/s$	$1/s$	$1/\sqrt{s}$	$1/S_{sp}$
Insulator thickness (t_{ox})	$1/s$	$1/\sqrt{s}$	$1/\sqrt{s}$	$1/S_{ox}$
Length (l_{loc})	$1/s$	$1/s$	$\cong 1/s$	$1/s$
Resistance (R_{int})	S	$\sqrt{s}$	1	$s_w s_H/s$
Capacitance to substrate	$1/s$	$1/s^{3/2}$	$\cong 1/s$	$s_{ox}/s\ s_w$
Capacitance between lines	$1/s$	$1/\sqrt{s}$	$\cong 1/s$	$s_{sp}/s\ s_H$
RC delay (t)	1	$1/\sqrt{s}$	$\cong 1/s$	$s_W\ s_H/s^2$
Voltage drop (IR)	1	$1/\sqrt{s}$	$1/s$	$s_w\ s_H/s^2$
Current density (J)	s	$\sqrt{s}$	1	$s_w\ s_H/s$

*Reprinted with permission: Addison Wesley, Reading, MA (H.B. Bakoglu, *Circuits, and Interconnections and Packaging for VLSI*). s_W and s_H are sacling factor for width and height.

Table 8.3 Scaling global interconnections

Parameters	Ideal scaling	Quasi-ideal scaling	Constant scaling	Generalized scaling
Thickness H_{int}	$1/s$	1	s_c	$1/s_H$
Width (W_{int})	$1/s$	1	s_c	$1/s_w$
Separation (W_{sp})	$1/s$	1	s_c	$1/S_{sp}$
Insulator thickness (t_{ox})	$1/s$	1	s_c	$1/S_{ox}$
Length (l_{loc})	s_c	s_c	s_c	s_c
Resistance (R_{int})	$s^2\ s_c{}^2$	s_c	$1/\ s_c$	$s_w S_H/\ s_c$
Capacitance to substrate	s_c	s_c	s_c	$\cong s_c$
RC delay (t)	$s^2\ s_c{}^2$	$s_c{}^2$	1	$s_W\ s_H/s^2$

Reprinted with permission Addison Wesley, Reading, MA; s_w, S_H, and s_c denotes scaling factor for width, height, chip size.

to articulate the routing hierarchy, with an ever-increasing disparity between the minimum pitches of the first and last metal levels [22].

The scaling down of the dimensions when applied to three-dimensional (isochronal delay surfaces) surfaces gives rise to a constant delay of 700 ps, which is equivalent to a 10 cm line with a dielectric having K value ~3.5 in the time of flight regime. As the lines get narrower, the location and the nature of the steps change because of the increasing line attenuation. When the source resistance $R_s >> Z_0$, the isochronal $R_s\ C$ charging curve shows a gradual fading nature due to multiple reflections. The source resistance affects the line performance by voltage division on signal launch and by its reflection coefficient. The length axis scale depends on the dielectric constant K and delay time t, as $(t/\sqrt{K})$, and the line attenuation is a function of line resistance and its characteristic impedance. Experimental evidence [23] shows that for $R_s = 10\ \Omega$ the minimum line widths for achieving time of flight performance are 5.3 and 1.9 μm with corresponding maximum line densities of 630 and 1750 per centimeter at 300 and 77 k, respectively. Experimental evidence also shows that when the interconnecting lines of ETA-10 and NEC SX2 supercomputers are changed from normal copper linings to superconducting linings there are no improvements in the speed of the machines. However, when the same experiment is conducted with fewer lines of superconducting lines [24–25], the interconnections yield higher signal amplitude (but give rise to undamped reflections and signal ringing).

8.1.3.4 Manhattan Design

Today, sophisticated computing resources and innovative development efforts have given rise to a fundamentally different approach to interconnect routing design. For the past 25 years, designers have laid out the interconnect routing primarily based on the *Manhattan* architecture. The *Manhattan* architecture is built on the basis of right angle layout, which requires shorter interconnect routing. When the interconnect routing consists of too many lines (e.g. in MPU) in *multilevel* interconnections metal deposition following the *Manhattan* architecture fails. As a result, design engineers had to come up with a different architecture for laying out the circuit, which is very popular and is known as the X-architecture. Figure 8.8 shows two architectures, namely the Manhatten and the X.

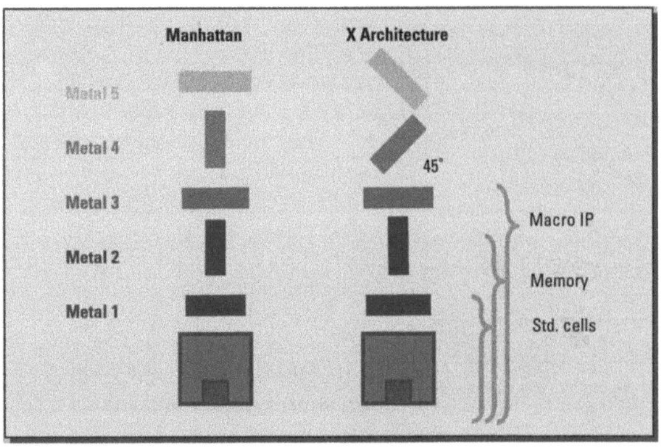

Fig. 8.8 Two architectures, namely the Manhattan and X

8.1.3.5 X-Architecture

It is true that the physical design procedure for a multilayer wiring with via holes shortens the length of the routing, but not without any penalty. Thus in complex multilayer design, the direction of the interconnections is rotated by 45° to reduce the metal line lengths. This multilayer architecture is reported to reduce power consumption within a circuit by 20%, and enhance chip performance by 10%. Figure 8.9 shows an X-architecture applied to a chip with five metal layers. The architecture is based on Simplex *liquid routing* technology.

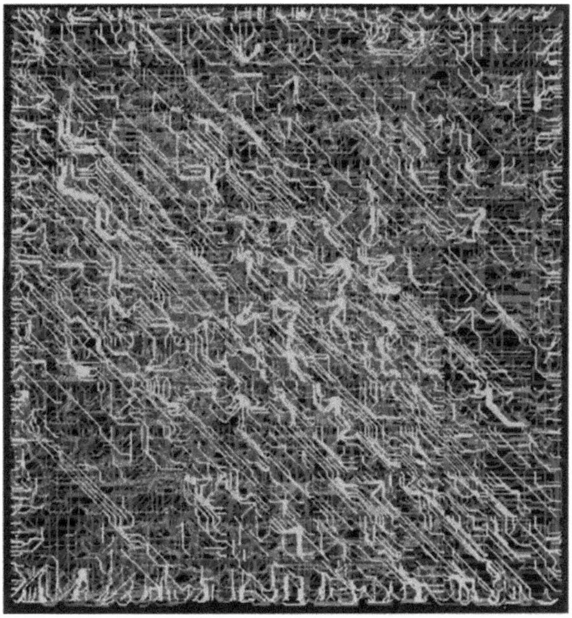

Fig. 8.9 X-architecture based on Simplex's liquid routing (Courtesy of Dr. K. Rygler, Rygler and Associates, Austin, TX)

In the Manhattan configuration, chip performance increases along the area, however, if the design is implemented using X, the result will be a chip that is of equal area but higher performance, or of equal performance but smaller area. Figure 8.10 shows a comparative analysis of the performances between the X versus Manhattan architecture.

It appears that the scalable X-architecture will be the mainstream design with a minimum of economic pain with 20% less interconnect and 30% less vias. However,

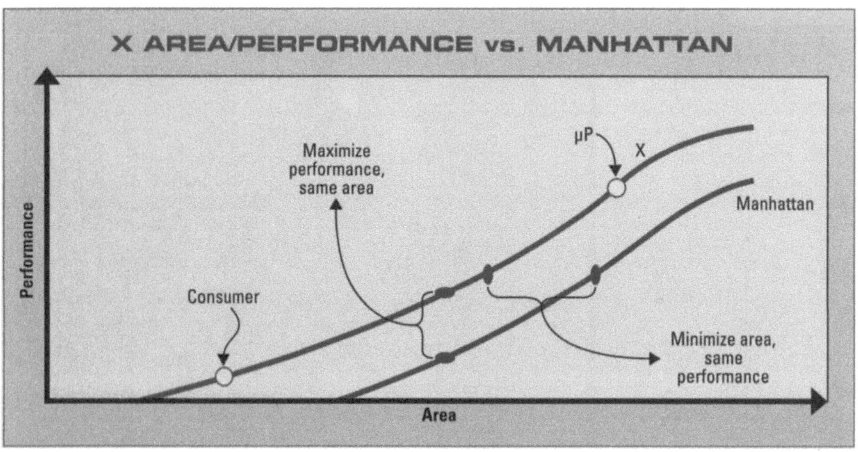

Fig. 8.10 Comparative analysis between X area performance versus Manhattan (Reprinted with permission, Semiconductor International April 2005, p. 44, Courtesy, Cadence Design Systems)

the main issue will be how to implement the architecture efficiently with CAD tools and make it as effortless for design as it is for manufacturing.

8.1.3.6 Multilevel Design

Multilevel interconnections with wider and thicker lines at the upper levels are suggested as one of the viable technologies. The scaled down first level metal is used as local interconnections. The global interconnections with wider and thicker lines, which yield shorter propagation delays, use upper level wiring in 3-D technology. Here, the low interconnectivity theme is given priority, with single package unit bonded with stacked die wire. However, packages with several interconnect layers that connect one or more embedded dies along the layer's edge are also very effective.

For some time interconnects through via tungsten (W) plugs between stacked active layers have been gaining popularity in some of the European trade markets. But the procedure is not very cost effective and ultra-thin-chip-stacking (UTCS) with three complex field effect transistors (FET) has been designed as an alternative. The ASIC die size has three stacks each measuring $15 \times 15\ \text{mm}^2$ with ~600 I/O and 100 μm pitch. It has an interlayer dielectric (ILD) 60–80 μm thick compatible with thermo-mechanical properties [26–27] and is expected to be more viable and cost effective particularly when a *high I/O* count die with small pitches is required [28].

Figure 8.11 shows a schematic of a transmission line distribution similar to a multichip module interconnect structure. The arrangement has sufficient headroom for chips to move into multi-GHz speed [29]. For ultra *high density* dynamic read access memory (DRAM) (Fig. 8.12a) and the typical logic multilayer metallization

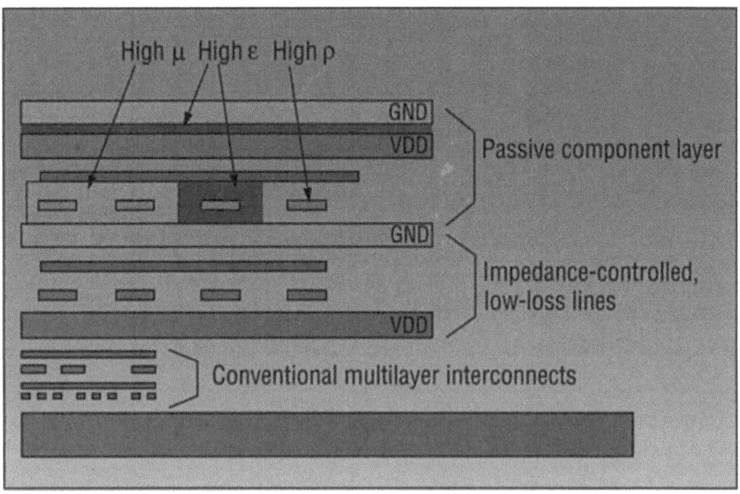

Fig. 8.11 Conception of a conventional mutilayer interconnection (Reprinted with permission, *Solid State Technol.* Sept. 1998, p. 88)

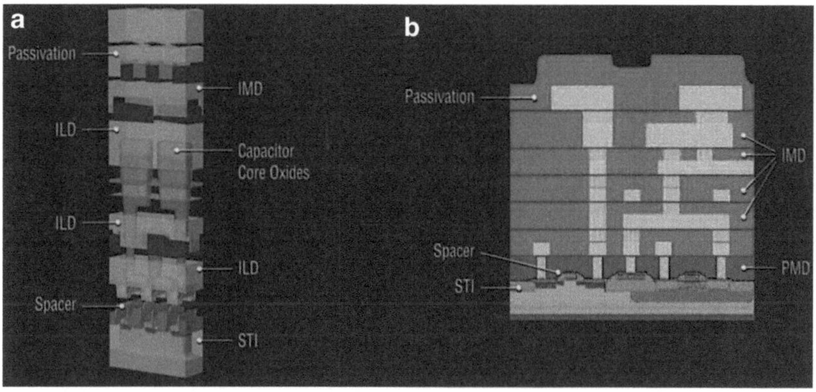

Fig. 8.12 (**a**) Ultra High Density DRAM and (**b**) Typical Logic Multilayer Metallization Structure (courtesy Wattkins-Johnson)

structure (Fig. 8.12b) with copper lines have been common practice in modern integrated circuits but not without penalties like noise, cross-talk, and related phenomena.

8.1.4 Challenges with High Density Routing

8.1.4.1 Noise

As the feature size decreases, the circuit element density and the *I/O* counts increase. As a result the self-inductance of the distributed lines from the power supply generates large amounts of chip current changes and introduces noise within the chip [30]. Different methods have been used to minimize the noise such as a decoupling capacitor, *multiple power* and *ground pins* and *tailor driver turn on* characteristics [31]. Beside transient noise, one has to take into account the resonance effect at the chip and board level. In high-speed systems, the distribution of signal and power and the power supply-level fluctuations play an important role, especially in faster CMOS, when the speed and I/O count become comparable to a bipolar transistor.

8.1.4.2 Cross-Talk

Deep sub-100 nm level feature size brings the power lines and the associated routing closer which results in *cross-talk* due to mutual capacitance and inductance between neighboring routes. The longer and closer the lines, the higher will be the cross-talk. Cross-talk is a noise-related phenomenon, which is associated with power distribution (inductive voltage fluctuations along the power lines). Noise generated by off-chip drivers and on-chip circuitry is a major discipline in package and integrated circuit (IC) design for high-speed systems. The cross-talk between package and chip interconnects can be modeled as coupled *microstrip* lines [32–33].

The number of modes of propagation through routing lines should have a different characteristic impedance and phase velocity. Any perturbation traveling through a coupled transmission line pair can be expressed as a superposition of even and odd modes, if the lines are symmetric [34]. As the number of metal interconnects increases inside a Chip, the spacing between metal lines decreases to accommodate more lines per unit area. Thus the substrate carrying the package with a close-spaced conductor will be a source of *cross-talk* [35] and will interfere with the integrated circuit's performance.

Recently, to minimize cross-talk, a meshed power system has been developed which consists of metal layers running parallel to each other along the *X* and *Y* directions. Via holes connecting ground layers are placed at each crossing point (Fig. 8.13). Using the power/ground mesh as a starting point, the power and the ground conductors are narrowed down in numbers. The design will shield each signal trace on all sides by power and ground traces. Thus it is expected that the signal traces routed through the mesh will have less cross-talk. Figure 8.14 shows the spectra of power distributed noise from an interconnected mesh power (IMPS) system.

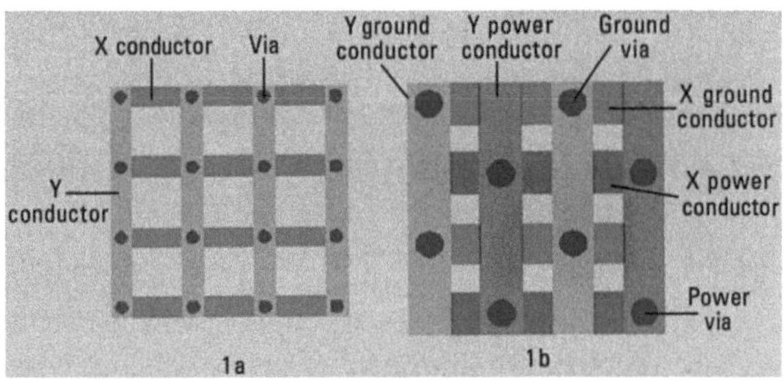

Fig. 8.13 Interconnected mesh power system (Photo courtesy, University of Arkansas and Arkansas (Fayetteville) and from Kyocera corporation (Kyoto, Japan)

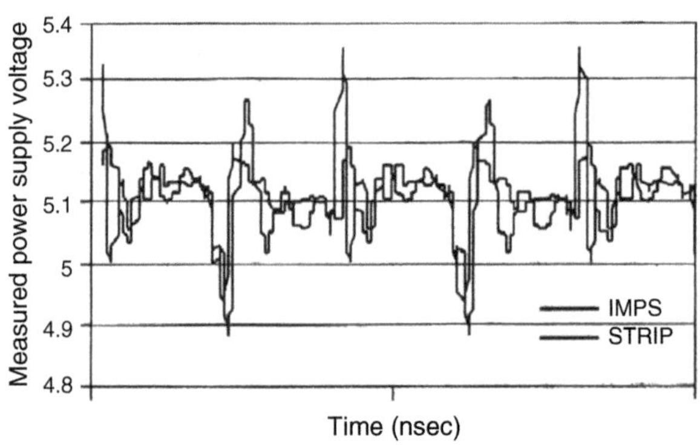

Fig. 8.14 Power distribution noise between strip line and IMPS topologies (Photo courtesy, University of Arkansas, (Fayetteville) and from Kyocera corporation (Kyoto, Japan)

8.1.5 Cascaded Driver

Just as repeaters are suitable for *RC* loads, cascaded drivers on the other hand are suitable for capacitive load, where, instead of a single minimum size driver, a chain of drivers is be used [36–39]. The method optimizes the delay of the interconnecting lines by charging the input capacitance of the cascaded drivers. Cascade drivers are preferred to repeaters when the global interconnection length (l_{int}) is short. It has been reported that optimal size repeaters with cascaded first stage obtain the shortest delay under all conditions.

8.1.6 Transmission Line Coupling

As the circuit density increases, interconnecting lines are laid very close to each other to minimize the space [40]. Two such interconnecting wires can support two modes of propagation with separate impedances and propagation speeds. If the lines are symmetric, the two modes are even and odd. In the even mode, both lines are positive, with respect to ground, while in other mode both lines are negative with respect to ground. Any signal traveling in the coupled transmission line system can be expressed as a superimposition of even and odd modes. The even mode impedance can be calculated but the odd mode impedance is harder to conceptualize. The even and odd mode parameters are related to line capacitance and inductance [41].

The total area occupied by transmission wires on a chip is dependent on the number of transistors. Rent's rule specifies that the number of wires that crosses the boundary of a block can be related to the number of transistors or nodes within the block, and the number of wires, connecting each transistor to other transistor within the block. It has been found that a memory chip requires fewer interconnection levels than a logic chip of roughly the same size [42].

The propagation velocity v through the transmission line will be affected due to the increase in the circuit density, which can be modeled mathematically as: $(v/c) = ((1/(\in_r \epsilon_r)^{1/2}) = ((1/(LC)^{1/2})$, where v is the propagation velocity through the transmission line, c the propagation velocity of the electromagnetic wave, L the inductance, C the capacitance per unit lengths, and $\in_r$ and ϵ_r the relative permeability and permittivity. It has been found that the low characteristic impedance in the transmission line requires a greater current drive capability from the clock driver to achieve transmission line speed.

8.1.7 Clocking of High-Speed System

In slow VLSI circuits the clock frequency is relatively low and minimum width lines are employed for clock distribution. Under these conditions clock skew is dominated by the difference in the *RC* time constants of the interconnections. Thus no matter how strong the clock driver is made, high clock frequencies will not be possible if the line has a large *RC* time constant. The time constant *RC* per unit length can only be determined by the interconnection and insulator thickness. At the same time it is not possible to improve the *RC* delay with wider interconnections by neglecting fringing fields. In other words, to enhance the *RC* performance the vertical dimensions of the interconnects are to be adjusted [43–45].

8.2 Reliability

8.2.1 Introduction

The reliability of Cu-interconnects poses a serious concern, especially as the feature size shrinks to 90 nm and smaller. As a matter of fact, beyond 0.18 μm feature size, interconnect delay contributes more to the overall device delay than the gate delay, forcing the industry to move from the well-known sputter-deposited and plasma etched Al-alloys to the new Cu-damascene procedure accompanied by chemical mechanical planarization/polishing (CMP).

Interconnections are now and will continue to be critical limitations to improving the performance and costs of integrated circuits as technology is moving towards smaller (sub-100 nm) feature size. However, there is no doubt that the fundamental change from aluminum (Al) to copper (Cu) as interconnecting material has increased the circuit *speed* and *performance*.

Figure 8.15 shows the superior behavior of a Cu-low-*K* system over a Al-SiO_2 system in global interconnecting lines. Copper-low-*K* has the advantage over Al-SiO_2 when the feature size is within 75 nm. However, beyond 75 nm feature

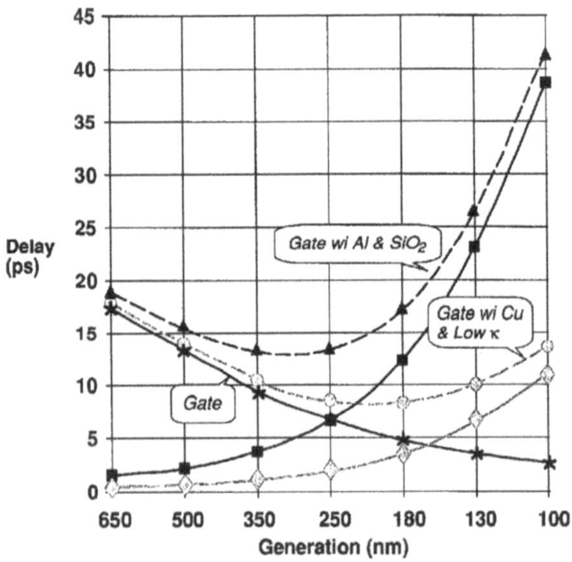

Fig. 8.15 Global interconnect delay improvement with Cu-low-*K* system (Reprinted with permission from IEEE, Ref. Mark Bhor IEDM Proc. 1995, IEEE)

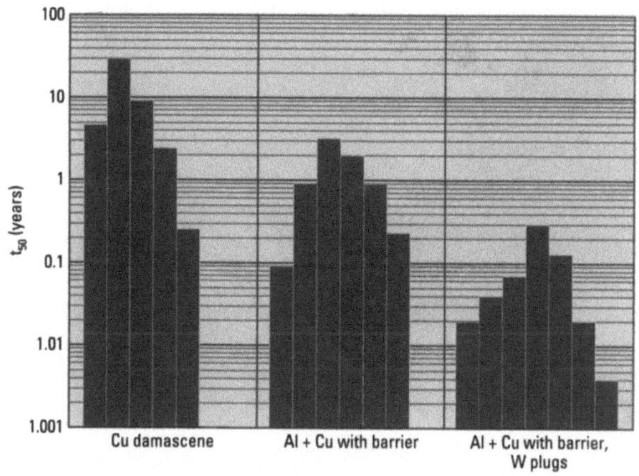

Fig. 8.16 The superior electromigration nature of copper (Reprinted with permission Semiconductor International April 2000, p. 98)

size the bulk properties of Cu change and there are some reliability issues that concern the use of copper (Cu) beyond 75 nm node technology.

Figure 8.16 shows the superior electromigartion resistance property of copper deposited by electroplated, and sputtered damascene etched copper, and doped etched copper with that of aluminum alloys with or without tungsten plug. In all cases the data are converted to conditions of temperature and current density of 125°C and 5×10^6 A/cm^2, respectively, using Black's equation [46] where the median time to failure is t_{50}.

Both the *speed* and the packing density of the chip steadily increase with greater integration of functionality. Downward scaling of physical feature size can perform this by a scale factor s, increasing the number of wiring levels and enlarging the chip size by a scale factor S_f. The smaller metal cross-sections and reduced wire spacing increased the R and C of the interconnect structures dramatically. In the sub-100 nm level, the interconnect crises (*RC* delay, scattering, surface wetting, stress, electromigration, etc.) are severe [47].

As the size and the internal structure changes, the resistivity (ρ) of a metal line no longer depends upon the size but is also affected by the surface and interface scattering, sidewall scattering (for Cu-damascene process) and grain boundary scattering (Fig. 8.17). As a matter of fact, grain boundary scattering is the second contribution to size effects [48].

Figure 8.18 shows the de-wetting problem with Cu-film upon annealing, when the feature size is comparable to the mean free path (~40 nm) of the electrons [49].

With smaller feature size (~100 nm), *the metal pitch* (width and spacing) decreases, and ultimately affects circuit performance. In order to maintain the same *RC* effects with an aluminum metal system, and with 100 nm design rules, one has to increase the number of metal level (n) [50], which will affect the cost of the circuit. In order to cope with number of levels of metallization, and to keep the metal pitch tight, one has to reduce R and C through the use of new materials. From Fig. 8.19,

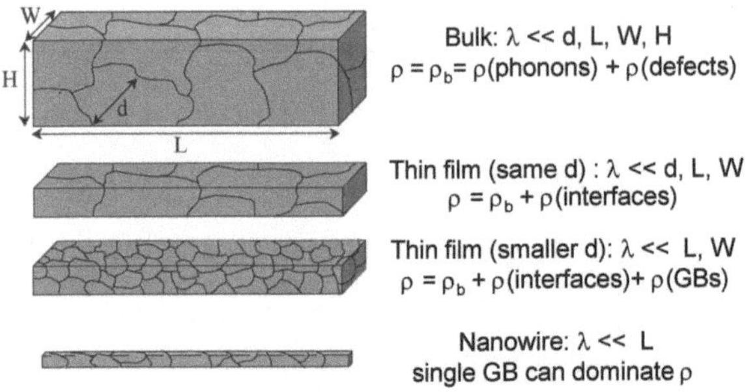

Fig. 8.17 The effect of down-sizing of the feature size and the importance of internal structure (Reprinted with permission, Drs. K.A. Dun and A.E. Kaloyeros, TRC Oct. 25–27, 2004)

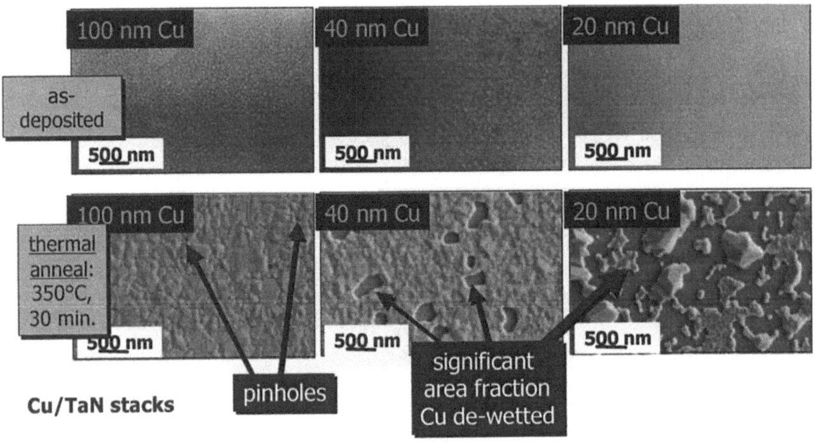

Fig. 8.18 The de-wetting problem of Cu-film when the feature size is very small (Reprinted with permission, Drs. K.A. Dun and A.E. Kaloyeros, TRC Oct. 25–27, 2004)

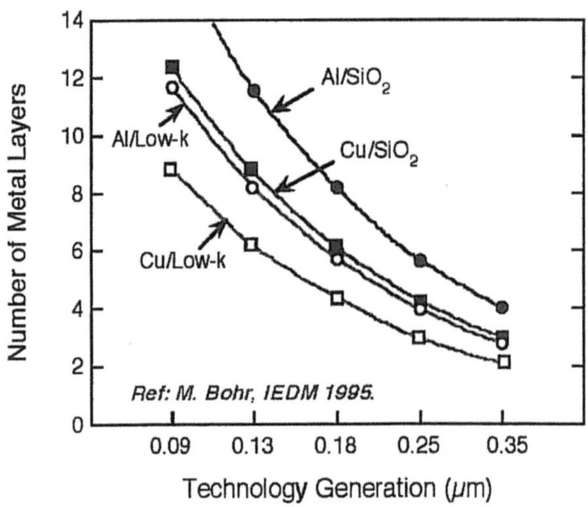

Fig. 8.19 Influence of materials on the number of required interconnect levels (Reprinted with permission from IEEE, IEDM Proc. 1995)

it is clear that replacing Al lines by Cu lines and dielectric (*K*) SiO_2 by low-*K* material one can reduce the number of metal layers and thereby reduce the cross-talk and *RC* delay [51–55].

8.2.2 Reliability Issues Related to Cu-Interconnects

Reliability is defined as the ability of a device or a material to conform to its electrical, mechanical, and visual specifications over a specified period of time under special conditions at a specified confidence level. Breakthrough in integrated circuit (IC) technology for smaller, and faster devices, and periodic improvements in circuit density and performance following scaling, has made aluminum (Al) interconnects obsolete, because it has reached the technological and physical limits. Copper (Cu) is becoming the metallization of choice for high performance integrated circuits (ICs), which offers 30% reduction in resistivity over Al, and more resistance to electromigration (EM). It is expected that Cu based chips are enabled to offer thinner and narrower interconnects without compromising yield and circuit performance.

At the same time, the replacement of silicon dioxide (SiO_2) by low-*K* material has further improved circuit speed and performance of a microprocessor. But in reality, as the feature size has scale down to 75 nm and below, replacement of Cu as interconnect has raised several questions related to reliability like *oxidation, adhesion, surface and interface scattering, Joule heating, stress, electromigration, diffusion,* and *void formation.*

8.2.2.1 Deposition Process Related Issues

Copper (Cu) is deposited by an electrochemical method over a thin layer of Cu (seed layer) serving as nucleation center. ECD-Cu is metallurgically a metastable film, which crystallizes even at room temperature [56] due to self-annealing and involves grain growth (from 0.1 to 1 μm) and change in textural behavior. It has been observed that when the initial grain sizes are small, there is substantial stress development within the film, which causes nucleation sites for stress voiding [57].

The crystallites of the Cu-film are observed to have a preferential growth along the <111> direction [58] with a drop in sheet resistivity about 30% after 180 h [59–60] which has also been noticed when a small amount of metallic impurity is added.

In order to maintain an *effective resistivity* of the copper interconnects well within 2.0 Ω-cm, thinner and conformal barrier and seed layers are needed. Considering the limitations of conventional PVD or ionized PVD, alternatives such as CVD, electroless deposition, and electochemical deposition (ECD) of barrier layers are considered [61–63]. On the other hand, a thin seed layer will increase film resistivity, which will ultimately change different process parameters during growth of the EP-Cu film. For example, a non-uniform current distribution from the edge to the center of the substrate (seed layer) causes non-uniform deposition and potential distribution in the electrolyte, i-E behavior at the metal-electrolyte interface, charge and mass transfer in the electrolyte and over all dynamics of the profile evolution [64].

The surface topography control inside a high aspect ratio via hole by the electroplating process is a key issue, since additives are required to ensure proper filling. But the EP-Cu filling often results in so-called *super-filling*, where the thickness of the overburden Cu is higher on the higher density patterned areas. Besides proper filling, grain growth and voids inside via holes and trenches continue to remain an issue due to aging of the electroplating bath and changes in the concentration of the additives. According to some observers, true bamboo structures in Cu are difficult to achieve by conventional deposition processes even after annealing of very narrow lines [65–66]. Experimental observations show that lines with bamboo structure, where the grains span the full width of the line, have increased EM lifetimes because of reduction of the fast atomic diffusion path [66].

8.2.2.2 Seed-Layer Related Reliability Issues

The resistivity of the Cu-seed layer will be influenced by electron scattering via the interface/surface layer as it approaches the electron mean free path (~40 nm). The electroplated copper film shows <111> growth on a PVD grown <111> textured seed layer. But deviation of the <111> orientation of the grains is noticed, when Cu is doped with tin (Sn) to improve EM resistance [67–69]. Moreover, addition of Sn shows reduced bond strength between the interface of the Cu and Ta-based barrier layer [70–71]. Figure 8.20 show the electrical resistivity of copper and copper doped alloys.

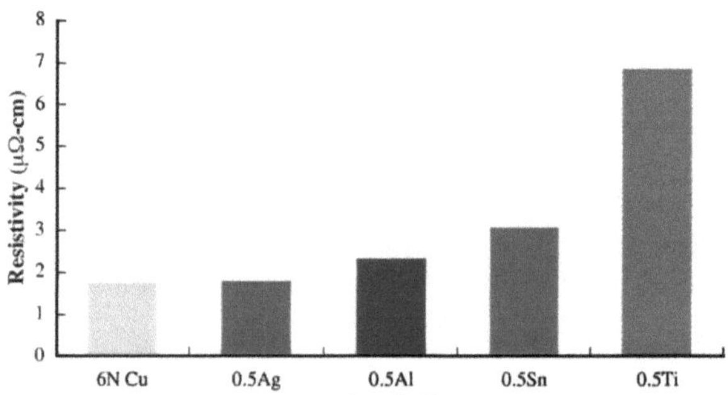

Fig. 8.20 Electrical resistivity of copper and copper doped alloys (Reproduced with permission Honeywell Electronic materials, V-EMT 1:16 (12 July, 2004)

The texture of an EP-Cu film on a vacuum-deposited Cu-seed layer is very different from a vacuum deposited Al-film. The grain structure in Al films is columnar and is identical to each other, whereas in Cu films grains can vary from bamboo-structure to polygranular clusters [72]. According to some observers, random seed structure on the trench wall is favorable to the formation of bamboo structure. The grain boundary (GB) planes in bamboo microstructures are normal to the direction of current flow and EM occurs via relatively slower surface diffusion, resulting in increased EM lifetime [73]. Bamboo structure can also be introduced in EP-Cu film by laser annealing.

Figure 8.21 shows the resistivity of copper film deposited by different methods. ALD deposited films are claimed to produce more conformal and void free thin films inside high aspect ratio (AR) via holes than CVD or PVD grown films. However, the complexity of controlling the ALD process due to a variable incubation time in the reactor can lead to fluctuations in film thickness. Both CVD and ALD are low-temperature processes that use precursors, and the selection of precursors is very critical. The PVD system, on the other hand, generally requires higher temperature or energetic ions for film deposition that might be detrimental to device fabrication.

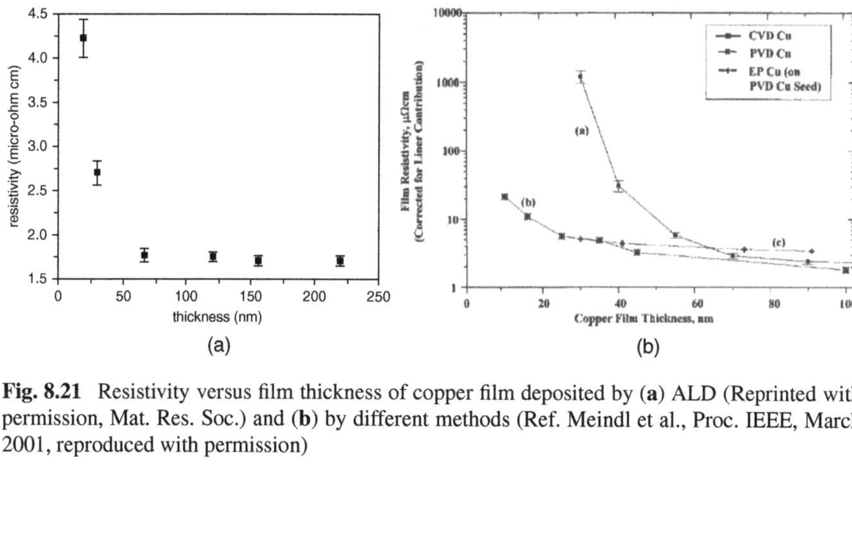

(a) (b)

Fig. 8.21 Resistivity versus film thickness of copper film deposited by (**a**) ALD (Reprinted with permission, Mat. Res. Soc.) and (**b**) by different methods (Ref. Meindl et al., Proc. IEEE, March 2001, reproduced with permission)

8.2.2.3 Oxidation Related Problems

Aluminum (Al) and copper (Cu) behave differently in the presence of oxygen. Al forms a protective layer, which prevents further oxidation of the metal, whereas oxide on Cu is a form of corrosion. It decreases the cross-sectional area and increases the total resistance of the metal line. However, deposition of a thin layer of Al over Cu contact pads has been used as a strategy to protect Cu-pads from oxidation [74–75].

During chemical mechanical polishing (CMP), the porous low-*K* dielectric materials can retain moisture and can corrode the metal (Cu). The acceleration factor

(A_F), for *corrosion* in Cu due to chemical or electrolytic reactions in the presence of moisture, and bias (V), can be modeled as:

$$A_F = (V)[R_{H\,test}/R_{H\,use})^3 \exp.(0.9/k)(1/T_{use} - 1/T_{test})], \tag{8.4}$$

where $R_{H\,test}$ and $R_{H\,use}$ are the relative humidity during test use and during use, T_{use} and T_{test} are the temperatures during use and test, respectively, and k is Boltzmann's constant. However, several procedures including deposition of a cap layer to seal the pores of the porous low-K materials may effectively solve the reliability problem.

8.2.2.4 Adhesion Problem

Adhesion is known to be one of the most important factors in microelectronics interconnects because device failures are often associated with stress and electromigration (EM) that are enhanced at poorly bonded interfaces. Therefore, there is no doubt that an understanding of the *adhesion strength* at a Cu/barrier interface and Cu-layer at different metal levels is necessary to develop a robust chemical mechanical polishing (CMP) process to avoid peeling and delamination. In the copper (Cu) damascene process (CDP) Cu-lines are encapsulated on the sides and bottom by barrier layer and on top by a barrier/etch stop layer. The process is completely different from conventional aluminum (Al) metallization. As a result, lack of adhesion between the Cu-dielectric interface is susceptible to Cu-diffusion at the top surface. The strongest adhesion strength between copper and the underlying dielectric layer is seen to exist when the seed layer is 500 Å thick and sitting on a tantalum (Ta) barrier. However, the bond between the seed layer and Ta is seen to deteriorate if the seed layer suffers from stress. Various surface treatments prior to dielectric deposition provide some near-term relief but no permanent solution. However, a cobalt tungsten phosphide (CoWP) cap with typical thickness between 100 and 150 nm has been shown to promote adhesion [76].

The adhesion property of Cu-lines has been estimated on the basis of the lattice misfit concept. It has been argued on the basis of the experimental results that hcp elements, especially ruthenium (Ru), as substrate material promotes the adhesion of fcc copper, better than bcc elements serving as substrate. According to one group of researchers, using a PVD capping layer of a refractory metal can minimize poor adhesion between the ALD TaN barrier and the PVD Cu. However, a composed layer of ALD TaN followed by ALD TiN has been found to eliminate Cu agglomeration and has promoted the adhesion property of copper.

Figure 8.22 shows the adhesion energy versus EM activation energy of different materials that are considered as potential candidates for the cap layer. From the graph it is clear that CoWP cap material supersedes both SiC and nitrogeneted Si/SiC materials. However, experimental evidence shows that an isotropically grown monolayer of CoWP cap layer increases the line to line leakage which has been interpreted due to the thickness dependent stress migration. So it is argued that low stress dielectric stacks will enhance adhesion strength and interfacial bonding between Cu-dielectric layers.

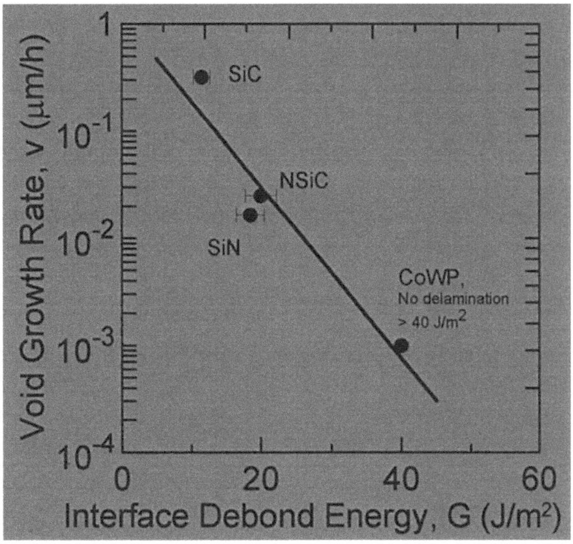

Fig. 8.22 The superior adhesion energy of CoWP compared to the other cap layers (Reprinted with permission IBM Research)

8.2.2.5 Scattering Related Reliability Issues

Scattering from a Cu-line is becoming a reliability issue as the dimensions of the lines are reduced. Indeed, as the lateral dimension of Cu-lines goes beyond 60 nm, the resistivity of copper changes drastically (roughly a factor of 2 higher than bulk resistivity value), due to scattering from the *sidewall* and *grain boundary* [77]. As a matter of fact, grain boundary scattering will make a significant change to the resistivity if the grain size of the Cu-lines is comparable to or smaller than the electron mean free path (~40 nm). As-deposited copper has a grain size of 10 nm which shows 20% higher resistivity than bulk copper. However, after post-anneal the grain size increases and the grain boundary scattering is reduced. But the grains within the trench do not grow with temperature due to the constraint of the sidewall. Thus the effective resistance of the Cu-line inside the trench will be different from the Cu-line on the blanket.

The other form of scattering that has been observed with small Cu-lines is interface scattering. In the presence of a diffusive reflection boundary, energy dissipation caused by scattering from a surface is characterized by a parameter p. On the basis of the free electron model the p value for surface scattering of conduction electrons is deduced from the diameter dependent resistivity. The value of p can vary from 0.18 to 0.66 depending on the surface condition of the film. As a result, surface scattering will contribute to the resistance as:

$$[\rho/\rho o][1 + \{k(1-p)\lambda\}/t] \quad (8.5)$$

where ρ is the resistivity of the film, ρ_0 the resistivity of the bulk material, k a constant, l the length, p the specularity parameter, and t is the thickness of the film. Thus we can see that the value of p appear to arise from differences in microscopically observed conditions. As a result, when the feature size of the Cu-lines reaches 45 nm and beyond the size effect becomes dominant because of electron–phonon scattering and presents a challenge for continuous device scaling. As a matter of fact, at high frequencies the mean free path λ and the Fermi velocity of the charge carriers in traveling distances between two scattering events on grain boundaries appear as the main factors effecting the resistance of nano-lines of copper.

8.2.2.6 Joule Heating and Reliability Issues

Joule heating has become more of a reliability issue on the scaled VLSI Cu-interconnect. Indeed Joule heating can strongly affect the maximum operating temperature of the global wires which in turn will constrain the scaling of current density to mitigate electromigration and consequently degrade the expected speed improvement from the use of low-K dielectric materials. Experimental evidence shows that a temperature rise in the Cu-interconnect with low-K is significant due to poor thermal conductivity of the low-K material. The observation further shows that Joule heating can delay the interconnect by as much as 15%. However, in terms of heat dissipation when a self-supporting Cu-line is considered and compared with an Al-line of the same dimension, the former is seen to dissipate more heat. To testify to this we present here a mathematical model. Let us consider that heat is being dissipated in integrated circuits (ICs) due to current in the metal interconnects and other resistive elements. The temperature rise ΔT within a self-supporting interconnecting line 100 μm long, and 1 × 1 μm in dimension carrying current 10^{-4} A/cm^2 for one second can be calculated as:

For Al-interconnect

$$\Delta T = (\Delta Q)/M\,C_p \approx 100°\text{C} \tag{8.6}$$

For Cu-interconnect

$$\Delta T = (Q)/M\,C_p \approx 48°\text{C} \tag{8.7}$$

where (ΔQ) is the heat supplied and is equal to $\{(I^2R \times 1)\ \text{J}\}$, M is the mass, and C_p is the specific heat of the metal line.

Table 8.4 shows the thermal and electrical properties of Al and Cu metals. From the above calculations we can see that the temperature rise in a self-supporting Cu-line is almost half compared to an Al-line. However, when the dimension of the conducting line is reduced to sub-100 nm the value of ΔT can go to several hundreds of degrees. The thermal energy F_{Th} transmitted across unit area per unit time is proportional to the temperature gradient (dT/dx) and can be written as:

$$F_{\text{Th}} = -K_{\text{Th}}\,(\mathrm{d}T/\mathrm{d}x)\ \text{W/cm}^2 \tag{8.8}$$

Table 8.4 Thermal and electrical properties of aluminum and copper

Materials	K_{Th} (W/cm.K)	C_p (J/g.K)	Resistivity (ρ) $\mu\Omega$.cm	Density (g/cm^3)
Aluminum	2.37	0.9	2.65	2.7
Copper	4.01	0.39	1.68	8.93

where K_{Th} is the thermal conductivity of the material. From the above equation if we calculate the value of F_{Th} for Cu and Al, we will see that the substrate carrying Cu-lines will have to handle approximately 20% less thermal energy than when it carries Al-lines.

8.2.2.7 Stress in Cu-Line and Its Effects on Reliability Issues

Stress migration or stress voiding has always been one of the most important reliability concerns in both Al-based and Cu-based interconnects. The blocking boundaries and the vacancy concentration in interconnecting lines introduce a *stress gradient* due to flow of excessive current. The *stress gradient* is seen to increase with time until the associated driving force becomes equal and opposite to the electromigration (EM) force [78–79]. Since Cu has larger grain boundary activation energy, which results in a lower mobility, Cu is expected to be more resistant to stress induced voiding (SIV) than Al under the same level of stress. However, Cu does not have a natural oxide protection like Al and its SIV can be process (EP-Cu, PVD-Cu-seed, and barrier layer, and CVD/SOD low-*K*, etc.) and structure dependent (damascene architecture where Cu is confined). For example when a nitride capping layer is replaced by tungsten (W) or CoWP, the line width dependency of SIV is seen to reduce, which indicates that the interface is really a fast diffusion path and the SIV of Cu-lines is process dependent.

On the other hand, the relative vacancy concentration of atoms is related to the hydrostatic stress (σ), activation volume (Ω), and the resistivity (ρ) of the material [80–82]. Experimental observation shows that the *stress gradient* represents a component of the chemical potential and the mass flux can be considered not only due to EM, but also as a joint effect of EM and *stress*. Finite element calculations based on Hull and Rimmer theory shows that void growth due to stress distribution confines around a via sandwiched between upper and lower lines. The void source results from the supersaturated vacancies that developed when Cu-lines are not properly annealed.

Sputter deposited thin films are generally found to be under stressed condition. The stress arising from the substrate attachment is composed of intrinsic stress σ_{in}, the thermal stress σ_{th}, and the transformation stress σ_{tr}. The first one arises from the deposition process, and the lattice mismatch between the film and the substrate. The second one originates from the difference between the coefficients of thermal expansions (CTEs) of the deposited film and the substrate. The transformation stress, however, refers to the induced stress change due to the phase transformation of the film during the deposition process.

The Cu/interlevel diffusion barrier interface is thought to provide a site void nucleation at relatively low *tensile stress* [83–84]. The low tensile stress forms voids in Cu-interconnects. There is another stress noticed in the damascene structure, which is known as *substrate induced stress* [85].

The *stress induced* EM in Cu-low-*K* interconnects is a serious reliability issue because low-*K* dielectrics (with a value ~2) are generally mechanically weak [86]. Indeed, the development of stress within interconnects will depend upon the architecture of the connecting wires together with the different components connected to the lines. Therefore, as a result of the induced stress, stress related de-lamination (lack of adhesion) is a frequent phenomenon with silicon nitride (Si_3N_4) when Cu-lines are laid on it.

8.2.2.8 Thermal Stress and Related Phenomena

Thermo-mechanical reliability has been a critical issue for both packaging and wafer level structures. Thermal deformation and thermal stress are directly related to the packaging and wafer reliability. Simulation results show that large residual stress can be induced from wafer processing which can significantly affect the wafer level structure's reliability. Moiré results show that thermal deformations of these packages are quite different from each other, leading to different reliability problems in these packages.

Experimental results show that thermal stress for Cu-lines depends on the barrier layer thickness. On the other hand, thermal stress in Cu-lines increases with a decrease of the aspect ratio of the Cu-lines. The diffusion barrier (Ta/TaN) has lower CTE and higher Young's modulus than Cu. During thermal cycling the diffusion barrier impedes the shrinkage of Cu-lines and induced tensile stress. However, the stress in the side line of the barrier is compressive and decreases with increasing barrier thickness.

The Cu-seed layer deposited over the barrier layer has different thermal coefficient of expansion (TCE). During thermal cycling expansion and contraction in the Cu-seed layer (which is a part of the copper interconnecting lines) and the barrier layer will be different. As a result the interconnecting Cu-lines will be left under *tensile stress*. Experimental observations show that this *tensile stress* is maximum, at the top of the trench (Fig. 8.23). Experimental observations further show that this tensile stress in the interstitial dielectric layer is largely hydrostatic [87–88].

Experimental observation shows further that the *thermal stress* has a profound effect on the *Blech length* (l_{Blech}). As a result, at high thermal stress, the Blech product ($J \times l_{Blech}$, where J is the current density) may vanish (which is known as *stress voiding)* causing device failure (even without any flow of current through the circuit).

Thermal cycling induces plastic deformation [89], which produces creep and interfacial sliding in thin Cu-film, and all of these have a profound effect on the reliability of Cu-lines forming interconnects. Due to thermal cycling the Al-film expands whereas the Cu-film shrinks, and thus the Cu-film will require a different film footprint on the substrate [90]. Thermal cycling related volume contraction in

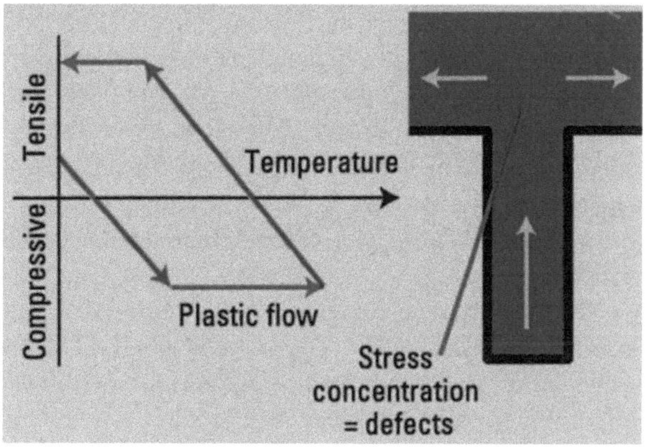

Fig. 8.23 Development of tensile stress after thermal cycling (Reprinted with permission, Novellus System, San Jose, CA)

Cu-film results in intergranular void formation and it has become an additional reliability problem in the space program. Figure 8.24 shows that the thermal stress-state in Al-lines over a passivated layer (TEOS decomposed SiO_2) is very different from Cu-lines with Silk (low-*K*).

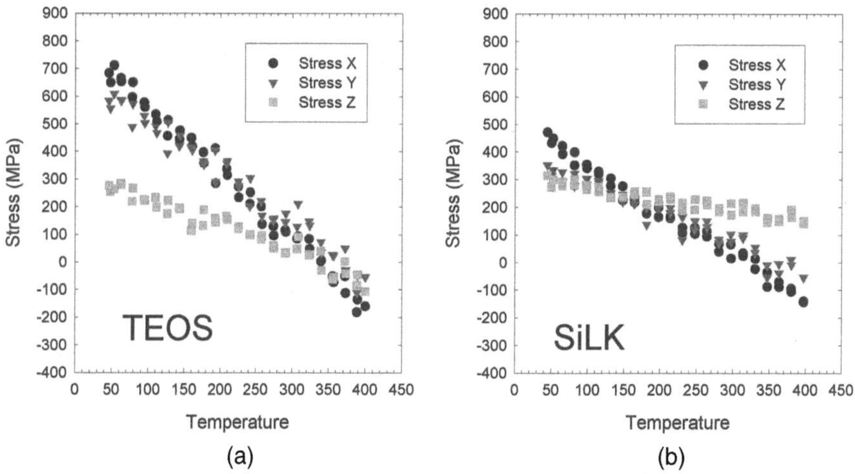

Fig. 8.24 Thermal stress: (**a**) Al-interconnect and (**b**) Cu-damascene interconnect (Reprinted with permission, Dr. P.S. Ho, The University of Texas, Austin)

8.2.2.9 Blech Effect and Its Consequences

In sub-micrometer level devices, Cu-interconnecting lines are subjected to *electromigration (EM)* induced extrusion failures due to the *Blech effect*. The extrusions and possibly corrosion can cause intra-line leakage breakdown. It has been theorized and experimentally proven that there is a critical lower limit of length of an interconnecting line known as the *Blech length*, and beyond that length Cu-interconnecting lines experience EM failure very often. It has been further observed that the poor adhesion between etch stop and Cu significantly reduces the critical stress for void nucleation. On the other hand, the use of low-K materials in Cu-interconnects has brought about an order of magnitude reduction in the modulus of the surrounding dielectric.

We know that the key to achieving immortality without voiding is a high critical stress for void nucleation and mechanically rigid surroundings to create a large back stress. Thus, in order to achieve immortality with voiding, it is essential to have an effective current shunt which is absent in Cu-dual-damascene interconnects.

8.2.2.10 Reliability Problems Due to Electromigration (EM)

Reliability is as much a key to success in the microelectronics industry as is performance. With the complexity of today's microelectronics, a phenomenal level of reliability must be maintained. As device features reduce in ultra-large-scale integrated circuits, current densities increase with the metallization layer complexity. The increased current density increases the temperature of the metal layers. As a result, the electromigration (EM), which is a *thermally activated* diffusion phenomenon, is very much affected by the increased current density.

Thermodynamic properties of copper films are very different from those of aluminum films, which are mechanically insensitive to artificial passivation. Passivated Cu films show distinct strain hardening due to its low stacking fault energy and highly twinned microstructure [91–94]. However, a cleaned Cu surface does not form a passivated oxide like Al. As a result, EM related phenomena are more prominent in Al (or its alloy) films. It has been observed that in polycrystalline Al lines, grain boundaries provide the highest diffusive paths for electromigration, but in Cu-metallization, the Cu/liner or Cu/interlevel diffusion barrier (usually Cu/Si_3N_4) interfaces provide even higher paths of diffusion than grain boundaries [95–99]. However, according to some observers, surface migration is the fastest diffusive pathway, compared to grain boundary (GB) and interface diffusions [100–102]. Whatever the true nature of the fastest pathway for EM in deep submicron level devices, microtextural effects might still be present and can alter the finest detail of EM failure [103–105].

Figure 8.25 shows the activation energies of different bulk metals and alloys versus line-width. For the damascene process the activation energy in Cu-interconnect is much higher than Al-alloy based interconnect which yields longer lifetimes in Cu-lines when extrapolating high temperature accelerated-life test data.

Figure 8.26 illustrates the differences in EM behavior of Cu-interconnects having different dimensions and with and without barrier layers. The EM failures consistently follow well-behaved long-normal distributions and demonstrate the extendibility of the barrier process for current and future generations of Cu-low-K interconnects. From the above figure it is also clear that with a re-sputtered barrier,

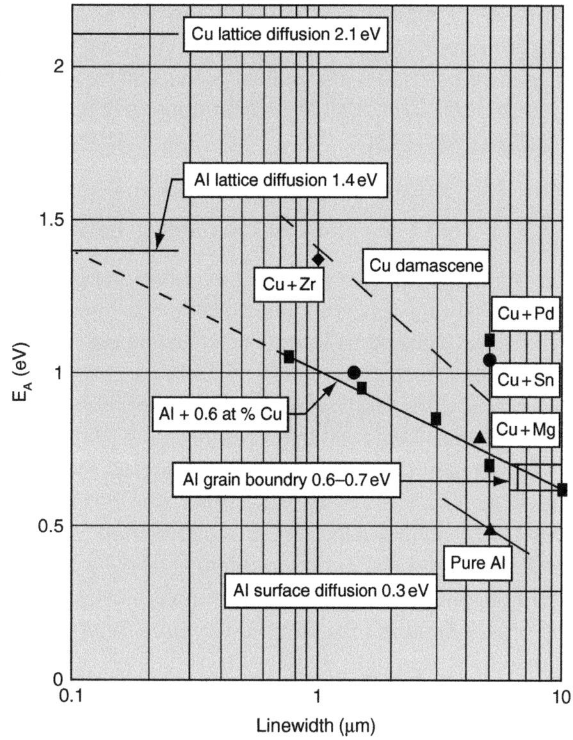

Fig. 8.25 Comparative analysis of the activation energies of different bulk metal and metal–alloys (Reprinted with permission *Semiconductor International* April 2000, p. 102)

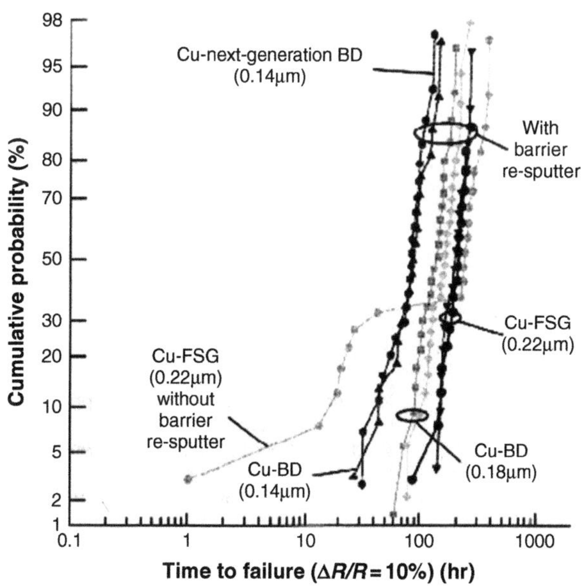

Fig. 8.26 Electromigration failure distributions with different technology nodes and associated interlayer dielectric (ILD) (Reprinted with permission from Applied Materials, CA). BD is black diamond

the film becomes monomodal with low log-normal sigma (σ) between 2.7 and 3.6. Re-sputtering the barrier at the via bottom not only creates a continuous path for copper diffusion, but also enhances the sidewall coverage through deposition of the re-sputtered barrier, thereby eliminating early modes of failure at the via/metal link interface [106–112].

8.2.2.11 Void Formation and Reliability Issues

The tendency of Cu to contract in volume and form inter-granular voids is an additional reliability concern with respect to Cu-interconnects that was not a major issue in better established Al-interconnect technology. Thermal cycle induced void formation would affect interconnect reliability and could accelerate EM by an early void nucleation time. Early onset of void formation would further degrade with EM, causing larger voids and open circuits.

Different microstructures of Cu-film result in different kinetics of void evolution. The wide dispersion of activation energy of EM (from 0.5 to 1.25 eV) can be related to that. For small grains having large curvature, surface diffusion leads to healing of the damage at the GBs and to thinning of the grains, whereas, for large grains with small curvature, it can redistribute material between the GBs and grooving results in voids. Tantalum (Ta) covered Cu films show no surface morphological changes, and voids started only at sidewalls of the line.

Like the Al-interconnect, a short segment of Cu-line forming an interconnect forms voids due to its failure to reach the back-stress which prevents the growth of voids. However, unlike Al-based interconnect, there is no apparent deterministic current density length product (*jL*) for Cu-lines which will determine the mortality rate [113].

Experimental observations show that voids (Fig. 8.27) preferentially nucleate at the Cu/Si_3N_4 surface near the cathode end of the Cu-line, since Si_3N_4 does not provide a conducting path to shunt [114]. In fact voids are seen to be prominent when the interconnect length is short and the current density is high.

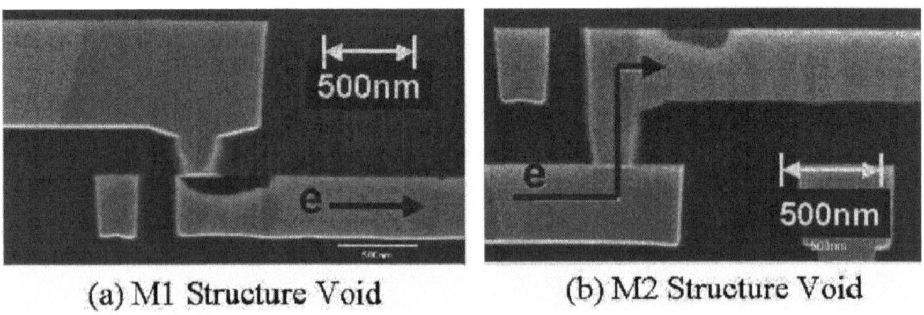

Fig. 8.27 The effect of interconnect line segments of different lengths and their consequences: (**a**) cross-section of FIB image of a small fatal void directly below a via; (**b**) formation of a non-fatal void at the Cu/Si_3N_4 top interface (Reprinted with permission, Materials Research Society)

From Fig. 8.27 one can see the impact of void formation when Cu-line segments have different lengths and the current density is high. It is true that the formation of voids requires nucleation and time. However, in the case of Cu-interconnect it is the growth not the nucleation that will take most of the time to fail the device [115].

8.2.2.12 Diffusion Related Problems

Cu-diffusion in silicon is very rapid (diffusivity $\sim 4.7 \times 10^3 \times \exp(-0.43\ \mathrm{eV}/kT)$ $\mathrm{cm^2./s}$), even at room temperature. Cu is a noble metal and, as such, readily diffuses through dielectrics. In order to obviate the problem, a liner or a barrier layer is fabricated to prevent Cu-diffusion. Diffusion of copper inside silicon is noticed by the precipitation of Cu in the formation of Cu-silicide (Cu_3Si) (Fig. 8.28). The process of Cu-precipitation is further complicated by the electrostatic interactions between the positively charged interstitial Cu-ions and charged Cu-precipitates or dislocations [116].

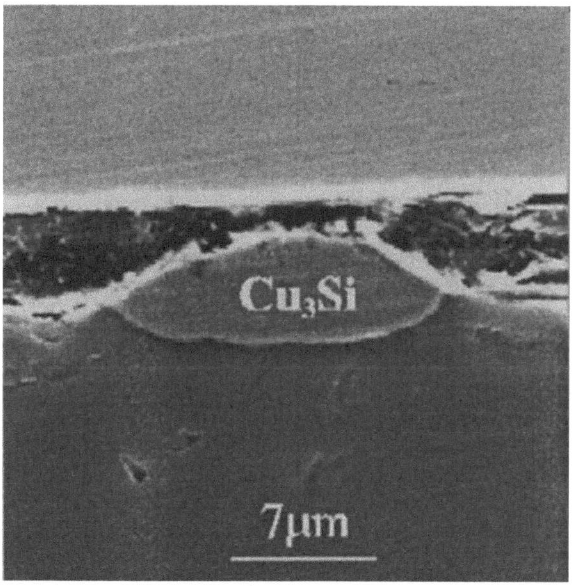

Fig. 8.28 The precipitation of Cu inside silicon and the formation of Cu-silicide (Cu_3Si) (Reproduced with permission, Elsevier Science, New York)

8.2.2.13 Reliability of the Barrier Layer

The use of Cu-interconnect, damascene architecture, low and high-*K* materials, and chemical mechanical process (CMP) for planarization has brought much hope and enthusiasm in integrated circuit fabrication. However, the use of new exotic materials, and the additional layers, has also become a reliability issue in terms of yield, and lifetime of the devices. For example, the barrier layer to minimize Cu-diffusion is an additional step required with Cu-interconnect, which was not necessary for Al-interconnect. The barrier layer controls the equivalent resistance, and adhesion of Cu-interconnects modifies the grain's microstructure during deposition, and finally controls the interface growth. Thus from the reliability standpoint deposition of a perfect barrier layer is very important. Figure 8.29 shows the patterned structure of the barrier layer serving as a substrate for the Cu-interconnect

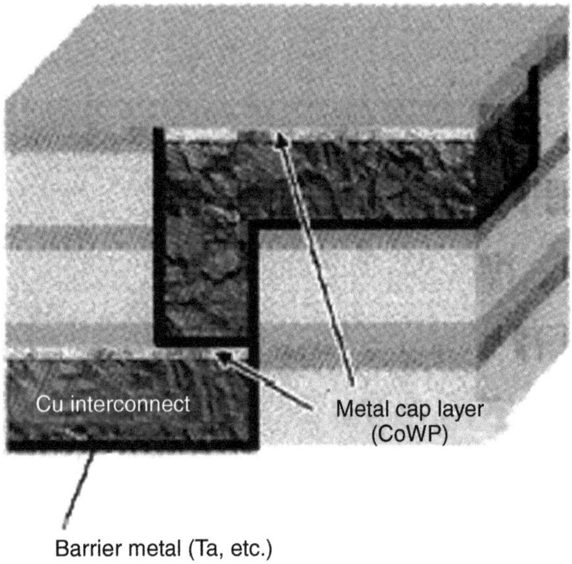

Fig. 8.29 The patterned structure of the barrier layer serving as substrate for the Cu-interconnect

Typical dielectric materials used in integrated circuits are not suitable as a barrier layer when Cu forms the interconnecting lines. At the same time, adhesion of copper to these materials is poor. The barrier layer should serve at least three purposes: (a) Cu-lines should adhere well to the barrier layer, (b) it should effectively encapsulate copper at high processing temperatures, and (c) the electrical resistivity of the barrier layer should be low [117, 118].

Experimental evidence shows that the microstructure of the deposited copper film will depend on the texture, and grain size of the substrate (barrier layer) [119]. Different methods are available to deposit the barrier layer and each method can produce film with different microstructures (especially Ta and its compounds) [120–121], which can change the composition, texture, grain size, and ultimately the electrical resistance, diffusion properties, interfacial adhesion, and film stress.

The most successful barrier layer that has worked well so far for Cu is a bi-layer stack of Ta/TaN. Ta has better adhesion, and seed properties and less stress induced voids due to its heteroepitaxial relationship with copper. TaN, on the other hand, has demonstrated excellent adhesion to SiO_2-based interlayer dielectric materials (ILDs). Being amorphous in nature it suppress the EM along GBs. Ta in β-phase has resistivity between 150 and 200 $\mu\Omega$. cm and PVD TaN film shows resistivity between 200 and 250 μΩ. cm. However CVD, and ALD deposited films show higher resistivities (500–1000 μΩ. Cm) compared to PVD films.

Different barrier materials show different annealing temperatures to form crystallites. The crystalline structure of a film will have different grain sizes and grain boundaries, which will form diffusion paths for the Cu-atoms. So, the reliability of the barrier layer is a function of temperature, microstructure of the layer, and the processing instruments and conditions. Barrier films are evaluated in terms of their ability to block copper diffusion at temperatures in the range 700–800 °C.

The barrier layer should be very thin (between 15 and 50 Å) and should provide adequate protection against Cu-diffusion into the dielectric layer. It has been found that when the film thickness is in between 200 and 400 Å, it works well at the 0.1 μm node level. But as the metal becomes thinner, filling the via holes and trenches at 0.1 μm level with barrier metal is a challenge especially when conformal and void free film is the primary objective. Moreover, the thinner barrier blanket will show higher electrical resistivity compared to a thicker blanket.

The barrier layer is generally selected from refractory metals and their compounds and can support current even after small void formation. Therefore, failure is not observed immediately after the formation of small isolated voids, until the current density reaches a maximum level where they can fuse. Moreover, since these barrier layers are composed of very low diffusing materials, EM will be very slow and the chance of open circuit will be minimum. EM may cause open circuit, discontinuity in current flow within the circuit, and increased electrical resistance of the route. Therefore, the failure times are not solely determined by void nucleation, but by combination of nucleation and growth [122].

8.2.2.14 Presence of Via Holes and Trenches and Reliability Related Problems

Via holes are used in multilayer circuits to reduce the interconnecting line length. The via holes are usually very small (~25 nm or less) and are metallized to make ohmic contacts between the upper and lower levels of the metal layers. To deposit metal layers within small via holes without voids is a real challenge, because of constraints in processing [123–124] (like shadowing) and poor adhesion of the metal on the walls of the via holes. Moreover, the finer dimensions and the higher current densities together with current crowding make these via holes much more vulnerable and raise some questions about their reliability.

Test structures designed with variations in reservoir area, the reservoir layout, and number of vias show that the current density through the different layouts is different. As a result, the lifetime of the EM will be different. As the current density increases, metal migration follows voids (due to tensile stress) at the vias closest to the line [126–127]. So, via layout should be designed to avoid *current crowding* (Fig. 8.30)[127]. The distances between via holes and their positions on the circuit can vary according to the requirements of the circuit layout. It has been observed that the via-above node has a much smaller mortality rate [128].

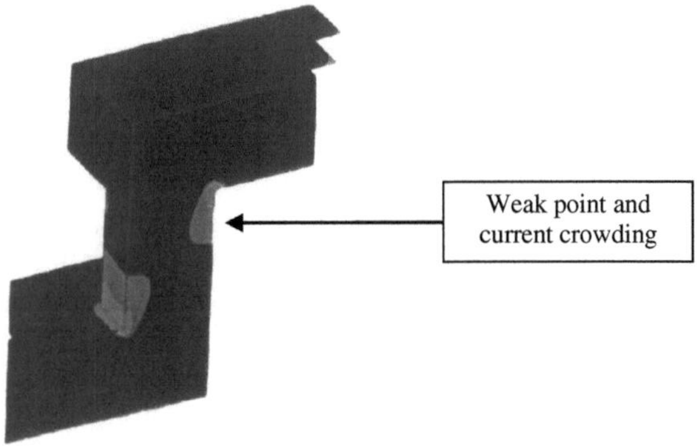

Fig. 8.30 The weak points and current crowding in the via hole

8.2.2.15 Reliability of the Low-K Materials

It is recognized that Cu/low-K is the interconnect system of choice for future device geometry. The critical need for low-K ILD in ULSI applications has already stimulated great efforts in the development of advanced low-K materials including fluorinated inorganic materials, porous silica, and porous organic materials. However, the lack of rigidity of low-K materials is an intrinsic property that will make EM more of a concern for all metal layers including the first few local interconnects that are usually less susceptible. As a result, the introduction of low-K materials to increase the speed of a device has added additional concerns of reliability, because the speed of an electrical signal in an IC is governed by two components – the switching time of an individual transistor, known as the transistor gate delay, and the signal propagation time between transistors, known as the RC delay. It is equal to $2\rho\varepsilon\ \{(4L^2/P^2)+(L^2/T^2)\}$, where ρ is the metal resistivity, ε is the permittivity of ILD, L is line length, P is the metal pitch, and T is the thickness of the metal. Figure 8.31 shows the low-K, cap layer and multilayer metals (M_1, M_2, M_3) inside a Cu-damascene structure.

Materials with K values lower than 3 are taken to be low-K materials, and they fall into three categories: organic, inorganic, and hybrid. Hybrid materials are typically inorganic materials doped with mostly carbon (C) or fluorine (F). The doped inorganic dielectric material, e.g. SiO_2 doped with C or F, reduces the effective K value from 4.2 to 3.0–3.5, but the introduction of C or F forces the material to change its volume, and reduces the *electronic polarizability*.

The other method which has been in practice to reduce the K value is to introduce pores into the material (air pockets, $K = 1$). The introduction of pores degrades the mechanical properties of the dielectric material, which becomes vulnerable during *chemical mechanical polishing* (CMP). The ITRS calls for $K = 2.2$ (for 65 nm generation) which means incorporation of ~20% porosity, which is actually above the porosity percolation threshold. At this porosity, low-K ILD will act like a sponge, and will make the dielectric absorb moisture and is susceptible to cracking. Moreover, percolated porosity provides a path for Cu ions to diffuse and possible

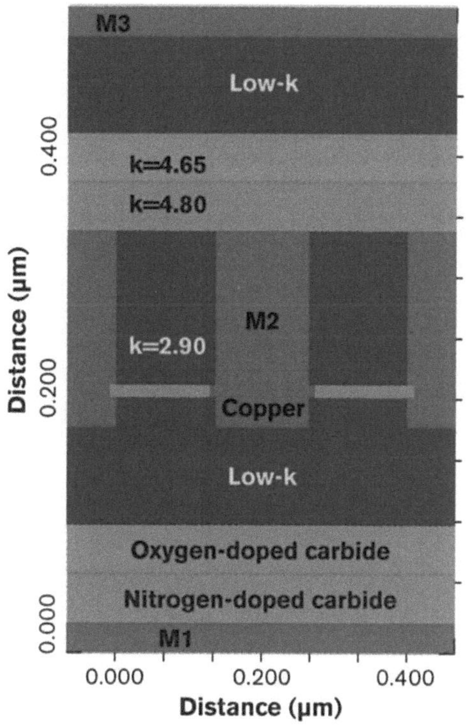

Fig. 8.31 The low-*K*, caplayer and multilayer metals (M_1, M_2, M_3) inside a Cu-damascene structure (Photo courtesy, *Semiconductor International*, Feb., 2007)

shortening of the metal lines. At the same time, highly porous materials are generally more ductile at room temperature, and a ductile-to-brittle transition occurs at lower temperature. However, when the pores are closed rather than open, and the pore sizes are less than 10% of the minimum feature size, the film shows better modulus to withstand CMP.

Adhesion is another issue related to low-*K* dielectric materials. Poor adhesion of the dielectric material to the barrier layer could create a fast diffusion path for copper along the low-*K*-barrier interface as well as delamination of the film under stress (Fig. 8.32).

These are not all. The thermal conductivity of low-*K* materials is poor compared to silicon dioxide (SiO_2). As a matter of fact, thermal conductivity decreases rapidly with the incorporation of porosity, as heat conduction is limited to the solid phase of a porous material. Some studies show a >10 times decrease in thermal conductivity when the material is accompanied with 20% porosity. As a result, the effective thermal resistance (R_{Th}) of the low-*K* materials will be higher than a conventional dielectric material like SiO_2. On the other hand, organic dielectric materials have high thermal coefficient of expansion, and low thermal conductivity leading to a high stress field at the interface, and will show delamination (Fig. 8.32) and Joule heating at the interconnects.

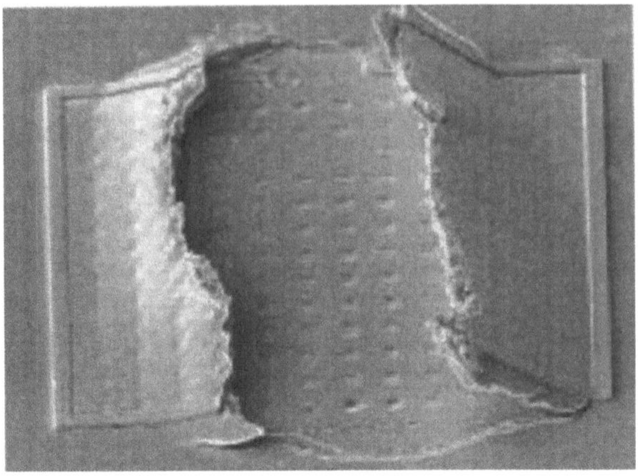

Fig. 8.32 Delamination along the FSG/barrier interface upon wire bonding, when FSG (fluorosilicate glass) is used as the final ILD (Reprinted with permission, Freescale Semiconductor Inc.)

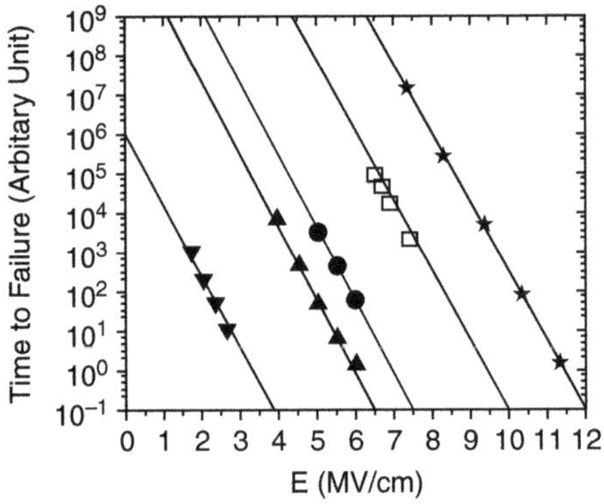

Fig. 8.33 Time dependent dielectric breakdown lifetime extrapolation curves for different dielectric materials used in the Cu-damascene process (Reprinted with permission, Dr. P.S. Ho, University of Texas). The *lines* from the *left-hand side* represent (1) P-MSQ, (2) SiOC, (3) SiOF, (4) PE-TEOS, (5) SiO_2, respectively. P stands for plasma and PE is plasma-enhanced

The time dependent dielectric breakdown (TDDB) lifetime extrapolation shows (Fig. 8.33) that the time to failure for SiO_2 is the highest compared to low-K dielectric materials. The integrity of the dielectric layer is a major overall reliability issue of copper low-K interconnects, which is not a reliability issue with SiO_2 in the conventional Al-interconnects. It has been observed that the breakdown strength and Weibull shape parameter β decrease as the porosity in low-K material increases.

Low-K films are more susceptible to *ash damage* (Fig. 8.34) and require a shift away from standard oxidizing chemistries. This is important for dual inlaid structures that require aggressive processing which affect low-K dielectric materials. It is sometimes advantageous to eliminate the middle stop layer (MSL), but without an MSL the trench bottom profile becomes more difficult.

During exposure the cross-linking of a low-K polymer film must be kept within some tolerable limits in order to keep the device integrity. Too little polymerization

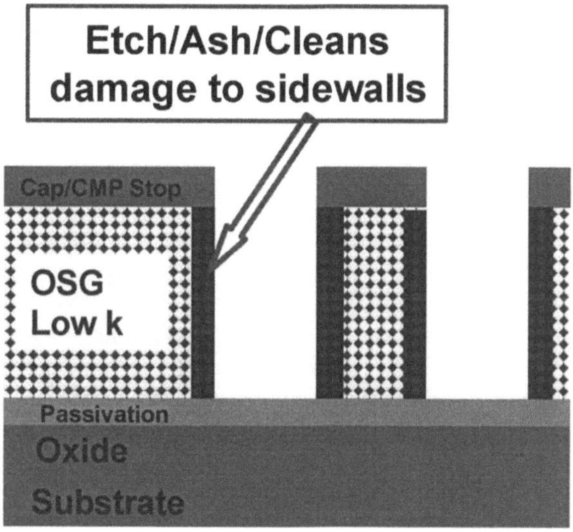

Fig. 8.34 Schematic of ash damage during plasma operation for low-*K* dielectric materials (Reproduced with permission, Dr. P.S. Ho, Univ. of TX. Austin, TX)

will degrade the hardness, whereas too much polymerization will make the material brittle. Control over electron dose is of imperative importance for the final structure of the low-*K* resist material. Another important criterion for these low-*K* materials is that they are susceptible to oxidation, which dramatically changes their *K* values. This can destroy the signal synchronization, ultimately rendering a processor inoperable.

8.2.2.16 Impact of Exotic Materials on Dual-Damascene Pattern Formation

Cu-metallization has introduced many challenges to interconnect process technology, among which three pertain to patterning : (a) control of substrate reflectivity; (b) selectivity to the etch-stop (ES) layer; and (c) control of the profile and etch bias of the patterned critical dimension. For patterning ILD, etch back (EB) and via flare are more critical due to the soft nature of the ILD. At the same time, utilization of a dual hard mask to define via and trench openings is not compatible with unlanded via holes and full via first opening.

Most of the time the ILDs that are used in the Cu-damascene process are composed of silicate glass (SG), fluorosilicate glass (FSG), and carbon doped silicate glass (CDO). These materials are very transparent at 248 and 193 nm lithography and the reflectivity of these materials serving as substrate are unacceptable. Therefore, in most of the cases use of an organic antireflective coating (ARC) is recommended. However, applying this technique to dual damascene (DD) patterning shows several defects that impact die yield and chip reliability [129]. As a matter of fact, during trench etching both the ARC layer and the ILD are etched simultaneously and that generates defects that are known as *shell* defects.

On the other hand, dielectric materials like silicon nitride (Si_3N_4) and silicon carbide (SiC) are used as the ES layer. Since the *K* values of these materials are relatively higher than that of an ILD, minimal ES layer thickness should be provided to avoid excessive increase in the effective *K* (K_{eff}) value of the overall integrated stack [130].

8.2.2.17 Etching Processes and the Reliability

The scaling of feature size and the growing complexity of the stacked materials have made traditional etching systems very difficult, e.g. control of profile, selectivity, critical dimensions (CDs), uniformity, and defects. Also of concern is the cost of ownership, which is impacted by parameters such as etch rate, the time between cleans, and the cost of consumables. Indeed, success in the etching process, whether it is wet or dry, depends upon successful control of the etch profile, the critical dimension (CD), damage, line edge roughness (LER) and line width roughness (LWR), uniformity and variability across the wafer, and wafer to wafer.

Dielectric etching traditionally uses highly polymerizing fluorocarbon gases to passivate the sidewalls for anisotropic vertical etching. However, the presence of a higher percentage of carbon and lower oxygen in SiOC films drives the etch into over-polymerization. This leads to cessation of SiOC etches – a condition referred to as *etch stop*. However, carbon tetrafluoride (CF_4) or H_2 based chemistry has been found to be helpful in removing the residue.

Dry anisotropic etching is very successful with dense dielectric materials. But when the dielectric is a porous one (to lower the *K* value) dry anisotropic etching is difficult to perform because of the etching rate. However, hydrocarbon gases are found to control anisotropy. On the other hand, during trench etching the presence of an intermediate stop layer has the advantage of controlling the depth of etching. But the presence of an additional layer has several disadvantages, like increase of the effective *K* value of the dielectric layer, potential delamination, formation of a facet at the top of the via inside the trench, and manufacturing cost. Facet formation becomes exacerbated when the density of the dielectric is very low. However, experimental study shows that low gas phase polymer with little or no fluorocarbon and low ion energy (low bias) are the key parameters to minimize faceting.

In the via first etching sequence, an *antireflecting coating* (ARC) is spun over the wafer and into the via holes prior to photoresist. It has been observed that ARC leaves a residue inside the via hole after photoresist and ARC etching are done. Such a residue is referred to as a *fence*. The presence of the fence causes problem during metal filling, which is a reliability issue.

Fencing and *faceting* around the via hole during the trench etch of the via first dual damascene integration scheme are particularly detrimental and can lead to problems during copper metallization and ultimately may cause device failure. During trench etching, the serious problem that has been noticed is the formations of fences, which are caused by the deposition of etch products of the dielectric. The other concern about the trench etch stop layer is that its removal can lead to increased capacitance and thickness variation of the metal layer.

As device dimensions shrink (from 90 to 32 nm generations), new exotic materials are being introduced, including low- and high-*K* dielectric materials and proprietary gate metals with fine tuned work functions. These new materials, particularly the metals, are creating new etch by-products that are sometimes difficult to remove. A similar situation is noticed with dual damascene patterning of via holes that are very small. In addition, etch stops, hard masks, antireflecting coatings (ARCs), and different process flows (via first or trench first) have introduced difficult situations during the etching process.

Another new and unique etch challenge is presented by 3-D integration, where via holes/trenches must be etched completely through a thin silicon wafer. These deep via etches require a specialized process, like fast switching between silicon etching and polymer deposition for silicon sidewall protection with SF_6 and C_4F_8. However, the switched approach (1–3 s) includes a better etch rate and selectivity but with scalloping of the sidewall.

Plasma induced low-*K* damage is also a critical concern, which affect the CDs. The etch process very often leaves some residues that must be removed along with the resist material. Since SiOC contains a higher percentage of carbon, fluorine-containing gas like carbon tetrafluoride or H_2 based chemistries is applied to enhance residue removal. The porosity of low-*K* films also affects the anisotropic etching rate. Hydrocarbon gases are found to control anisotropy. Figure 8.35 shows a scanning electron micrograph of the cross-section showing the formation of a fence during etching.

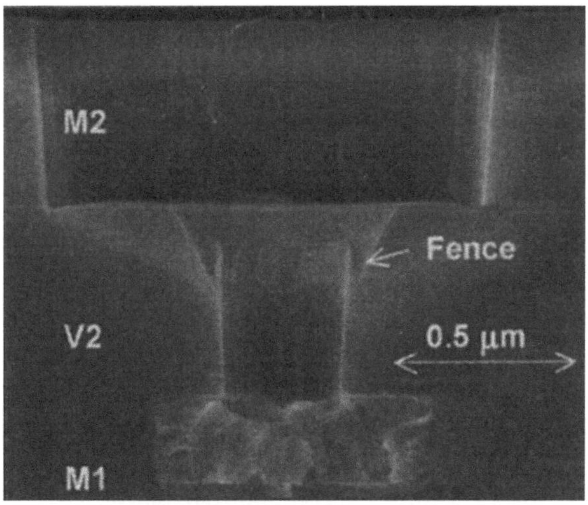

Fig. 8.35 Scanning electron micrograph of the cross-section showing the formation of a fence during etching (Reproduced with permission, IBM Research)

8.2.2.18 Reliability of Chemical-Mechanical Polishing (CMP)

Both dry and wet etching systems are not suitable for the copper damascene process. As a result, the deposited metal is polished back using chemical mechanical polishing (CMP). Conventional CMP requires a long time for planarization and a large quantity of consumable materials, especially when the interlayer dielectric is a low-*K* material. Furthermore, it presents a tremendous challenge for post-CMP cleaning. Remaining particles, erosion and dishing are the major factors that affect production yield. Recently, a current controlled electro-polishing CMP system has been able to produce a stress free surface of the substrate. According to the report, the system is capable of removing atomic level particles from the surface of the substrate with minimum damage (Fig. 8.36) [131].

Fig. 8.36 Scanning electron microphotograph of a patterned copper (Cu) wafer showing a clean surface with no damage and corrosion after post Cu-CMP cleaning (Reprinted with permission, Conexant Syst. Inc. Newport Beach, CA)

Etching and planarization methods include mechanical milling, reactive ion etching (RIE), wet chemical etching, parallel polishing, and chemical mechanical polishing (CMP). It is possible that RIE is able to produce a more uniform anisotropic feature size with better selectivity, whereas fluorine based inductively coupled plasma (ICP) can balance chemical etching and ion assisted etching.

Reliability in CMP and post-CMP processes involves addressing plasma damage, contamination, thermal budgets, cleaning of high aspect ratio features, defect tolerant processes, and elimination or minimization of complicated processes. Acceptable reliability will only come through the control of interfaces, more emphasis on modeling, and the elimination of failure mechanisms.

8.2.2.19 Cleaning Process and Reliability

As features are becoming smaller, and specs get tighter, most cleaning challenges are becoming evolutionary. Although dry processes continue to evolve and offer unique advantages for some cleaning operations, wet cleaning with a mixture of chemicals such as hydrofluoric (HF), hydrochloric, (HCl), sulfuric (H_2SO_4), or phosphoric acid (H_3PO_4), or hydrogen peroxide (H_2O_2), or ammonium hydroxide (NH_4OH) diluted with de-ionized (DI) water, is still considered to be the most valuable cleaning solutions for modern integrated circuits. As a matter of fact,

RCA cleaning with more dilution remains a production workhorse. However, all of the wet chemistry cleaning in some way or other involves high-temperature (~160 °C) operation and it creates significant reliability and contamination issues for submicron size devices. Thus, the ultimate solution will be to come up with a low-temperature solvent.

The most critical and challenging job is to keep the Cu-surface clean without corrosion. Compatibility with the ILD and ES layers is also required for dual damascene (DD) cleans to be effective. Use of oxygen ashing is a dilemma in the Cu-damascene process. Use of the method will degrade the ILD's *K*-value but not using it will require more cleaning, especially to remove hardened via polymer and antireflecting coating.

In the back-end-of-line (BEOL) copper cleaning seems to be under control, but the main challenge is cleaning porous low-*K* dielectric materials. As the industry moves to 32 and 22 nm node technologies, low-*K* will pose new challenges for wet processing (fluoride ions of HF can change the effective *K*-value). Here, upstream processing will prove to be critical to the effectiveness of the cleaning and some post-clean processing may be essential to render the films stable for the down stream steps.

Post-etch clean strategies for high-*K* and metal gate materials require an integrated strategy for implementation into a transistor flow. Preventing corrosion of metal gate and new materials also needs to be considered. Cleaning photomasks are becoming more critical for 65 nm technology and it requires a special wet etching formula as conventional wet cleaning processes with HF/H_2O_2 followed by NH_4OH/H_2O_2 are not adequate. On the other hand, at 193 nm, sub-pellicle defects and photon induced haze resulting from sulfur residues that remain after cleaning are becoming the key challenges.

Another big challenge facing wet processing equipment and materials supplies is the removal of photoresist after high dose ion implantation. The obvious challenge is to remove the resist completely, while at the same time keeping post-strip defectivity within specifications and not doing any damage to the film and structures. At the same time, dry stripping in and of itself is not the best solution because of substrate loss and other challenges. As a result the wet clean formulation should be such that the solution becomes very aggressive in removing particles but should be gentle to the device structures. For example, the compatibility of tungsten to oxidizers, in particular things like hydrogen peroxide, is a concern, selectivity being the biggest challenge.

According to some observers, with submicron devices, surface cleaning after etching or ashing, or CMP, is virtually to apply physical force to clean particles as well as residues. Thus research is focused on megasonic brushing and cleaning with non-megasonic solutions. However, the brushing and cleaning of the particles including residues have to be performed efficiently without any damage to delicate structures. Right now the particle size is about 60 nm with particle density typically on the order of 0.04 particles per square centimeter on a 300 mm wafer. It is expected that by the end of 2010 the particle size will drop to 40 to 30 nm with a particle density of 0.06 particles per square centimeter on a 300 mm wafer. These particles are smaller than an airborne virus and the reliability of the metrology tools to detect these *defects* leaves a lot to be considered.

8.2.3 Measurements

(a) *Film thickness, roughness and density measurements*: Recently X-ray reflective (XRR) and X-ray fluorescence (XRF) techniques have been used for non-destructive *in situ* measurements of thickness, surface roughness, and density of the multilayer stack during the damascene process. The XRR and XRF measurements can detect any change in density (due to interactions between the layers) of the interfacial layer by measuring different incidence angles of the critical angle of the X-ray reflectivity. This interaction involves changes in effective resistance, and mechanical behavior of the adjacent film. It is also possible to calculate the percentage porosity of a low-K material from the density measurements of the material. Besides, XRR and XRF are used to measure the thickness of a metal or dielectric layer in Cu-damascene process. XRR and XRF techniques have also been used for the measurements of the film roughness. However, the XRF technique cannot be applied when the layer thickness or roughness is 250 nm or higher because of X-ray absorption.

(b) *Stress measurements*: Stress in Cu-interconnects is a complex function of a variety of parameters, including interconnection width. As film stress competes with adhesion strength of the film, loss of adhesion may occur for a high modulus material for a given thermally induced strain and identical adhesion strength [132].
Several methods have been used to measure strains, and hence to infer stress within thin metal interconnects. One such method is X-ray micro-diffraction, where a focused beam of approximately 1 μm [133] is produced from synchrotron radiation. The other method which has been successfully used to measure stress is a narrow (~10 nm) convergent electron diffraction (CBED) beam in a transmission electron microscope (TEM). A cone-shaped electron beam illuminates the sample and the diffracted spot looks like a disk from parallel illumination. The central disk is used for strain analysis [134].

(c) *Electromigration (EM) measurements in Cu-damascene line*: To detect mass migration of copper (Cu) from the Cu-interconnect line, the test structure is made out of Cu line 10 μm wide. Gallium metal is placed on the surface of the film. The test structure is put under current stress under a specific current load and X-ray fluorescence mapping is obtained by an X-ray scanning beam 1 μm wide. Figure 8.37 (a) shows a schematic of the test structure of the Cu-line with a small lump of gallium (Ga). Figure 8.37(b) shows the X-ray fluorescence (XRF) mapping obtained by X-ray micro-beam scanning. Figure 8.37(c) shows the distribution paths of Ga and Cu atoms along the grain boundaries (GBs) [135].
Electromigration (EM) in Cu-damascene interconnecting lines has also been studied by time resolved X-ray microscopy [136]. Multiple X-ray (1.8 k eV.

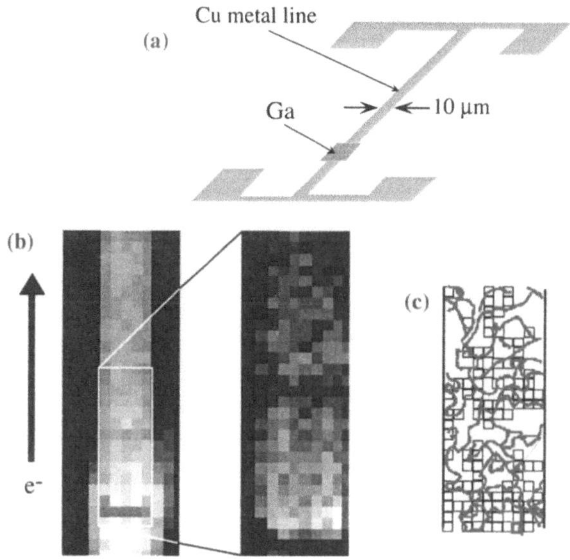

Fig. 8.37 (**a**) Initial distribution of Ga atoms; (**b**) Ga distribution due to current stress; and (**c**) coincidence between Cu and Ga atom paths through Cu-interconnects (Reproduced with permission, Central Res. Lab. Hitachi, Japan)

photon energy) images are recorded when the lines are under current stress. The image sequences show void formation, the way the voids are nucleated, and the mass migration of Cu-atom towards the via sidewall.

(d) *Measurement of Cu-damascene line texture*: The texture of the damascene Cu-line is complicated due to high aspect ratio trench and via lines, where nucleation during deposition occurs on the sidewalls and at the bottom of the interconnect features [137]. Electron back scattering diffraction (EDSD) scans are carried out in a small area in the longitudinal plane (RD-ND plane where RD is the rolling direction and ND is the normal direction). Later, EBSD map-points are imaged in every 0.1 µm intervals (corresponding to the map step size). After suitable image processing, evaluation of the microstructure is performed by means of automated acquisition and indexing of Kakuchi patterns. Afterwards, pole figures and disorientation distributions are determined for each mapped region. In order to compare the texture between two adjacent grains in the vicinity of the grain boundary, the grains are scanned and imaged with a field emission scanning electron microscope (FE-SEM) [138].

(e) *Line edge roughness (LER) or line width roughness (LWR):* The rapid shrinking of feature size and the introduction of new exotic materials continue to challenge metrology. The shrinking feature size has shrunk the critical dimension (CD) to a few tenths of a nanometer. As a result the line edge roughness (LER) or line width roughness (LWR) becomes a critical issue because it can degrade resolution and line-width accuracy [139–140]. According to the *International Technology Road Map for Semiconductors* (ITRS) 2004, the line width roughness (LWR, nm) $<\delta\%$ of CD for 90 nm node technology should be 3.0 nm. From

ITRS requirements, one can see that there is a real need for 3-D metrology in various fields.

Both CD-AFM (atomic force microscopy) and CD-SEM (scanning electron microscopy) have been used to measure (a) sidewall angle (SWA) height, (b) the feature's profile, and (c) LER and LWR. CD-AFM studies have been extended to measure top rounding of the etching profile. The commercial instrument available in the market (Veeco, Dimension X3RD) has shown static repeatability under 0.7 nm (3σ) and dynamic repeatability around 1 nm (3σ). For accurate and dependable measurements the tip width and tip shape of the flared silicon are precisely determined. Once these measurements are done, the equipment is set in CD-mode and further measurements are carried out [141]. For LER or LWR measurements, different light scattering measurements are performed based on an optical Fourier transform (OFT) using different simulations. The measurement is capable of detecting simultaneously the specular and non-specular diffraction patterns coming from different structures having periodic roughness [142].

Similarly, using CD-SEM (Hitachi), the LWR and LER for resist material can be measured. The measurement repeatability is 1.0 nm (3σ), the beam energy ranges from 300 to 1600 V, resolution 2.0 nm, throughput 55 wafers per hour, MAM (make acquire, and measure) time is <5 s, and visual field position accuracy is $\pm$ 1 μm. Experimental observation shows that the 3σ value increases as the length (L) of the measurement increases. The best workable values for length and edge measurements are found to be 2 μm and 200 edge points, respectively. A value of 2 μm can also be applied for CD measurements. The CD variations are caused by roughness of the surface. The frequency of roughness smaller than 'L' is called as LWR. On the other hand, when the frequency of roughness is greater than 'L' it is called CD variations [143].

(f) *Time dependent dielectric breakdown (TDDB)*: The integrity of the dielectric layer is a major overall reliability issue of copper low-K interconnects. Low leakage current for assessing isolation properties of back-end-of-line (BEOL) dielectric materials and time dependent dielectric break down (TDDB) measurements have gained more acceptance for assessment of BEOL reliability. TDDB is not an issue for SiO_2 based dielectric interconnects but becomes increasingly important when low-K materials with fine pitch are considered.

TDDB is found to accelerate by the drift of Cu^+ ions in a dielectric layer under the influence of an electric field [144] and can be modeled after McPherson and Mogul as:

$$\tau_{BD} = \{B \exp(E_a/kt)\}/\{[\exp(q\lambda E/kT) - [\exp(-q\lambda E/kT)]\} \qquad (8.9)$$

where τ_{BD} is the time to form Cu^+ passage from anode to cathode, B is a proportionality constant, E_a is the diffusion activation energy of Cu^+, E the electric field, λ the jump distance between two adjacent Cu^+ positions in dielectric, k the Boltzmann constant and T is absolute temperature.

It has been observed that for a particular field E, and temperature T, τ_{BD} is higher in a dielectric material having higher activation energy (e.g. τ_{BD} for a particular temperature, field, and λ, is higher in Si_3N_4 compared to SiO_2) [145–146]. Thus one can assume that low-K materials will give lower τ_{BD} values compare to silicon dioxide, which has a higher K value compared to these low-K materials. When the oxide layer is very thin, the breakdown mechanism is widely accepted to be the result of neutral defect and TDDB no longer provides a suitable method to assess the reliability of the oxide layer. As a matter of fact, as the oxide thickness continues to scale down towards the intrinsic limit, plasma-charging damage has become a serious problem [147].

To measure the scaled-down feature size of copper, barrier, and the dielectric materials, a series of analytical tools is used to characterize Cu-damascene processing. A list of such analytical tools is given in Table 8.5 .

Table 8.5 A list of the analytical tools used to characterize copper damascene processing

Analytical tools	Applications
Focused Ion Beam (FIB)	Void Exposure, grain size, defect cross-section
Scanning Electron Microscope (SEM)	Grain size, defect imagimg
Atomic Force Microscope (AFM)	Etch depth, contact geometry, surface morphology
In Lens Field Emission Scanning Electron Microscopy (FE-SEM)	Layer thickness, step-coverage, grain size, defect imaging
Time of Flight Secondary Ion Mass Spectroscopy (TOF-SIMS)	Post-CMP cleaning, surface contamination
Dynamic SIMS	Cu-diffusion, in-film metal contamination
Surface SIMS	Surface and back side contamination of a wafer
Total Reflection X-Ray Fluorescence (TXRF)	Surface and back side contamination of a wafer
Field Emission Auger Electron Spectroscopy (FE-AES)	Killer particle identification
X-Ray Photoelectron Spectroscopy (XPS)	Silicon oxide thickness, surface contaminant

Figure 8.38 shows an image of a cleave cross-section of a trench region. The image has been taken by using the backscatter mode of an immersion lens FE-SEM. The barrier layer shows excellent uniformity while the seed layer shows variation in thickness along the length.

Figure 8.39 shows a SEM image of a FIB prepared cross-section of Cu film over the trenches. The formation of voids during electrochemical deposition of Cu is also shown in the figure. Figure 8.40 shows an image of the grain structures of the Cu-seed and barrier layer inside a via hole. The image has been taken with the help of an immersion lens FE-SEM.

Figure 8.41 shows an overlay of two secondary ion mass spectroscopy (SIMS) depth profiles. From the figure it is clear that the barrier layer of sample B is not affected as much as sample A in the same experimental environment.

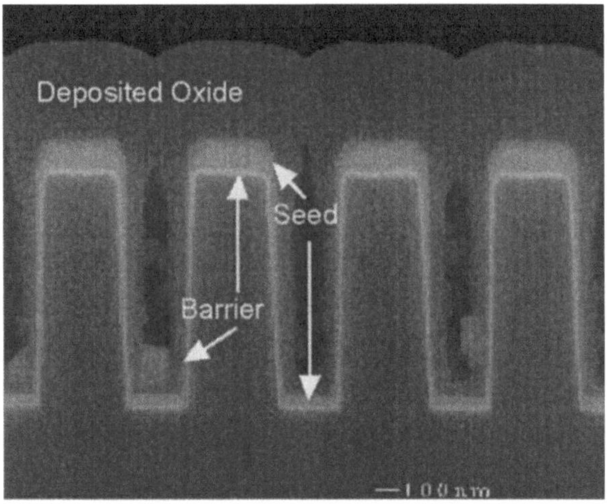

Fig. 8.38 The deposition of an oxide layer to retain the structural integrity of the seed layer. Imaging is done using FE-SEM (Photo courtesy, Charles Evans, & Associates, Sunnyvale, CA)

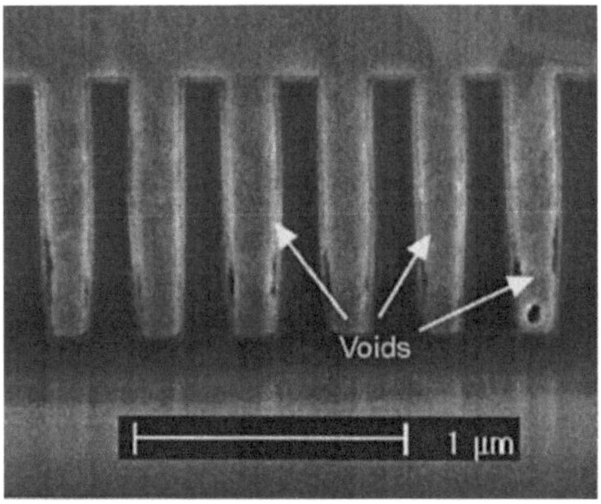

Fig. 8.39 The secondary of voids in Cu-filled trenches (Photo courtesy, Charles Evans, & Associates, Sunnyvale, CA)

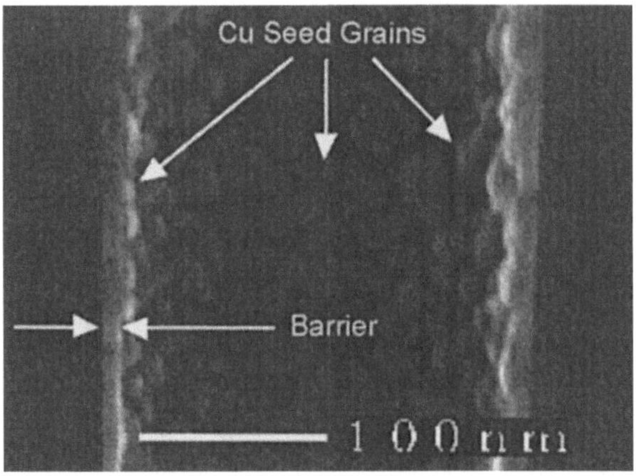

Fig. 8.40 The grain structure of a Cu-seed deposited over a via hole (Photo courtesy, Charles Evans, & Associates, Sunnyvale, CA)

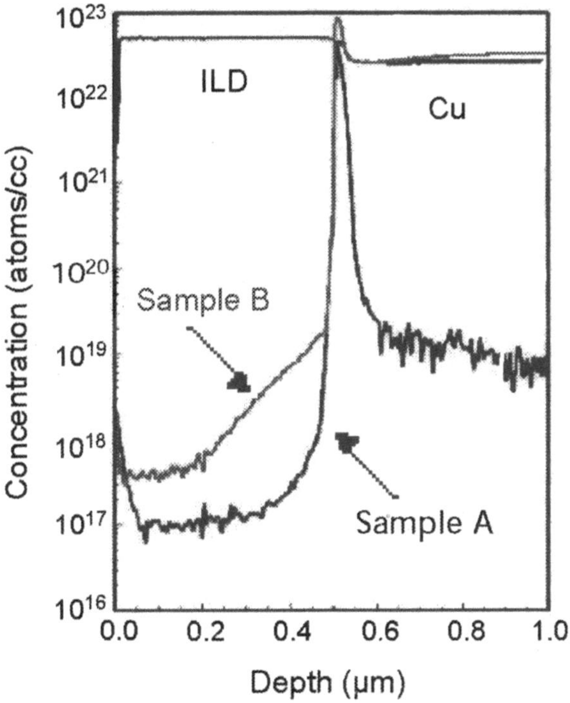

Fig. 8.41 SIMS profiles of two samples A and B encapsulated in the barrier layer (Photo courtesy, Charles Evans, & Associates, Sunnyvale, CA)

8.3 Summary

It is widely known that the performance of interconnect routing systems in sub-100 nm node integrated circuits (ICs) will be highly influenced by the metal used for the interconnect. Scaling of interconnect routing is also a critical determinant for future circuit density, especially for logic and microprocessor products. As higher levels of interconnect routing for future logic and microprocessor products are needed, the interconnecting routing system has to provide a cost-effective solution for yield management and wafer manufacturing. The materials, the process architecture of dual damascene (DD) technology, and the adoption and reliability of emerging Cu-routing, will decide the future of Cu-DD processes. The major line monitoring and yield management in the future are high aspect ratio trenches and *via holes*, new process integration of Cu-low-*K* systems, and integrity of the thin *barrier layer*. The number of contacts and *via holes* that can be used is also constrained by the routing area due to the via blockage effect [148]. Hence, efficient design of contacts and via holes is pivotal to robustness and reliability of sub-micrometer scale Cu-interconnects. In addition the highly resistive diffusion barrier layer needed for Cu-interconnects occupies nearly 20% of the drawn wire width at each technology node [149]. Therefore, the barrier layer has added an additional challenge to the Cu-damascene routing system [150]. It is also true that due to the deep submicron (reduced geometry) Cu-interconnecting system, the line resistances have decreased further and the exiting routing system has introduced additional inductive cross-talk which is being considered as the most serious problem in VLSI interconnects, specially for large busses [151]. Unfortunately, the self, mutual and loop inductances of a Cu-interconnecting system do not depend on their geometry, but also depends on the current return path. Thus the routing is becoming a reliability issue with the present interconnect scaling scheme.

Deposition of a thin barrier layer on to a wafer that exhibits surface topology variation results in a non-uniform film thickness after chemical mechanical polishing (CMP). This non-uniformity leads to resistivity variations resulting in the reliability problem of the device performance. Therefore, stringent control of the starting material has to be established especially when there is a variation in the surface topology of the blanket. Despite the higher conductivity, the aspect ratio of Cu-interconnects remains high to allow finer pitch. To lower the capacitance, the permittivity of the intermetal dielectric (IMD) and interlevel dielectric layer (ILD) films should have lower values than silicon dioxide (SiO_2). However, the low-K dielectric materials that are being used to lower the RC effect have not yet reached the manufacturing stage. Thus more work and research is going on for a successful implementation of Cu-interconnects and their routing in the sub-micron level devices.

In the near future more margins are required for electromigration with cobalt tungsten capping layers and alloying of copper which is not without risk, because both come with an increase in the RC product, which ultimately will affect the speed of the device. As a matter of fact, due to increasing complexity and clock frequency, temperature has become a major concern in IC design. Higher temperatures not only degrade system performance, raise packaging costs, and increase leakage power, but they also reduce system reliability via temperature enhanced failure mechanisms such as gate oxide breakdown, interconnect fast thermal cycling, stress migration, and electromigration (EM) [152]. In order to minimize temperature related reliability problems, several low-temperature deposition technologies (CVD, ALD, etc.) have been either in use or at the research stage for submicron film deposition in the Cu-damascene process.

In the area of thin film deposition, PVD deposited copper and tantalum are unique because thermodynamic deposition of pure tantalum (Ta) by the ALD method is difficult. Moreover, during metal organic and ALD depositions, precursors are used and there is always the possibility of the presence of carbon and nitrogen in the film, which can degrade adhesion. The biggest issue is scaling, especially with low-K dielectric materials, where water absorption and mechanical integrity become severe if not properly sealed. However, it is worth mentioning that joint research efforts by the industries and universities for the implementation of Cu-low-K system to replace the Al-SiO_2 system have made so much progress within ten years that the future of manufacturable 65 nm node devices with Cu-low-K is bright.

References

1. R.L. Geiger, P.E. Allen, and N.R. Strader, VLSI design techniques for analog and digital circuits, McGraw Hill, p. 853 (1990) and also P. Hofstee et al, A 1 GHz single issue 64 b power PC processor, Proc. ISSCC, p. 92 Feb. (2000) and M. Forshaw et al., A short review

of nanoelectronic architecture, 15, S220 (2004) and also K. Fife, A. El Gamel and H-S-P Wong, IEEE ISSCC Tech. Dig. Feb., 48 (2008)

2. M.T. Bohr, Interconnect scaling—the real limiter to high performance ULSI, in IEDM Tech. Dig., 241–244 (1995) and V. Mehrotra and D. Boeing, Technology scaling impact of variation on clock skew and interconnect delay, Proc. Intl. Interconnect Technol. Conf. June (2001) and T. Hoffmann, IEEE IEDM Tech Dig., 269 (2006) and also S-Y Lee, C. Zhu, J.D. Cressler, and S-H. Lee, Jpn. J. Appl. Phys., 47, 5309 (2008)
3. K.C. Saraswat and F. Mahammadi, Effect of interconnection scaling time delay of VLSI circuits, IEEE Trans. Electron. Dev., ED-29, 645–650 (1982) and also J. Davis, V. De, and J. Meindl, A stochastic wire-length distribution for gigascale integration, IEEE Trans. Electron. Dev., 45 (3), 580 (1998) and also K. Ishimaru, Solid Sate Electron, 52 (9), 1266 (2008)
4. H.B. Bakoglu, Circuits, interconnects and packaging for VLSI, Addison, Reading, MA, Chapter 2, p. 26 (1990) and also K. Yamashita, and S. Odanaka, Interconnect scaling scenario using chip level interconnect model, Proc. Symp. VLSI Technol., 53 (1997) and S. Thompson et al, 130 nm logic technology featuring 60 nm transistors, Intel. Tech. J., (May 2002) and also A. Piontek et al., Semicond. Sci. Technol., 22, S9 (2007)
5. Y.V. Rajput, Modeling distributed RC lines for the transit analysis of complex networks, Int. J. Electronic, 36 (5), 709–717 (1974) and also T. Sakurai, Closed form expression for interconnect delay, coupling and cross talk in VLSI's, IEEE Trans. Electron. Dev., 40 (1), 118 Jan (1993) and P. Kapur et al., Technology and reliability constrained future Cu-interconnects: Resistance modeling: IEEE Trans. Electron Dev., 590, April (2002) and also C-C Ho and B.S. Chiou, Microelectron. Eng. 84 (4), 646 (2007)
6. W.E. Engeler and D.M. Brown, Performance of refractory metal multilevel interconnection system, IEEE Trans. Electron. Dev., ED-19 (1), 54–61, Jan (1972) and also X.W. Lin, and D. Pramanik, Future interconnect technologies and copper metallization, Solid State Technol., Oct. (1998) and also A. Knorr, B. Kastenmeier, and N. Moumen, US Patent, 7125782, Nov. (2006)
7. C. Alpert, A. Devgan, and S. Quay, Buffer insertion for noise and delay optimization, IEEE Trans. Computer-aided design of integrated circuits and systems, 18 (11), 1633 Nov. (1999) and also K.C. Saraswat and F. Mohammadi, Effect of scaling of interconnections on the time delay of VLSI circuits, IEEE J. Solid State Circuits, SC-17, 275–280, April (1982) and also H. Kudo et al., IEEE Conf. IITC, San Francisco, CA, June (2008)
8. C.P. Yuan and T.N. Trick, A simple formula for the estimation of the capacitance of two-dimensional interconnects in VLSI circuits, IEEE Electron. Dev. Lett., EDL-3, 391–393, Dec. (1982) and also O. S. Nakagawa, D. Sylvester, J. McBride, and S.Y. Oh, On chip cross talk noise model for deep submicrometer ULSI interconnect, The Hewlett Packard J., 39, May (1998) and L. Lang et al., IEEE IEDM Tech. Dig., Issue 1, 46 (2000)
9. D.B. Jarvis, The effects of interconnections on high speed logic circuits, IEEE Trans. Electron. Circ. Comp., 12, 476–487, Oct. (1963) and also J. Davis, V. De, and J. Meindl, A stochastic wire-length distribution for gigascale integration, Part-II, IEEE Trans. Electron. Dev. 45 (3), 590, March (1998) and also K. Fife, A. El Gamal and H-S.P. Wong, IEEE IEDM Tech Dig., 1003, Dec (2007)
10. I. Catt, Cross talk in digital systems, IEEE Trans. Electronic Comp., EC-16 (6), 743–763, Dec. (1967) and also S. Takahashi, M. Edahiro, and Y. Hayashi, Interconnect design strategy: Structures, Repeaters, and Materials toward 0.1 μm ULSI with giga-hertz clock operation, Proc. IEDM, 833, Dec. (1998) and also K. Fife, A. El Gamel, and H-S-P Wong, IEEE ISSCC Tech. Dig., 48, Feb. (2008)
11. B. Amrutur and M. Horowitz, A replica technique for word line and sense control in low power SRAMs, IEEE J. Solid State Circ., 33 (8) Aug. (1998) and A. Sakata et al., IEEE Conf. IITC, San Francisco, CA, June (2008)
12. J.A. Lowersson and P. Anderson SRAM cells for low power write in buffer memories, 1995 Symp. on low power electronics, San Jose, CA, (Oct 9–11, 1995) and K. Kim, H. Mahmodbi,

and K. Roy, Proc. Int. Symp. On Low power Electronics and Design, ACM, New York, USA, (2007)

13. M. Matsumiya et al., A15-ns 16 Mb CMOS and SRAMs with interdigital bit line architecture, IEEE J. solid state circuits, 27 (11), Nov. (1992) and US Patent 20070245065 Oct (2007)
14. H.B. Bakoglu and J.D. Meindl, Optimal interconnect for VLSI, IEEE Int. Solid State Circ., Conf. ISSCC 1984, San Francisco, 164–165, Feb. (1984) and IEEE Trans. Electron Dev., ED-32, 903–909, May (1985) and also J. Davis and J. Meindl, Compact distributed RLC interconnect models Part-II: Computed line transient expressions and peak cross talk in multilevel networks, IEEE Trans. Electron. Dev., 47 (11), 2078 Nov. (2000)
15. D. Edelstein et al., Full copper wiring in a sub-0.25 μm CMOS ULSI technology, Proc. IEEE IEDM, 773–776 (1997) and also R. Venkatesan, J. Davis, K. Bowman, and J. Meindl, Optimal repeater insertion for N-tier multilevel interconnect architecture, Proc. Intl. Interconnect Technol. Conf., 132, June (2000) and also A. Knorr, B. Kastenmeier, and N. Moumen, US Patent, 7125782, Nov. (2006)
16. P. Kapur, P. McVittie, and K.C. Saraswat, Technology and reliability constrained future copper interconnects, IEEE Trans. Electron Dev., 49 (4) April (2002) and T. Murata et al., Jpn. J. Appl. Phys., 47, 2488 (2008)
17. C.K. Hu and J.M.E. Harper, Copper interconnects and reliability, Mater. Chem. Phys., 52 (5) (1998) and also S.P. Jenget et al., Process integration of low-dielectric constant material, Mater. Res. Soc. Symp. Proc., 381, 197 (1995) and T. Nogami et al., Characteristics of Cu/barrier metal interface for copper interconnects, Proc. IEEE Int. Interconnect. Tech. Conf., 298 (1998) and also C-L. Chang, C-H. Hsieh, and S-L. Shue, US Patent, 20060113675 June (2006)
18. G.B. Sorkin, A symmetrically perfect trivial global routing, IEEE Trans. Computer-aided Des. CAD 6 (9), 820 (1987) and also D. Boeing, and S. Nassif, Models of process variations in device interconnect, in Design of High Performance micrprocessor circuits, In A. Chandrakasan, W. Bowhill, and F. Fox (eds.). IEEE Pub., (2000) and also N. Toon, Lecture Notes in Comp. Sci., Springer, Berlin, Vol. 849, p. 292, Jan (2006)
19. R.K. Watts, Submicron integrated circuits, in interconnects, Wiley, New York, p. 270 (1989) and also Z. Lin, C. Spanos, L. Milor, and Y.T. Lin, Circuit sensitivity to interconnect variations, IEEE Trans. Semi. Manuf., 11 (4), 557 (1998) and X-J. Yuan, J.M. hart, G-Z Ling, and P.S. Ypung, US Patent, 7109734, Sept 19 (2006)
20. L.A. Glasser and D.W. Dobberpuhl, The design analysis of VLSI circuits, Addison Wesley, Reading, MA (1985) and also B. Amrutur, M. Horowitz, A replica technique for worldline and sense control in low-power SRAMs, IEEE J. Solid State Circ., 33 (8) (1998) and also J. Allowerson, P. Anderson, SRAM cells for low-power write in buffer memories, 1995 Symp. On low-power electronics, San Jose, CA, Oct. 9–11 (1995)
21. H.B. Bakoglu, Circuits, interconnections and packaging for VLSI, Addison Wesley, Reading, MA, p. 196 (1990) and also K. Rahmat, O.S. Nakagawa, S.Y. Oh, and J. Moll, A scalling scheme in deep submicron process, Proc. IEDM., 241, Dec. (1995) and R. Ho et al., The future of wires, Proc. IEEE, 490–504, April (2001) and J. Srinivasan, Proc. Int'l. Conf. On dependable systems and networks (DSN-04), (2004) and S. Borakar, IEEE MICRO, July–Aug. (1999)
22. T.N. Theis, The future of interconnection technology, J. IBM Res. Devp. 44 (1), 379, May (2000), and C.L. Gan et al., Adv. Maters.for Micro 7 Nano Systems (AMMNS) 7th May (2008)
23. C. Hilbert, D.A. Gibson, and D. Herrel, A composition of lossy superconducting interconnect for computers, IEEE Trans. Electron. Dev., 36 (9), 1830, Sept. (1989) and also Y. Liu, L. Pileggi, and A. Strojwas, Model order reduction on RC(L) interconnect, including variational analysis, Proc. DAC., 201 (1999)
24. D.R. Resnick, Hardware technologies from ETA-10 in Proc. 2nd Int. Conf. supercomputing, 153–157 (1987) and also A. Zorin, S.V. Lotkhov, H. Zangerle, and J. Niemeyer, Coulomb

blockade co-tunneling in single electron circuits with on chip resistors, J. Appl. Phys., 88 (5), 2665 (2000)
25. T. Furukatsu, T. Watanabe, and R. Kondo, NEC super computer SX system, Nikki electron, 11–19, 237–272 (1984)
26. S. Leseduarte et al., IEEE-CPMT., 23 (4), 673–679, Dec. (2000) and In P. Garrou, C. Bower, and P. Ramm (eds.) Hand book of 3-D integration, J. Wiley, NJ, (2008)
27. S. Pine et al., Euro. Symp., Paris, France (2001) and also D. Boning, T. Maung, J. Chung, K.J. Chang, S.Y. Oh., and D. Bartelink, Statistical metrology of interlevel dielectric thickness variation, Proc. SPIE Symp. On Microelectronics manufacturing, Austin, TX., Vol. 2334, p. 316 (1994)
28. T.K. Gupta, Handbook of thick and thin film hybrid microelectronics, Wiley, NJ, Chapter 11, p. 363 (2003) and also E. Beyne, High bandwidth interconnect via a novel chip stack package, Solid State Technol., s19 (2002)
29. K Lee, On chip interconnects—gigahertz and beyond, Solid State Tech., 41, Sept. (1998) and M. Nawaz, W. Molzer, S. Decker, L.F. Giles and T. Schulz, Microelectron. J., 38 (12), 1238 (2007)
30. G.A. Katopis, Delta noise specification for a high performance computing machine, Proc. IEEE, 75 (9), 1405–1415, Sept.(1985) and also K. Tang and E. Friedman, Peak cross talk noise estimation in CMOS VLSI circuits, Proc. Intl. Conf. On Electron. Circ. Syst., 3, 1539 (1999)
31. H.B. Bakoglu, Circuits interconnections and packaging for VLSI, Addison Wesley, Reading, MA, p. 311 (1990) and also A. Devgan, Efficient coupled noise estimation for on chip interconnects, Proc. Intl. Conf. On Computer aided design, p. 147 (1997)
32. K.C. Gupta, R. Grag, and I.H. Bahl, Microstrips and slot lines, Artech House, Dedham, MA (1979) and also N. Chang, V. Kanevsky, O.S. Nagawa, K. Rahmat, and S.Y. Oh, Fast generation of statistically based worst case modeling of on chip interconnect, Proc. IEEE Intl. Conf. Computer-aided design: VLSI in computers and processors, p. 720 (1997) and also K. Minoglu et al., IEEE Conf. IITC, San Francisco, CA, June (2008)
33. T.K. Gupta, Hand book of thin and thin film hybrid microelectronics, Wiley, NJ, Chapter 11, p. 364 (2003)
34. H.B. Bakoglu, Circuits Interconnections and packaging for VLSI, Addison Wesley, Reading, MA, (1990) and also C. Kozyrakis et al., Scaleable processors in the billion transistors era: IRAM, IEEE Computer, 75–78, Sept. (1997)
35. R. Ansari and B. Liu, Multi rate signal processing, In S.C. Mitra and J.J.F. Kaiser (eds.), Hand book for digital signal processing, John Wiley, New York, p. 1016 (1993)
36. H.C. Lin and L.W. Linholm, An optimized output stage for MOS integrated circuits, IEEE J. Solid State Circuits, Sc-10 (2), 106–109, April (1975) and also J.H. Chen, J. Huang, L. Arledge, P.C. Li., and P. Yang, Multilevel metal capacitance models for CAD design synthesis systems, IEEE Electron Device Lett., 13 (1), 32, Jan. (1992)
37. C. Mead and M. Rem, Minimum propagation delays in VLSI, IEEE J. Solid State Circuits, Sc-17, 773–775, Aug. (1982) and also K. Nabors and J. White, Fast cap: A multiple accelerated 3-D capacitance extraction program, IEEE Trans. On Computer Aided Design of Integrated Circuits and Systems, 10 (11), 1447 (1991)
38. E.T. Lewis, Optimization of device area and overall delay for CMOS VLSI design, Proc. IEEE, 72, 670–689, June (1984)
39. H.B. Bakoglu, Circuits, Interconnections, and Packaging for VLSI, Addison Wesley, Reading, MA, p. 215 (1990)
40. W.R. Heller, W.E. Donath, and W.F. Mikhail, Prediction of wiring space requirements for LSI, Proc. 14th design automation Conf., New Orleans, pp. 32–43 (1977), and also W.E. Donath, Placement and average interconnection lengths of computer logic, IEEE Trans. Circuits, Systems, CAS-26, 272–277, April (1979)
41. H.B. Bakoglu, Circuits, interconnections, and packaging for VLSI, Addition Wesley, Reading MA, p. 290 (1990) and also L. He, N. Chang, S. Lin, and O.S. Nagawa,

An efficient inductance modeling for on chip interconnects, Proc. CICC, 457–460 (1999)

42. M. Bohr et al., A 90 nm generation logic technology, Proc. IEDM, (2002) and also T. Theis, The future of interconnection technology, IBM Res. Dev., 44 (3), May (2000)
43. S.H. Unger and C.J. Tan, Clocking schemes for high speed digital systems, IEEE Trans. Comput., C-35 (10), 880–895, Oct. (1986) and also B. Stine, D. Boning, and J. Chung, Analysis and decomposition of spatial variation in integrated circuit processes and devices, IEEE Trans. Semi. Manuf. 10 (1), 24 (1997) and also A. Knorr, B. Kastenmeir, and N. Moumen, U.S. Patent, 7125782 Nov. (2006)
44. J.A. Marques and A. Cunha Clocking of VLSI circuits, Chapter 15 of VLSI architecture, B. Randell and P.C. Treleaven (eds.) Prentice Hall Inc. Englewood Cliffs, NJ, (1983) and also R. Streiter, H. Wolf, Z. Zhu, X. Xiao, and T. Gessner, Thermal and electrical simulation of deep submicron interconnection system, Microelectron. Eng. 60, 39 (2002)
45. K.D. Wagner, Clock system design, IEEE Design Test Computers, Oct. (1988) and also M.T. Bohr, Interconnect scaling—a real limiter of high performance ULSI, IEEE International Electron Device Meeting (IEDM), 95, 241 (1995)
46. J.R. Black, Proc. IEEE, 57, 1587 (1969) and also J.R. Black. Proc. 6th annual Int. Reliab. Symp., Piscataway, NJ, 148 (1967) and also Z. Lu et al., IEEE Trans. VLSY Syst., 15 (920), 159 (2007)
47. H.B. Bakoglu, Circuits, interconnections, and packaging for VLSI, Addition Wesley, Reading MA, p. 375 (1990) and also X.W. Lin, and D. Pramanik, Future interconnect technologies and copper metallization, Solid State Technol., 631, Oct. (1998) and V. Sukharev, E. Zschech, and N.D. Nix, J. Appl. Phys., 102, 3505 (2007)
48. W. Steinhoegl, G. Schindler, G. Steinlesberger, M. Traving, and M. Engelhardt, Comprehensive study of copper wires with lateral dimensions of 100 nm and smaller, J. Appl. Phys., 97, 23706, (2005)
49. N.W. Ashcroft and N.D. Mermin, Solid State Physics., Holt, Rinehert and Winston Pub., New York, p. 52 (1962) and also W.R. Hunter, IEEE Trans. Electron Dev., 44 (2), 304 (1997) and M. Rossnagel and T.S. Kuan, Alteration of Cu-conductivity in size effect regime, J. Vac. Sci. Technol., 22 (1), 240 (2004) and also O. Van der Straten, Y. Zhu, J. Rullan, K. Dunn, and A.E. Kaloyeros, MRS Symp. (JMR), Jan. (2006)
50. P.A. Hart, T.V. Hof, and F.M. Klaassen, Device down scaling and expected circuit performance, IEEE Trans. Electron. Dev., ED-26 (4), 421–429, April (1979) and L. Peters, Semicond. Int., 56, Nov. (1999) and also A. Gouldstone et al., Acta Maerialia, 55, 4015 (2007)
51. C. Ryu et al., Microstructure and reliability of Cu-interconnect, IEEE Trans. Electron. Dev., 46, 1113–1120 (1999) and also L. Peters, Yield management, Semicond. Int., 34, Jan. (2003) and K. Mistry et al., IEEE IEDM Tech Dig., 247 (2007)
52. C. Lingk, M.E. Gross, and W.L. Brown, Texture development of blanket electroplated copper films, J. Appl. Phys., 87, 2232–2236 (2000) and also J.A. Cunningham, improving copper interconnects, Semicond. Int., 98, April (2000)
53. S. Lagrange et al., Microelectron. Eng., 50, 449 (2000) and A.M. Minor et al., Phil. Mag., 85 (2–3), 323 (2005)
54. D. Edelstein et al., Tech Dig. IEEE Int. Elect. Dev. Mtg., 773 (1997) and also A. Gouldstone et al., Acta Mater., 55, 4015 (2007)
55. D. Edelstein et al., High performance liner for copper damascene interconnects, Conf. IEEE Interconnect Int'l Tech. Conf. (IITC) (2001) and M. Jin, A.M. Minor, E.A. Stach, and J.W. Morris Jr., Acta Mater., 52, 5381 (2004)
56. C. Cabral Jr. et al, Room temperature evaluation of microstructure and resistivity in electroplated Cu-films, Proc. Advanced metallization Conf. 1998, Pub. Mat. Res. Soc. PA, p. 81 (1999) and J.H. An and P.J. Ferreira, Appl. Phys. Lett., 89, 151919 (2006)
57. T. Sullivan and D. Pierce, Interconnect reliability with a focus on Cu, Int'l Reliab. Workshop (IRW) Final report (1998) and C.K. Hu, L. Gignac, and R. Rosenberg, Microelectron. Relib., 46, 213 (2006)

58. S. Lagrange et al., Micrelectron. Eng., 50, 449 (2000) and also L. Lu, R. Tao, L.B. Wang, B.Z. Ding, and K. Lu, Grain growth and strain release in nanocrystalline copper, J. Appl. Phys., 88 (11), 6408 (2002)
59. T. Ohba and T. Nakamura, Current status and issues for Cu-interconnect technology, Proc. Sci. Technol. Div. (IV) Symp., Jap. Int'l. Metals, Tokyo, p. 5 Jan. 26, (2001)
60. T. Ohba, A study of current multilevel interconnect technologies 90 nm nodes and beyond, Fujitsu Sci. Tech. J., 38, 1, June (2002)
61. J. Flake, Y. Solomentsev, J. Cooper, and K. Cooper, Wafer scale profile evolution of electrochemically deposited copper films, J. Electrochem. Soc.,150 (4), C195 (2003) and P.R. Besser et al., U.S. Patent, 6992004 Jan. (2006)
62. D. Wang and U. Landau, J. Electrochem. Soc., 142, 2598 (1995)
63. Y. Lantasov, R. Palmans, and K. Maex, Microelectron. Eng., 50, 441 (2000) and S-H. Kim et al., Electrochem. Solid State Lett., 11 (5), H-127 (2008)
64. E.K. Broadbent, E.J. McInerney, L.A. Gochberg, and R.L. Jackson, J. Vac. Sci. Technol., B-17, 2584 (1999) and also H. Deligiani, J.O. Dukovic, E.G. Walton, R.J. Contolini, J. Reid, and E. Patton, Electrochemical processing in ULSI fabrication of semiconductor/metal deposition II, and In P.C. Andricacos, P.C. Sarason, P. Allongue, J.L. Stickney, and G.M. Oleszek (eds.) PV99-9, Electrochem. Pub. Pennington, NJ, p. 83 (1999) and also P.S. Ho, K.D. Lee, J.W. Pyun, X. Lu, and S. Yoon, Dielectric and scaling effects of electromigration for copper interconnect, Springer, London, 225 (2006)
65. C.K. Hu, K.Y. Lee, L. Gignac, and R. Craters, Thin Solid Films, 443, 338 (1977) and also C.S. Hau-Riege and C.V. Thompson, Appl. Phys. Lett., 77 (3), 352 (2000)
66. J.K. Jung, N.M. Hwang, and Y.C. Joo, J. Koren Phys. Soc., 40 (1), 90 (2002) and also J.M. Paik, K.C. Park, and Y.C. Joo, Relationship between grain structures and texture of damascene Cu-lines, J. Electronic. Mater., 33 (1), 48 (2004)
67. F.M. D'heurel and A. Gangulee, Thin Solid Films, 25, 531 (1975)
68. Y. Igarashi and T. Ito, J. Vac. Sci. Technol., B-16 (5), 2745 (1998)
69. T. Tonegawa, M. Hirori, K. Motoyama, K. Fuzi, and H. Miyamoto, Proc. IEEE Interconnect Int'l Tech. Conf. (IITC), 216 (2003)
70. E. Lee et al., Evaluation of PVD-deposited Cu-alloy interconnect applications, Adv. Electronic Manufacturing Technol., V-EMT 1:16, 12th July, (2003) and J. Gambino et al., Proc of IEEE IITC, 22 (2007)
71. N. Iwamoto, N. Truong, and E. Lee, Proc. Int. Conf. Metallurgical Coatings and Thin Films, (2004)
72. J.M. Paik, K.C. Park, and Y.C. Joo, Relationship between grain structures and texture of damascene Cu-lines, J. Electronic. Mater., 33 (1), 48 (2004) and also C.E. Kalnas, P.R. Keller, and D.P. Field, Crystallographic characterization of interconnects by orientation mapping, Int. Conf. Characterization and metrology for ULSI Tech. Am. Inst. Phys. New York., 849 (1998)
73. C.S. Hau-Reige and C.V. Thompson, Appl. Phys. Lett., 77 (3), 352 (2000)
74. E.T. Ogawa, K.D. Lee, V. Balaschke, and P.S. Ho, Electromigration reliability issues in dual damascene Cu-interconnections, IEEE Trans. Reliab., 51, 403 (2002)
75. L.S. Darken and R.W. Gurry, Physical Chemistry of Metals, McGraw Hill, New York, (1956) and also A.A. Volinsky, Mechanical reliability and characterization of modern microelectronic interconnect structures, Nano-engineering world forum, 2003, Int'l Engineering consortium (IEC), Royal Plaza Hotel and Trade Center, Marlbrough, MA, June 23–25 (2003)
76. A. Kohn et al., Characterization of electroless deposited CoWP thin films for encapsulation of Cu-metallization, Mat. Sci. Eng., A302, 18, (2001) and C.K. Hu, Appl. Phys. Lett., 84, 4986, (2004) and C-C Chiu, H.H. Chang, C.C. Lee, C. Chsia, and K.N. Chang, Microelectron. Relib., 47 (9) 1506 (2007)
77. G. Steinlesberrger et al., Proc. the advanced metallization Conf., p. 397 (2002) and also p. 213 (2003)
78. J.R. Lloyd, Electromigration in integrated circuit conductors, J. Phys. D, Appl. Phys., 32, R109 (1999) and also Y. Liu and R.J. Diefendorf, On stress evolution and interaction during

electromigration in near bamboo structure lines, Appl. Phys. Lett., 71, 3171 (1997) and M.S. Yeh et al., IEEE IITC, 113 (2006)
79. I.A. Blech and C. Herring, Appl. Phys. Lett., 29, 131 (1976) and also I.A. Blech, Critical length in electromigration, in AIP Conf. On Proc. Stress induced phenomena in metallization: Fourth Int'l. Workshop, 418, 3–13 (1998)
80. J.J. Clement and C.V. Thompson, J. Appl. Phys., 2, 900 (1995) and also R.P. Vinci, E.M. Zielinski, and J.C. Bravman, Thin Solid Films, 262, 142 (1995)
81. N.L. Beverly, G.B. Alers, and J.A. Prybyle, Appl. Phys. Letts., 68 (17), 2372 (1996)
82. R.G. Filippi, G.A. Biery, and R.A. Wachine, J. Appl. Phys., 78 (6), 3756 (1995)
83. C.L. Gan, C.V. Thompson, K.L. Pey, W.H. Choi, H.L. Tay, B. Yu, and M.K. Radhakrishan, Appl. Phys. Lett., 79, 4592 (2001) and Y. Hou et al., Semicond. Sci. and Technol, 23, 075023 (2008)
84. S.P. Hau-Reige, J. Appl. Phys., 91, 2014 (2002)
85. Y. Liu and X. Huang, Substrate induced stress and transformation characteristics of a deposited Ti-Ni-Cu thin film, Phil. Mag., 84 (19), 1919 (2004)
86. K.N. Tu, Recent advances on EM in very large scale integration of interconnects, J. Appl. Phys., 94 (9), 1 (2003) and J. Gimbino et al., IPFA, 59 (2007)
87. R.P. Vince, E.M. Zielinski, and J.C. Bravman, Mat. Res. Soc. Symp., (1994) and A. Wei et al., Symp. On VLSI Tech. Dig., 216 (2007)
88. M.A. Marcus, W.F. Flood, R.A. Cirelli, R.C. Kristler, N.A. Ciampa, W.M. Mansfield, D.L. Berr, C.A. Volkert, and K.G. Steiner, Mat. Res. Soc. Symp. Proc., 338, 203 (1994) and also M.P. Surh, Threshold stress behavior in thin film electromigration behavior, J. Appl. Phys., 85, 8145 (1999)
89. E.S. Ege and Y.L. Shen, Thermomechanical response and stress analysis of copper interconnects, J. Electronic. Mats., 32 (10), 1000 (2003) and Z. Lu et al., IEE Trans on VLSI Syst. 15 (20), 159 (2007)
90. D. Weiss, O. Kraft, E. Artz, Grain boundary voiding in self-passivated Cu and Al alloy films in silicon substrates, J. Mat. Res., 17 (6), 1363–1370, June (2002)
91. S.P. Baker, A. Kretchmann, and E. Aret, Acta Mater., 49, 2145 (2001)
92. V. Weihnacht and W. Bruckner, Acta Mater., 49, 2365 (2001)
93. Y.L. Shen and V. Rammurthy, J. Appl. Phys., 93, 1806 (2001)
94. T.N. Ning, P.W. Cook, R.H. Dennard, C.M. Osburn, S.E. Shuster, and H.N. Yu, 1 μm MOSFET VLSI technology, Part-IV, Hot electron design constraints, IEEE Trans. Electron. Dev., ED-26, 346–353 April (1979) and also K.N. Tu, Electromigration in stressed thin films, Phys. Rev., 45, 1409 (1992)
95. C.L. Gan, C.V. Thompson, K.L. Pey, and W.K. Choi, Experimental characterization and modeling of the reliability of three terminal dual damascene interconnect trees, J. Appl. Phys., 94^{TM}.(2), 1222 (2003)
96. J. Lloyd and J.J. Clement, Thin Solid Films, 262, 135 (1995) and also J.C. Ling and C. Leez, A study of grain boundary diffusion of copper in titanium nitride thin films, Electrochem. Solid State Lett., 2 (4), 181 (1999)
97. E. Kolawa, P.J. Pokela, J.S. Reid, J.S. Chen, and M. Nicolett, Amorphous Ta-Si-N diffusion barrier in Si/Al& Si/Cu metallization, Appl. Surf. Sci., 53, 373 (1991) and also C.K. Hu, R. Rosenberg, and K.Y. Lee, Appl. Phys. Lett., 74, 2945 (1999)
98. L. Arnaud, G. Tartavel, T. Berger, D. Mariolle, Y. Gobil, and T. Touet, Microelectron. Relib., 40, 77 (2000)
99. N.D. McCusker, H.S. Gamble, and B.M. Armstrong, Microelectron. Relib., 40, 69 (2000)
100. C.K. Hu, R. Rosenberg, and K.Y. Lee, Electromigration path in thin film lines, Appl. Phys. Letts., 74, 2945–2947 (1999) and C.K. Hu, L. Gignac, and R. Rosenberg, Microelectron. Relib., 46, 213 (2006)
101. C.S.H. Reige and C.V. Thompson, Electromigration in Cu-interconnects with very different grain structures, Appl. Phys. Letts., 78, 3451–3453 (2001)

102. C.K. Hu, K.Y. Lee, L. Gignac, and R. Carruthers, Electromigration in 0.25 W wide Cu line on W, Thin Solid Films, 308–309, 443–447 (1997)
103. K. Abe, Y. Harada, and H. Onoda, Cu-damascene interconnects with crystallographic texture controlled electromigration performance, 36th Annual IEEE Int. Reliability Phys. Symp. Proc., 342–347 (1998) and J.R. Loyd et al., IEEE Trans. Electron Dev. Mat. Reliab., 5 (1), 113 (2005)
104. C.K. Hu et al., Mechanisms of very long electromigration lifetime in dual damascene Cu interconnections, Appl. Phys. Letts., 78, 904–906 (2001) and M.Y. Meyer, Ph.D. thesis, Brandenburg University of Technol. Cottbus, Germany, (2007)
105. M.P. Surth, Thresh hold stress behavior in thin film electromigration, J. Appl. Phys., 85, 8145–8154 (1999) and M.A. Meyer and E. Zschech Proc. Int. Workshop on stress induced phenomena in Metallization, AIP Conf. Proc. Kyoto, Japan, (2007)
106. H. Ono, T. Nakano, T. Ohta, Appl. Phys. Lett., 64 (12), 1511–1513 (1994) and also J.C. Slit, D. Vick, K. Robbie, and M.J. Brett, Thin film microstructural control using glancing angle deposition by sputtering, J. Mater. Res., 14, 1197 (1999)
107. K. Holloway et al., J. Appl. Phys., 71 (11), 5433–5443 (1992) and also J.F. Whitacre, Z.U. Rek, J.C. Bilello, and S.M. Yalisov, Surface roughness and in plane texturing in sputtered film, J. Appl. Phys., 84 (3), 1346 (1998)
108. M. Takeyama et al., J. Vac. Sci. Technol., B-14 (2), 674–678 (1996), and D. Pandi and G.A. Dixit, Solid State Technol., 28, Nov. (2003)
109. J.R. Lloyd and J.J. Clement, Electromigration in copper conductors, Thin Solid Films, 262, 135 (1995)
110. C.K. Hu, R. Rosenberg, and K.Y. Lee, Electromigration path in Cu thin film lines, Appl. Phys. Lett., 74 (20), 2495 (1999)
111. L. Arnaud, G. Tartavel, T. Berger, D. Mariolle, Y. Gobil, and I. Touet, Microstructure and electromigration in copper damascene lines, Microelectron. Relib., 40, 77 (2000)
112. J. Proost, T. Hirato, T. Furuhara, K. Maex, and J.P. Celis, Microstructure and electromigration induced drift in electroplated damascene Cu, J. Appl. Phys., 87 (6), 2792 (2000)
113. F. Wei et al., Lengths effect on the reliability of dual damascene Cu interconnects, Mat. Sci. Soc. Symp., 716, B13.3.1 (2002)
114. C.L. Gan et al., Appl. Phys. Lett., 79, 4592 (2001) and also E.T. Ogawa et al., Proc. 39th Int. Reliab. Phys. Symp. P. 341 (2000) and S.P. Hau-Riege, J. Appl. Phys., 91, 2014 (2002)
115. K.N. Tu, J. Appl. Phys., 94 (9), 5451 (2003) and also K.N. Tu, C.C. Yeh, C.Y. Liu, and C. Chen, Effect of current crowding on vacancy diffusion and void formation in electromigration, Appl. Phys. Lett., 76, 9881 (2000) and also M.A. Meyer and E. Zschech, Proc. Int. Workshop on stress induced phenomena in Metallization, AIP Conf. Proc. Kyoto, Japan, (2007)
116. A.A. Istratov and R. Weber, Physics of copper in silicon, J. Electrochem. Soc., 149 (1), G-21 (2002) and also S.P. Hau-Reige, New methodologies for interconnect reliability assessments of integrated circuits, Ph.D. dissertation, MIT, April, (2000)
117. K. Holloway et al., J. Appl. Phys., 71, 5433 (1992) and also P. Kapur, G. Chandra, J.P. McVitte, and K.C. Saraswat, Technology and reliability constrained future copper interconnects, Part II, IEEE Trans. Electron. Dev., 40 (4), 598–604 (2002)
118. E.E. Davidson, B.D. Mc Credie, and W.V. Vilkelis, Long lossy lines (L3) and their impact upon chip performance, IEEE. Comp. Packaging Manufac. Tech. No.1, Part B (4), 361–375 (1997) and also K.H. Min, K.C. Chun, and K.B. Kim, Comparative study of Tantalum and Tantalum nitrides as a diffusion barrier for Cu metallization, J. Vac. Sci. Technol., B-14 (5), Sept–Oct (1996) and also N-H. Kim, S-Y. Kim, W-S. Lee, and E. Chang, Microelectron. Eng., 84 (11), 2663 (2007)
119. J.M. Paik, K.C. Park, and Y.C. Joo, Relationship between grain structures and texture of damascene Cu line, J. Electronic. Mats., 33 (1) (2004) and J. Gambino et al., IEEE IITC, June (2007)

120. H.B. Nie et al., Appl. Phys., A73, 229 (2001) and J.Y. Cho, K. Mirpuri, D.N. Lee, J.K. An and J.A. Szpunar, J. Electron. Mats., 34 (1), 53 (2005)
121. A.E. Kaloyeros et al., J. Mater. Res., 15 (12) 2800 (2000) and B. Predel, In O. Medelung (ed.) Electronic Mats. Semiconductors, Landolt Bornstein, Springer, Verlag, (2006)
122. J.D. Torre et al., J. Appl. Phys., 94 (1), 263 (2003) and Y. Kakuhara, Jpn. J. Appl. Phys., 47, 4475 (2008)
123. R.N. Tait, S.K. Dew, T. Smy, and M.J. Brett, J. Appl. Phys., 70, 429 (1991) and also S. Krongelb, L.T. Romankiw, and J.A. Tornello, IBM J. Res. Dev., 42, 575 (1998) and K.D. Lee et al., IEEE 94th Int. Pelib. Phys. Symp. April (2008)
124. P. Muller and T. Kamins, Device electronics for integrated circuits, 2nd. Ed. Wiley, pp.1–56 (1986) and also L. Maggnin, R. Maboudian, and C. Carraro, Selective deposition of thin films on to silicon with improved adhesion, Electrochem. Solid State Lett., 4 (1) C-5 Jan. (2001)
125. H.A. Lee, L. Ting, N.C. Tso, and C.U. Kim, Analysis of the reservoir length and its effect on electromigration, J. Mats. Res. (1), 167 (2002) and C.S. Hau Riege and A.P. Marathe, IEEE Int. Symp. Reliab. Phys., Phoenix, AZ, April (2008)
126. E.C. Yeh and K.N. Tu, Numerical simulation of current crowding phenomenon and their effects on the electromigration in vary large scale integration interconnects, J. Appl. Phys., 88 (10), 5680 (2000)
127. N.V. Hieu, Ph.D. Thesis, Multilevel interconnect reliability, Univ. Twente, Natherland, March, (2004) and K.N. Tu, Recent advances ion electromigration in very large scale integration interconnects, J. Appl. Phys., 94 (9) (2003) and J.A. West, D.M. Barth, and P.S. Zuhoski, US Patent, 730965 (2007)
128. K.D. Lee, E.T. Ogawa, H. Matsuhashi, P.R. Justison, K.S. Ko, and P.S. Ho, Electromigration critical length effect in Cu/oxide dual damascene interconnects, Appl. Phys. Lett., 79, 3236 (2001)
129. S. Lin et al., An optimized integration scheme for 0.13 μm technology node dual damascene Cu interconnect, Proc. Int. Interconn. Tech. Conf. (IITC, IEEE) 273, San Francisco, CA, June 2–4 (2000)
130. T. Kajita et al., Highly reliable Cu-low dual damascene interconnect technology with hybrid (PAE/SiOC) dielectrics for 65 nm–node high performance eDRAM, in Proc. Int. Interconn. Tech. Conf. (IITC, IEEE) 273, San Francisco, CA, June 2–4 (2003)
131. D.H. Wang, M. Afnan, and S.S. Chiao, Stress free polishing advances use of copper interconnects on silicon wafers, Semiconductor Fab. Tech. 13th ed. 255 (2003) and Y. Toma et al., Adv. Mat. Conf. Sept 26–27, The Univ. of Tokyo, Japan, (2006)
132. J.J. Toomy, S. Hymes, and S.P. Murarka, Stress effects in thermal cycling of copper (magnesium) thin films, Appl. Phys. Lett., (1995)
133. N. Tamura, J.S. Ching, G.E. Ice, and B.C. Larson, Mater. Relib. Microelectron., IX In D.D. Brown, A.H. Verbruggen, and C.A. Volkert (eds.) Mater. Res. Soc. Proc. 563, Pittsburgh, PA, 175 (1999) and P. Goudeau, O. Sicardy, N. Tamura, and O. Thomas, Int. J. Mat. Prod. Technol., 26 (3–4), 354 (2006)
134. J.A. Nucci, R. Keller, S. Kramer, C.A. Volkert, and M.E. Gross, Localized measurement of strains in damascene copper interconnects by convergent beam electron diffraction, Mater. Res. Soc. Symp. Proc. 617, D.8.5.1 (2000) and also E.S. Ege and Y.L. Shen, Thermomechanical response and stress analysis of copper interconnects, J. Electron. Maters., 23 (10), 1000 (2003) and K. Kudo et al., IEEE Symp. Int. Reliab. Phys. Symp., Phoenix, AZ, April 27 (2008)
135. H. Hasegawa and Y. Hirai, J. Appl. Phys., 90, 2792 (2001)
136. G. Schneider et al., Electromigration in passivated Cu-interconnects studied by transmission X-ray microscopy, J. Vac. Sci. Technol., 20 (6), 3089 (2002)
137. J.M.E. Harper and K.P. Rodbell, J. Vac. Sci. Technol., B-15, 763 (1997), and also K. Nagao, J.B. Neaton, and N.W. Ashcroft, Phys. Rev., B68, 125403 (2003)

138. D. Rabbe, Mat. Sci. Technol., 11, 461 (1995) and A.D. Rollett, L. Mitra, and B.S. ElDasher, 9th Int. Symp. On plasticity and its current applications, In A.S. Khan and O.L. Pamies (eds.) Neat Press, MD, (2002)
139. C. Diaz, H-J. Tao, Y-C. Ku, A. Yen, and K. Young, IEEE Electron Dev. Letts., 22 (6), 287 (2001) and F. Chen et al., IEEE Int. Symp. Reliab. Phys., Phoenix AZ, April (2008)
140. J.A. Croom et al., Line edge roughness characterization modeling and impact on device behavior, Proc. IEDM (2002)
141. Y. Martin and H.K. Wickramasinghe, Method for image side walls by AFM, Appl. Phys. Letts., 64, N19 (1994)
142. P. Boher et al., Optical fourier transform spectrometry for LER & LWR metrology, SPIE, 5752, 192 (2005)
143. A. Yamaguchi et al., Characterization of line edge roughness in resist patterns and estimation of its effect on device performance, Proc. SPIE, 5038, 689 (2003)
144. J. Noguchi et al., Impact of low-K dielectrics and barrier material on TDDB life time of Cu-interconnects, IEEE-IRPS Proc., 353 (2001) and also G. Raghaban et al., Diffusion of copper through dielectric films under bias temperature stress, Thin Solid Films, 262, 168 (1995)
145. M. Vogt, M. Kachel, M. Plotner, and K. Drescher, Dielectric barriers for Cu-metallization system, Microelectron. Eng. 37/38, 181 (1997) and also T.P. Ma, IEEE Trans. Electron. Dev., 45, 680 (1998)
146. J. Noguchi et al., Effect of NH_3-plasma treatment and CMP modification on TDDB improvement in Cu-metallization, IEEE Trans. Electron. Dev., 48 (7) 1340 (2001)
147. K.P. Cheung, IRPS, 432 (2003) and also D.J. Dumin, Int. J. High speed electron, Syst. 11, 617 (2001) and J. McPherson and H. Mogul, Underlying physics of the thermomechanical E model in describing low field time dependent dielectric breakdown in SiO_2 films, J. Appl. Phys., 84 (3), 1513 (1998) and N. Suzumura et al., IEEE Int. Symp. Reliab. Phys., Phoenix AZ, April (2008)
148. Q. Chen, J.A. Davis, P. Zarkesh Ha, and J. Mendil, A compact physical via blockage model, IEEE Trans VLSI Sys., 8 (6), 689 (2000)
149. ITRS, (2003)
150. N. Srivastava and K. Banerjee, A comparative scaling analysis of metallic and carbon nanotube interconnections for nanometer scale VLSI technologies, Proc. 21st Int. VLSI multilevel Interconnect Conf. Sept. 29-Oct. 2, Wikoloa, HI, 393–398 (2004)
151. K. Banerjee and A. Mehrotra, Analog Integrated Circuits and Signal Processing, Kluwer Academic Pub. Netherland, p. 97 (2003)
152. Z. Lu, W. Huang, M.R. Stan, K. Skadron, and J. Lach, Interconnect lifetime prediction for reliability-aware systems, IEEE Trans. VLSI systems, 15 (2), 159 (2007)

Glossary (Copper Interconnects)

ACE A (aqueous), C (chemically) E (enhanced), Post-CMP hybrid cleaning technology consisting of a combination of CO_2 cryogenic along with conventional wet etching.

Adatom A single atom on the surface of a material taking part in adsorption during nucleation.

Agglomeration Small particles (less than 5 μm) when sticking together or the action or process of collecting small particles in a mass.

ALD A (atomic) L (layer) D (deposition) is performed by sequential atomic layer deposition of material/s at a low temperature.

Aligner Tool used in photolithography for adequate positioning of the photomask relative to the wafer.Ashing: Removal of organic material by volatilization.

Amorphous A material which has random atomic structure with short-range order, indicated by a diffuse diffraction pattern.

Ångstrom Unit of a length, and 1 Å is equal to one ten billionth (10^{-10}) of a meter.

Anisotropic etch Etching normal to the surface of the wafer to be etched.

Anneal Thermal cycling (heating and cooling) to change the morphology and stress related phenomena.

APCVD A (atmospheric) P (pressure) C (chemical) V (vapor) D (deposition), meaning that the CVD reactor will operate at atmospheric pressure. CVD is defined as the formation of a non-volatile solid film by vapor phase chemical reactions.

AR A (aspect) R (ratio), meaning the ratio of height to width of a damascene structure (either trench or via hole).

ARC A (anti) R (reflecting) C (coating), is a thin layer of material used to prevent reflection of light from the surface. It can be used on the front surface or the back (BARC) of a wafer.

Aromatic thermosetting polymer Cross-linked polymer that provides a three-dimensional structure, so that the polymer forms rigid, completely insoluble material. (Aromatic means in general the organic material "having pleasant odors" and possesses a benzene-like structure.)

ASIC A (application) S (specific) I (integrated) C (circuit), meaning an IC customized for a particular use.

Attenuated PSM A phase shift mask which forms its pattern through adjacent areas of quartz or MoSi.

Barrier layer A sandwich layer designed to prevent any interaction between layers above and below the barrier layer.

Bandgap The energy gap between the conduction and valence bands of solids (metal, conductor, and insulator). The electron energy in a crystal corresponds to an energy band whose width is determined by the corresponding frequencies of the neighboring atoms.

Bandgap engineering The process or the technology of altering the energy gap of a solid material.

Bandwidth The frequency range of the messages handled by an information processing system.

BEOL B (back) E (end) O (of) L (line), the process involves the wiring of the wafer during integrated circuit (IC) fabrication.

BiCMOS The device is packed with Bi (bipolar transistor) and CMOS (complimentary MOS) transistors.

Blech length A critical length of a metal line, which allows mass migration of the metal.

Buckling When a structure undergoes visibly large displacements due to large stress. It is an indication of the growth of bulges or ripples.

CD C (critical) D (dimension), referred to dimensional control or line-width control of the imaging pattern during lithography.

CDO C (carbon) D (doped) O (oxide), involves intentional introduction of carbon impurities inside a metal oxide to reduce the normal dielectric constant of the host material.

Clock One of the most important signals in synchronous digital circuits, which controls the timing and throughput of the entire system.

Clocking/clock frequency The fundamental rate in cycles per second of a computer. In the CPU (central processing unit) the clock frequency determines the rate of data processing.

CMOS C (complimentary) M (metal) O (oxide) S (semiconductor), a monolithic switching device comprising a p-channel (PMOS) and an n-channel (NMOS) field effect transistor connected in series.

CMP C (chemical) M (mechanical P (polishing/planarization), a method to polish/planarize a surface by mechanical methods aided by chemicals.

Coherent light Light waves having the same wavelength, phase, and direction.

Cohesive energy The strength or the energy with which the atoms are bound within a metal. The strength of this binding energy can conveniently be measured by the enthalpies of atomization.

Complimentary PSM A phase shift mask, which utilizes multiple reticles to enhance the images printed on the wafer.

Conduction The process by which heat or electric current is transferred through matter without transfer of the matter itself.

Conductivity The power of conducting heat, electricity, sound, or other forms of energy.

Conductor Material which transmits heat, electricity, light, sound, or other forms of energy.

Conformal Leaving unchanged the size of all angles.

Contact printing The process of printing the replica of a mask placed directly in contact with a substrate.

Contrast Related to the quality of image formation during lithography. For negative resist, the contrast is related to the rate of cross-linked network formation, and for positive resist, it is related to the rate of chain scission at a constant irradiation source.

CP-4 etch An etch solution comprising HF (hydrofluoric acid): HNO_3 (Nitric acid): CH_3COOH (acetic acid glacial) in 3:5:3 proportion.

CPU C (central) P (processing U (unit), is the heart of a computer system. Three main sections are generally located within CPUs of all sizes. A CPU has two components: (i) arithmetic logic unit (ALU); and (ii) the control unit (CU).

Creep High temperature related progressive deformation of a material at constant stress.

Cross-talk Undesired noise level due to capacitive reactance between the metal lines in a multilayer metal structure.

Crystalline In crystalline solids, the atoms are spatially arranged in a periodic fashion. When this periodicity exists throughout the entire solid it forms a single crystal.

CTE C (coefficient) of T (thermal) E (expansion), measures the increase in length (or volume) per unit length (or volume) of a solid per unit rise of temperature. It is also expressed as TCE.

CVAL C (curvilinear) V (variable) A (axis) L (lens), a special lens in an optical system for achieving large distance beam scanning at a resolution of <100 nm. It has been introduced for the optics of projection reduction exposure with variable axis immersion lens.

CVD C (chemical) V (vapor) D (deposition), involves chemical reactions between gaseous reactants at elevated temperature and consequent condensation and deposition of the gaseous material over a wafer.

Damascene A process practiced in Damascus to decorate a sword. At present in Cu-interconnecting systems the damascene process means inlaying copper within the dielectric channels to form a metal interconnect.

Dangling bond A broken chemical bond associated with an atom in the surface layer of a solid extending to the solid's exterior.

Detector A device used to detect light or nuclear particles. A detector can also be meant as a device to be used in the demodulation stage of an AM and FM receiver.

Dielectric Insulating material (organic/inorganic) mostly used in integrated circuits to form capacitors, masks, and crossover, and to create structures to inlay metals to form interconnects.

DIGM D (diffusion) I (induced) G (grain) B (boundary) M (motion), stress related diffusion of the solute and consequently movement of the grain boundary in a polycrystalline material.

Dishing Over-polishing the burden layer during CMP operation, making a depression that looks like that of a dish.

Dissolution inhibitor A chemical which can reduce the dissolution rate of a photoresist.

DLC D (diamond) L (like) C (carbon), laboratory synthesized hydrogenated amorphous material resembling the physical and mechanical properties of carbon.

Dopant Intentional impurity administered in a semiconductor to change the electrical behavior of the host material.

Double (dual) damascene The modified version of the damascene process where two interlayer dielectric patterning requires two metal CMP steps.

DRAM D (dynamic) R (random) A (access) M (memory), meaning a solid state memory device based on MOS technology.

Dry-etching Etching performed in dry conditions by the volatilization of atoms by chemical reactions or stripping of atoms with high energetic ions.

DUV D (deep) U (ultra) V (violet), meaning light having short wavelength (~200–300 nm).

e-beam evaporation Where evaporation of a material is performed by the bombardment of high-energy electrons generated in an electron gun inside an evacuated chamber.

EBL E (electron) B (beam) L (lithography), the focused beam of electrons used to perform photolithography operation with special type of resist material.

ECR E (electron) C (Cyclotron) R (resonance), used to generate high-density plasma.

EDS E (energy) D (dispersive) S spectroscopy, a method to determine the energy spectrum of X-ray radiation. Mainly used in chemical analysis.

Electromigration Movement of the atoms in the direction of electron flow. It is the mass transport of metal atoms due to an electric current in a bulk specimen.

Electronegativity An empirical measure of the tendency of an atom in a molecule to attract electrons.

Electroplating Deposition of a metal layer from an electrolyte when an electric current is passed through the electrolyte.

Enthalpy The measure of heat exchange of a system, related to the heat and work related thermodynamical studies of materials.

EOT E (equivalent) O (oxide) T (thickness), the optimum thickness of oxide necessary for speed, high channel mobility, and low leakage current.

Erosion Excess removal of material during chemical mechanical polishing (CMP).

EUV E (extreme) U (ultra) V (violet), a wavelength of light shorter than 180 nm.

EUVL Extreme ultraviolet lithography, means a photolithography operation, where the resist is exposed to EUV to create sub-micron structures.

Excimer laser Chemically interactive (generally fluorinated argon or krypton) laser (light amplified by stimulated emission of radiation) capable of generating very short wavelengths of 248 and 193 nm.

FEOL F (front) E (end) O (of) L (line), a process involving the first step of IC fabrication where the individual active devices are patterned.

FIB F (Focused) I (ion) B (beam) generally means an instrument used to perform lithography, and is characterized by its spot size, spot current, deflection frequency, field size, and writing speed. These instruments are less prone to back-scattering and better resolution limiting proximity effects than electron beam lithography.

FET F (field) E (effect) T (transistor), a special type of transistor where the electric field controls the channel conductivity.

Galvanic couple Two dissimilar metals in electrical contact inside an electrolyte.

GB G (grain) B (boundary), the interface between two grains in a polycrystalline material.

GDOES G (glow) D (discharge) O (optical) E (emission) S (spectroscopy), a tool for quantitative analysis of multi-elemental depth profiling.

Glazing Sluggishness in the removal rate of a pad during chemical mechanical polishing/planarization (CMP).

g-line lithography Photolithography operation using 436 nm exposure light.

Global planarization A planarization process in which the relaxation distance is in the rage of millimeters.

High-*K* Dielectric materials having dielectric constant (*K*) higher than 3.0.

Hydrophilic The property of a material, which has a natural affinity to bond with water through the hydrogen bond.

Hydrophobic The property of a non-polar molecule that likes to self-associate itself in the presence of moisture.

IC I (integrated) C (circuit), A circuit comprising diodes, transistors, resistors, and capacitors formed on a single semiconductor chip (monolithic IC, meaning IC built on a single (mono) stone (lithos)).

ICP I (inductively) C (coupled) P (plasma), a process where the plasma created inside the reactor is inductively coupled for uniform and highly ionized flux.

ILD I (inter) L (layer) D (dielectric), insulating layer separating two metal lines.

i-line lithography Photolithography operation using 365 nm exposure light.

IMD I (inter) M (metal) D (dielectric), insulating layer between two adjacent metal lines.

Immersion lithography Lithography that involves the immersion of the optical system in a higher refractive index medium than air.

Interconnect Metal lines formed from a deposited film of metal to connect different parts of the integrated circuit (IC).

Joule heating The increase in heat due to excessive electric current through a conductive material.

K Used to denote heat conductivity or the dielectric constant of an insulator.

k Used to denote a unit of temperature in kelvins.

Lattice Periodic arrangements of atoms in a crystal specific to a solid material.

LED L (light) E (emitting) D (diode), a junction diode. When properly biased, the injected electron takes part in radiative recombination to give rise to an emitted photon (quantum of light energy).

Local planarization A planarization process in which the relaxation distance is in the rage of < 100 μm.

Low-*K* Dielectric materials having dielectric constant (*K*) lower than 3.0.

LSP L (low) S (stand by) P (power).

LSTP L (Low) ST (standby) P (power) A term applicable to logic devices in which the prime concern is to keep the current in the "off" state as low as possible. Mostly meant for portable gadgets.

LTCC L (low) T (temperature) C (co-fired) C (ceramic), a low-temperature co-fired ceramic material used as a hybrid circuit substrate. The substrate material has low dielectric constant.

MBE M (molecular) B (beam) E (epitaxy), a versatile technique to grow an oriented single crystal layer on a substrate by impinging molecular or atomic beams of the constituent material inside a high vacuum system.

Modulator A device that modulates (amplitude, frequency, etc.).

Monomodal failure The failure statistics in all the lines are due to either polygranular or transgranular failure mechanism. When the same failure statistics for some lines are polygranular and for some transgranular, the failure mode is bimodal. The failure time statistics for a given population are sensitive to the value of the critical stress.

MPSA M (Marcapto) P (propane) S (sulfonic) A (acid) Chemical used as an additive during copper plating.

MTF M (modulation) T (transfer) F (function), a parameter which when approximately plotted provides a rapid convenient indication of the capability of an optical projection.

Multulayer metallization More than one metal layer deposited on the top of an another layer without short circuit.

NA N (numerical) A (aperture), defines the geometry of an objective lens and is a measure of a lens's capacity to collect diffracted light from an object in an optical system.

Nucleation Phase transition in stable condition.

OLED O (organic) LED, made out of organic materials.

Oswald ripening The process by which larger particles grow at the expense of smaller particles due to higher solubility of the smaller particles and to molecular diffusion through the continuous phase.

PAC P (photo) A (active) C (chemicals/compounds), chemicals (or compounds) that are sensitive to certain radiation of light, specially meant for additives in a high polymer, e.g., photoresist.

PAG P (photo) A (acid) G (generator), an organic compound with the radical (R) associated with sulfonic group ($-SO_3$). It generates acid (sulfonic RSO_3H on homolysis and dissociation) when it is being exposed to light and significantly changes the thermal decomposition temperature of the polymer.

PCB P (printed) C (circuit) B (board), which carries the hybrid circuit, consists of a thin copper sheet laminated on to some kind of resin bonded material.

PEB P (post) E (exposure) B (bake), heat treatment of a resist material after it has been exposed to a particular radiation of light.

PECVD P (plasma) E (enhanced) CVD, a special CVD process which uses rf-induced glow discharge to transfer energy into the reactant gases inside a vacuum chamber allowing the substrate to remain at a lower temperature than thermally energized reactant gases.

Photolithography The process of producing an optical image through a mask and an optical system using a photosensitive material (photoresist).

Photoresist (resist) A light sensitive high molecular weight polymer, very much in use in the formation of pattern in modern VLSI circuits. The main ingredients are resin (binder), the sensitizer, and additives in a proper solvent. In a positive resist irradiated regions are soluble and it produces a positive image of the mask. On the other hand, a negative resist behaves almost opposite to the positive resist.

Pitch The distance between center to center spacing between pads, rows of bumps, pins, posts, leads.

PLED P (polymer) LED, an LED fabricated from polymeric materials.

Plug Linkage provided by a metal inside a via hole between two conducting layers of a multilayer metal system.

Polycrystalline When a solid is made of several smaller crystallites quasi-periodic in nature.

Polymer A large molecule made by the addition of many molecular units to each other.

Polymerization The starting units from which the polymer is formed are called monomers, and the reaction that joins the monomers together is called polymerization.

Porogen A sacrificial polymeric material, which is thermally decomposable and the decomposition products are volatilized and permeated through the matrix leaving pores in their initially occupied sites.

Positron annihilation spectroscopy Powerful tool for microstructure investigation of condensed matter. The high momentum part of the Doppler broadening spectra can be used to distinguish different elements.

PREVAIL P (projection) R (reduction) E (exposure) with V (variable) A (axis) I (immersion) L (lens), an optical system which combines high exposure efficiency of massively parallel projection with a scanning probe to correct dynamic aberrations.

Projection printing A process that involves projection of the mask image on the surface of a wafer placed at a distance from the mask/reticle.

PSM P (phase) S (shift) M (mask), a non-conventional mask provided with material/s to introduce a phase shift of an exposure light by utilizing interference to enhance resolution.

RBS R (Rutherford) B (back) S (scattering). The technique is a non-destructive method to determine concentration profiles in the outer 0.5–3.0 μm of a material by backward scattering of positively charged α particle from a positively charged nucleus.

SCALPEL SC (scattering with) A (angular) L (limitation) P (projection) E (electron) B (beam) L (lithography), a new lithographic technique which offers high resolution and wide process latitude.

SEM S (scanning) E (electron) M (microscope), an important tool for surface morphology analysis of a material. It has the capability of providing higher magnification, resolution and depth of field than optical microscope.

Step coverage Covering of steps left by the previous layer. It is the ratio of the minimum thickness of a film as it crosses a step to the nominal thickness of the film on flat regions.

Superconduting material The resistivity, the reciprocal of the conductivity of a material, is temperature dependent, and at low temperatures, the resistivity of certain materials drops abruptly to an immeasurably low value. This phenomenon is called superconductivity and the material undergoing such a change is called superconducting material.

Supercritical fluids Special fluids with minimum or no surface tension that possess mass transport properties comparable to a gas.

TDDB T (time) D (dependent) D (dielectric) B (breakdown), typical measurements to evaluate the quality of the dielectric of a transistor. The breakdown of the dielectric is the failure mode for integrated circuit (IC).

TEOS Tetra-ethyl orthosilicate, which is tetraethosiloxane Si $(OC_2H_5)_4$ and is used to form silicon dioxide at low temperature during integrated circuit device fabrication.

TFT T (thin) F (film) T (transistor), a transistor prepared by the deposition of thin film.

Transition elements Elements that have partly filled d and f shells. They are all hard, strong, high melting metals that conduct heat and electricity well.

Triple point Intersection on a phase diagram where three phases are in equilibrium.

USJ U (ultra) S (shallow) J (junction), generally a junction in the sub-100 nm depth. To cope with the ever-decreasing feature size of a device ultra-shallow junctions are important to improve the short channel performance of the device.

UTQUANT A quasi-static CV (capacitance-voltage) simulator for one-dimensional silicon MOS structures. Originally written to study quantization effects in the MOS inversion layer.

UTCS U (ultra) T (thin) C (chip) S (stacking), an approach for high density and silicon based chip packaging, (SiP) integration as well as 3-D interconnectivity by embedding of thin die in thin film interconnect layers.

VLSI V (very) L (large) S (scale) I (integration), integration of more than 106 active and passive components.

Voids Formation of pits due to stress.

Zeta potential The measure of electrostatic forces that exist between particles and the wafer surface.

Index

MIX
Papier aus verantwortungsvollen Quellen
Paper from responsible sources
FSC® C105338

If you have any concerns about our products,
you can contact us on
ProductSafety@springernature.com

In case Publisher is established outside the EU,
the EU authorized representative is:
Springer Nature Customer Service Center GmbH
Europaplatz 3, 69115 Heidelberg, Germany

Printed by Libri Plureos GmbH
in Hamburg, Germany